Albrecht's Foundation Concepts

The Albrecht Papers, Volume I

by William A. Albrecht, Ph.D.
Edited by Charles Walters

William A. Albrecht, Ph.D.

Albrecht's Foundation Concepts

The Albrecht Papers, Volume I

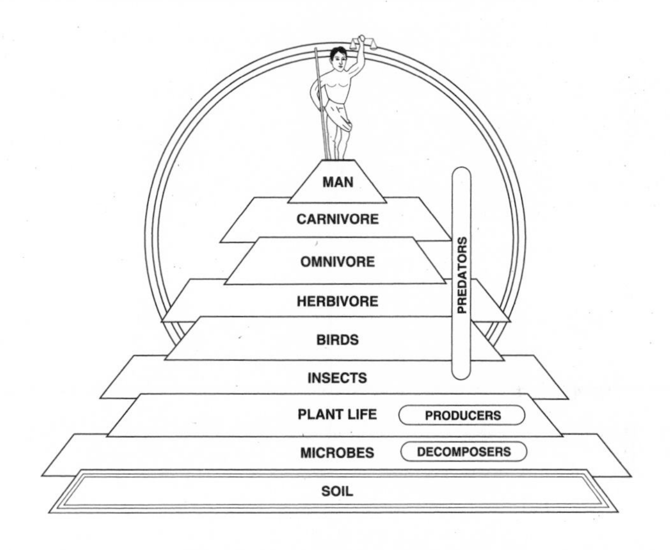

by William A. Albrecht, Ph.D.
Edited by Charles Walters

Acres U.S.A.
Austin, Texas

Albrecht's Foundation Concepts

Fourth printing, 2011

The information in this book is true and complete to the best of our knowledge. All recommendations are made without guarantee on the part of the author and Acres U.S.A. The author and publisher disclaim any liability in connection with the use or misuse of this information.

Acres U.S.A.
P.O. Box 91299
Austin, Texas 78709 U.S.A.
(512) 892-4400 • fax (512) 892-4448
info@acresusa.com • www.acresusa.com

Printed in the United States of America

Publisher's Cataloging-in-Publication

Albrecht, William A., 1888-1974
Foundation concepts / William A. Albrecht. — 2 ed., Austin, TX, ACRES U.S.A., 2011
x, 515 pp., 23 cm.
First edition published as The Albrecht Papers, Volume I
Includes Index
Includes Bibliography
ISBN 978-1-601730-27-5 (trade)

1. Soil fertility. 2. Soils and nutrition. 3. Soil science.
4. Agriculture — crops and soils. 5. Soil-plant relationships.
I. Albrecht, William A., 1888-1974 II. Title.

SF96.7 A42 2011 631.4

Tart Albrechtisms

Don't lime to fight soil acidity. Use lime to feed the plant.

•

Insects and disease are the symptoms of a failing crop, not the cause of it.

•

The use of sprays is an act of desperation in a dying agriculture. It's not the overpowering invader we must fear but the weakened condition of the victim.

•

The excessive use of chemical salts in fertilizers is upsetting plant nutrition.

•

Manure forms an organic shield around the salts. It is a buffer against salt injury. As soils become lower in organic matter we will not be able to use salts so directly.

•

Fertilizer placement is the art of putting the salts in the ground so the plant roots can dodge it.

•

To help them maintain their soils, farmers should be given a depletion allowance on their income tax the same as owners of mines, oil wells and timber tracts.

•

The South will always be a good bull market, just as it was a good mule market. The soils of the South do not have the virgin fertility to produce a good breeding stock.

•

We are exhausting the quality of our soils. As we do so the quality of our plants goes down. And we are accepting this.

•

Those who teach must constantly hold up the challenge to study nature, not books.

•

Students today throw themselves down in front of teachers like a pile of boards to be turned into furniture. They don't come to the university to educate themselves.

ABOUT THE AUTHOR

Dr. William A. Albrecht, the author of these papers, was chairman of the Department of Soils at the University of Missouri College of Agriculture, where he had been a member of the staff since 1916. He held four degrees, A.B., B.S., M.S. and Ph.D., from the University of Illinois. During a vivid and crowded career, he traveled widely and studied soils in Great Britain, on the European continent, and in Australia.

Born on a farm in central Illinois in an area of highly fertile soil typical of the cornbelt and educated in his native state, Dr. Albrecht grew up with an intense interest in the soil and all things agricultural. These were approached, however, through the avenues of the basic sciences and liberal arts and not primarily through applied practices and their economics. Some teaching experience after completing the liberal arts course, with some thought of the medical profession, as well as an assistantship in botany, gave an early vision of the interrelationships that enrich the facts acquired in various fields when viewed as part of a master design.

These experiences led him into additional undergraduate and graduate work, encouraged by scholarships and fellowships, until he received his doctor's degree in 1919. In the meantime, he joined the research and teaching staff at the University of Missouri.

Both as a writer and speaker, Dr. Albrecht served tirelessly as an interpreter of scientific truth to inquiring minds, as this volume amply illustrates. Dr. Albrecht retired in 1959 and passed from the scene in May 1974 as his 86th birthday approached.

ABOUT THE EDITOR

Charles Walters was the founder and executive editor of *Acres U.S.A*, a magazine he started in 1971 to spread the word of eco-agriculture. A recognized leader in the field of raw materials-based economic research and sustainable food and farming systems, this confirmed maverick saw one of his missions as to rescue lost knowledge. Perhaps the most important were the papers of Dr. William A. Albrecht, whose low profile obscured decades of brilliant work in soil science. Albrecht's papers, which Walters rescued from the historical dustbin and published in an initial four volumes, continue to provide a rock-solid foundation for the scientific approach to organic farming. Additional volumes of Albrecht's papers were organized and edited by Walters for later publication — the result is shown here with this book.

During his life, Walters penned thousands of article on the technologies of organic and sustainable agriculture and is the author of more than two dozen books (and co-author of several more), including *Eco-Farm: An Acres U.S.A. Primer*, *Weeds — Control Without Poisons*, *A Farmer's Guide to the Bottom Line*, *Dung Beetles*, *Mainline Farming for Century 21* and many more.

Charles Walters generously shared his vision, energy and passion through his writing and public speaking for more than 35 years and made it his lifelong mission to save the family farm and give farmers an operating manual that they couldn't live without. The Albrecht Papers are an important part of this message. Charles Walters passed on in January 2009 at the age of 83.

Dedicated to the pioneers of the Price-Pottenger Nutrition Foundation and to the eco-farmers of America.

CONTENTS

FOREWORD

Dr. William A. Albrecht, Professor of Soils and Chairman of the Department of Soils at the University of Missouri College of Agriculture, was less an enigma than a recap of his career might suggest. He was, first, a student of nature, and his greatest credential was that nature accepted his findings, worked with them, and delivered much of American agriculture from the bondage of ignorance. In the fullness of time, he earned four academic degrees—A.B., B.S., M.S. and Ph.D., all from the University of Illinois. He did not consider credentials an education, or an education in terms of credentials.

Dr. Albrecht traveled widely. He studied the soils of Great Britain, the European continent, and Australia on-scene, and he always drew conclusions seasoned by a farm boy's upbringing.

He was born on a farm in central Illinois, and as a son of the cornbelt he retained an abiding interest in all things agricultural. His education in this field, however, was primarily through the basic sciences and liberal arts rather than the applied practices. "It was during a period spent in teaching as Assistant Professor of Botany that he first visualized a master design in the chemistry of soil and life," wrote A. A. Jeffrey as agricultural editor of the Missouri Agricultural Experiment Station. "This experience led him into additional undergraduate and graduate work that was encouraged by scholarships and fellowships. He received his doctor's degree in 1919, having in the meantime joined the research and teaching staff at the University of Missouri."

The bibliography contained in this volume tells something about Dr. Albrecht's production of scientific and popular papers. In all cases his contributions emphasized the fundamental necessity of feeding plants, animals and human beings through ministrations to the soil itself, correcting deficiencies of diet at their point of origin—that is, in soils a scientific readout found wanting.

Scholars and farmers alike have flavored the memory of Dr. Albrecht with vignettes that merit telling now. E. M. Poirot, a farmer at Golden City, Missouri, first met Dr. Albrecht in 1926. "We began research for making artificial manure out of straw in order to have a bit of organic fertilizer. This, we hoped, would encourage the growth of nitrogen-fixing bacteria on sweet clover in soil where the organic level was too low to support the bacterial life in then available laboratory cultures of inoculation materials. Later I was granted two patents on the process which was made available to farmers by the Capper Foundation of Kansas."

It was on Poirot's acres that Albrecht proved the contention that calcium was the prince of nutrients. The year was 1928. Later observations of cows chewing bones, grazing where the soil was high in magnesium and where extra phosphate and lime had been applied, and still later—when cows walked through what appeared to be good grass to eat 17 different kinds of weeds where calcium limestone, magnesium and phosphate had been applied—caused Albrecht to think in terms of fertilizing crops so they would be more nutritious to animals eating them.

The Albrecht Papers tell this story. They tell more, but not all. For it is a fact that Albrecht's findings enabled the nutrient-deficient Ozarks to produce good calf crops until now Missouri is second only to Texas in calf production. Dr. Albrecht lived his life as a scientist, a writer and a speaker, always serving tirelessly as an interpreter of scientific truth so that inquiring minds could use this knowledge in service to mankind. His papers provide a glimpse of that story.—*Charles Walters Jr., Publisher/Editor, Acres U.S.A.*

• • •

Disease has been defined as *"an illness or sickness. . .a disturbance in function or structure of any organ or part of the body."* Over the centuries, scientists have been attempting to uncover the causes of disease, hoping thereby to find specific forms of treatment. Fortunately, we have come a long way from the miasmas, exorcisms, purges and bloodletting of former times.

The germ theory led to the concept of specific infectious agents as responsible for many diseases. The discovery of the malarial parasite, the tubercle bacillus, the treponema of syphilis, the leprosy bacillus, and the organisms responsible for tonsillitis, pneumonia and other infectious states, was an important step forward. So was the recognition of still smaller organisms such as PPLO's and the viruses as causative agents. However, these concepts narrowed the thinking of many investigators.

A logical approach to treatment seemed to be to find chemicals or other substances, such as antibiotics, that would kill the germs responsible for various diseases. The discovery of the sulfa drugs, followed by penicillin, and the family of mycins did save untold millions from suffering and death.

However, development of potent "germ killers" and their widespread use has been accompanied by a marked increase in viral infections. Degenerative diseases such as arthritis, cardiovascular diseases and cancer are on the increase. There is evidence that a strange organism, named *"progenitor cryptocides"* by Virginia Livingston, M.D., is found in abundance in the blood of those developing cancer.

Susceptibility to "cryptocides," to cancer, to the toxic effects of petro-chemicals and to chronic degenerative diseases of all types may well be primarily the result of inadequate nutrition. Only recently has this been realized.

One of the first to stress this concept in the 20th century was a modest scientist who for many years was Chairman of the Department of Soils at the University of Missouri—Dr. William A. Albrecht. Extensive experiments with growing plants and animals substantiated his theory and observation that a declining soil fertility, due to a lack of organic material, major elements and trace minerals—or a marked imbalance in these nutrients—was responsible for poor crops and in turn for pathological conditions in animals fed deficient feeds from such soils. Obviously mankind is no exception.

Dr. Albrecht was a member of the Board of Directors of the Price-Pottenger Nutrition Foundation—a non-profit corporation dedicated to spreading the knowledge of good nutrition via the works of Dr. Weston A. Price, the late F. M. Pottenger, Jr., M.D., Dr. Albrecht and others. He is being sorely missed not only by those who have had the privilege of knowing him, but eventually by thousands who learn of his work. His findings are immortal. The Price-Pottenger Nutrition Foundation hopes this timely book will have the widest possible circulation.

How long will it take for agriculture, physicians and scientists in all fields to realize—in the words of Dr. Price—*"Life in all its fullness is mother nature obeyed,"* or as succinctly stated by Dr. Albrecht, *"To be well fed is to be healthy."—Granville F. Knight, M.D., President, Price-Pottenger Nutrition Foundation.*

Bibliography

Here is a listing of the papers Dr. William A. Albrecht accounted for during a working lifetime. There may be some items not contained in this list, and there are certainly duplications of content from one paper to the next. But, withal, this is a comprehensive statement of fact. Almost all of these papers and manuscripts are on file at *Acres U.S.A.* Dr. Albrecht was a prolific writer and a conscientious publicist. Often he would recycle articles under several titles. In this compilation, due note has been made of such information when available. Many of the publications cited here no longer exist. Finally, use has been made of Dr. Albrecht's own notes and citations in compiling this bibliography.

1918

Changes in the Nitrogen Content of Stored Soils. *Journal of the American Society of Agronomy,* volume 10, number 2, pp. 83-88, 1918.

1919

Symbiotic Nitrogen Fixation as Influenced by the Nitrogen in the Soil. Thesis, submitted in partial fulfillment of the requirements for the degree of Doctor of Philosophy in Agronomy in the Graduate School of the University of Illinois, 1919. Also published in *Soil Science,* volume 9, number 5, pp. 275-328, 1920.

Soil Inoculation for Legumes. *Missouri Agricultural Experiment Station Circular 86,* 1919.

1921

Bat Guano and its Fertilizing Value. *Missouri Agricultural Experiment Station Bulletin 180,* 1921.

1922

Viable Legume Bacteria in Sun-Dried Soil. *Journal of the American Society of Agronomy,* volume 14, numbers 1 and 2, pp. 49-51, 1922.

1924

Inoculation for Legumes. *Missouri Agricultural Experiment Circular 121,* 1924.

1925

Nitrate Accumulation Under the Straw Mulch. *Soil Science,* volume 20, number 4, pp. 253-353, 1925.

1926

Nitrate Accumulation in Soil as Influenced by Tillage and Straw Mulch. Paper read as part of the symposium on Soil Bacteriology-Nitrification Studies at the meeting of the American Society of Agronomy, November 17, 1925. Also published in *Journal of the American Society of Agronomy,* volume 20, pp. 841-853, 1926.

1927

Artificial Manure Production on the Farm. *Missouri Agricultural Experiment Station Bulletin 258,* 1927.

1928

Farm Trials of Artificial Manure. *Journal of the American Society of Agronomy,* volume 20, pp. 123-132, 1928. Co-author, E. M. Poirot.

Calcium as a Factor in Soybean Inoculation. *Soil Science,* volume 25, number 4, pp. 313-325. Student paper by R. W. Scanlan [W. A. Albrecht, advisor].

1929

A Protector for Graduated Cylinders. *Journal of Chemical Education,* p. 336, February 1929.

Relation of Calcium to the Nodulation of Soybeans on Acid and Neutral Soils. *Soil Science,* volume 28, number 4, pp. 261-279, 1929. Co-author, Franklin L. Davis.

Physiological Importance of Calcium and Legume Inoculation. *The Botanical Gazette,* volume 88, number 3, pp. 310-321, 1929. Co-author, F. L. Davis.

1930

Dry Inoculants for Alfalfa. *Journal of the American Society of Agronomy,* volume 22, number 11, pp. 916-918, 1930.

Legume Bacteria with Reference to Light and Longevity. *Missouri Agricultural Experiment Station Research Bulletin 132,* 1930. Co-author, Lloyd M. Turk.

Fractional Neutralization of Soil Acidity for the Establishment of Clover. *Journal of the American Society of Agronomy,* volume 30, pp. 649-657, 1930. Co-author, E. M. Poirot.

Nitrogen Fixation as Influenced by Calcium. *Proceedings and Papers of the Second International Congress of Soil Science,* Leningrad, Moscow, U.S.S.R., volume 3, pp. 29-39, 1930.

Local Variation of Soil Acidity in Relation to Soybean Inoculation. *Soil Science,* volume 30, pp. 273-287, 1930. Student paper by George Z. Doolas [W. A. Albrecht, advisor].

1931

Changes in Composition of Soybeans Toward Maturity as Related to their Use as Green Manure. *Soil Science,* volume 32, number 4, pp. 271-282, 1931. Co-author, W. H. Allison.

Available Soil Calcium in Relation to "Damping Off" of Soybean Seedlings. *Botanical Gazette,* volume 92, pp. 263-278, 1931. Co-author, Hans Jenny.

1932

Calcium and Hydrogen-Ion Concentration in the Growth and Inoculation of Soybeans. *Journal of the American Society of Agronomy*, volume 24, number 10, pp. 793-806, 1932.

The Composition of Soybean Plants at Various Stages as Related to their Rate of Decomposition and Use as Green Manure. *Missouri Agricultural Experiment Station Bulletin 1973*, 1932. Student paper by Lloyd M. Turk [W. A. Albrecht, advisor].

Physiochemical Reactions Between Organic and Inorganic Soil Colloids as Related to Aggregate Formation. *Soil Science*, volume 5, number 44, pp. 331-358, 1937. Student paper by H. E. Myers [W. A. Albrecht, advisor].

Physiology of Root Nodule Bacteria in Relation to Fertility Levels of the Soil. *Proceedings of the Soil Science Society*, volume 2, pp. 315-327, 1937. Co-author, F. L. Davis.

Nitrification of Ammonia Adsorbed on Colloidal Clay. *Proceedings of the Soil Science Society of America*, volume 2, pp. 263-267. Co-author, T. M. McCalla.

A New Culture Medium for Rhizobia. *Journal of Bacteriology*, volume 34, number 4, pp. 455-457, 1937. Co-author, T. M. McCalla.

1933

Drilling Powdered Agricultural Limestone. *Agricultural Engineering*, volume 14, number 4, April 1933. Co-author, M. M. Jones.

Inoculation of Legumes as Related to Soil Acidity. *Journal of the American Society of Agronomy*, volume 25, number 8, pp. 512-522, 1933.

1934

Declining Nitrate Levels in Putnam Silt Loam. *Journal of the American Society of Agronomy*, volume 26, number 7, pp. 569-574, 1934.

1935.

Relation of the Degree of Base Saturation of a Colloidal Clay by Calcium to the Growth, Nodulation and Composition of Soybeans. *Missouri Agricultural Experiment Station Research Bulletin 232*, 1935. Student paper by Glenn M. Horner [W. A. Albrecht, advisor].

Discrimination in Food Selection by Animals. *Scientific Monthly*, volume 60, pp. 347-352, 1935.

1936

Methods of Incorporating Organic Matter with the Soil in Relation to Nitrogen Accumulations. *Missouri Agricultural Experiment Station Research Bulletin 249*, 1936.

Nodulation and Growth of Soybeans as Influenced by Calcium and Hydrogen-Ion Concentration in Putnam Silt Loam Soil. *Greek Journal*, volume 3, number 48, 1936. Student paper by George Z. Doolas [W. A. Albrecht, advisor].

A Study of the Uniformity of Soil Types and the Fundamental Differences Between the Different Soil Series. *Alabama Agricultural Experiment Station Bulletin*, 1936. Student paper by F. L .Davis [W. A. Albrecht, advisor].

Artificial Manure Production on the Farm. *Missouri Agricultural Experiment Station Bulletin 369*, 1936.

Drilling Fine Limestone for Legumes. *Missouri Agricultural Experiment Station Research Bulletin 367*, 1936.

1937

The Nitrate Nitrogen in the Soil as Influenced by the Crop and the Soil Treatments. *Missouri Agricultural Experiment Station Research Bulletin 250*, 1937.

Variant Forms of Rhizobia (Root Nodule Bacteria) in Relation to the Calcium of the Soil. *Proceedings of the Soil Science Society of America*, volume 1, number 2, 1937. Co-author, T. M. McCalla.

Behavior of Legume Bacteria (Rhizobium) in Relation to Exchangeable Calcium and Hydrogen-Ion Concentration of the Colloidal Fraction of the Soil. *Missouri Agricultural Experiment Station Research Bulletin 256*, 1937. Student paper by T. M. McCalla [W. A. Albrecht, advisor].

Longevity of Legume Bacteria (Rhizobium) in Water. *Journal of the American Society of Agronomy*, volume 29, number 1, pp. 76-79, 1937. Co-author, T. M. McCalla.

1938

The Colloidal Clay Fraction of Soil as a Cultural Medium. *American Journal of Botany*, volume 25, pp. 403-407, 1938.

Variable Levels of Biological Activity in Sanborn Field after Fifty Years of Treatment. *Proceedings of the Soil Science Society of America*, volume 4, pp. 77-82, 1938.

Magnesium as a Factor in Nitrogen Fixation by Soybeans. *Missouri Agricultural Experiment Station Research Bulletin 288*, 1938. Student paper by Ellis R. Graham [W. A. Albrecht, advisor].

Nitrate Production in Soils as Influenced by Cropping and Soil Treatments. *Missouri Agricultural Experiment Station Research Bulletin 294*, 1938.

Loss of Soil Organic Matter and Its Restoration. *Soils and Men, Yearbook of Agriculture*, pp. 347-360, 1938. Also published as There Is No Substitute for Soil Fertility, *Yearbook Separate No. 1626*, 1939.

Effect of Different Nitrogenous Fertilizers on the pH and Available Phosphorus of Soils and Their Relation to the Yield of Cotton. *The American Fertilizer*, May 1938.

Fertility Reflected in Feeding. *The Fertilizer Review*, p. 11, September-October 1938.

1939

Plant Growth and the Breakdown of Inorganic Soil Colloids. *Soil Science*, volume 47, number 6, pp. 455-458, 1939. Co-authors, Ellis Graham and Carl Ferguson.

Some Soil Factors in Nitrogen Fixation by Legumes. *Transactions of III Commission of International Society of Soil Science*, New Brunswick, New Jersey, volume A, pp. 71-84, 1939.

Limestone Mobilizes Phosphates into Korean Lespedeza. *Journal of the American Society of Agronomy*, volume 31, number 4, April 1939. Co-author, A. W. Klemme.

Calcium in Relation to Phosphorus Utilization by Some Legumes and Non-Legumes. *Soil Science Society of America*, volume 4, pp. 260-265, 1939. Co-author, N. C. Smith.

Colloidal Clay Culture for Refined Control of Nutritional Experiments with Vegetables. *Proceedings of the American Society for Horticultural Science*, volume 37, pp. 689-692, 1939. Co-author, R. A. Schroeder.

The Soil and the Times. *Technical Supplement to Program*, USDA, Soil Conservation Service, Region 5, Milwaukee, Wisconsin, 1939.

Evaluating Productivity. Foreword, *Bulletin 405*, 1939.

Dangerous Grass. *Capper's Farmer*, volume 50, p. 9, 1939.

Value of Timothy Hay as Sheep Feed in Response to the Soil Treatment. *Missouri Agricultural Experiment Station Bulletin 444*, June 30, 1939. Co-authors, A. G. Hogan and George Norwood.

Doubles Protein Yield. *The Farmer's Digest*, pp. 4-6, August 1939. [Condensed from Capper's Farmer.]

1940

Nitrogen Fixation and Soil Fertility Exhaustion by Soybeans Under Different Levels of Potassium. *Missouri Agricultural Experiment Station Research Bulletin 330*, May 1941. Co-author, Carl E. Ferguson.

Land Classification in Relation to the Soil and its Development. *Proceedings of the First National Conference on Land Classification, Agricultural Experiment Station Bulletin 421*, pp. 45-53, December 1940.

Calcium and Phosphorus in ihrem Einflusz auf die Manganaufnahme durch die Futterplfanzen. *Bodenkunde and Pflanzenernahrung*, volume 21, 22, pp. 757-767. Co-author, N. C. Smith.

Saturation Degree of Soil and Nutrient Delivery to the Crop. *Journal of the American Society of Agronomy*, volume 32, number 2, February 1940. Co-author, N. C. Smith.

Calcium-Potassium-Phosphorus Relation as a Possible Factor in Ecological Array of Plants. *Journal of the American Society of Agronomy*, volume 32, number 5, pp. 411-418, June 1940.

Adsorbed Ions on the Colloidal Complex and Plant Nutrition. *Soil Science Society of America Proceedings*, volume 5, pp. 8-16, 1940.

The Soil as a Farm Commodity or a Factory. Conservation Conference, University of Missouri, Columbia, Missouri, pp. 38-42, June 1940. Also in *Journal of the American Society of Farm Managers & Rural Appraisers*, April 1941.

Good Horses Require Good Soils. *Horse and Mule Association of America Booklet No. 256*, December 1940

Missouri Soils and Their Grazing Crops (abstract). Conference on Pasture Farming and Grazing Management, University of Missouri, College of Agriculture, July 1940.

Organic Matter—The Life of the Soil. *Farmer's Week*, Ohio State University, 1940.

Making Organic Matter Effective in the Soil. *Annual Report of The Ohio Vegetable and Potato Growers Association*, 1940.

1941

Nitrogen Fixation and Soil Fertility Exhaustion by Soybeans Under Different Levels of of Potassium. *Missouri Agricultural Experiment Station Research Bulletin 330*, May 1941. Co-author, Carl E. Ferguson.

Calcium Saturation and Anaerobic Bacteria as Possible Factors in Gleization. *Soil Science*, volume 51, number 3, pp. 213-217, March 1941.

Biological Assays of Soil Fertility. *Soil Science Society of America Proceedings*, volume 6, pp. 252-258. Co-author, G. E. Smith.

Soil Organic Matter and Ion Availability for Plants. *Soil Science*, volume 51, number 6, pp. 487-494, June 1941.

Calcium and Phosphorus as they Influence Manganese in Forage Crops. *Bulletin of the Torrey Botanical Club*, volume 68, pp. 372-380, 1941. Co-author, N. C. Smith.

Drilling Limestone for Legumes. *Missouri Agricultural Experiment Station Research Bulletin 429*, May 1941.

Plants and the Exchangeable Calcium of the Soil. *American Journal of Botany*, volume 28, number 5, pp. 394-402, May 1941.

Interrelationships of Calcium, Nitrogen and Phosphorus in Vegetable Crops. *Plant Physiology*, volume 22, pp. 244-256, 1941. Co-authors, S. H. Wittwer and R. A. Schroeder.

Calcium as a Factor in Seed Germination. *Journal of the American Society of Agronomy*, volume 33, number 2, pp. 153-155, February 1941.

Limestone (A Fertilizer). *Capper's Farmer*, p. 18, 1941.

Bread From Stones. *Grain and Feed Review*, p. 12, November 1941.

It's the Calcium Not the Alkalinity. *Soybean Digest*, 1941.

Sperm Production as a Guide to the Adequacy of a Diet for Farm Animals. *Missouri Agricultural Experiment Station Bulletin 477, Report*, June 30, 1941. Co-authors F. F. McKenzie, J. F. Lasley, G. T. Easley and George Smith.

Calcium-Bearing Versus Neutral Fertilizers. *Commercial Fertilizer Year Book*, 1941.

Helps in Transition by Gamma Alphans. *The Gamma Alpha Record*, volume 40, pp. 7-10, 1941.

1942

Surface Relationships of Roots and Colloidal Clay in Plant Nutrition. *American Journal of Botany*, volume 29, pp. 210-233, 1942. Co-authors, E. R. Graham and N. R. Shepard.

Plant Nutrition and the Hydrogen Ion: I. Plant Nutrients Used Most Effectively in the Presence of a Significant Concentration of Hydrogen Ions. *Soil Science*, volume 53, pp. 313-327, 1942. Co-author, R. A. Schroeder.

The Development of Loessial Soils in Central United States as it Reflects Differences in Climate. *Missouri Agricultural Experiment Station Research Bulletin 345*, 1942. Co-author, H. B. Vanderford.

Plant Nutrition and the Hydrogen Ion: II. Potato Scab. *Soil Science*, volume 52, pp. 481-488, 1942. Co-author, R. A. Schroeder.

The Fertility Problem of Missouri Soils. *Proceedings, Missouri State Horticulture Society*, December 1940-November 1942.

Feed Efficiency in Terms of Biological Assay of Soil Treatments. *Proceedings of the Soil Science Society of America*, volume 7, pp. 322-330, 1942. Co-author, G. E. Smith.

Plant Nutrition and the Hydrogen Ion: III. Soil Calcium and the Oxalate Content of Spinach. *Bulletin of the Torrey Botanical Club*, volume 69, pp. 561-568, 1942. Co-author, R. A. Schroeder.

Plant Nutrition and Hydrogen Ion: IV. Soil Acidity for Improved Nutrient Delivery and Nitrogen Fixation. *Proceedings of the Soil Science Society of America*, volume 7, pp. 247-257. Co-author, C. B. Harston.

Lime for Backbone. *Business of Farming*, September-October 1942.

The Business Called Plant Growth. *Office of War Information*, pp. 9 and 10, 1942.

Sound Horses are Bred on Fertile Soils. *Percheron News*, pp. 15-22, 1942.

Health Depends on Soil. *The Land*, volume 4, number 2, pp. 137-142, 1942.

Neglect of Soil Fertility Reflected by Farm Animals. *Kansas City Daily Drovers Telegram*, 1942.

Sound Bones Basis for Healthy Horses—Fertility of Soil Helps Develop Healthy Animals. *Farm Topics*, 1942.

Buying Fertilizers Wisely. *Missouri Agriculture Experiment Station Circular 227*, 1942. Co-author, L. D. Haigh.

Forest Trees Require Soil Fertility. *American Forests*, volume 48, pp. 328, 1942.

Growth Processes Explained, Part Played by Clay, Humus, Silt Minerals. *Southern Florist and Nurseryman*, 1942.

Think of the Soil as a Factory—Not as a Commodity. No date or source, probably 1942.

1943

The Soil is the Basis of the Farming Business. *Better Crops with Plant Foods*, volume 27, 1943.

Why Do Farmers Plow? *Better Crops with Plant Foods*, volume 27, June-July; volume 27, August-September, 1943. Also *The Organic Farmers*, 1950. Also *Missouri Agricultural Experiment Station Research Bulletin 474*.

Nitrate Adsorption by Plants as Anion Exchange Phenomenon. *American Journal of Botany*, volume 30, pp. 195-198, 1943. Co-author, E. R. Graham.

Potassium in the Soil Colloid Complex and Plant Nutrition. *Soil Science*, volume 55, pp. 13-21, 1943.

Magnesium Depletion in Relation to Some Cropping Systems and Soil Treatments. *Soil Science*, volume 55, pp. 445-447, 1943. Co-authors, J. J. Pettijohn and E. D. McLean.

Pattern of Wildlife Distribution Fits the Soil Pattern. *Missouri Conservationist*, pp. 1-4, 1943.

Soil and Livestock. *The Land*, II: pp. 298-305, 1943.

Hogs Benefit from Crops Grown on Fertile Soils. *Weekly Kansas City Star*, 1943.

Soil Fertility and the Human Species. American Chemical Society, *Chemical and Engineering News*, volume 21, p. 221, 1943.

Calcium. *The Land*, III: 1943

Fertilizers and Soil Management in Wartime. *Missouri Agricultural Experiment Station Bulletin 474*, 1943.

Soil Fertility, An Important Factor in Horticultural Crops. *Transcript Illinois State Horticulture Society*, pp. 434-440, 1943. Also, *Missouri Agricultural Experiment Station Circular 345*, 1950.

Fertility of Soil Measures the Protein in the Crop. *Weekly Kansas City Star*, 1943.

Feed the Soil to Feed Yourself. *The Furrow*, Second Quarter, 1943.

Soils and Future Agricultural Engineering. *Missouri Shamrock*, volume 10, pp. 8-19, 1943.

Our Soil Fertility—One of the Allied Powers. *Weekly Kansas City Star*, 1943.

The Soil is Big Farming Business. *Better Crops with Plant Food*, pp. 26, 1943.

Make the Grass Greener on Your Side of the Fence. *The Business of Farming*, volume 2, number 6, 1943.

The Fertility Problem of Our Soils. *Agricultural Education Magazine*, 1943.

Our Soils in Selected Service. *Farmers Digest*, pp. 1-10, 1943.

Beef Yields, Too, Measured by Fertility of the Soil. *Weekly Kansas City Star*, 1943.

For Best Crop Yields, The Farmer Must Mobilize the Soil. *Weekly Kansas City Star*, 1943.

Bugaboo of Soil Acidity Dispelled. In *Farm Topics, Buda* [Illinois] *Plain Dealer*, 1943.

"Grow" Foods or Only "Go" Foods According to the Soil. *School of Science and Mathematics*, 1943.

Better Seed or Better Soil. No date or source. Probably 1943.

Biological Assays of Some Soil Types Under Treatments. *Soil Science Society Proceedings*, pp. 282-286, 1943. Co-authors, Eugene O. McLean and G. E. Smith.

Double Cropping for Double Profits. . .Requires Good Soil Management. No date or source. Probably 1943.

1944

Nitrogen Fixation, Composition and Growth of Soybeans in Relation to Variable Amounts of Potassium and Calcium. *Missouri Agricultural Experiment Station Research Bulletin 381,* 1944. Co-author, Herbert Hampton.

Nodulation Modifies Nutrient Intake from Colloidal Clay by Soybeans. *Soil Science Society of America Proceedings,* volume 8, pp. 234-247. Co-author, Herbert Hampton.

Soil Granulation and Percolation Rate as Related to Crop and Manuring. *Journal of American Society of Agronomy,* volume 36, pp. 646-648. Co-author, Jacob Sosne.

Science on the March. Soil Acidity—A Nutrient Deficiency. *Science Monthly,* volume 58, p. 237, 1944.

Soil Fertility and Soybean Production. Soybean Digest, February 1944.

Proves Weedy Pastures Lack in Plant Food. *Weekly Kansas City Star,* 1944.

Soil Fertility Food Source. *Technical Review* (Massachusetts Institute of Technology), volume 46, pp. 3-7, 1944.

Is the Plow on Its Way Out? *The Farmer Stockman,* April 1944.

Soil Fertility and National Nutrition. *Journal of American Society of Farm Managers and Rural Appraisers,* volume 8, pp. 45-66, 1944.

Go Ahead and Plow! *Farm Journal,* 1944.

Mobilizing the Fertilizer Resources of Our Nation's Soil. *Rock Products Magazine,* 1944. Also presented at 28th Annual Convention of National Crushed Stone Association, New York City, 1945.

Taking Our Soils for Granted. *Philfarmer,* 3rd Quarter, 1944. Also in *The Ranchman,* November 1944.

School of the Soil: Our Soil-Dirt or Design. *Philfarmer,* 4th Quarter, 1944.

Only Fertile Soil Can Grow Healthy Plants, Sound Virile Animals and Most Civilized Justice Loving Nations. *American Banker,* 1944.

Feed Efficiency in Terms of Biological Assays of Soil Treatments. *Farm for Victory,* 1944.

Soil Fertility in its Broader Implications. *The Fertilizer Review,* 1944. Also in *Florists Exchange and Horticultural Trade World,* November 1944; *The Farmer-Stockman,* May 1944; *Extension Bulletin 66,* General Extension Service, University of New Hampshire, Durham, New Hampshire.

T.B.A. Deficiency Disease, *West Plains* [Missouri] *Journal,* 1944.

Fertility: The 4th Dimension. *The Land,* volume 3, number 2, pp. 185-189, 1944.

Soil Fertility and Food Quality. *Proceedings of the Soil Science Society of Florida,* volume 6, pp. 108-122, 1944.

Missouri University Makes Interesting Soil Survey; Wool Quality Depends on Soil Fertility. *Midwest Wool Growers News,* volume 13, number 1, 1944.

Soils Take A Rest. *Science on the March,* volume 59, number 3, pp. 235-236, 1944 [?].

Better Pastures Depend on Soil Fertility. *The Fertilizer Review,* 1944.

Is Plowing Folly? *Better Crops with Plant Food,* pp. 31-32, 1944.

Plowman's Wisdom. *The Home Garden,* volume 3, pp. 13-16, 1944.

Soil and Human Health. *Garden Club of America Annual Report,* pp. 22-26, 1944.

Soil Improvement. *Better Crops with Plant Food,* 1944.

Sweet Clover Responds to Potash Fertilizer. *Better Crops with Plant Food,* 1944.

A New Emphasis on Plant Food. *Better Crops with Plant Food,* 1944.

Soil Fertility and Wildlife—Cause and Effect. *Transactions of the Ninth North American Wildlife Conference,* American Wildlife Institute, pp. 19-28, 1944.

1945

Soil Fertility and Food Quality. *45th Annual Report of the Indiana Corn Growers Association,* January 10, 1945.

Vegetable Crops in Relation to Soil Fertility: II. Vitamin C and Nitrogen Fertilizers. *Soil Science,* volume 59, pp. 329-336, 1945. Co-authors, S. W. Wittwer and R. A. Schroeder.

Our Soil Holds Our Future. *Family Herald and Weekly Star,* Montreal, Canada, 1945.

Why Plow? *91st Annual Report,* Ontario Association of Agronomists Society, Toronto, Canada, pp. 72-74, 1945. Also in *The Furrow,* 1945.

War Is Result of Global Struggle for Soil Fertility. No source [South Haven, Michigan, 1945].

Fertile Soil Makes Better Livestock—Healthier Humans. *Hormel Farmer,* volume 8, number 2, 1945.

They Like Grass from Top-Dressed Fields. *The Jersey Bulletin,* p. 496, 1945.

Soil and Livestock. *Your Farm,* pp. 97-105, 1945.

Soil Fertility and its Health Implications. *American Journal of Orthodontics and Oral Surgery,* volume 31, pp. 279-286, 1945.

Our Soil An Active Body. *Philfarmer,* 1st Quarter, 1945.

How Long Do the Effects of Fertilizer Last? *Better Crops with Plant Foods,* p. 14, 1945. Also in *The Farmer's Digest,* volume 8, number 11, 1945.

Applied Nitrogen as Possible Assistance for Legumes. *Farm for Victory,* 1945.

Animals Recognize Good Soil Treatment. *The Berkshire News,* 1945.

How Soil Determines Human Growth. *Southwest Review,* Southern Methodist University, pp. 320-323, 1945.

Plants Vary Widely in Mineral Composition. *Farmer's Digest,* volume 9, p. 9, 1945.

Our Soils—Under Construction. *Philfarmer,* 3rd Quarter, 1945.

Potash Deficiency Follows Continuous Wheat. *Better Crop with Plant Food,* volume 29, p. 24, 1945. Co-author, N. C. Smith.

Agriculture Limestone—A Life-Giving Grist. *Rock Products*, pp. 92-94, 1945.

The Farmer and the Rest of Us. Book review by W. A. Albrecht. *Soil Conservation*, 1945.

Food is Fabricated Soil Fertility. Chapter 23, *Nutrition and Physical Degeneration*, [by Weston A. Price], pp. 435-436 and 461-469. Also published by W. A. Albrecht, 1945.

Better Soil—Better Sections. *Turtox News*, volume 23, number 7, 1945.

Discrimination in Food Selection by Animals. *Science Monthly*, volume 50, pp. 347-352, 1945.

Red Clover Suggests Shortage of Potash. *Better Crops with Plant Foods*, 1945.

Our Soils, Too, Must Be Fed. *Missouri Farm Bureau News*. No date or citation. Probably 1945.

Protein Production by Plants. No date or source. Probably 1945.

Dr. Albrecht Says We Should "Treat Trees as a Crop." *West Plains* [Missouri] *Journal*, 1945.

Some Nutritional Aspects of Green Leafy Vegetables. *The Scientific Monthly*, volume 61, pp. 71-73, 1945.

1946

Colloidal Clay Cultures. Preparation of the Clay and Procedures in its Use as a Plant Growth Medium. *Soil Science*, volume 62, pp. 23, 31, 1946.

Plant Nutrition and the Hydrogen Ion, V: Relative Effectiveness of Coarsely Ground and Finely Pulverized Limestone. *Soil Science*, volume 61, pp. 265-271, 1946.

Vegetables Crops in Relation to Soil Fertility, III: Oxalate Content and Nitrogen Fertilization. *Food Research*, volume 2, pp. 54-60, 1946. Co-authors, S. H. Wittwer and H. R. Goff.

Soil Fertility and Farm Security. *Farmer's Digest*, volume 10, pp. 40-43, 1946.

Our Soil—Under Destruction. *Philfarmer*, 1st Quarter, 1946.

The French Don't Dare Wear Out Their Farms. *Missouri Ruralist*, 1946. Also *Farmer's Digest*, 1946.

Soil Fertility and Farm Security. *The Farmer's Digest*, 1946.

Soil Fertility and Nutrition. *New Agriculture*, p. 10, 1946.

How Soils Nourish Plants: Clay Holds the Active Supply of Nutrients. *Philfarmer*, 2nd Quarter, 1946.

Cater to Cow's Taste by Soil Treatments on Pastures. *Guernsey Breeders Journal*, 1946.

Agricultural Limestone for Better Quality of Foods. *Pit and Quarry*, 1946.

Lime-Rich Soils Give Size and Vigor to French Stock. *National Livestock Producers*, volume 24, number 9, 1946.

What Fineness of Limestone. *Pit and Quarry*, 1946.

Extra Soil Fertility Lengthens Grazing Season. *Guernsey Breeders Journal*, 1946.

How Soils Nourish Plants: Mineral Reserves Restock Clay with Nutrients. *Philfarmer*, 3rd Quarter, 1946.

Why Be A Friend of the Land? *Friends of the Land. The Land Letter Series VI*, 1946.

The Soil Nitrogen Supply is Still a Big Problem. *Victory Farm Forum*, pp. 1-4, 1946.

Protein Takes More Than Air and Rain (A Key to Failing Fertility). *The Land*, 1946.

How Soils Nourish Plants: Soil Fertility is Needed, So is Soil Acidity. *Philfarmer*, 4th Quarter, 1946.

Growing Legumes on Acid Soils. *The Rural New Yorker*, 1946.

The Cow Ahead of the Plow. (Older Soils Under Older Civilizations). *New Agriculture*, 1946.

Older Soils Under Older Civilizations. (More Permanent Soils Under More Permanent Civilizations.) *New Agriculture*, 1946.

As Animals Judge Your Crops. *The Furrow*, 1946.

Bigger Bones Demand Better Soils. No date or source. Probably 1946.

Scientific Answer is Given to the Question: Why Plow? *The Goderich Signal Star*, Goderich, Ontario, p. 8, 1946.

The Soil as the Basis of Wildlife Management. Address at the banquet of the 8th Midwest Wildlife Conference, Columbia, Missouri, 1946.

Is Nitrogen Going West? *The Fertilizer Review*, 1946.

1947

Soil Fertility and Alfalfa Production. *Feedstuffs*, 1947.

Soils and Livestock Work Together, I: The Protein Problem and the Pattern of Soil Fertility. *Meat*, pp. 192-194, 1947.

Dr. Albrecht Tells How Livestock and Soils Are Related. *The National Provisioner*, pp. 261-269, 1947.

Some Late Developments in Fertility Thinking. Outline of Ideas Presented to the Meeting of the Itinerant Teachers of Agriculture, Columbia, Missouri, 1947.

Plant Nutrition and the Hydrogen Ion, VI: Calcium Carbonate, a Disturbing Fertility Factor in Soils. *Soils Science Society of America Proceedings*, volume 12, pp. 342-347, 1947. Co-author, D. A. Brown.

Healthy Soils for Healthy Plants, Animals and Humans. *Bulletin*, Council of State Garden Clubs, volume 17, 1947.

Soil Fertility—The Basis of Agriculture Production, Annual Meeting Western Colorado Horticulture Society, p. 10-10, 1947.

Older Soils Under Older Civilizations. Bigger Horses on Better Soils. *New Agriculture*, 1947.

Old Soils Under Old Civilizations. The European Compost Bespeaks Soil Conservation. *New Agriculture*, 1947.

Use Extra Soil Fertility to Provide Protein. *Guernsey Breeders Journal*, volume 71, number 6, p. 738, 1947.

Feed Values are Soil Values. *The Nation's Agriculture* [Swift and Company]. No date. Probably 1947.

Better Soils for Better Grass. *Rural New Yorker*, 1947.

Limestone—The Foremost of Natural Fertilizers. *Pit and Quarry*, May 1947.

Better Soils Make Better Hogs. *Hampshire Herdsman*, 1947. Also in *The Practical Farmer*, 1948.

Soil Builders Build Better Cattle. *The American Hereford Journal*, 1947.

Buy More Fertilizer But Less Feed. *Hoard's Dairyman*, 1947.

Fertilize the Soil then the Crop. *Commercial Fertilizer Year Book*, 1947.

Too Much Nitrogen Puts Plants on "Jag." *National Livestock Producer*, 1947.

Better Soil Management Makes Better Wheat. News Service, College of Agriculture, University of Missouri, 1947.

Hidden Hungers Point to Soil Fertility. *Victory Farm Forum*, pp. 1-4, 1947. Also in *The Practical Farmer*, p. 8, 1948.

How Soils Nourish Plants—Plants Barter for their Nutrients. *Philfarmer*, 2nd Quarter, 1947.

Soil and Survival. *The Land*, volume 6, pp. 383-386, 1947.

The Basic Need for Correcting Mineral Deficiencies in our Soils. 25th Annual Southern Regional Conference, U.S. Office of Education, Vocation Division, p. 2, 1947. Miscellaneous publication 3250, Alabama State Board, Montgomery, Alabama, 1947.

Soil Fertility Needs High Levels. *Farm Bureau News*, 1947.

We Can Grow Legumes on Acid Soils. *Hoard's Dairyman*, 1947. Also in *The Farmer's Digest*, pp. 56-61, 1947.

Is There A Livestock Crisis in the United States? *Victory Farm Forum*, p. 5, 1947.

Soil Fertility and Animal Production. 58th Annual Report of the Indiana State Dairy Association, Purdue University, pp. 35-52, 1947.

Soil Conservation in its Broader Implications. Conservation and a Stable Society. *Ecological Society of American Natural Resources Council*, 3rd Annual Meeting, 1947.

How Soils Nourish Plants: Balanced Diets Required by Plants. *Philfarmer*, 1st Quarter, 1947.

How Soils Nourish Plants: Plants "Select" Their Diets from the Soil. *Philfarmer*, 3rd Quarter, 1947.

How Soils Nourish Plants: Soil is a Collection of Nutrient-Bearing Mineral Centers Undergoing Mobilization by Acid Clay. *Philfarmer*, 4th Quarter, 1947.

Our Teeth and Our Soils. Annals of Dentistry, number 6, 1947. Lee Foundation for Nutritional Research, Milwaukee, Wisconsin, Reprint Number 37, 1947. Also *Missouri Agricultural Experiment Station Circular 333*, 1948.

"Tomorrow's Food"—The Coming Revolution in Nutrition. Book review by Dr. Albrecht. *New Agriculture,* pp. 57-58, 1947.

Soil Fertility as a Pattern of Possible Deficiencies. *Journal of the American Academy of Applied Nutrition,* volume 1, pp. 17-28, 1947.

Soils in Relation to Human Nutrition. Address given at the Special Women's Meeting of the 4th Annual Meeting of the Western Colorado Horticulture Society, Grand Junction, Colorado, 1947.

[An Illustrated Lecture by Ollie E. Fink with information provided by Dr. Albrecht.] No date. Probably 1947.

The Quality of Our Food—Where it Comes From and What it Does. A Review of *"The Soil and Health* by Sir Albert Howard by W. A. Albrecht. *Journal of the New York Botanical Gardens,* 1947.

1948

Nutrition and the Climatic Pattern of Soil Development. *American Association for the Advancement of Science Central Volume,* 1948.

Abstract of a paper presented at the Centennial Celebration of the American Association for the Advancement of Science, 1948, Washington, D.C. Published in Science 108:599, 1948. Also in the *Journal of American Dietetic Association,* 1949.

Diversity of Amino Acids in Legumes According to the Soil Fertility. *Science,* volume 108, pp. 426-428. Co-authors, William Blue and V. L. Sheldon.

Microbiological Assays of Hays for Their Amino Acids According to Soil Types and Treatments, Including Trace Elements. *Proceedings Soil Science Society of America,* volume 13, pp. 318-322, 1948. Co-authors, William G. Blue and Victor L. Sheldon.

Composition of Alluvial Deposits Viewed as Probable Source of Loess. *Proceedings of Soil Science Society of America,* volume 13, pp. 468-470, 1948. Co-author, Alvin Beavers.

Carbohydrate-Protein Ratio of Peas in Relation to Fertilization with Potassium, Calcium, and Nitrogen. *Proceedings Soil Science Society of America,* volume 13, pp. 352-357, 1948. Co-authors, R. A. Schroder and C. G. Vidalon.

Potassium Helps Put More Nitrogen Into Sweet Clover. *Journal American Society of Agronomy,* volume 40, pp. 1106-1109, 1948.

Climate, Soil and Health, I: Climatic Soil Pattern and Food Composition. *Oral Surgery, Oral Medicine and Oral Pathology,* volume 1, pp. 199-214, 1948.

Climate, Soil and Health II: Managing Health Via the Soil. *Oral Surgery, Oral Medicine and Oral Pathology,* volume 1, pp. 214-, 1948.

Soil Acidity is Beneficial, Not Harmful. *Guernsey Breeders Journal,* 1948. Also in *Southern Florist and Nurseryman,* 1948.

Building Soils for Better Herds. *The Polled Hereford World,* 1948.

What's New in Soil Husbandry. *Better Farming Methods,* 1948.

How Soils Nourish Plants. The Nitrogen Problem—Its Climatic Aspects; What is "Hard" or "Soft" according to Pattern of Soil Nitrogen; Drought Damage May be Starvation for Nitrogen. *Philfarmer,* 1st, 2nd and 3rd Quarters, 1948.

National Pattern of Tooth Troubles Points to Pattern of Soil Fertility. *Journal Missouri State Dental Association,* volume 106, p. 8, 1948.

Quantity or Quality. *The Grain and Feed Review,* 1948.

Quality of Crops Also Depends on Soil Fertility. *Victory Farm Journal,* 1948.

Soil and Proteins. *The Land,* volume 7, 1948.

Food—What's In It? *The Land,* p. 55, 1948.

Pasture Grasses Need Additional Nutrients to Furnish Livestock Ample Protein Supply. *Dairy and Poultry News,* volume 3, 1948.

Diseases as Deficiencies via the Soil. *The Iowa State College Veterinarian,* volume 12, number 3, 1948.

There's A Cure for Sick Soils. *The Shorthorn World,* volume 32, p. 21, 1948.

Soil Fertility and the Nutritive Value of Foods. *Agriculture Leaders Digest,* 1948.

Soil Fertility and the Nutritive Value of Foods. *Agriculture Leaders Digest,* 1948.

Some Rates of Fertility Decline. Better Crops with Better Foods, 1948.

Fertilizers Reduce Root Rot of Sweet Clover. *The Furrow,* volume 53, 1948.

Lime Your Soils for Better Crops. *Missouri Agricultural Experiment Station Bulletin 566,* 1948.

Soil Microbes Get Their Food First. *Victory Farm Forum,* 1948.

Our Soils, Our Food and Ourselves. *Farmer's Digest,* 1948. Also *Organic Farming,* pp. 9-13, 1950.

Root Rot of Sweet Clover Reduced by Soil Fertility. *Better Crops with Plant Food,* 1948.

Is the Cure in the Soil? A Report on "Trace" Elements as Nutrients for Plant and Animal. *The Furrow,* 1948.

Soil Fertility and Nutritive Value of Foods. *Agricultural Leaders Digest,* 1948. Also in *The Land,* volume 7, number 3, 1948.

When and What Do We Eat? *The Practical Farmer,* 1948.

Animals Appreciate Good Soil Treatment. *The Practical Farmer,* 1948.

The Earth Is Ours. Book review by W. A. Albrecht. *Soil Conservation,* volume 13, number 12, p. 264, 1948.

1949

Nutrition and the Climatic Pattern of Soil Development. *Journal of American Dietetic Association,* 1949.

Nutrition Via Soil Fertility According to the Climatic Pattern. Commonwealth Scientific and Industrial Reserach Organization, Melbourne, Australia, pp. 1-30, 1949. Published by Thurston Chemical Company as Plant and Animal Nutrition in Relation to Soil and Climatic Factors, 1951, Proceedings Specialist Conference, London.

Brucella Infections. *The Merck Report*, pp. 13-14, 1949. Also *Mother Earth, Journal of the Soil Association*, volume 7, pp. 49-50. Co-authors, F. M. Pottenger, Jr. and Ira Allison.

Are We Going to Grass? *The Shorthorn World*, 1949.

When is a Legume not a Legume? *News and Views*, volume 4, number 1, pp. 112. Coke Oven American Research Bureau, 1949.

The Fundamentals of Soil Survival. *Missouri College Farmer*, 1949.

What's New in Soil Husbandry? *Better Farming Methods*, 1949.

Cows Are Capable Soil Chemists. *Guernsey Breeders Journal*, 1949. Also in *The Organic Farmer*, 1950.

Gearing the Soil to our Economy. *The Mail* [Adelaide, Australia], 1949.

Soil the Assembly Line of Agriculture. *Queensland Country Life* [Australia], 1949.

Soil as the Basis of Wildlife Management. *Missouri Conservation Commission Circular 134*, pp. 3-9, 1949.

Trace Minerals in Relation to Animal Health. *Stock and Land* [Australia], 1949.

Nitrogen for Proteins and Protection Against Disease. *Victory Farm Forum*, 1949.

Erosion Measuring Ideas Spread to Australia. *Missouri College Farmer*, p. 8, 1949.

Paper for the Fifth Seminar, or Trace Element Clinic, 1949.

Declining Soil Fertility, Its National and International Implications. An Address by Dr. Albrecht at the 4th Annual Convention of the National Agricultural Limestone Association, Inc., Washington, D.C., 1949.

1950

Soil Fertility Pattern: Its Suggestion About Deficiencies and Disease (Part II). Soil Fertility: Its Climatic Pattern (Part I). *The Journal of Osteopathy*, volume 57, number 11, pp. 19-25, 1950; volume 57, number 12, pp. 23-26, 1950; volume 58, pp. 12-16, 1951.

Soils and Their Minerals. *Rural New Yorker*, 1950.

Soil is the Key to Good Food, Good Health. *Pacific Northwest Cooperator*, Walla Walla, Washington, 1950.

Quality of Food Crops According to Soil Fertility. *The Technology Review*, volume 52, pp. 432-436, Massachusetts Institute of Technology, Cambridge, Massachusetts, 1950.

Cows Refuse Grass to Eat Weeds. *Capper's Farmer*, 1950.

Weed Killers and Soil Fertility. *Rural New Yorker*, 1950.

Variable Levels of Biological Activity in Sanborn Field After Fifty Years of Treatment. *Bio-Dynamics*, pp. 6-15, 1950.

"Deep-Rooting" Depends on Your Soil and How Your Treat It—Not What You Plant. *Flying Plowman*, pp. 4-5, 1950. Also in *The Practical Farmer*, 1952.

Plenty of Moisture, Not Enough Soil Fertility. *Better Crops with Plant Food*, volume 34, pp. 20-21, 1950. Missouri Farm News Service, 1950.

Phosphorus Helps Grass Make Quick New Growth. Missouri Farm News Service, 1951.

Too Much Nitrogen or Not Enough Else? *National Livestock Producer*, 1950.

Helps in Transition by Gamma Alphans. *The Gamma Alpha Record*, Missouri Issue, volume 4, pp. 7-10, 1950.

Soil Scientist Theorizes on Reasons for Cutback in Sheep. *The Weekly Kansas City Star*, July 26, 1950.

Soils and Proteins, The Declining Protein Content of Wheat and Corn is a Sign of Decreasing Soil Fertility. *The Organic Farmer*, p. 16, 1950.

Our Soils and Our Health (High Time to Learn About). Rogers Grains & Feed Company, Ainsworth, Nebraska, 1950.

Effect of Soil Deficiencies on National Health. National Council of Farmer Cooperatives, Chicago, Illinois, 1950.

Chemicals in Food Products. Testimony by W. A. Albrecht to House of Representatives, Select Committee to Investigate the Use of Chemicals in Food Products, Washington, D.C., 1950.

Soil Microbes. . .Eat First. *Let's Live*, January 1950.

Health Is Born in the Soil. *Let's Live*, February 1950.

European Composts Really "Make" Manure. *Let's Live*, March 1950.

Soil Is Key to Good Food, Good Health. *Let's Live*, April 1950.

Our Teeth and Our Soils. *Let's Live*, May 1950.

Soil Nitrogen Still a Big Problem. *Let's Live*, June 1950.

Pass-along Hungers. *Let's Live*, July 1950.

Live Soil and Pass-along Hungers. *Let's Live*, August 1950.

We Take Our Soil for Granted. *Let's Live*, September 1950.

Calcium and Soil-borne Nutrients. *Let's Live*, October 1950.

Soil Fertility Pattern. *Let's Live*, November 1950.

Quality Garden Crops. *Let's Live*, December 1950.

1951

Pattern of Caries in Relation to the Pattern of Soil Fertility in the United States. *Dental Journal of Australia*, volume 23, pp. 1-7, 1951. Also in *Journal of Applied Nutrition*, volume 10, pp. 521-524, 1957.

Plant Nutrition and the Hydrogen Ion, VII: Cation Exchange between Hydrogen Clay and Soils. *Missouri Agricultural Experiment Station Research Bulletin 477*, 1951. Co-author, D. A. Brown.

Biosynthesis of Amino Acids According to Soil Fertility, I: Tryptophane in Forage Crops. *Plant and Soil* (Holland) III, number 1, pp. 33-39; II. Methionine Content of Plants and the Sulfur Applied. *Plant and Soil* (Holland) III, number 4, pp. 361-365.

Complementary Ion Effects in Soil as Measured by Cation Exchange Between Electrodialized Hydrogen, Clay and Soils. *Proceedings of Soil Science Society of America*, volume 15, pp. 133-138, 1951. Co-author, D. A. Brown

Reconstructing the Soils of the World to Meet Human Needs. *Chemurgic Papers*, National Farm Chemurgic Council, number 5, pp. 1-12, 1951.

Phosphorus Gives Quick Recovery in Grasses that are Cut or Grazed. *Better Crops and Soils*, 1951. Also, Missouri Farm News Service, 1951.

Diseases as Deficiencies via the Soil. *The Iowa State College Veterinarian*, number 12, 1951. Also in *Natural Food and Farming Digest*, pp. 106-109, 1957.

War, Some Agricultural Implications. *The Organic Farmer*, volume 2, number 8, pp. 19-23, 1951. Also in *The Practical Farmer*, 1951. Also in *Health for You*, volume 1, number 8, p. 4, 1953.

What's New in Soil Husbandry? *Better Farming Methods*, p. 70, March 1951.

Soil Fertility and Our National Future. Hoblitzell Agriculture Laboratory, Texas Research Foundation, Special Series No. 1, 1951.

Our Soils and Our Foods. Annual Report, Nebraska Crop Improvement Association, pp. 16-24, 1951.

Soil and Democracy. *Journal of American Academy of Applied Nutrition*, pp. 14-17, 1951.

Why Roots Grow Deep? *Farm Quarterly*, pp. 66-68, Winter 1951.

Fairy-Ring Mushrooms Make Protein-Rich Grass. *Bulletin Torrey Botanical Club*, volume 78, number 1, pp. 83-88, 1951.

Managing Nitrogen to Increase Protein in Grains. *Victory Farm Forum*, pp. 16-18, 1951.

Soil Fertility in Relation to Animal and Human Health. *Milk Industry Foundation Convention Proceedings*, volume 5, Detroit, Michigan 1951.

Starter Fertilizer or Sustaining Fertility. *Commercial Fertilizer*, 82:6, June 1951.

Science, Thinking Naturally. No date or source. Probably 1951.

Nitrogen for Proteins and Protection. *Let's Live*, January 1951.

Health is Born in the Soil. *Let's Live*, February 1951.

Soil. . .to Feed Us or to Fail Us. *Let's Live*, March 1951.

A Weak Soil Body. *Let's Live*, April 1951.

Lime—Nutritional Service to Plant Growth. *Let's Live*, May 1951.

Roots Don't Go Joy-Riding. *Let's Live*, June 1951.

Getting Our Minerals. *Let's Live,* July 1951.

Fertile Soils Make Big Plants. *Let's Live,* August 1951.

Feed Your Plants Well. *Let's Live,* September 1951.

Protein Is Protection. *Let's Live,* October 1951.

Animals Prefer Nutritional Values. *Let's Live,* November 1951.

Wildlife Looks for Better Nutrition. *Let's Live,* December 1951.

1952

Potassium-Bearing Minerals as Soil Treatments. *Missouri Agricultural Experiment Station Research Bulletin 510,* 1952. Co-author, E. R. Graham.

Protein Deficiencies Via Soil Deficiencies. I: Ecological Indications. *Oral Surgery, Oral Medicine and Oral Pathology Journal,* volume 5, pp. 371-383, 1952. II: Experimental Evidence, volume 5, pp. 483-499, 1952.

Correcting Soil Deficiencies for More and Better Forage from Permanent Pastures. *Missouri Agricultural Experiment Station Bulletin 582,* 1952.

Soil Acidity as Calcium (Fertility) Deficiency. *International Society of Soil Science Joint Meeting Commission II and IV,* volume 1, Dublin, 1952. Also in *Missouri Agricultural Experiment Station Research Bulletin 513,* 1952. Co-author, G. E. Smith.

The Cow Ahead of the Plow. *Milking Shorthorn Journal,* pp. 10, January 1952. Also in *Guernsey Breeders Journal,* pp. 1173-1177, April 1952. Also *Shorthorn World,* pp. 70 ff., August 10, 1952.

Health of Man and the Soil. *Certified Milk,* volume 27, p. 10, March 1952.

Poison Weeds or Pamper Crops? *New Agriculture,* volume 34, pp. 6-7, March-May 1952.

The Value of Organic Matter. *The Rural New Yorker,* May 3, 1952. Also in *Modern Nutrition,* pp. 13-15, 1952.

The Use of Mulches. *Bulletin Garden Club of America,* volume 40, number 3, pp. 83-89, May 1952.

Soil Fertility—A Weapon Against Weeds. *The Practical Farmer,* Allentown, Pennsylvania, 1953. Also in *The Organic Farmer,* volume 3, number 11, 1952.

Pastures and Soils. *Cornbelt Livestock Feeder,* volume 4, pp. 6-9, number 10, 1952.

Proteins and Reproduction. *The Land,* volume 11, number 2, 1952.

Better Proteins Grow on Better Soils. *Commercial Fertilizer,* 1952.

Soil Science Looks to the Cow. *The Polled Hereford World,* volume 6, number 9, 1952.

Nitrogen Increases Protein in Grains. *Nitrogen News and Views*, volume 7, number 5, 1952.

Educating Potential Teachers of Soils. *School of Science and Mathematics*, volume 52, pp. 617-621, November 1952.

More and Better Proteins Make Better Food and Feed. *Better Crops with Plant Food*, volume 36, 1952.

How Smart is a Cow? *Missouri Ruralist*, volume 93, number 21, 1952.

Hidden Hunger & Soil Fertility. *Victory Farm Forum*. Also in *The Challenge*, 1952.

Taking Our Soils for Granted. *The Philfarmer*, 1952. Also in *The Challenge*, 1952.

Education Versus Training. A Request Presentation made at the Annual Meeting of the American Society of Agronomy, Davis, California, 1955; also at the Honor's Convocation, University of Missouri, 1952.

Nitrogen for Proteins and Protection Against Disease. No date and no source. Probably 1952.

Report of the Subcommittee on Trace Elements, submitted before the General Committee on Fertilizer Application, Cincinnati, Ohio, November 1952.

How Much Can Your Soil Do? *Missouri Ruralist*, volume 93, November 22, 1952.

Now We Know Lime is a Plant Food—Not Merely a Treatment for Acidity. *Missouri Ruralist*, volume 94, number 1, 1952.

Will We Have Enough to Eat? *Missouri Ruralist*, volume 93, December 13, 1952.

La fertilidad del suelo y el valor nutritivo de los alimentos. *Campo*, volume 5, number 71, pp. 25 ff., La Paz, Bolivia, 1956.

Soil Treatment and Wool Output. *Let's Live*, January 1952.

Diseases As Deficiencies Via the Soil. *Let's Live*, February 1952.

How Smart is a Cow? *Let's Live*, March 1952.

Quality vs. Quantity Crops. *Let's Live*, April 1952.

"Deep Rooting" Depends on Soil. *Let's Live*, May 1952.

Soil Fertility and Nutritive Food Values. *Let's Live*, June 1952.

Soil and Proteins. *Let's Live*, July 1952.

Our Teeth and Our Soils. *Let's Live*, August 1952.

Mineral Hunger. *Let's Live*, September 1952.

Soil Acidity is Beneficial. *Let's Live*, October 1952.

Too Much Nitrogen? *Let's Live*, November 1952.

Protein Deficiencies. . .through Soil Deficiencies. *Let's Live*, December 1952.

1953

Biosynthesis of Amino Acids According to Soil Fertility, III: Bioassay of Forage and Grain Fertilized with "Trace" Elements. *Plant and Soil* (Holland), volume 4, pp. 336-343, 1953. Co-author, Fred E. Koehler.

Soil Fertility and Amino Acid Synthesis by Plants. *Proceedings National Institute of Science* (India), volume 19, pp. 89-95, 1953. Co-authors, William G. Blue and V. L. Sheldon.

Some Rates of Fertility Decline. *Journal of the Soil Association,* pp. 47-48, 1953.

Manage Nitrogen Fertilizer Application for Best Protein Production. *Missouri Ruralist,* volume 94, January 24, 1953.

Legumes Take Nitrogen from the Air, But Need Some from the Soil. *Missouri Ruralist,* January 24, 1953.

In Defense of the Cow. *Livestock Weekly,* March, 1953.

How Can We Best Use Chemical Soil Conditioners? *Better Farming Methods,* p. 14, 1953.

High Time to Learn About Our Soils and Our Health. *Agriculture Leaders Digest,* volume 34, 1953.

On Soils and Cattle. *Rural New Yorker,* March 16, 1953.

Proteins Are Becoming Scarcer as the Soil Fertility Goes Lower. *The Polled Hereford World,* volume 7, number 7, pp. 20-27, 1953.

The Fertility Pattern (Human Ecology). *The Land,* volume 12, pp. 217-220, 1953.

Soil Fertility and Plant Nutrition—Some Basic Principles. No date, no source. Probably 1953.

Soil Fertility and Nutritive Value of Foods. *The Practical Farmer,* volume 14, number 3, 1953.

Some Aims of Soil Research. Speech before agronomist and fertilizer meeting, Chicago, 1953.

Chemicals for the Improvement of Soils. No date, no source. Probably 1953.

Soil and Nutrition (Via Complete Proteins). Presented at 30th Annual Convention, California Fertilizer Association, Carmel, California, 1953.

Less Water Required Per Bushel of Corn with Adequate Fertility. Missouri Farm News Service, 1953.

Soil and Nutrition. *Let's Live,* January 1953

The Role of Nitrogen. *Let's Live,* February 1953.

More and Better Proteins. *Let's Live,* March 1953.

Proteins and Reproduction. *Let's Live,* April 1953

Protein Protection. *Let's Live,* May 1953.

Better Proteins. *Let's Live,* June 1953.

Soil Fertility for Proteins. *Let's Live,* July 1953.

Soil Conservation. *Let's Live,* August 1953.

The Importance of Soil. *Let's Live,* September 1953.

High Time to Learn About Our Soils and Our Health. *Let's Live,* October 1953.

The Use of Mulches. *Let's Live,* November 1953.

Protein Service in Nutrition. *Let's Live,* December 1953.

1954

"Let Rocks Their Silence Break." *11th Annual Meeting American Institute of Dental Medicine, Annual Volume,* Palm Springs, California, November 1, 1954.

Soil and Nutrition. *American Institute of Dental Medicine, Annual Volume,* Palm Springs, November 2, 1954.

Droughts—The Soil as Reasons for Them. *Institute of Dental Medicine, Annual Volume,* Palm Springs, California, November 3, 1954.

How to Sell More Fertilizer. *Commercial Fertilizer,* volume 88, pp. 20-23, January 1954.

Do We Overlook Protein Quality? *What's New in Crops and Soils,* volume 6, number 5, p. 9, 1954.

Soil Acidity (Low pH) Spells Fertility Deficiencies. *Pit and Quarry,* 1954.

Lime the Soil to Correct its Major Deficiencies. *Rock Products,* volume 57, number 4, pp. 116-117, 1954. Also in *Natural Food and Farming,* volume 2, number 1, 1955.

Some Aims of Soil Research. *Better Crops with Plant Food Magazine,* volume 38, number 15, 1954.

Statement before the Senate Committee. Hearings before the Senate Appropriations Committee on Agriculture, 83rd Congress, Second Session, HR 8779, pp. 1161-1174, 1954.

Drought. *Better Crops with Plant Food,* volume 38, pp. 6-8, 1954. Also in *Modern Nutrition,* volume 7, pp. 10-13, 1954.

Fertilizer's Service in Plant Nutrition. *Proceedings 4th Annual Convention,* Agricultural Ammonia Institute, December 6-8, 1954.

A Specialist Questions Use of Phenothiazine for Sheep. *Weekly Star Farmer,* January 6, 1954.

Boron Improves Alfalfa Quality. Missouri Farm News Service, May 19, 1954.

Agronomists in National Defense Activities. *Agronomy Journal,* volume 46, number 1, pp. 36-47, 1954.

Lime Soil to Feed Crops—Not to Remove Soil Acidity. Missouri Farm News Service, 1954.

Are We Poisoning Our Sheep? *Organic Gardening and Farming,* volume 1, number 4, pp. 30-33, 1954.

Amino Acids in Legumes According to Soil Fertility. *Let's Live,* January 1954.

The Soil Fertility Pattern. *Let's Live,* February 1954.

The Little Things Count in Nutrition. *Let's Live,* March 1954.

Animals Know Good Food! *Let's Live,* April 1954.

Agriculture is Biology First and Foremost. *Let's Live,* May 1954.

The Upset Biological Processes. *Let's Live,* June 1954.

Consider the Soil—Not Technologies! *Let's Live,* July 1954.

Some Soils Analyzed. *Let's Live,* August 1954.

The Role of Clay in Plant Nutrition. *Let's Live,* September 1954.

Root Chemistry and Clay Chemistry. *Let's Live,* October 1954.

The Sustaining Fertility of the Soil. *Let's Live,* November 1954.

The Importance of Soil Economics. *Let's Live,* December 1954.

1955

The Influence of Soil Mineral Elements on Animal Nutrition. *Michigan State University Centennial Symposium on Nutrition of Plants, Animal and Man,* 1955.

How Good Is Grassland Farming? *The Organic Farmer,* volume 2, pp. 56-59, 1955.

Capital Is No Substitute for Soil Fertility. *Rock Products,* 1955.

Why Your Cattle Break Through the Fence. *Western Livestock Journal,* volume 33, pp. 35-38, 1955.

The Living Soil. *Parks and Sports Grounds,* volume 20, number 9, pp. 577-578; number 11, pp. 728-752, 1955. Also in *Journal of Applied Nutrition,* volume 9, pp. 382-386, 1955.

Make Tax Allowances for Fertility Depletion. *Agriculture Leader's Digest,* volume 6, p. 34, 1955.

Trace Elements and Agricultural Production. *Journal of Applied Nutrition,* volume 8, pp. 352-354, 1955.

Soil Areas, Medical Rejectees Give Similar Maps. Missouri Farm News Service, 1955.

Quality as Feed, Not Only Quality as Crop, Demonstrates Effects of Soil Fertility. It's the Soil That Feeds Us. Part 5. No date, no source. Probably 1955.

Weeds, As the Cows Classify Them. It's the Soil That Feeds Us. Part 6. No date, no source. Probably 1955.

Education Versus Training. Presented before the American Society of Agronomy Section on Teaching, 1955. Also in *Man's Role in Changing the Face of the Earth,* pp. 648-673. William L. Thomas, Editor, University of Chicago Press, 1956. Also Wenner-Gren Foundation International Symposium, *Man's Role in Changing the Face of the Earth,* Princeton, New Jersey, 1955.

It's the Soil That Feeds Us. Climates Make Soils to Feed or to Fail Us. Part I. *Natural Food and Farming,* 1955.

Fighting Drought on the Farm. *Organic Gardening and Farming,* volume 20, number 9, pp. 56-60, 1955.

More Thoughts on Availability. Extracts from an article, *Natural Food and Farming, Mother Earth, Journal of the Soil Association,* London, 1955.

Consider the Soil—Not Only Technologies! *Natural Food and Farming,* volume 2, number 1, 1955.

Soils, Plant, Animal and Human Health. Manuscript, April 21, 1955.

Proteins, The Struggle for Them by all Forms of Life, Premised on the Fertility of the Soil. Manuscript, 1955.

Should Farmers Receive Tax Allowance for Soil-Building. Missouri Farm News Service, volume 44, number 51, 1955.

Not How Much but How Good. *Missouri Conservationist,* 1955.

Soil Fertility. . .Its Climactic Pattern. *Let's Live,* January 1955.

Problems of the Small Farm. *Let's Live,* February 1955.

The Pattern of the Small Farm. *Let's Live,* March 1955.

Income Problems of the Small Farm. *Let's Live,* April 1955.

Use of Resources of the Small Farm. *Let's Live,* May 1955.

Why Small Farms Are Popular. *Let's Live,* June 1955.

Creating Good Soil is a Challenge. *Let's Live,* July 1955.

Biology vs. Technology in Growing Things. *Let's Live,* August 1955.

Organic Soils and "Good Constitutions." *Let's Live,* September 1955.

The Soil's Assembly Line. *Let's Live,* October 1955.

Some Aims of Soil Research. *Let's Live,* November 1955.

Make Tax Allowance for Fertility Depletion. *Let's Live,* December 1955.

1956

Soils, Nutrition an Animal Health. *Journal of American Society of Farm Managers and Rural Appraisers,* volume 20, pp. 24-37, number 7, 1956.

Agricultural Limestone for the Sake of More than its Calcium. *Pit and Quarry,* 1956.

Boden und Ernahrung. Planzen qualitat-Nahrungs-grundlage. Zeitsctrift Landiwirtschaftliche Forchunge, Frankfurt, Germany, 1956.

Proteins. *Modern Nutrition,* volume 9, number 7, pp. 16-18; number 8, pp. 4-6; number 10, pp. 4-7, 1956.

White Clover Years in Cycles of Soil Changes. *Better Crops with Plant Food,* volume 11, p. 17, 1956.

Commercial Fertilizers. *Missouri Agricultural Experiment Station Research Bulletin 607,* 1956.

Trace Elements and Agricultural Production. *Let's Live,* January 1956.

Our Soils Affect Nutrition. *Let's Live,* February 1956.

Basic Facts of Soil Science. *Let's Live,* March 1956.

A New Book—"Our Daily Poison." *Let's Live,* April 1956.

Soils, Plants and Nutrition. *Let's Live,* May 1956.

Soil Calcium and the Quality of Leafy Greens. *Let's Live,* June 1956.

Soil Alters Calcium Digestibility in Leafy Greens. *Let's Live,* July 1956.

Is Commercial Urea an "Organic" Fertilizer? *Let's Live,* August 1956.

Compost for the Sea or the Soil? *Let's Live,* September 1956.

Food Quality—as Physiology Demands It. *Let's Live,* October 1956.

Blast Furnace Slag—a Soil Builder. *Let's Live,* November 1956.

Plants Struggle for their Proteins, Too. *Let's Live,* December 1956.

1957

Soil Fertility and Quality of Seeds. *Missouri Agricultural Experiment Station Research Bulletin 619,* 1957. Co-author, Robert L. Fox.

Soil Fertility and Biotic Geography. *The Geographical Review,* volume 47, number 1, pp. 86-105, 1957.

Fertilite du Sol et Geógraphic Biotique aux Etats Unis. *L'Agronomic Tropicale,* volume 12, number 3, pp. 355-356, 1957.

Trace Elements and the Production of Proteins. *Journal of Applied Nutrition,* volume 10, pp. 534-543, 1957.

Fertilize for Higher Feed Value. Joint meeting of Agronomists-Fertilizer Industry, 1957.

Rhododendrons. . .a Problem of Soil Fertility, Not Acidity. *Let's Live,* January 1957.

Is Soil Fertility via Food Quality Reported in Your Varied Pulse Rate? *Let's Live,* February 1957.

Breeding Out Plant Proteins—Bringing in Diseases. *Let's Live,* March 1957.

Cycles of Soil Changes In White Clover Years. *Let's Live,* April 1957.

Cycles of Soil Changes in White Clover Years. *Let's Live, May 1957.*

Fertilizing with Nitrogen: The Cow Makes Her Suggestions. *Let's Live,* June 1957.

Fertilizing with Nitrogen: Rabbits Testify by Experiments. *Let's Live,* July 1957.

Fertilizing with Nitrogen: We May Use Too Much Salt. *Let's Live,* August 1957.

Fertilizing with Nitrogen: Fertility Imbalance and Insect Damage. *Let's Live,* September 1957.

Too Much Salt for the Soil. *Let's Live,* October 1957.

Blood Will Tell. *Let's Live,* November 1957.

What Texture of Soil is Preferred? *Let's Live,* December 1957.

1958

Calcium—Boron Interactions. Demonstrated by Lemna Minor on Clay Suspensions. *Missouri Agricultural Experiment Research Bulletin 663*, 1958. Co-author, Robert L. Fox.

Better Soil Fertility, Less Plant Pests and Diseases. *Better Crops with Plant Food*, 1958. Also in *American Agricultural Reports*, 1958.

Growing Our Own Protein Supplements. *Polled Hereford World*, volume 12, number 7, pp. 28, 1958.

Soil Fertility and Plant Nutrition. Some Basic Principles. *Natural Food and Farming*, volume 5, number 4, pp. 6-11, 1958.

Soil Fertility and Animal Health. Fred Hahne Printing Company, Webster City, Iowa, 1958.

Some Significant Truths About the Good Earth. *Natural Food and Farming*, volume 5, number 6, pp. 17ff., 1958.

Fertile Soils Lessen Insect Injury. No date, no source. Probably 1958.

Silt Loams—Nutritional Blessing of the Winds. *Let's Live*, January 1958.

What Texture of Soil is Preferred? II. Clay—The Soil's Jobber. *Let's Live, February 1958.*

Water—Major Mineral of Soil Nutrition. *Let's Live*, March 1958.

Water—Nature's Major Biochemical Reagent. *Let's Live*, April 1958.

Cows Know Nutrition. *Let's Live*, May 1958.

Soil Organic Matter, Builder of Climax Crops. *Let's Live*, June 1958.

Soil Organic Matter—Fertility and Crop Needs. *Let's Live*, July 1958.

Soil Organic Matter—"Constitution of the Soil." *Let's Live*, August 1958.

Soil Organic Matter—Mobilizer of Inorganic Soil. *Let's Live*, September 1958.

Soil Organic Matter—Includes Much "Et Cetera." *Let's Live*, October 1958.

Soil Organic Matter—Includes Much "Et Cetera." *Let's Live*, November 1958.

Soil Organic Matter—Farm Manures Help Maintain It. *Let's Live*, December 1958.

1959

Human Health Closely Related to Soil Fertility. *School and Community*, volume 46, pp. 20-21 (Missouri State Teachers Association, Columbia, Missouri), 1959. Co-author, Charles Boyles.

Soil and Health. Some Imbalances, Deficiencies and Deceptions via Soil and Crops. *Natural Food and Farming*, volume 5, p. 6, 1959.

Nature Teaches Health via Nutrition (Gueste Editorial), *Journal of Applied Nutrition*, volume 12, number 4, p. 162, 1959.

Diagnoses or Post-Mortems. Declining Soil Fertility Brings Pests and Diseases. *Natural Food and Farming*, volume 6, p. 6, 1959.

Nitrogen, Proteins and People. *Agricultural Ammonia News*, numbers 1 and 2, pp. 18-20, 1959.

Studying Nature. *Natural Food and Farming*, volume 6, pp. 10, 1959.

Book Review, Encyclopedia of Organic Gardening. *Modern Nutrition*, volume 12, number 6, pp. 25-26, 1959. Also in *Natural Food and Farming*, volume 5, number 11, 1959.

Book Review. Ear, Nose and Throat Dysfunctions Due to Deficiencies and Imbalances. *Journal of Applied Nutrition*, volume 12, number 4, pp. 178-179, 1959.

Soil Organic Matter—Possible Poisons Naturally. *Let's Live*, January 1959.

Soil Organic Matter—Possible Poisons of our Own Make. *Let's Live*, February 1959.

Soil Organic Matter—And Man-Made Poisons. *Let's Live*, March 1959.

Soil Organic Matter—Man-Made Poisons. *Let's Live*, April 1959.

Natural Organic Matter—Man-Made Organic Supplements. *Let's Live*, May 1959.

Soil and Plant Compositions: 1. Too Much Nitrogen or Not Enough Else? *Let's Live*, June 1959.

Different Soils, Different Plant Compositions (Both Soil, Plant Responsible). *Let's Live*, July 1959.

Different Soils, Different Plant Compositions (Both Soil, Plants Responsible). *Let's Live*, August 1959.

Different Soils, Different Plant Compositions (Phosphorus in Crop Varies with Nitrogen Applied). *Let's Live*, September 1959.

Different Soils, Different Plant Compositions (Soil Nitrogen and Vitamin C in Plants). *Let's Live*, October 1959.

Different Soils, Different Plant Compositions (Varied Soil Potassium Means Varied Organic Values). *Let's Live*, November 1959.

Different Soils, Different Plant Compositions. *Let's Live*, December 1959.

1960

Trace Elements, Allergies, and Soil Deficiencies. *The Journal of Applied Nutrition*, volume 13, pp. 20-32, 1960. Co-authors, Lee Pettit Gay and G. S. Jones.

Boden und Gesundheit. Boden, Pflanzen, Pestizide; Fruchtbarer Boden—widerstandsfahige, planzen, Germany, 1960.

Kleine Lebenskunde (I). Gesundheit Gedeiht nur auf Fruchtbaren Boden, number 35, pp. 2-4, 1960.

Kleine Lebenskunde (II). Frage der Kalkdungen, number 36, pp. 22-23, 1960.

Kleine Lebenskunde (III). Calcium: eine Notwendigkeit von Anbegin, number 37, pp. 9-10, 1960.

Kleine Lebenskunde (IV). Calcium: Biochemisch Wichtiger als Chemisch, number 39, pp. 8-9, 1960.

Kleine Lebenskunde (V). Gesteinminerale and Pflanzerernahrung, number 40, pp. 7-8, 1960.

Kleine Lebenskunde (VI). Uberdie vielfaltigen indirekten Nutzen des Kalziums in Boden, number 41, pp. 5-7, 1960.

Policies Regarding Agricultural Chemicals (Special Report). *Natural Food and Farming,* volume 8, number 7, 1960.

Soils Must Nourish Plants Promptly. No date, no source. Probably 1960.

Give Your Soil a Balanced Diet. No date, no source. Probably 1960.

Book Review of *Hunza Land* [by Dr. Allen E. Banik and Renee Taylor]. *The Weekly Kansas City Star,* 1960.

Book Review of *Soil, Grass and Cancer* [by Andre Voisin]. *Weekly Star Farmer,* 1960.

It's the Soil That Feeds Us. A collection of 12 articles outlining basic principles in the relation of soil to health. Natural Food Associates, Atlanta, Texas, 1960.

Book Review of *The Poisons in Your Food; You Are What You Eat* [by William Longgwood], 1960.

Different Soils, Different Plant Compositions (Bacteria Help Legume Roots Mobilize Fertility). *Let's Live,* January 1960.

Different Soils, Different Plant Compositions (Microbes Give Legumes Their Protein Power). *Let's Live,* February 1960.

Different Soils, Different Plant Compositions (Fertility Effects Show Early in Plants). *Let's Live,* March 1960.

Different Soils. . .(Big Yields of Bulk—Low Phosphorus Concentration). *Let's Live,* April 1960.

Different Soils. . .(Boron Interrelated with Potassium). *Let's Live,* May 1960.

Different Soils. . .(Boron Helps Maintain Potassium Balance). *Let's Live,* June 1960.

Different Soils. . .(Soil Exhaustion—Variable Organic and Inorganic Composition of Plants). *Let's Live,* July 1960.

Different Soils. . .(Sulfur Deficiency in Soils). *Let's Live,* August 1960.

Different Soils. . .(Chemical Composition of Plants and the "Feeding Power" of Their Roots). *Let's Live,* September 1960.

Different Soils. . .(Chemical Composition of Plants and the "Feeding Power" of Their Roots). *Let's Live,* October 1960.

Different Soils. . .(Vegetable Quality Reveals its Connection with Soil Organic Matter). *Let's Live,* November 1960.

Different Soils. . .("Chelation")—Nature's Emphasis on Soil Organic Matter. *Let's Live,* December 1960.

1961

Healthy Plants are Resistant to Disease and Infestation. *Modern Nutrition,* volume 14, number 12, pp. 16-18, 1961.

Soils, Their Effects on the Nutritional Values of Foods. *Consumer Bulletin,* volume 44, number 1, pp. 20-23, 1961.

Wastebasket of the Earth. *Bulletin of the Atomic Scientists,* volume 17, number 8, pp. 335-340, 1961.

Insoluble Yet Available. *Farm and Garden* (England), volume 4, number 2, pp. 29-36, 1962. Originally presented at Arkansas Plant Food Conference, Little Rock, Arkansas, 1961.

The Living Soil. *Natural Food and Farming Special Report,* pp. 23-30, 1961.

Soil Fertility. The Basis for Formulating an Agricultural Policy. For Brookside Consultants Association, New Knoxville, Ohio, October 1, 1961.

Boden und Nahrung: Wie die Boden so die Inhaltsstoffe der Pflanzen, I. Stickstoff im Boden und Vitamin C in den Pflanzen. *Organischer Landbau,* volume 4, number 6, 1961.

Different Soils, Different Plant Compositions (Natural Laws Regarding Soils and Plant Compositions). *Let's Live,* January 1961.

Different Soils. . .(Depleted Soils—Species Extinction). *Let's Live,* February 1961.

Different Soils. . .(Soil Organic Matter Mobilizes the Phosphorus for Plants). *Let's Live,* March 1961.

Different Soils. . .(Soil Organic Matter Mobilizes the Phosphorus for Plants). *Let's Live,* April 1961.

Different Soils. . .(Depleted Soils Change Sugar, Starch, Proteins and Yields of Crop). *Let's Live,* May 1961.

Mobilizing the Natural Soil Potassium. *Let's Live,* June 1961.

Nitrates. . .Possible Poison Grown into Foods. *Let's Live,* July 1961.

To Keep the Soil a "Living" One. *Let's Live,* August 1961.

Soil's Resurrection in Three Years. *Let's Live,* September 1961.

Schedule of Soil Fertility Delivery and Crop Growth. *Let's Live,* October 1961.

Schedule of Soil Fertility Delivery and Crop Growth. *Let's Live,* November 1961.

An Old Problem—Loss of Applied Nitrogen. *Let's Live,* December 1961.

1962

Are We Committing Soil Suicide: *The Cotton Trade Journal 29th International Edition,* 1961-1962. Also, A Policy for Preventing Agricultural Suicide, *Natural Food and Farming,* volume 10, 1963.

Organic Matter for Plant Nutrition. *Clinical Physiology,* volume 4, number 3, pp. 212-224, 1962. Also in *Journal of Applied Nutrition,* volume 17, pp. 168-178, 1964.

The Healthy Hunzas. A Climax Human Crop. *Journal of Applied Nutrition,* volume 15, numbers 3 and 4, pp. 171-279, 1962.

An introduction to *Our Synthetic Environment* [by Lewis Herber], Alfred A. Knopf Publishers, New York, 1962.

A review of *The Ecology of Man and His Earth*, by George D. Scarseth, Iowa State University Press. Published in *Newsletter on Human Ecology of Elsah, Illinois*, volume 4, number 5, 1962.

Boden und Nahrung: Wie die Boden so die Inhaltsstoffe der Pflanzen. II. "Chelatbildung"—Die Natur hat der organischen Masse im Boden wichtige Aufgaben ubertragen. *Organischer Landbau*, volume 5, number 2, 1962.

Boden und Gesundheit: Vermehrte Fruchtbarkeit. *Boden unde Gesundheit*, volume 42, number 6, p. 6, 1962-1963.

"A Few Facts About Soils." *Let's Live*, January 1962.

"Agricultural Education." *Let's Live*, February 1962.

"The Natural" vs. "The Artificial." *Let's Live*, March 1962.

Salt Damage to Seedlings. *Let's Live*, April 1962.

Salt Damage to Seedlings. *Let's Live*, May 1962.

Purpose of Liming Soil an Enigma. *Let's Live*, June 1962.

Immunity Against Leaf-Eating Insects via Soil as Nutrition. *Let's Live*, July 1962.

Immunity Against Leaf-Eating Insects via Soil as Nutrition. *Let's Live*, August 1962.

Garden Soils and Bio-geochemistry. *Let's Live*, September 1962.

Bio-Assays Rather than Chemical Analyses. *Let's Live*, October 1962.

Their Questions—My Answers. *Let's Live*, November 1962.

Less Soil Organic Matter Spells Lower Form of Vegetation. *Let's Live*, December 1962.

1963

Soil and Health via Nutrition. *American Vegetarian Hygienist*, volume 20, number 2, 1963.

A Policy for Preventing Agricultural Suicide. *Natural Food and Farming*, volume 10, 1963.

Soils Needs Living Fertility. *Western Livestock Journal* (Anniversary Yearbook Feature), volume 41, number 9, pp. 174-175, 1963.

Organic Matter Balances Soil Fertility. *Natural Food and Farming*, volume 10, pp. 8-10, February 1963.

Soil and Survival of the Fit. *Journal of Applied Nutrition*, volume 16, numbers 2 and 3, pp. 83-100, 1963.

Biosynthesis of Amino Acids According to Soil Fertility: IV. Timothy Hay Grown with Trace Elements. *Plant and Soil*, volume 18, number 3, pp. 298-308, 1963. Co-author, John de Jonge.

Weeds Suggest Low Nutritional Values. *Let's Live*, January 1963.

Weeds Suggest Low Nutritional Values. *Let's Live*, February 1963.

Animals Choose Feed for Quality—Not for Tonnages Per Acre. *Let's Live*, March 1963.

"Rule of Return." *Let's Live,* April 1963.

Soil Fertility: First Concern for Human Survival. *Let's Live,* May 1963.

Organic Matter Makes "Healthy" Soils. *Let's Live,* June 1963.

Soil Organic Matter Under Time and Treatment. *Let's Live,* July 1963.

Changes in Quality of Soil Organic Matter. *Let's Live,* August 1963.

Soil Humus. . .Chelator of Inorganic Elements. *Let's Live,* September 1963.

Humus. . .Soil Microbial Product. *Let's Live,* October 1963.

"Mycorrhiza," I. Mobilizers of Organic Plant Nutrition. *Let's Live,* November 1963.

"Mychorrhiza," II. Misconceptions Persist. *Let's Live,* December 1963.

1964

A review of *Grass Tetany* (by Andre Voisin). *The Land Bulletin Number 85,* Ontario, Canada, 1964.

The Story of a Living Soil. *The Missouri Farmer,* volume 56, number 2, pp. 10-11, 1964.

Grow Self-Protection via Soil as Nutrition. *Clinical Physiology,* volume 6, number 1, pp. 45-53, 1964.

Pflanzenschutz durch Boden-fruchtbarkeit. Kleine Lebenskunde VIII. *Boden und Gesundheit,* volume 4, 1964.

Livestock Can Teach Us a Lesson on Nutrition from the Ground Up. *Breeder's Gazette,* volume 129, number 4, pp. 13-15, 1964; pp. 11-15, number 5, 1964. Also, Let's Look at Nutrition from the Ground Up, *Polled Hereford World Magazine,* volume 18, p. 40 ff., 1964.

Our Livestock—Cooperative Chemists. *Brangus Journal,* volume 12, number 5, pp. 4-10, 1964.

Langbeinite—Its Use as a Natural Potash-Mineral Fertilizer. Manuscript, 1964.

The Future of Langbeinite in the Maintenance of Soils. *Agricultural Chemicals,* pp. 18-20, 1964.

Fluoridation of Public Drinking Water. Manuscript, 1964.

Organic Matter for Plant Nutrition. *Journal of Applied Nutrition,* pp. 168-177, 1964.

Die Lebensfunktion des Bodens. Kleine Lebenskunde. IX. Pflanzenschutz durch Boden-fruchtbarkeit, number 45, 1964.

Let's Look at Nutrition from the Ground Up. *Polled Hereford World,* volume 18, pp. 40 ff., 1964.

"Mychorrhiza," II. Misconceptions Persist. *Let's Live,* January 1964.

"Mycorrhiza," III.Facts About Their Magnitude. *Let's Live,* February 1964.

"Mycorrhiza," III. Facts About Their Magnitude. *Let's Live,* March 1964.

"Mycorrhiza," IV. Revelations of Species. *Let's Live,* April 1964.

"Mycorrhiza," V. Parasite or Symbiont According to Soil as Nutrition. *Let's Live*, May 1964.

"Mycorrhiza," V. Parasite or Symbiont According to Soil as Nutrition. *Let's Live*, June 1964.

"Mycorrhiza," VI. Some Field Observations. *Let's Live*, July 1964.

"Mycorrhiza," VI. Some Field Observations. *Let's Live*, August 1964.

"Mycorrhiza," VII. Proteins, Amino Acids and Benzene Rings. *Let's Live*, September 1964.

"Mycorrhiza," VII. Proteins, Amino Acids and Benzene Rings. *Let's Live*, October 1964.

"Mycorrhiza," VIII. Early Beliefs Lately Confirmed. *Let's Live*, November 1964.

Magnesium. . .Balance in Soil, Plants and Bodies. *Let's Live*, December 1964.

1965

Correcting Borderline Soil Conditions. *Modern Nutrition*, volume 18, number 3, pp. 9-10, 1965.

Soils and Chemistry (Biochemistry): Frontiers of Knowledge for All. Symposium, Oklahoma City, Oklahoma, 1965. *Frontiers of Science Foundation of Oklahoma Text*, pp. 1-8, inclusive. Appendix, illustrations, "Soil as Nutrition," pp. 19-28, inclusive.

Soil Management, By Nature or By Man? *Natural Food and Farming*, volume 12, number 7, pp. 31-34, 1965.

Book review by W. A. Albrecht of *Marshall's Volume I on Soil Materials. The Missouri Soil Builder Official Publication*, The Soil Fertility and Plant Nutrition Council of Missouri, volume 13, 1966.

Trace Elements and Soil Organic Matter. Manuscript, dated 1965-1967.

Magnesium. . .Its Relations to Calcium. *Let's Live*, January 1965.

Magnesium. . .Its Relation to Calcium in Plants. *Let's Live*, February 1965.

Magnesium. . .Its Relation to Calcium in Body Tissues. *Let's Live*, March 1965.

Magnesium. . .Its Relation to Potassium. *Let's Live*, April 1965.

Magnesium. . .Its Excess, According to Plant Species. *Let's Live*, May 1965.

Magnesium. . .Indirect Modifications via Mixed Flora. *Let's Live*, June 1965.

Magnesium. . .Imbalances Among Companion Elements. *Let's Live*, July 1965.

Magnesium. . .Biochemically, So Little is so Important. *Let's Live*, August 1965.

Magnesium. . .Relation of Soil Test to Crop Analyses. *Let's Live*, September 1965.

Balanced Soil Fertility, Requisite for Nutritional Quality of Crops. *Let's Live*, October 1965.

Quality Becomes More Quantitative. *Let's Live*, November 1965.

Nutritional Quality of Vegetables via Plant Species and Soil Fertility. *Let's Live*, December 1965.

1966

The "Half-Lives" of Our Soils. *Natural Food and Farming,* volume 13, number 4, pp. 7-10, 1966.

Boden und Ernahrung. Planzenqualitat-Nahrung Sgrundlage, 1966.

Hearings before the Subcommittee on Foreign Aid Expenditures of the Committee on Government Operations, 89th Congress, Second Session on S. 1676, Part 5-A, 1966.

Plant, Animal and Human Health Vary with Soil Fertility. *Modern Nutrition,* volume 19, 1966.

Education versus Training in Christian Living. Manuscript, 1966.

Self-Protection by Plants, Lined to Nutritional Values. *Let's Live,* January 1966.

Variable Quality Production by Food Plants. *Let's Live,* February 1966.

Nutritional Values of Vegetables Change Rapidly. *Let's Live,* March 1966.

Variable Nutritive Values in Carrots. *Let's Live,* March 1966.

Variable Nutritive Values in Carrots. *Let's Live,* April 1966.

Nutritional Values of Apples, According to Chemical Analysis. *Let's Live,* May 1966.

Nutritive Value of Apples, According to Biological Assay. *Let's Live,* June 1966.

Size and Weight versus Nutritional Quality. *Let's Live,* July 1966.

Magnesium in the Soils of the United States. *Let's Live,* August 1966.

Health as Different Soil Areas Nourish It. *Let's Live,* September 1966.

Health as Different Soil Areas Nourish It. *Let's Live,* October 1966.

Soil Phosphorus—Activated via Soil Organic Matter. *Let's Live,* November 1966.

Balanced Soil Fertility—Better Start of Life. *Lets Live,* December 1966.

1967

Soil Reaction (pH) and Balanced Plant Nutrition. Mimeographed booklet distributed by W. A. Albrecht, 33 pp., 1967.

What Animals can Teach Us About Nutrition. *Here's Health,* volume 11, number 130, pp. 113-115, 1967.

Healthy Soils Mean Healthy Humans. *Here's Health,* volume 11, number 128, pp. 115-120, 1967.

Magnesium Integrates with Calcium. *Natural Food and Farming,* volume 14, number 4, 1967.

Problems of Quality in the Productivity of Agricultural Land. *Natural Resources.* Papers presented before a faculty seminar at the University of California, Berkeley, 1967. Also reprinted in *Journal of Applied Nutrition,* volume 20, numbers 3 and 4, pp. 65-78, 1968.

1968

Calcium Membrances in Plants, Animals and Man. *Journal of Applied Nutrition*, volume 20, numbers 1 and 2, pp. 10-19, 1968.

Nitrogen and Organic Matter Turnover. *Modern Nutrition*, volume 21, number 5, 1968.

1970

Concerning the Influence of Calcium on the Physiological Function of Magnesium. Translated by Dr. Albrecht from the original German as it appeared in "Die Ernahrung der Pflanze," volume 27, number 3, pp. 97-101; number 6, pp. 121-122, 1931. Translation dated 1970.

Plants Protected by Fertile Soil. *Journal of Applied Nutrition*, volume 22, numbers 1 and 2, pp. 23-32, 1970.

Nutritional Role of Calcium in Plants. I. Prominent in the Non-Legume Crops, Sugar Beets. *Plant and Soil*, volume 33, number 2, pp. 361-382, 1970.

1971

Conversations with Dr. William A. Albrecht. Interview in *Acres U.S.A.*, August 1971.

Conversations with Dr. William A. Albrecht. Interview in *Acres U.S.A.*, September 1971.

Waste Basket of the Earth. *Acres U.S.A.*, Part I, November 1971.

Fertilizing Soils with Nitrogen. *Acres U.S.A.*, December 1971.

Diseases as Deficiencies via the Soil. *Acres U.S.A.*, December 1971.

1972

Nodulation Modifies Nutrient Intake from Colloidal Clay by Soybeans, Supplements and Modifications. Originally presented in *Soil Science Society of America Proceedings*, volume 8, pp. 234-237, 1944. Co-author, W. E. Hampton, 1972.

Fertilizing Soils with Nitrogen, Part II. *Acres U.S.A.*, January 1972.

Fertilizing Soils with Nitrogen, Part III. *Acres U.S.A.*, February 1972.

Less Water Required per Bushel of Corn with Adequate Fertility. *Acres U.S.A.*, June 1972.

Fertile Soils for Self-Protection by Crops. *Acres U.S.A.*, September 1972.

In Defense of the Cow. *Acres U.S.A.*, November 1972.

1973

School of the Soil. *Acres U.S.A.*, February, March, April, May, June, July, November, December 1973; January, February, March, April and July 1974.

1974

Good Horses Require Good Soil. *Acres U.S.A.*, April 1974.

Enter Without Knocking. . .and Leave the Same Way. *Acres U.S.A.*, September 1974.

Insoluble Yet Available. *Acres U.S.A.*, September and October 1974.

Quality Crops also Depend on Soil Fertility. *Acres U.S.A.*, December 1974; January and February 1975.

LABORATORY PAPERS

1935

Laboratory Instructions for the Course in Soils I. Also Laboratory Exercises for Soils I.

Soils. A Laboratory Study. No date. Probably 1935.

1946

Laboratory Instruction for the Course in Soils 25.

1951

Our Soils and Ourselves. Briefs of lectures of Soils I course at the University of Missouri [copyrighted in 1954].

1956

Understanding Our Soils by Clarence M.Woodruff, associate professor of Soils, University of Missouri. Represents the development of work organized originally by W. A. Albrecht.

1957

Study Guide for Basic Soils. Published by Lucas Brothers, Columbia, Missouri, 1957.

CLASSROOM MATERIALS

Laboratory Instructions for the Course in Soils I, *1935*.

Laboratory Instruction for the Course in Soils 25, *1946*.

Our Soils and Ourselves. Lucas Brothers, Publishers, Columbia, Missouri, *1951*.

Our Soils and Ourselves. Lucas Brothers, Publishers, Columbia, Missouri, *1954*.

Understanding Our Soils. Probably *1956*.

Study Guide for Basic Soils. Lucas Brothers, Publishers, Columbia, Missouri, *1957*.

Outline of Laboratory Exercises for the Course Soils I. No date.

Soils: A Laboratory Study. No date.

Soils: A Laboratory Study. No date.

FILM

The Other Side of the Fence

OVERVIEW

Foundation Concepts

1

When William A. Albrecht was a young lad growing up—one of eight farm children to receive a college education—he had a good doctor friend. His friend, Bill Albrecht later recalled, spent a lifetime teaching people about health, and yet when he died he had nearly 72% of his business on the books. "So I got discouraged as a boy," was the way Dr. William A. Albrecht recalled it near the end of his long life. "I said, I am afraid that I don't have enough association with the medical profession to make a go of it. Having been a country boy with a lot of curiosity, and interested in the physiology of plants, animals and man, I decided I'd better stay with plants and agriculture. So I took soil fertility and soil microbiology for my major, and they brought me here [to the University of Missouri, Columbia] because at that time soybeans were new and there were no cultures." By that time Albrecht was a rare educator, indeed. He was a plant microbial nutritionist. "In other words," he would say, "you get down to the single cell." In agriculture and in soil microbiology and in medicine, he discovered what the country boy said when he came home from the college of agriculture. ["Dad, they teach so much that ain't so."] "As I learned," Albrecht often summarized, "I wrote everything out and studied it out, and put it into manuscript form." When he first came to the University of Missouri as a microbiologist, he had completed his doctor's thesis, Symbiotic Nitrogen Fixation as Influenced by the Nitrogen in the Soil. *Now the university wanted him to grow cultures. "And they thought I could grow a bacteria that would make a plant fix nitrogen and be inoculated. And I was here six months before I discovered that was what they believed, and I was terribly disgusted. I said, I'll have to tell those people that when a bull and a cow get together, the cow has to do her part, too, not just the bull. All of my research here is merely the conviction that when my cultures do not make nodules on legumes, I've got a plant that is too sick to carry its half. But I haven't got that across so far." Nevertheless, Albrecht did his work. The paper reproduced here came styled* Soil Inoculation for Legumes. *It was published in March 1919 as University of Missouri Agricultural Experiment Station Circular 86, and represents the anchor item in that legume bacteria study department. A later paper,* Viable Legume Bacteria in Sun Dried Soil, *was published in 1922.* Inoculation for Legumes *was issued as an Experiment Station Circular 121 in 1924.*

Soil Inoculation for Legumes

The present high prices for farm products have caused a widespread interest in those practices which promise to increase the productiveness of farm lands. Soil inoculation for legumes may be classed as such a practice, but its adoption has not been general because the beneficial effects of the practice are not yet fully appreciated. This is indicated by the many inquiries received by the Experiment

Station. Farmers intending to practice inoculation for the first time wish to know when and why it is necessary, how it is carried out, and what beneficial effects it may have. It is the purpose of this circular to answer these questions.

1. what is soil inoculation?

Inoculation of any kind deals with the introduction of bacteria, and the term "soil inoculation" refers to the introduction of certain desirable bacteria into the soil. Many years ago it was learned that certain bacteria have beneficial effects on the soils, and that to realize these benefits in new fields it was necessary only to scatter on them small amounts of soil brought from a field containing the bacteria. The good effects of such treatment were marked, especially when leguminous crops such as lupines, cowpeas and beans were being grown. Small amounts of soil in which a legume had been grown, transferred to very poor soil newly seeded to the same legume, made the crop do well. It has long been known that beans will grow on soil too poor for other crops, and if the soil is inoculated, that the next crop to follow the beans will do better than if beans had not been grown. The reason for this has not been known so long, and was learned only when it was discovered how legumes differ from other crops.

2. legumes differ from non-legumes

Leguminous crops are very rich in protein. This is true of both their seed and hay. Alfalfa hay, for example, is as rich in protein as rye bran, while the soybean is richer in this respect than any cut of beef. For this reason the legumes have a very high feeding value, which makes them especially desirable as part of the ration for young animals. To produce this protein the plant must have large amounts of nitrogen, which is one of the plant-food elements commonly lacking in most soils, and the most costly of the fertilizer constituents. It is the ability of the leguminous plants to take these amounts of nitrogen from the air that makes them so valuable.

Besides being richer in protein, there is another significant difference between legumes and other plants. If a cowpea, bean, or any other leguminous plant is carefully dug up and its roots washed free of dirt, many wart-like growths will be seen. These are called nodules. They vary with the different legumes in color, size, shape and location on the roots. Red clover, for instance, bears small club-shaped, flesh-colored nodules about the size of a pin-head which are distributed at random on all the roots; on the soybean and cowpea the nodules may be round and as large as a common pea, and are usually, although not always, located on the upper part of the main root. Those on alfalfa often occur as clusters of club-shaped growths appearing on any part of the root system.

If the nodules are opened and a part of the inside examined under a microscope, numerous small bodies of rod-like form will be seen. Others may look like the letters "Y" and "T." All these are nitrogen-gathering or nitrogen-fixing bacteria, and are not found on common crop plants other than the legumes. The nodules must not be confused with the nodular or warty growths called crown-gall, which appear as a disease on a great variety of vegetable plants and fruit trees. Crown-gall is a disease injuring the plant and stunting its growth, while the nodules of legumes are normally found on the roots, increasing rather than reducing the size of the plant.

The fact that the bacteria inhabiting the nodules gather nitrogen and aid plant growth is the greatest difference between legumes and non-legumes. Through some unknown means the bacteria living within the nodules are able to take nitrogen from the soil air and give it to the plant. In return for this, the bacteria feed on the plant juices. The plant and bacteria work cooperatively, helping each other in a close relationship, called symbiosis. This close relation is not absolutely necessary for the existence of both, since either the plant or bacteria can live alone, but the growth of each one is greatly benefited by the presence of the other.

Legumes growing without the bacteria feed on the nitrogen of the soil in the same manner as corn, oats, wheat and grasses. By means of the bacteria living so helpfully in the nodules, the leguminous plants get a good share of their nitrogen from the bountiful supply of free or uncombined nitrogen in the soil air. It is this property which makes it possible for legumes to grow well on soils poor in nitrogen and yet contain a large percentage of nitrogen in their seeds, leaves and stems.

3. is inoculation always necessary?

"Must I inoculate my field to get a good crop?" is a question often asked and one not easily answered. To inoculate a leguminous crop at every seeding is both laborious and expensive, and if not needed it is a waste of inoculating material. On the other hand, the failure to inoculate, when inoculation is needed, is gross neglect and a significant loss in money. It is important to know when a leguminous crop should be inoculated, or when the bacteria must be introduced into the soil.

Unfortunately, no chemical test or other rapid means can be used to determine when bacteria are needed on a field. The only certain method of answering this question is to grow the leguminous crop. If root nodules do not develop at all, or develop on only a few scattered plants, then that crop will be improved by inoculation. If, however, some few nodules develop on each plant, inoculation is not necessary. Under the latter conditions the growth of one crop of legumes will enable the few bacteria in the soil to multiply and produce numerous nodules during the next season of the crop, providing this legume is again seeded on the land within a few years. To test a soil in this manner requires time, and many farmers prefer to inoculate a leguminous crop rather than wait for the results of such an experiment. Although no other test will decide the question with certainty, yet there are some general facts that may aid in the decision.

Such crops as red clover and cowpeas which have been grown in this state for many years probably need no inoculation, except in rare cases. In some isolated sections where neither of these two crops has ever been produced it may be possible that even red clover and cowpeas will do better if inoculated. No direct evidence for such statements is available, but the ease with which these two crops are started in the general farming districts of the state where they have been widely grown indicates that the inoculation for these crops is not the most important requisite in establishing them. Such, however, is not the case with sweet clover, alfalfa, soybeans, or with any other leguminous crop that is new to the district and has never been raised on the farm in question. None of the three crops last mentioned is native to the state [Missouri], nor have any of them been extensively grown. On this account the bacteria adapted to each of them have not been widely scattered over the farming sections, and when a crop of alfalfa, sweet clover or soybeans is planted on a field for the first time it will need inoculation. The labor and expense required for this simple treatment are too small to trust to inoculation by chance, with the risk of a poor crop. Occasionally some seeds carry a few bacteria into the field, which inoculate a few scattered plants and slowly spread to other parts of the field through distribution by animals, flood waters, winds or other chance agencies, but this is too slow when artificial inoculation is so easy and costs so little.

The need of inoculation is not always shown by poor plant growth. Inoculation is necessary when the legumes make a fair growth but fail to form nodules. Under such conditions the plants are using the nitrogen of the soil instead of drawing on the unlimited supply in the air. On very rich soils, uninoculated legumes may make a good growth by feeding on the soil nitrogen the same as other crops, and inoculation is the one treatment necessary to enable them to secure a part of their nitrogen from the air, and thus save the supply in the soil.

4. conditions may be unfavorable to thorough inoculation

Sometimes the first crop growth of a new legume is poor with only a few nodules on each plant, even though it was given thorough inoculation. This may possibly be due to one of two causes: the bacteria may be present in small numbers because they have not had time to multiply; or some unfavorable soil condition keeps the legume plant and bacteria from doing better. In the former case, cultivation of the soil to distribute the bacteria, or a second crop of the legume to allow them to increase, will give the soil good infection. If the soil condition is at fault, this must first be remedied by special soil treatment before the legume will do well and the bacteria be of benefit. The most common conditions unfavorable to legumes are those of poor drainage and lack of lime. Under such conditions inoculation must be used again after the soil condition has been remedied.

5. legume bacteria do not live in the soil indefinitely

These beneficial bacteria do not live in the soil indefinitely in the absence of the leguminous plants on which they feed. For this reason inoculation may be needed when a legume is grown again on land after several years of cropping by non-legumes. However, the length of time the bacteria remain active in the soil in the absence of the legume on which they live is not definitely known. Experience indicates that reinoculation should be practiced if the desired legume has not been grown for eight or ten years. Short lapses of time between reseeding of the same crop do not permit the bacteria to die, and inoculation is unnecessary. Its practice under such conditions would not increase the crop yields and would be unprofitable.

6. cross inoculation

Inoculation for one legume crop will not necessarily serve for all legumes. If one legume crop has been grown on a soil and was thoroughly inoculated with many nodules on its roots, this gives no assurance that any other kind of legume will grow without the special treatment of inoculation. Some different kinds of legumes inoculate each other, but legumes do not cross-inoculate universally. Some have a single kind of bacteria which will grow on no other plant, and they must be inoculated with that specific kind of bacteria for their best growth, while others have one kind of bacteria common to two or more types of plants. It is a well-established fact that alfalfa can be inoculated with the bacteria from sweet clover, and that the cowpea may be cross-inoculated with the wild partridge pea. Likewise, red clover, white clover and all other true clovers will inoculate each other. Soybeans, however, have a particular kind of bacteria not yet found on any other legume. According as the common legumes will cross-inoculate, or as they have the same kind of bacteria in common, they may be separated into the following groups:

1. The true clovers, including red, white, alsike, crimson and mammoth red.
2. Alfalfa, yellow and white sweet clovers, bur clover, and black medic or yellow trefoil.
3. Cowpea, partridge pea, peanut, velvet bean and Japan clover.
4. Garden, field and sweet peas, and vetches.
5. Garden beans.
6. Soybeans.

Because sweet clover and alfalfa inoculate each other many persons are apt to think that simply seeding sweet clover will inoculate the soil for alfalfa. Such is not the case. If proper bacteria have never been introduced artificially or by chance they must be applied when the first of these two crops on the soil is sweet clover as well as when it is alfalfa. Sweet clover often gets into alfalfa and smothers it, because conditions favorable for alfalfa are also favorable for sweet clover. The bacteria that nourish the alfalfa also nourish the sweet clover, and the sweet clover makes a much more rank growth, so that the alfalfa is smothered out. The soil is, however, no better supplied naturally with the bacteria for sweet clover than it is with those for alfalfa the first time one of these crops is grown; and if they must be added to the soil for the alfalfa, the need of inoculation cannot be met by substituting sweet clover. If the soil is to be put into alfalfa eventually, there may be some advantages in first seeding it to sweet clover for a season or two, since the latter crop seems to be less delicate than alfalfa. It establishes itself over weeds more readily and helps distribute the bacteria for the alfalfa. It cannot, however, generate the bacteria needed for the alfalfa if they are not put there either by chance or by artificial inoculation, and the venture of establishing inoculation for sweet clover is as great as establishing it for alfalfa.

7. *how to inoculate*

Soil may be inoculated by two methods: by transferring soil from a field where the same legume has been growing with plenty of nodules, and by the use of pure cultures, or artificial cultures, grown especially for this purpose.

8. *the soil method*

The use of inoculated soil was the first means of introducing the proper legume bacteria into a field, and this was accompanied with such good results that it has become established as good procedure. By this method, from 300 to 500 pounds of inoculated soil are scattered over each acre of the field and disced or harrowed in before seeding. The soil so distributed is collected from the surface six inches of a field on which the same legume—or one which cross inoculates with it—has been grown recently with many nodules. Extensive exposure of this soil to the sun before scattering is thought to be detrimental to the bacteria, but definite evidence fails to prove that this is as harmful as might be thought. Useless exposure to the sun, however, should be avoided.

In case only small amounts of soil are available, it may be applied to the seed rather than to the field. Soil collected as just mentioned is dried and sifted over the seed which has been moistened (not wet) with a ten per cent glue solution (one pound of liquid furniture glue to one gallon of water, or one pound of dry glue to three gallons of water). Dry, powdered soil is added while the moist seed is stirred until every grain has become coated and the seed has a dirty appearance. After drying rapidly to prevent molding and loss of vitality, the seed is ready to be sown. The smaller seeds, such as alfalfa and sweet clover, will need to be screened before seeding in order to break up the clusters of seeds held together by the glue.

Some objections have been made to the soil inoculation method, but under proper precautions the objections are not serious. The method may be laborious, and rather expensive when no thoroughly infected soil is available in the immediate vicinity. It can also introduce noxious weed seeds or dangerous plant diseases, but these can be avoided by using uninfested soil. When any of these undesirable conditions prevail, the pure-culture method might well be used. But whenever good soil can be had for the same cost as pure cultures, the soil method is doubtless preferable. It has given a greater degree of success in past experiences.

9. *pure or artificial cultures.*

The pure-culture method, like the soil method, is based on the principle of scattering the bacteria on the field. The bacteria are grown especially for such use. When bacteria were discovered to be the cause of the nitrogen-gathering power of legumes, attempts were soon made to grow pure cultures of them to avoid the difficulties of the soil method. For several years the pure cultures were in ill repute, but recently improved methods of propagation and distribution have overcome the former opposition and tests of commercial cultures by many experiment stations have found them generally reliable.

The desired bacteria are separated from the nodule and grown on neutral jelly in the laboratory, where they increase rapidly. These are sent to the farmer in various ways, sometimes in solution, sometimes on vegetable jelly, but more often on sterilized soil or sand. The method used is immaterial, provided the bacteria are alive. In using the culture, it is diluted with water, sprinkled on the seed and allowed to dry.

The advantages of this method are its simplicity and ease of performance. No great expense is involved in making the cultures. With pure cultures there is no liability of introducing dangerous diseases, destructive insects or noxious weed seeds. They must not be used carelessly, however, since the bacteria are living organisms and die when the cultures are stored too long or subjected to excessive heat or cold. Directions for using are always supplied and if followed with reasonable care, this method of inoculation should be successful.

10. small acreages of legumes advisable for beginners.

Whenever a new legume is seeded for the first time it is good policy to attempt it on a small acreage. Such procedure offers a chance to become acquainted with the habits of the crop and the best methods of handling it, without the chance of a heavy loss of money and labor in case of failure. If this small area is inoculated, soil may be taken from it to inoculate larger fields on years following. It is, however, well to leave a part of this small area uninoculated and if the treated part has dark green plants with many nodules on the roots while the untreated has poorer plants or fails to develop nodules, the soil needs inoculation for the crop. Such procedure with the small acreage of a new legume has the following advantages. First, it involves no great risk of a heavy loss in labor and money; second, it determines at a very small cost whether the legume crop will do well; third, it tests the soil to see if inoculation is necessary; and fourth, if such treatment is necessary, it furnishes the soil from the inoculated part of the field as good material for inoculating the rest of the farm. Venturing on a small scale is far better policy than trying a large one, and anyone who is seeding a new legume for the first time will do well to be guided by this principle, trying a few acres to find out how successful the crop is in the locality, rather than investing large sums in seed and labor to be lost in possible crop failure.

11. benefits from inoculation.

As previously mentioned, inoculation enables legumes to make a good growth on poor soil by taking nitrogen from the air. It also enables the plants to take up larger amounts of potash and phosphorus, and to give larger yields. By this practice, the total crop is increased and its percentage of protein becomes higher, with a resulting higher feeding value.

Besides larger yields, and a higher protein content in the crop, inoculation causes greater root growth. This with the nodules included gives the root system also a high total nitrogen content; and since the root system remains in the soil, legumes do not exhaust the soil nitrogen as rapidly as other crops. The legume, by taking nitrogen from the air, is a soil enricher and when plowed under serves as a nitrogen fertilizer. How much nitrogen a single crop of legumes plowed under will add to an acre of soil varies widely, and figures varying from twenty to one hundred and fifty pounds have been given. It is generally believed that about three-fourths of the nitrogen in the roots and tops of leguminous plants, grown on average soil, may come from the air. Regardless of what this amount may be, the fact remains that properly inoculated legumes use the nitrogen of the atmosphere without depleting the nitrogen in the soil, while in the absence of the bacteria they may drain the soil of its nitrogen the same as any other crop.

12. possible failure of inoculation.

Inoculation, while giving striking benefits, is by no means a cure-all for soil troubles. The bacteria concerned require satisfactory soil conditions if they are to flourish. In a very sour soil, they do poorly and their growth can readily be improved by lime. If a farmer is uncertain regarding the acidity of his soil, it is well to have the soil tested before seeding a leguminous crop.

For a legume, as for any other crop, the seed bed should be well prepared, the soil well drained, and its acidity neutralized by the use of lime. When all these requirements are met and the inoculation is given in addition, the legume crop should be successful. Well inoculated legumes should find a place in every system of rotation in order to assist in keeping the soils permanently fertile.

2.

The old farmer's expression that when a bull and a cow get together, the cow has to do her part, too, not just the bull, *had special significance for Dr. William A. Albrecht. It also meant that when a seed and soil got together, the soil had to do its part. Speaking into an* Acres U.S.A. *tape recorder, Albrecht put it this way. "Professor M. F. Miller [then head of the Department] thought I should grow bacteria that would make the cow have a calf whether she wanted to or not. And I had to politely show the points I wanted to make." As usual, Albrecht had his "vision" to lean on—that is, his informed conceptualization. He liked to tap foreign scholars, and so he knew about Russia's S. N. Winogradsky and Holland's M. W. Beijerinck, who had proved that nitrogen fixation took place in the soil without any legumes whatsoever. The essentials of rapid fixation were found to be absence of readily available nitrogen and the presence of carbohydrates, phosphate and lime. He pondered the findings of Thomas Way of England. It was Way, it will be recalled, who in 1852 discovered that when soil absorbed ammonia, a corresponding amount of calcium was released to the drainage water. The exchange mechanism, he found, was seated in the clay. Continued study revealed that such an exchange did exist, and that it involved all cation (positively charged) elements. Moreover, it had been developed that the total exchange capacity of soil depended on both the colloidal clay and the organic matter. With these few points in mind, Albrecht continued observing as nature revealed herself. The titles of early papers suggest the usual method of approach. "We took things one leg at a time," Albrecht said, "and wrote down the results." Often as not, significant findings went out under the imprimatur of the Department, students being the authors, special credit being given to "Dr. Wm. A. Albrecht, under whom the author worked."* Calcium as a Factor in Soybean Inoculation *read one paper by Robert W. Scanlan.* Physiochemical Reactions Between Organic and Inorganic Soil Colloids as Related to Aggregate Formation *was another, H. E. Myers, author, Albrecht, godfather. The paper reproduced here floated to the surface in the Albrecht collection as a manuscript dated 1953. It has never been published in this form before. In easy-to-understand language, it tells something about the classical colloidal clay experiments at the University of Missouri. A precise statement of the objective and the method used has been published as Agricultural Experiment Station Research Bulletin 60,* The Chemical Nature of a Colloidal Clay, *June 1923, Richard Bradfield, author, Drs. M. F. Miller, W. A. Albrecht and F. L. Duley (of University of Missouri) and Drs. F. E. Bear, Edward Mack and E. N. Transeau (Ohio State University), godfathers. Additional work on the subject by one of Dr. Bradfield's students, Leonard D. Bauer, has passed into the literature as Missouri Agricultural Experiment Station Research Bulletin 129,* The Effect of the Amount and Nature of Exchangeable Cations on the Structure of Colloidal Clay, *1929. Indeed, there were many papers, all highly technical. Albrecht described his own role in these studies as follows: "I separated the finest part of the clay out of Putnam silt loam by churning in a centrifuge running at 32,000 r.p.m. after the clay had been suspended and settled for three weeks. At the bottom, that clay finally plugged up the machinery. But we had thinner and thinner, smaller and smaller clay until about halfway up in that centrifuge, there we had it as clear as vaseline. We took the upper half of that clay. We made pounds and pounds of it. We put it into an electrical field and made it acidic. We took off all the cations so it was acid clay. That was how we studied plant nutrition. We put on different elements in different orders. We mixed them, balanced them." Standing on the shoulders of giants, Albrecht knew he had to begin with calcium. Extensive research projects served up this working code for balanced plant nutrition: hydrogen, 10%; calcium, 60-75%; magnesium, 10-20%; in some plants, 7-15%; potassium, 2-5%; sodium, 0.5-5.0%; and other cations, 5%. "While the above ratios are guidelines," wrote Albrecht in* Soil Reaction (pH) and Balanced Plant Nutrition, *1967, "they have been found most helpful for humid soil treatments as more nearly balanced plant nutrition for legumes. They are also a sound reasoning basis for better growth of non-legumes. Those same ratios between the nutrient cations, emphasizing calcium almost ten times higher, and more, than others among the five, should make us believe that one is apt to find the calcium the more commonly deficient nutrient element for crops, too long covered in our belief that soil acidity calls for a cheap carbonate, being taken as limestone." The nutrient code for cations expressed above is being used by the important laboratories serving eco-agriculture. The logic of Albrecht's search caused him to look at anion (negatively charged elements) next, at sulfur and the trace elements (which suggested themselves as reasons for the low protein values of crops being grown). The conclusions seemed obvious. "The plant's struggle is one for its synthesis of proteins," Albrecht summarized in* Soil Reaction (pH), *"its living tissues giving growth, self-protection and reproduction. Humans and all other warmblooded bodies are struggling similarly for adequate proteins. Those are synthesized from the elements as the starting point of creation by only plants and microbes; hence the soil, its microbes (centered about soil organic matter) and its crops, are the quality control of man's nutrition and of all the animals supporting him. In our management of the soil we are not yet the equal of what nature was before man's advent on the scene to take over that responsible part of creation."*

Soil Fertility and Plant Nutrition

In our research in the Department of Soils of the Missouri Experiment Station, we have labored for a number of years to work the combination so that we can get *soil fertility* and *plant nutrition* linked together in a language which we all understand. In opening this broader subject of soil fertility and plant nutrition, we may well be reminded that it is much broader than what is commonly included under that title. It will eventually encompass more than any of the rest of us might now include.

Much that is said about our scientific progress, in which we are about to believe that we have reached the pinnacle, deserves some rather critical examination. It seems that we are allowing ourselves to be so easily deceived by our success. There is a terrific danger in over-confidence. Even though we can fertilize the soil, we are not quite yet controlling nature at that point of her activities. The burden of the thought is in those few words, namely, we do not yet create crops.

We are delighted in the technical progress which we have built up and which raises our standard of living when you consider technologies. Technologies apply only, however, when we consider our control of dead matter. As for me, I am not quite ready to put agriculture in the class of the science under control, like a technology is, for example, when you look at the assembly line of an airplane or an automobile. It is more nearly correct to talk about agriculture as an art, plus some science. It is not yet a case in which the science is in complete control. Agriculture is still an art, which we study by deduction, that is, we look at it as a natural behavior. We take a fragment out of it and put a little science into that portion. But when we take complete control we must have the science so well organized that we can put all the parts together and run the whole process from creation to death. Nobody as yet has been able to do that with agriculture.

Agriculture is biology first and foremost. It is technology and management second. We need only to remind ourselves of the last two seasons to be reminded how readily we use the weather as a scapegoat, when the crops didn't behave as we thought we would like to control them. It is fitting, therefore, to provoke your thinking about the sciences applied to agriculture in a technological viewpoint only as a possible or even serious danger. Some of the agricultural troubles for which we are apologizing came about because we used technologies to upset the biology of agriculture. Much that is apt to be called agricultural science has upset the biology and we are coming now to reap the bad harvest. We are beginning to realize that the matter of agricultural production has been largely nature's performance. Very often we have not had very much to do with it. We have been copying and memorizing agricultural practices, but have not been comprehending the basic principles that operate under nature.

1. some biological processes we have upset

As illustrations, we may well list several cases in which we have upset the biology rather than helped it. We have been taught to believe that crop rotations build up the soil in fertility. Yet nature uses continuous cropping and doesn't rotate the crops when she builds up the soils in fertility. We have upset the biology in that case very decidedly. We have had to use 65 years of trials on Sanborn Field to discover what nature's truth about crop rotations in relation to soil fertility really is.

We are now trying to put a grass agriculture over much of the country where nature had tree culture. Then we are going to let the cow make up the difference between our ignorance of quality grass and the cow's knowledge of it. We put the plow ahead of the cow. The cow is about to be extinguished if, as a matter of legal procedure, we follow the philosophy of killing cows to get rid of diseases like brucellosis, hoof and mouth disease, and others. How we ever got such a belief to prevail is strange, namely, that we can have cows when we keep killing them because we make them sick.

Primitive man lived on the dry lands. More recently in human history we began to farm those lands which have high rainfalls. Then when rainfall goes back on us we run to the federal treasury, as though that were the place where one gets any biological help.

As another upset of biology, instead of leaving plant residues on the top of the soil, as nature does, we bury them as deeply as we can.

Nature used different soils to make grasses on the prairies than she uses in making trees in the forest. Yet here are some folks who believe that the prairie grasses make prairie soils and the forest trees make forest soils.

Nature uses plant roots to make soils acid in order to get the fertility into the plants. We want to put a carbonate in the soil so that the root cannot put its acid out any farther than barely off itself. Yet plants nourish themselves by making soils acid. We fight the soil acidity by means of the carbonate instead of feeding the plant with the calcium and the magnesium in the limestone.

Nature washes the soluble fertility out of the soil into the sea. We make the fertilizers soluble before we put them on the soil. If they remained there in that form they would soon be in the sea, too. Or, we mine the sea salts as soluble fertilizers, like potash salts, and put them through another round of going from the soil and into the sea.

We put cattle and grass into the plains areas of ever-threatening drought, and succeed because that is where nature had successfully put a similar beast in the bison. But we put them on the grass amongst the forest tree soils under high rainfall and then wonder why their reproduction is failing.

Nature doesn't have animals live to get fat. Experiments point out that our animals are searching for anything else but fattening feeds. They are searching for those which help them protect themselves against diseases and encourage them in their reproduction. But when we feed animals we cut the amount of protein in their ration down to the limit, because we want to make cheap gains rather than to give the animals the help they need to be healthy and to multiply. Instead of letting the animal live long, we cut their lifespans down to the most early maturity we possibly can and then call it "cheap gains."

Nature placed the animal's matings in the spring of the year. We try to have them mating all the year round.

We use artificial insemination to manage animal reproduction in that we limit the supply of sperm per animal. We omit thereby much of the hormone treatment which the bull uses when he deposits the sperm. We have given little thought to the fact that in the case of the so-called "hard-breeder" cows, the bull doesn't give up. He succeeds eventually where the costs of the repeated failing artificial inseminations would condemn the cow to slaughter. Here nature's biology succeeds and the species multiplies, but our technologies fail and the species becomes extinct. We upset the biology, but cling to our technology.

Some significant economic aspects might well be considered critically when the manipulated economics have manipulated machinery, money and technologies into agriculture, but have almost manipulated biology out. Bankers are about to believe that we can substitute capital for land. Certainly all the capital in all the banks cannot substitute for the soil of the land. We know of no bank with all its money that could by means of that wealth have a litter of pigs, lay an egg, or give birth to a calf. And yet we have folks believing that one can manipulate biology by means of economics. You can't do that any more with money alone than you can with machinery and technology.

We have not yet understood, nor appreciated, agriculture as a collection of complex, but well-integrated, biological processes. We have comprehended some fragments of this natural art which science has studied deductively, but we have not yet constructed the whole of a scientific agriculture inductively. We have not seen the soil as plant nutrition and thereby as animal and human nutrition, or the soil as the very foundation of all agriculture.

2. confusion will prevail until the soil is considered

Because we turned away from much of the art of agriculture in the absence of a complete science of it, we have a serious confusion. That is all the more serious now and at the moment we discover that we have rapidly mounting numbers of people and are soon running out of ample food for them. We are confused about the natural performances or about the biology in agriculture. We have permitted ourselves to be led astray and are asking the science of agriculture now to bring us back to where we can understand the basic principles rather than merely mimic any practice. We dare not be mere followers of traditions. We must face the problems and solve them. All of that calls for rather clear diagnoses. Let us try and comprehend the fact then that soil fertility properly coupled with plant nutrition is a form of creation, a form of outdoor biology, and not a matter merely of scientific technology. In that combination wisely used there may be some solution for our food problem.

Now what are some of these confusions about the basic facts of soil fertility and plant nutrition? First of all, we seem to have lost sight of the fact that the creative business of agriculture has always started in the soil. That great truth was told us about six or more thousand years ago, but we didn't take that remark very seriously. We are beginning to appreciate it now. We shall face it more seriously when we have the least of creative capacity left in the soil and when we need to know most about it.

In terms of wise fertilizer use, the most shocking confusion prevails when we talk about soluble fertilizers, considering water as the agency for solution, and then we make laws requiring that fertilizers must be water-soluble and thereby so-called "available." In fact and in nature, these soluble fertilizers are never taken out of the soil because the plant takes them into itself along with water it takes from the soil. The use of the major amount of water by the plant is that of keeping the respiring leaf tissues moist for the exchange of the gases, namely carbon dioxide and oxygen. That escape of water from the leaf is what we call "transpiration," and it is in that service where most of the absorbed water goes from the soil into the atmosphere. That use of soil water is controlled by the meteorological situation inviting water to evaporate from the leaves of the plants against the forces holding the water in the soil. The plant is an innocent connection between those two opposing forces acting on the water. Does the moisture in your breath move nutrients from your bloodstream into the tissues, or from your stomach into your bloodstream? But yet we take to the concept that the transpiration by the plant has something to do with the movement of nutrients from the clay of the soil into the roots. The transpiration stream of water from the soil, through the plant and into the atmosphere is independent of the nutrient stream from the soil into the roots. That may not be true for nutrients moving within the plant's conducting tissue. The water uptake by the roots is the result of atmospheric conditions favoring evaporation from the leaves with a set of dynamics which are more than a match against the forces holding the water on the surfaces within the soil.

Nutrient intake by crops is a function of three colloids, or possibly four, in contact. First of all, there are the nutrients on the clay colloid, or on the organic colloid of the soil. The soil colloid is in contact with the root membrane which is another colloid. That root membrane is in contact with the contents of a cell on the inside, namely the protoplasm, or the cytoplasm. Then, in turn, that cell is in contact with another cell. In that you have the combination of the three or four colloids in contact. The movement of the nutrient ions from the clay into the root membrane and into the cells follows the chemical laws controlling their traverse there because of the differences in activities, adsorption capacities, interfering ions and other factors along that line.

That movement of nutrients into the root is independent of the transpiration of water. We have demonstrated transpiration going forward regularly, or water moving from the soil through the plant to the atmosphere when the nutrient ions were moving in the reverse direction, namely, going from the plant back to the soil. We have demonstrated the ions going into the plants regularly when there was no transpiration. You can demonstrate this when you put a bell jar with atmosphere saturated with CO_2 and with water over that plant. In that case, you can stop the transpiration but you don't stop the ionic nutrient movement into the plant. Some recent work at the California Technological Institute has shown that the desert plants put water back into the soil while they are growing, therefore the water can be going back into the soil while the nutrients are going in the opposite direction. *We must get rid of this water-soluble fertilizer bugaboo in considering soil fertility and plant nutrition, because transpiration runs independently of our control and we need to concentrate our efforts on keeping the stream of fertility flowing more regularly into the plants.*

Let us not cover either our ignorance or our responsibility toward maintaining the soil fertility by trying to blame the water situation in the soil and the rainfall. The idea that the *drought* is responsible for the failure of plant nutrition persists. But what is commonly called *drought* isn't trouble in terms of water only. It is apt to be due to the fact that the upper layer of the soil, where the fertility is, dries and the roots must go down through a tight clay layer which has almost no fertility. Then, because of the crop failure in the absence of plant nutrition in that soil layer of stored water, we try to blame the drought or the bad weather. Drought may be merely that soil situation in which we have no soil fertility deep enough to feed the plants when they are compelled to have their roots go deeper to get stored water. We have emphasized the water so much that the situation suggests itself as a relic of the old "saloon" days, when men thought they had to stay in a saloon and drink, but forgot to take some groceries home for the family. Plants will scarcely emphasize drink to that much neglect of food. Our confused thinking about drink for plants emphasizes the water facts as an alibi for our ignorance of plant nutrition and the soil fertility factor where the emphasis properly belongs. During the drought we don't use the water to the best of our ability. We neglect to remind ourselves that the plant is about 95% air, water and sunshine, and only about 5% fertility. We are too indifferent to that fact to consider carefully how we can use that 5% as the requirement to produce the other 95% of plant growth, a performance which offers chances as a gamble better than one would scarcely anticipate.

We blame the water. We blame the weather. The water of transpiration from the plants is like the water going over the millwheel, only a part of that coming down the millstream. The amount of grist that one grinds in the mill is determined not so much by the amount of water going over the millwheel, the amount of which is fixed or limited, as by the diligence with which wheat is kept going into the millstones for 24 hours a day at full capacity. We haven't been keeping the soil fertility well and properly supplied to the crop plant and are therefore in error when for disturbed yields we blame the drought.

3. the problem analyzed

The problem of relating soil fertility to the plant's nutrition as well as to the plant's drink was approached and put under study at the Missouri Experiment Station many years ago. It seems fitting to review here the history of the mental procedures by which we attacked this problem of adequate soil fertility and its services in the growth of plants—at least in terms of this soil aspect when others were adequate to the best of our knowledge. Many years ago we made a kind of problem analysis. Our first division of the problem into its parts divided it according to the soil texture, namely, the sand, the silt and the clay. We decided on the clay as the part of the soil we should study first. That choice was expectable because in Missouri we have almost 4 million acres of claypan soils. About those it was commonly said, "Oh, they will make only about 20-30 bushels to the acre. Your crops drown out in the spring and you dry out in the summer." The Putnam claypan soil is a nice silt loam. It has a level topography and it would be fine but it just doesn't deliver by production. Farmers on it have some silly ideas and ways about handling it. They *bar* off the corn in the spring, as they say, and they *hill it up* in the summer. It is a terribly acid or sour soil. You can't grow legumes on it. You can't build up its nitrogen. Yet in spite of all the kind and unkind things said about this silt loam as a claypan soil, it is still a good one in contrast to the 10½ million acres of stoney land we have in Missouri. As an outcome we picked first on this Putnam silt loam and worked on its clay until now that clay of this claypan soil is known all over the world by its technical name, Putnam clay. This basic research in the laboratory, in the greenhouse, and later in the field has moved this soil into the corn growers' contests for the winning high yields per acre.

Let us follow with the next separate and inquire, "What is in the sand as fertility for plant nutrition?" You may well reply, "It is largely quartz, it contains only silicon and oxygen. Those minerals grains do not weather down." That is the reason the quartz crystals are still big grains. It never has weathered. It is a kind of soil skeleton. "Consequently for the time being," we said, "let's throw that soil separate out of our mind and concentrate on the silt." It also can be quartz. But then, too, it can be other-than-quartz. Its composition depends on the place where it is. The farther east one goes in the United States, and to the higher rainfalls, the more quartz there is in the silt of the soil. Silt doesn't have much capacity to hold fertilizer—neither does the sand. Because of their large particle size, these two soil separates have little capacity to absorb and to exchange nutrients to plants. But yet there is much silt blown in here from the floodplains of the Missouri River and from the west. It piles up along the river bluffs to give what is fairly good soil. Thus the soil separates were catalogued for their order of impotence for research attention. The silt fraction was set aside for later study when the initial study took to the clay separate.

4. clay research laid the foundation

We began with the claypan soil and its high content of clay since almost everybody wanted to enlist himself in what might be a fight with that tight clay. The early researchers bought and used much dynamite on it. They dug ditches of various kinds in it. They pushed it around with powerful machines, but about the time they would have the treatment complete, the soil was behaving just about the way it was before.

Whether fortunately or unfortunately, this clay has little or nothing inside of its crystal form of significant fertility contribution. We tried it some time in some of the early work we did with it in the laboratory. We tried bubbling carbon dioxide through it only to discover that if you really treated it long and hard with carbonic acid you could break out of it no more than about the iron that would be required to grow a crop. You might get a little magnesium out of it, but as a contributor of fertility we might credit it with iron, and one would be generous in doing that. From those early studies and the light they shed on the importance of the clay separate of the soil, there developed the research studies at this Station leading to our better understanding of soil fertility and plant nutrition.

"But what about the organic matter as a colloid similar to the clay?" you might well ask next. We found some organic matter in the clay. About 1½% carbon and about .15% of nitrogen are found in our Missouri clay. They are still very tightly linked into the clay molecule even after you have oxidized the clay with hydrogen peroxide, and after electrodialysis to take out everything that you possibly can. These clay studies brought carbon and nitrogen in a ratio of 10:1 right in the colloidal clay itself. That is the carbon-nitrogen ratio commonly given for well-weathered soils. But we can't get much nitrogen or much carbon out of it. These organic elements seem to be a highly fixed part of the clay.

Because of these discoveries, the clay had much later research consideration. A good number of men on the Station staff have contributed to this whole project. We began by taking the soil separates out and directed concentrated study to the clay. Its separation and preparation were no small task. We dispersed it in water. The larger particles were settled out, the coarsest clay was thrown out of the supernatant water by centrifuging. The remaining opalescent suspension was then electrodialized and the intense study of this clay began. It was on the basis of that attack and our increasing knowledge resulting therefrom that we are bold enough to talk about soil fertility and plant nutrition.

Because we have studied clay chemistry now for a long time, we have moved to study plant chemistry in combination with that clay. In his first experience of bringing colloidal clay chemistry into combination with the biochemistry of the plant, Dr. Hans Jenny met with disappointment in the plants so often that he was about to give up. But when near complete disgust with our theories, he finally caught the vision that the clay might be the dynamic center of the soil. He had thrown out two or three sets of plants before we rescued the situation. He had demonstrated the fact that it is a hard task to load the clay with enough fertility to feed a crop for its good health and growth. Subsequent trials in goodly numbers demonstrated quite clearly that the clay is the major center of the chemical dynamics in the soil which deal with the speedy process we might envision when we talk about growing a plant.

It was also discovered that the clay holds nearly the season's supply of plant nutrients ready and exchangeable for the root when it comes along. We have in the silt of the soil some of the reserve fertility that can be broken out when we consider that the soil is resting. The research found that the clay within the plant root zone must hold approximately a season's supply of fertility. That supply can be increased by adsorbing more nutrients on the clay or by adding more clay. It is in that factor of a higher saturation of the clay or of more clay where the different fertilizer use on soils of different textures is determined. One can put a tremendous amount of fertilizer on a heavy clay soil and not see much difference in terms of any tests of the soil you use, but yet the plants will register it pronouncedly. And that fact holds true whether you are considering fertilizers in the form of nitrogen, phosphorus and potassium, or in the form of magnesium and calcium as limestone.

The fertility held within the root zone is not leachable by water. It is not in the free water of the soil, and is therefore not in the water-soluble condition when it moves to nourish the plant. *We make fertilizers soluble so they will be speedily adsorbed on the clay rather than be held in water and be sucked in with the water by the plant.* We have learned from these clay studies that calcium is major nutrient for most any crop we grow. It must be ready for the plant early in the plant's life. It does more than appear in the plant ash. It is not hitchhiking, nor are any other nutrient elements coming out of the soil as we find them when we burn or ash the plant. Calcium is the major one among the elements in the ash, while nitrogen is the foremost, when all the elements, both combustible and non-combustible, are considered.

Our agricultural soils in general have less calcium when they are more acid, that is, have more hydrogen, and conversely, as they have more calcium they have less hydrogen. This simple nutritional situation of the importance of calcium, and the way the soil behaves under acidity as mainly a calcium deficiency has kept us in ignorance about what soil acidity really is, namely, a fertility deficiency rather than a bad environment.The roots make the clay more acid when the plants grow, since the hydrogen from the root and the calcium from the clay are exchanging places. The legumes are taking tremendous amounts of calcium off the clay. It was discovered that the degree of the calcium saturation on the clay determined whether the nutrients moved from the soil into the plant root or vice versa, from the plant root back to the soil. We grew some legume hay crops of beautiful appearance, even when they were putting back into the soil a good share of what fertility originally was in the seed. *When we grew a legume crop that was a so-called "hay" crop, but not a "seed" crop, we had less nitrogen, less phosphorus, and less potassium in the hay and the roots combined than was in the seeds initially planted.* It was from that day forward that we connected plant nutrition with the soil and also animal nutrition with the soil, because we couldn't see the reason for trying to fool the cow and expect her to be a hay baler in terms of eating enough of that vegetative stuff to represent nutrition for herself when it had less of those three nutrient elements in it than the seed we planted.

The clay, then, is the seat of all of these activities. First, there is the absorption of the nutrients from any solution. That activity is involved in our use of soluble fertilizers. Acceptance of hydrogen from the plant root by the clay, and the exchange from the clay to the plant root of some of the cations that are nutrition from that source when the hydrogen accepted there is not the major clay activity. The clay is also the seat of the breakdown of the reserve silt minerals as this decomposition serves to restock the clay, which is especially noticeable while the soils are commonly said to be "resting." Those are some of the basic facts, in summary about the soil's clay factor as it plays a significant role in plant nutrition.

5. *root biochemistry connected with clay chemistry*

Let us visualize next the happenings when the plant root comes in touch with the clay. We need to remind ourselves that the plant is carrying on a biochemical operation. It is not merely standing out of doors without being influenced very definitely by the soil. We discovered that when the same amount of fertility was adsorbed on less clay or the clay given a higher degree of saturation, the plant root in contact with that clay experienced an increasing efficiency with which that fertility moved into it and into the crop. In that simple fact there is the basis for the practice of banding the fertilizer application in the soil in place of mixing it throughout a large soil volume. In limited soil volumes saturated with fertility rather than having it distributed all through the soils to have the clay as a competitor colloid against your plant root is the basis for the efficiency in applying fertilizer in bands. The plant root finds the soil areas where it feeds itself to advantage. The clay has always been serving to remove a good deal of the fertilizer hastily from solution by absorbing it and thus getting rid of the dangerous salt effects, which the plants always suffered when we drilled fertilizer, with the exception of superphosphate, along with the seeding. But as long as we used ordinary superphosphate or others containing a large amount of calcium, we could drill much more fertilizer with the seeding than one could without it. If potassium salts, or sodium salts were used with the seed its germination and emergence were quickly disturbed. However, when plenty of calcium salts were mixed with them, then there was safety. Gypsum was that safety factor often without our recognition of that saving service to the plant's biochemical activities in the soil.

There is another interesting and significant fact we learned. We could study one ion and vary it on the clay to get a certain behavior with increasing saturation of the clay by that ion, provided the accompanying ions didn't disturb that phenomenon. We learned that these different ions adsorbed on the clay behave differently even when the amounts exchangeable are constant; that they are held with different forces; that they move off variably into the atmosphere of their colloid; and that they exhibit different effects and different energies. Consequently, consideration had to be given to the interrelationships of the different ions in the suite of them adsorbed on the clay. If we had a big divalent or a trivalent, or usually less soluble ion, its behavior was much different than if it was a monovalent one. Consequently, the monovalent ions have different effects on others in the suite than have the divalent ones. The monovalent ions as variables in the fertilizer can upset one's thinking seriously if that has been more confined to the divalent and the trivalent ions required in such larger amounts in the plant root environment. The hydrogen is a particularly active one because that is nature's tool which the root sends in close to the clay molecule to hustle the rest of the ions out as they might be clinging so closely to the clay and intending to stay there. Monovalent ions are disturbers to the rest of them, particularly those of higher valences, in their movement into the plant roots as compared to the amounts exchangeable.

That disturbed behavior relative to the amounts of the inorganic elements is very significant in case we bring a big, organic molecule into contact with the clay molecule or into the suite of adsorbed ions. When you take a soil which is low in the inorganic nutrient ions on the clay and then can get a big, organic molecule adsorbed there, that phenomenon changes decidedly the whole situation of relative activities of ions for movement into the plant root. Those inorganic ions then come into action of a higher degree than the exchangeable amounts to suggest in the absence of the organic ion. A decided difference in the behavior of calcium was demonstrated as the clay was more highly saturated by it but accompanied by a little methyl blue. The degree of calcium saturation didn't make much difference in the relative amount taken by the plant root. This suggests that when we have been using fertilizers and trying to explain the effects, the organic matter in the soil has been the saving grace for the plants in many cases. A low degree of saturation of the clay by the inorganic ions in the less fertile soils may be improved decidedly when we get more organic matter into those soils. When we put organic matter back into the soil, we upset this more commonly considered set of inorganic chemo-dynamics of the soil, which must have initially dominated there. With less organic matter in the soil, we have a different rate of fertility deliveries and a rearrangement of the suite of ions on the clay. Organic matter, then, shifts the ration of the crop about very decidedly. Adsorbed on the clay, it moves different ions into and out of action in different proportions, and therefore the soil is feeding the crop different diets. We ought then to expect different crop compositions in terms of some of the food compounds the crop manufactures.

Considering the plant root as a factor in modifying soil fertility in plant nutrition, we have found that if the root was a protein-rich one, it exhausted the inorganic fertility of the soil to a higher degree than any non-protein root. In other words, a legume is the quickest way we can take the inorganic fertility off the clay and deplete its fertility to a lower level than we can by any other crop. Sanborn Field tells us that fact with decided emphasis. The legume crops take the potassium off the clay more speedily than most other elements. On Sanborn Field, continuous clover cropping was tried beginning with the year 1888. After 14 years, this effort to grow the red clover was abandoned under both manure and no manure treatments. After abandonment of these plots, for some time they were planted to continuous cowpeas for about 20 years. Then they were left to grass for a period and later went to brome sedge, as a nurse crop for Korean lespedeza. This combination prevailed under the plot neglect until in 1950, when $70 an acre was spent for fertilizers and the plots put into continuous corn. The yields were 125 bushels of corn in 1950, 110 in 1951, 57 in 1952, and 73 bushels in 1953 when there was much talk about a drought. During those four years, the corn yield gave an average of 91 bushels per acre.

On the basis of those yields as a sequel to the preceding history, one might well raise the point whether legumes have been soil-improvers. It suggests that we never really knew (we just had a kind of blind faith) that the growing of legumes improved the soil. We know now from laboratory research and from such field records just cited that they can take the fertility out of the soil faster than the non-legumes. The first failures in the crops in that field era showed up in the continuous legumes. They couldn't save themselves. The question may well be asked, "Shall we go back to that plot and try to build up its fertility by seeding legumes, or shall we study the sciences of the soil and plant nutrition to work with the biology the best we can by applied nitrogen and other fertilizers?" With present high costs of production and high taxes, what solution have we for some of these problems except higher yields per acre via fertilizers for economy of production?

6. the crops' physiology becomes a deciding factor

If the soil fertility is to function efficiently in plant nutrition, we must give attention to the physiological requirements of the crop we expect to grow. Do we know the proper ratios of the inorganic nutrients that ought to be moving off that clay into the plant root? As yet we do not, but researches are giving suggestions. By means of the colloidal clay, we are trying to work out a concept of a blanced ration of calcium, magnesium, potassium, etc., on that clay, possibly for each different crop because different crops are synthesizing different organic compounds. We know that the legumes require more of the different elements than are required by the non-legumes. Legume tops and roots are running a bigger factory. They are creating collections of different proteins about which we don't know much. What do we know about the plant's nutrient needs during the different phases of the growing season? While all of the different phases of this problem are confronting us, our declining fertility is slipping down faster and the problem is becoming much larger.

7. the soil's silt fraction may serve as sustaining fertility

The research on the clay at the Missouri Experiment Station suggests that the restocking of the clay, after exhaustion of its fertility supply by cropping, resulted from mineral breakdown in the silt fraction. After we have studied the clay so much we are now studying the silt as a mineral reserve for crop nourishment. This separate is of more service in areas of low rainfall and of less of soil development in the west than it is in the high rainfalls of the east. Probably the silt breakdown in many of the midwestern soils and in others on coming east farther, has been the major supply of fertility. Therefore, our crop production is holding up to what that annual silt breakdown allows. The clay is a weathering agent for soil minerals. Dr. E. R. Graham has done some very clever work to show that we can use these reserve minerals in contact with the dynamic clay and thus restock the clay. That is what limestone has been doing when we use it on the soil. That is what rock phosphate does, too. When we have the Missouri River hauling lots of unweathered silt in from the west; with rather dry winters; and with the wind blowing that unweathered silt out of the river bottom and depositing about a thousand pounds per acre of Missouri every year, it would seem to be good agricultural foresight to think about this reserve material as a fertilizer which nature is giving us very generously each year.

This silt fraction, composed as it is of minerals other than quartz, should bring all of us to consider more seriously the *sustaining fertility* in the soil. While we have learned much about soils, we scarcely have knowledge enough to maintain production by starter fertilizers only, particularly when we put them at the top of the soil where that is dried out most of the time and then blame the drought for the failure of the fertilizers in doing what some folks commonly expect of them.

8. organic matter—the "constitution" of the soils

The most neglected and most important chemo-dynamic factor of the soil is the organic matter. Organic matter may be said to be the *constitution* of the soil. As a definition of the word *constitution* in that usage, we take its meaning when the doctor consoles the friends of a patient in serious illness by reminding them that the patient has a good constitution. According to its meaning, as used in medical practice, a good constitution is the capacity of the individual to survive in spite of the doctors rather than because of them. The organic matter in the soil has been the capacity for our soils and our crops to survive in spite of the soil doctors, rather than because of them.

Your attention has already been called to the importance of the organic molecule when it is on the clay. There is also the tremendous significance of the organic matter as a season's release of plant nutrition. This release is timed to increase during the growing season or become larger as the temperature goes higher. The microbial activities follow Vant Hoff's Law and double their rate of decay of the organic residues with every 10 degree rise in centigrade temperature. Nature has always been fertilizing with the organic matter which is dropped back to the soil from the previous plant generations which have died in place. Organic matter is still the most reliable fertilizer in terms of the nutrient ratios and of the time when maximums must be delivered.

Another aspect of organic matter about which we probably haven't thought much is the value of some organic compounds in cycle, that is, they may be dropped back as crop residues and the next crop's roots may be taking them up, using them and dropping them back again. Plants need the various ring compounds in very small amounts to make some of the essential amino acids. They need the phenol ring in phenylalanine, one of the essential amino acids, essential for plant growth as well as for animals and ourselves. They need the indole ring, which is a phenol ring plus a side ring. It is the compound which gives the odor to feces when the digestion acts on the tryptophane of which that ring is a part. Tryptophane is the most commonly deficient amino acid, and is one of marked complexity.

Then there are also the sulfur compounds and the sulfur-containing amino acids. We might well wonder whether man and his flocks have not been geared together so closely in their past history because some of those excreted organic compounds were put back into the soil, and were going through the cycle over and over again as a help in the survival of both man and beast. Now we are trying to divorce ourselves from animals, but perhaps we haven't found the basis of safety for it. Organic matter must find a new and more important place in our minds as the neglected half of plant nutrition and soil fertility. In terms of the inorganic half of that responsibility we have partly understood about one-half of that phase, namely, the major cations. We don't know much yet about the cations of the trace elements. When it comes to those major inorganic ions which are cations, we have a good concept of their chemo-dynamics for plant nutrition. As for the anions like sulfates, nitrates, carbonates and others, we do not yet know how they are handled by the microbes and the plant roots in the soil. We have much yet to learn when we have scarcely one-fourth of the field of soil fertility interpreted in terms of the basic principles of absorption, exchange, solubility and what have you, when it comes to the problem of soil fertility and plant nutrition. There is enormous opportunity for a big research program ahead. However, we have charted our course now and believe we have analyzed the problems, though by no means outlined the solutions for all of them.

9. conservation of soil fertility demands revision of some economics

We need help from observing and thinking minds to take these concepts about soil fertility and plant nutrition out into the fields for test, whether our concepts are on the right or the wrong track. Those who supply fertilizers have not discharged their responsibility completely when a carload of fertilizer has been delivered on the farm. They have the responsibility of making those goods serve properly in crop nutrition. Fertilizers must serve not just for increasing the crop bulk, but in terms of the necessary chemo-dynamics of soil fertility and plant nutrition which mean better nutrition for animals and man as well.

Our concern about soil fertility and plant nutrition naturally emphasizes their biological aspects. But even under demand for more food and need for a national agricultural policy, one dare not disregard some of the economics involved. Unfortunately, agriculture uses soil fertility as its biological capital. To date, such capital is still an unknown in the money marts, the bankers' vaults, and the political areas. According to present economics applied to agriculture, soil fertility capital is thrown into the bargain when we make a sale of agricultural products. Such values are not interpreted in dollars. The depletion of soil fertility, that is, the foundation for real food values in land values, is not yet considered. Consequently, the agricultural business does not have an accounting system for taxes on income, on lands, etc., set up to include fertility depletion allowances, allowed labor income, guarantee of perpetuity of capital assets, etc., to make agriculture, and the soil under it, self-perpetuating. Soil exploitation and land ruin with time are therefore inevitable.

In spite of this lop-sided kind of economics for agriculture, when economics looks toward guaranteeing self-perpetuation for most other forms of making one's livelihood—even for the laborer by means of strikes—we expect increasing food delivery from the soil. There is no economic alternative except that the soil must be mined. That must be the result if food production by agriculture is to continue and to increase under the present economic disregard of fertility as the factor giving real value to the acres of land. No national agricultural policy for survival under high standards of living can come forth unless we finally realize that our national strength lies in the fertility of the soil and our future survival in the wise management and utmost conservation of it.

3.

The research articles listed in the Albrecht bibliography would take a dozen volumes to reprint. Each details a factor or two from the grand mosaic of the whole. Working with graduate students—drawn, in the main, from Brigham Young University—Albrecht took the factors one leg at a time. The starting points were real enough. "Once the soil is exhausted, you're on an ash heap," Albrecht told Acres U.S.A. "You don't know what's missing except that nothing grows. The mysteries of creation haven't all been put under button pushing technology." When it became obvious that a culture wouldn't fix nitrogen on legumes by itself,

Albrecht's attention was given to finding answers. In isolation, these answers became entries in the technical literature. Once in a while, however, Albrecht paused to summarize. The first great summary became a matter of record in the Proceedings and Papers of the Second International Congress of Soil Science, *Leningrad-Moscow, U.S.S.R., July 20-31, 1930, and is presented below.*

Nitrogen-Fixation as Influenced By Calcium

The legume plants are distinctly different from other plants in consequence of their utilization of large amounts of nitrogen by means of symbiotic bacteria on their roots. As a result of this capacity to utilize atmospheric nitrogen and build comparatively large amounts of this element into their tissues, the variation in the nitrogen supply to the plant manifests wide differences in the plant growth and plant constitution. Running parallel with this growth variation is a correspondingly wide range in degree of nodulation on the roots of the plant, since, in general, extensive nodule production is associated with extensive growth. The study of the nutrition of the legumes offers possibilities beyond those of ordinary plants, because this wide variation in nitrogen metabolism may serve as an index of the irregularities in the nutrition of elements other than nitrogen. Through fluctuations in this nitrogen metabolism measurable by differences in plant growth, degree of nodulation, nitrogen content, and nitrogen-fixation, one can determine the influence of the other elements such as calcium upon the plant's behavior. In the following study, the above were used as criteria to determine the importance of calcium in legume growth and its activity in nitrogen-fixation.

In undertaking the study of calcium in connection with nitrogen-fixation by legumes, there is recognized the widely spread experience of legumes failing on soils which are acid, or the general recognition of improvement of legumes through the use of limestone. It has long been known that the majority of legumes on sour soils are improved when limestone is applied. As to the particular factor responsible for this improvement, a great number of suggestions have been made. The lowering of the degree of acidity through limestone application is widely held as the responsible agent. The importance of the calcium as an element of nutrition, however, has not received much consideration. Even if the excessive degree of acidity, or the deficiency in calcium is responsible, the question still remains as to the mechanism, or method through which either of these manifests itself. The questions are still open as to whether one or the other of these inhibits growth by disturbing the microorganisms infecting the plant, or by disturbing the plant either directly, or indirectly through irregular nutrition. Likewise, we may ask whether the behavior of bacteria is disturbed and they never attempt to enter the plant, or whether the plants are disturbed so as never to attain the conditions suitable for the bacteria to associate with them.

Since a high degree of acidity, particularly in humid soils, is usually accompanied by a corresponding deficiency in calcium, it is readily possible that the deficiency of the calcium rather than the presence of the acidity is the factor responsible for the failure of the legume. There is a danger of committing the mistake of attributing causal value to the wrong factor of these two concomitantly associated with this crop disturbance. In the following study an attempt has been made to study the importance of calcium in contrast to the degree of soil acidity in connection with the nodulation of the soybean, and to determine, if possible, whether this importance is manifested through influences on the bacteria or on the plant itself.

The literature on the importance of lime in connection with legumes and nitrogen-fixation is extensive and includes a wide range of phases of the question. Much attention has been given to the significance of the degree of acidity in relation to the legume bacteria and to legume plants. No attempt will be made to review this extensive list of publications. Stress has not commonly been laid, however, on the element calcium, as independent of the degree of acidity. Only recently has this phase of liming received attention as shown by literature reviews already published. It is the importance of calcium, more particularly, that has received attention in the following study.

1. experimental work

The soybean was used for this study of legumes in relation to calcium of the soil. Observations on the inoculation of soybeans had pointed out that there were wide variations in the degree of nodulation on soybeans, independent of significant variations in the acidity of the soil. This observed variation coupled with the difficulty of obtaining successful nodulation on lime deficient soils suggested the relation of the nodulation of the crop to some particular soil condition. This plant was selected also because it is not commonly considered as sensitive to soil acidity. Therefore, irregularities manifested in this crop through calcium deficiency might be more readily expected to manifest themselves in the crops more sensitive in this respect.

2. fertilizer treatments influence inoculation

The inoculation of soybeans was studied in soils from different experiment fields, located on soils commonly planted by soybeans, and given various soil treatments. The irregularity in nodulation on these soils is given on the next page. A careful study of the results in these pots emphasized that growth and nodule production were decidedly improved by lime, by lime and phosphate, and by lime phosphate and potash. The improvement through these treatments was decidedly significant, giving as much as 100% increase in growth, or more than such increase in nodule production on some of the soils. The application of phosphate as acid phosphate which is a carrier of both phosphorus and soluble calcium may render calcium responsible for the effects manifested apparently by both of the elements, calcium and phosphorus. The potash was not used alone but in combination with other treatments.

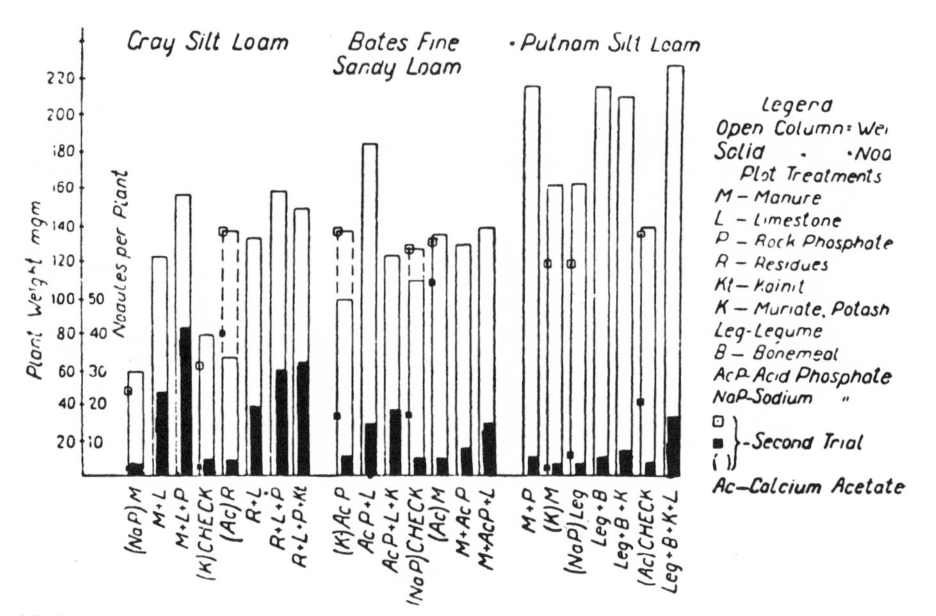

Nodule numbers and plant weights as influenced by soil treatments. (Second trial treatments are enclosed in parenthesis, results are indicated as column heights by open and solid squares.)The above graph has been reproduced from the Russian proceedings.

3. fertilizer effects on center treatments using lime or calcium

In order to determine which of the separate elements in the field treatment was responsible, to the same untouched soils were applied phosphorus as a soluble sodium phosphate, calcium as an acetate and potassium as a chloride, and these soils were replanted to the soybean crop. By these treatments the potassium chloride would manifest its importance. The phosphate would establish the importance of this element as independent of the calcium commonly associated with it in acid phosphate. The calcium acetate would determine the importance of the element calcium independently of the phosphorus. In addition, the acetate anion has little effect on the soil reaction, and consequently eliminates, in part, this item. The results of the plant growth and nodulation under these treatments are given as the second trial.

The most noticeable results of this trial were the failure of soluble phosphate to give decided improvement in the nodulation and growth of the crop, and the significant increase in nodulation which resulted in consequence of the treatment with calcium acetate. The calcium acetate was distinctly pre-eminent in its effects on nodule production in all three soils.

4. possible effects by calcium

The increased nodulation and growth may result from an increased supply of fixed nitrogen offered to the plant by the bacteria. This hypothesis suggests that the calcium may exercise its action by causing more legume bacteria to produce nodules and thus speed up the entire plant growth. Increased nodulation may result through greater vitality and activity of the legume bacteria in the soil as brought about by the effects of calcium on the bacteria. It may result also through the effect of calcium on the plant giving greater and greater efficiency to their function there. Experiments were undertaken, and the ease with which the bacteria already present in the soil enter into the plant root determines the importance of the calcium in these two respects.

5. calcium and longevity of legume bacteria

In order to test the importance of calcium in connection with vitality, soybean bacteria were stored for 60 days as cultures in water alone, and in water with calcium chloride at the rate of 1 part calcium to 1,500 of water. Cultures were likewise made in soil extract alone, and in soil extract with calcium. These cultures were used as inoculation for soybeans grown in soil extract and in soil extract plus calcium, both maintained at a pH of 5.7. They were continued until the plants developed several sets of leaves, when the nodule production was noted.

Poor inoculation resulted when the plants were grown in solutions free from calcium and inoculated by bacteria kept in the absence of calcium. The intervention of calcium either in the keeping of the bacteria, or in the growth of the plant favored the degree of inoculation. The bacteria, which had been stored in the absence of calcium for 90 days and inoculated into extracts with calcium, were able to produce effective nodulation. This suggests that the influence of the calcium is not wholly exercised through the bacteria, since the culture of bacteria either with or without calcium gave the same effect when introduced to the plant root in the calcium media.

6. susceptibility of plant to inoculation

Associated with the question of the longevity of the bacteria for an increased number of nodules is the question whether the legume plant remains susceptible to inoculation during its entire life history. Such extended susceptibility would make increased nodulation possible since the previous tests establish the fact that extensive longevity of the bacteria in the solution, and doubtless in the soil, may be expected to run parallel with it. Observations were made on the ability of the plant to be inoculated artificially at various ages, by counting the nodule numbers on soybean plants in pots inoculated the succeeding days after planting. Counts of nodules were made in the field at successive periods to note the increase of nodules with age. Results are given in table 1.

Table 1

Susceptibility to inoculation according to age, and nodulation according to growth stage

Time of inoculation (days after planting)	Stage of maturity	Total nodules	Average per plant
	Plants grown in pots		
5		237	23.7
7		125	12.5
9		79	7.9
11		5	0.5
13		5	0.5
15		30	3.0
17		0	0.0
19		0	0.0
21		10	1.0
23		0	0.0
	Plants grown in the field		
Age of plants			
33	Full growing period .	80	1.6
47	Beginning to bloom .	680	13.6
61	Full bloom — pods forming	885	17.7
75	Pods formed — not filled	910	18.2
89	Pods filled, ½ leaves dropped	935	18.7
103	Seeds beginning to mature, ⅔ leaves dropped	365	7.3
117	Plants mature, all leaves dropped . . .	270	5.4

These results suggest that the plant remains susceptible to inoculation during the greater portion of its growth. Since the inoculation on the pots was introduced, however, on the surface of the soil, it is readily possible that the susceptible roots had penetrated to such depths that the surface application of bacteria failed to reach them. The results suggest, nevertheless, the ready susceptibility of the young plant to go through inoculation by means of artificial cultures. The nodule increase in the field continued until the beginning of maturity. Hence the plant itself is susceptible to inoculation, though with a varying degree, during the greater portion of its growth period.

These trials suggest that the plants growing normally remain readily susceptible to inoculation, and that the bacteria remain sufficiently vital to serve as inoculation under a wide range of conditions. They suggest also that the influence of the calcium emphasizes itself possibly more in the presence of the plant than through any effects on the bacteria.

7. *other effects beyond neutralization of soil acidity*

To isolate further the effects of the calcium through the plant in contrast to its effects on the bacteria, some acid soil (pH 5.4) was used in which soybeans had been grown for 3 years and in which nodulation was apparently well established. One portion of this soil was treated with calcium carbonate at the rate of 4 tons per acre, while the other portion was given no calcium. Sterile soybean seeds were planted in both portions and no additional inoculation was supplied. Examination after 5 weeks' growth revealed an increase of over 300% according to part of table 2, in the inoculation on the portion given calcium carbonate in contrast to the soil given no treatment. This suggests that the effect of calcium cannot be one of influencing the longevity of the organisms since the flora of bacteria in this case was already living in the soil.

Another test on a sterile acid soil (pH 5.4) was undertaken by using calcium chloride. This was applied as solution equivalent in calcium to only 200 pounds of

Table 2
Nodulation of soybeans on acid soils as influenced by calcium compounds

Soil Treatment		No. of pots	Total No. of plants	Range in nodules per pot	Average nodules per plant	Per cent increase
Soil already inoculated given calcium carbonate						
None		30	130	10-81 [1]	12.0	—
Calcium carbonate-4t.		30	133	64-247 [2]	40.2	336
Sterile soil given calcium chloride with inoculation						
First Crop	None	30	130	—	—	—
	Calcium chloride	30	133	—	— [3]	—
Second Crop	None	16	72	—	— [3]	—
	Calcium chloride	16	80	1-26	1.8	—

[1] 3 pots exceeded this range greatly, having 110, 140 and 142 nodules.
[2] 2 pots exceeded this range greatly, having 297 and 380 nodules.
[3] 3 plants had 1 nodule each in these trials.

calcium carbonate for 2 million pounds of soil. The beans were inoculated with artificial cultures and examination was made at the end of 5 weeks. No nodules were formed. A second planting without reinoculation was made and excellent nodulation occurred on the calcium treated soil, while no nodules were produced on the untreated soil, as given in table 2.

Evidently the bacteria introduced as inoculation in this case remained viable in the acid soil. The influence of the calcium as demonstrated in this trial can scarcely be one of neutralization of the soil acidity, since the calcium application was in the form of chloride and in a quantity equivalent to only 200 pounds calcium carbonate per acre.

8. small amounts of calcium influential in field trial

The influence on nodulation by small applications of calcium was also tested in field trials using small amounts of calcium compounds either in the row or as salts applied on the seeds at planting. This soil was unlimed, and apparently sterile so far as soybean bacteria were concerned, hence artificial cultures were used. The salts were applied as solutions to the dirty seeds, which had been given enough dry soil to take up the moisture after applying the inoculation. Again the influence of small amounts of calcium salts was evident, as shown in table 3, giving increased numbers of nodules and the percent of plants inoculated when the calcium applications were made.

Table 3
Nodulation of soybeans as influenced by various treatments applied with the seed

Treatment	No. of plants examined	Nodules per plants	Plants nodulated
			per cent
Not inoculated. No treatment	40	0.1	14
Inoculated. Limestone[1] 52 pounds per acre	95	1.3	44
Inoculated. Limestone 183 pounds per acre	113	3.4	67
Inoculated. Limestone 183 pounds per acre, phosphate 78 pounds	97	4.1	68
Inoculated. Lime hydrate 52 pounds per acre	117	2.6	68
Inoculated. Lime hydrate 183 pounds per acre	125	2.6	71
Inoculated. Lime hydrate 183 pounds per acre, phosphate 78 pounds	149	6.4	77
Inoculated. Calcium chloride solution[2]	165	5.7	86
Inoculated. Calcium nitrate solution	110	11.8	97

[1] Applied in the row at seeding
[2] Applied on the seed at seeding.

A later field trial on an acid silt loam (Marion) used inoculated seeds, and had various calcium salts as chloride, nitrate, and hydroxide, applied to it through the fertilizer attachment of the seeding machinery. On this particular soil type the influence of the calcium salts was most marked in conjunction with the artificial cultures. When used in connection with an inoculated soil it enlarged the nodules but gave no increase in nodule numbers as shown in table 4.

Table 4
Nodulation of soybeans on acid soils in field treatments of calcium

	pH	Nodules per plants	Per 100 infected plants	Nodule Volume cc. per 1000 per nodule	Nodule Weight mg. per nodule
None [1]	6.4	0.2	20.0	—	—
Culture	6.1	0.6	6.6	900.0	133.0
Culture and calcium chloride	6.0	26.6	100.0	24.1	24.4
Culture and calcium nitrate	6.0	22.9	100.0	28.6	28.9
Culture and calcium hydroxide	6.0	22.9	100.0	21.1	23.0
Inoculated soil	6.4	16.0	100.0	31.2	30.0
Soil and calcium chloride [2]	6.4	13.4	100.0	45.5	44.4
Soil and calcium nitrate [3]	6.4	11.6	100.0	53.5	58.3

[1] About 30 plants were examined in each case. They contained 1.23% nitrogen in the summits and 0.76% in the roots.
[2] The plants in this treatment contained 2.39% N in the summits, 2.04% in the roots.
[3] The plants in this treatment contained 2.16% N in the summits, 2.13% in the roots.

9. calcium within the plant influences nodulation

In order to discover more fully the role played by the plant in the improved nodulation, through calcium treatment, a test was conducted on the influence which could be exerted by the calcium when within the plant tissue. Sterile soybeans were planted into calcium deficient sand and the same given calcium carbonate at the rate of 5 tons per 2 million. After a growth of 10 days, the seedlings were taken up, washed and replanted into other soils which had grown soybeans in the field and which were well inoculated. Half of each of these soils was planted with calcium starved seedlings and the other half with calcium fed seedlings. In about 5 weeks the plants were examined for their nodulation with the results given on table 5. On the soil which was not acid, and was rich in electrodialyzable calcium, there was little difference in the nodulation in consequence of the extra calcium carried within the soybean seedling. In the acid soil with a lower content of dialyzable calcium, the degree of nodulation was increased more than four times, in consequence of the extra calcium taken up by the seedlings in its early growth. This suggests that calcium exerted its influence upon nodulation by its effects on the plant or on the plant tissue which increased nodulation.

Table 5
Nodulation of soybeans in neutral and acid soils as influenced by the calcium in the seedlings

Soil character	pH	Seedling treatment	No. of plants	Nodule Nos. per plant		Calcium content		
				Range	Avg.	Per 100 seed-lings	Electro-dialyz-able per 10 g. soil	Per 100 seeds
						mg.	mg.	mg.
Neutral	7.8	None......	60	12-77	36.6	17.07		
		calcium....	67	9-67	38.9	30.14	24.07	6.85
Acid	5.5	None......	69	1-7	3.4	17.07		
		calcium....	79	2-25	15.1	30.14	11.78	6.85

10. calcium influences nitrogen fixation

The increased nodulation and better growth due to calcium within the plant raises the question whether increased nitrogen-fixation and nitrogen nutrition are brought about by the calcium under these conditions. Seeds were again planted into the calcium-laden and calcium-deficient sands and transplanted into a well inoculated acid soil (pH 5.5) after 10 days. At intervals of 10 days growth in the soil, 50 plants were removed, their length and weight taken, the nodules counted and analyses made for nitrogen content with the results given on table 6.

There was a decidedly greater growth, and a greater nitrogen harvest in consequence of the treatment of the seedlings with calcium. It is interesting to note that the calcium in the seedling encouraged its lengthening but not necessarily its increase in weight. This is in accord with recent observations by Miss Dorothy Day, *Some Effects of Calcium Deficiency on Pisum sativum.* It also encouraged an earlier increase in nodule numbers and an earlier increase in nitrogen content. That the increased nitrogen in the plant is not wholly due to the nodular activity in nitrogen-fixation, however, is shown by the fact that after growing in the soil for 10 days without producing nodules, the calcium-fed plants had already increased their nitrogen content 18.4% over that in the plants lacking of nitrogen in their seedling period. With the advent of nodules after 20 days the total nitrogen of the calcium-fed plants increased faster than that of those not given calcium. The percent increase in nitrogen of the former over the latter varied, but maintained an average of 17.3% for the five different determinations. This increase in total nitrogen occurred mainly through increased growth, rather than through increased concentration of nitrogen within the plants, since the percent of nitrogen in the dry matter was regularly higher in these plants grown from calcium-starved seedlings. According to these results, the whole plant activity is modified decidedly, its lengthening of the tissues, its metabolism of nitrogen, its production of chlorophyll and consequent greener color, its nitrogen fixation and other physiological phenomena seem to be increased by the presence of but a small increase of calcium in the young plant tissue.

11. calcium modifies plant structure

Since the influence of the calcium manifests itself through the plant, such an influence must of necessity have some fundamental basis in the plant, possibly purely functional and possibly structural. If structural, it is readily possible that these variations in structure caused by the presence and the absence of calcium may be responsible for freedom or failure of entrance, respectively of the nodule forming organisms. A study of the plant structure was undertaken to discern the structural differences associated with calcium irregularities, as these differences might be causally associated with the ready entrance and function of the legume organisms.

Further studies in conjunction with nodular structure are contemplated with a view of learning of the structural irregularities in that portion of the plant as possibly responsible for nodulation irregularities under deficient calcium supply. Studies on a chemical difference within the root cells may also be necessary to fully understand the mechanism through which calcium plays such a significant role in the inoculation and growth of the legume crop.

The following points are significant in this connection:

Table 6
Nitrogen content of soybeans as influenced by calcium in the seedlings

Age of plants	Previous Treatment	Avg. height cm.	Percent increast over no lime	Oven dry weight 50 plants	Percent increase over no lime	Nodule number per plant	Percent Nitrogen	Nitrogen content 50 plants g.	Increase over no lime	
									mg.	%
Seedlings	None	7.0		4.77		0	6.57	0.3134		
10 days . . .	Lime	12.0	71	4.77		0	6.51	0.3105		
Soil	None	13.5		6.64		0	5.58	0.3703		
10 days . . .	Lime	20.0	49	10.03	50.1	0	4.33	0.4387	68.4	18.4
Soil	None	21.7		10.90		1.09	4.82	0.5247		
20 days . . .	Lime	31.8	47	17.87	63.9	9.06	3.44	0.6147	90.0	17.1
Soil	None	30.4		23.9		6.04	3.40	0.7799		
30 days . . .	Lime	35.6	17	29.7	24.2	16.30	2.85	0.8467	66.8	8.5
Soil	None	31.0		24.8		13.81	2.92	0.7243		
40 days . . .	Lime	39.0	25	38.4	54.4	17.51	2.46	0.9472	222.9	30.7
Soil	None	31.0		35.2		17.30	3.13	1.1020		
50 days . . .	Lime	41.0	32	41.0	16.4	17.02	3.00	1.2340	132.0	11.9

1. The liming of soils for legumes is essential for the establishment of these crops not solely through the effect of neutralizing soil acidity.

2. It is clearly evident that lime plays an essential role in supplying the soil with the nutrient element calcium, often so deficient in the soil as to demand its application for a success of the crop.

3. The influences of the calcium are manifested through improved nodulation of the crop, its greater nitrogen-fixation and consequently better growth.

4. Results suggest that the improved nitrogen-fixation activities owing to calcium not necessarily result through the effects of this element on the legume bacteria, but possibly through the action of calcium within the plant.

5. It is highly probable that plant structural conditions, and possibly also its functional conditions, may be modified by a liberal calcium supply so as to facilitate better symbiosis between the legume bacteria and their host plant.

6. The use of lime in connection with legumes thus seems to stimulate nitrogen-fixation by legume bacteria, indirectly through the influences of the element calcium on the plant make up by favoring plant growth, bacterial entrance and multiplication, and thus the entire process of nitrogen-fixation.

When Is A Legume Not A Legume?

"Legumes are not soil builders when grown on exhausted, 'tired' soils," stated Dr. William Albrecht in a State Research Bulletin.

On exhausted soils, deficient in nitrogen, phosphorus, potash and calcium, legumes take little if any nitrogen from the air. Dr. Albrecht pointed out that although fairly good tonnage may be produced per acre the growth consists largely of carbohydrates and consequently is not high protein feed.

Such legumes are not even efficient producers of organic matter which could be used to increase soil humus. According to Dr. Albrecht, under declining soil fertility the plant's protein manufacturing processes are evidently the first to decline.

In conclusion, Albrecht stated that grasses grown on mineral and nitrogen rich soil and pastured or cut for hay at the right stage have much higher percentages of protein than legumes grown on soils with a low level of "bases and phosphorus." Thus, legumes as well as nitrogen fertilizers pay greatest dividends in conjunction with adequate minerals.

4.

The sweep of thinking that was to characterize Dr. William A. Albrecht's output for several decades of his life came to the fore at the Specialist Conference in Agriculture, Melbourne, Australia, May 1949. For the occasion, Albrecht prepared a paper that codified a great deal of the work he and his associates had been doing. In the Albrecht bibliography, these backup details come styled Calcium-Potassium-Phosphorus Relation as a Possible Factor in Ecological Array of Plants; Surface Relationships of Roots and Colloidal Clay in Plant Nutrition; Plant Nutrients Used Most Effectively in the Presence of a Significant Concentration of Hydrogen Ions; Potassium in the Soil Colloid Complex and Plant Nutrition; Soil Fertility in its Broader Implications; Colloidal Clay Cultures, Preparation of the Clay and Procedures in its Use as a Plant Growth Medium; Our Teeth and Our Soils; Diversity of Amino Acids in Legumes According to the Soil Fertility; Feed Efficiency in Terms of Biological Assays of Soil Treatments; *on and on. Here, Albrecht deals with these findings in wide brush strokes. Later, in papers that will lace together the rest of this volume, his entries fill out the details, much as would an artist with a fine pointed brush.*

Nutrition Via Soil Fertility According to the Climatic Pattern

The human species seems to be slow in learning how to feed itself effectively. Its nutrition is not guided wholly by instinct, as is true of the lower animals, nor yet by design of its own sufficiently reliable to guarantee its survival. Degenerative diseases, as causes of death in the United States, have risen from 39% of the population in the decade 1920-1929 to 60% in the year 1948. During the same period the infectious and general diseases decreased from 41 to 17%.

"Disease is so generally associated with positive agents—the parasite, the toxin, the *materies morbi*—that the thought of the pathologist turns naturally to such positive associations and seems to believe with difficulty in causation prefixed by a minus sign," [wrote the Medical Research Council, Special Report Series, London, 1924].

We are slow to believe that the entrance into the body by microbes is not necessarily by their overwhelming attack, but probably their initiatory part of a task of disposal under the beginnings of death recognized earlier by them than by their pseudo-victim; that much of what we call disease is nutritional deficiency; that we do not retain in our foods and feeds all their nutritional values possible between their production and their consumption; that the soil may be deficient in the contribution of the 14 or more indispensable inorganic elements; and that the foods cannot have in them those essential inorganic elements absent in the soil growing them. Man's place at the top of the biotic pyramid may be impressive through its loftiness; but it is correspondingly hazardous in nutrition and thereby in survival in terms of the myriads of compounds and their chemical complexity, because of man's dependence on the many services of collection and synthesis coming from below him through the animals, the plants, the microbes and finally the soil fertility. The soil is the point at which the assembly lines of all life take off.

1. deficiencies are the expectable, not the unusual

If the evolution of higher life started in the seawater, its physiology represents an ionic complexity in equilibrium with that growth medium. And if soil is a temporary rest stop of the rocks on their way to solution and to the sea, then man's evolutionary migration from out of the sea on to the land is one going against the current of soil development. Consequently, it is a migration toward nutrition on the level of the single rock contents. It represents the hazard of deficiencies certainly in the inorganic elements. These deficiencies are expectable when rocks consist of limited minerals and these in turn of limited numbers of elements. When of the 18 or more elements required by our bodies, only four originate in air and water while 14 or more must be supplied by the soil, and when any rock going to make soils seldom contains half that many, then man's evolutionary movement to land suggests the necessity of maintaining some life lines reaching from there back to the sea, or its fertility equivalent, for prevention of deficiencies of inorganic elements alone. Movement from the sea's varied collection of elements suggests that deficiencies should be expectable in nutrition even in just the ash constituents of the body.

Equally as indispensable as the 14 inorganic elements are the 15 or more vitamins going back to animals, plants, and microbes for their synthesis. These complex compounds have been too small in amounts present, too unstable in structural nature, and too baffling in chemical composition to become a part of the list of requisites for ourselves before but recent years. They render their services, not as parts of the body structure, but rather as tools and catalyzers in the many processes bringing about body construction. As catalysts, should we expect them to stand out before the reactions, of which they increase the speed, fail us and thereby bring about some biotic disasters? Recent research suggests that the concentration of some vitamins in vegetables increases as the fertility of the soil growing them moves toward imbalance and shortages in the elements. If these are the facts, shall we be hesitant in postulating that vitamins climbed to recognizable concentrations and biochemical prominence because of the increasing inorganic deficiencies coming via the soil? Like the whip for the team of horses, which comes into more prominent use as the team is less equal to the load, shall we not see vitamins coming into prominence because of deficiencies and imbalances in the substances reacting under the stimulation of these catalysts?

Beside this significant number of 29 indispensables already listed as inorganic, soil-borne elements and vitamin compounds, we must have also ten specific amino acids out of a known list of more than twice that number. These ten are requirements as constructive parts of our body proteins. Proteins are the only compounds of the body that can reproduce and that can transmit life. The chemical reactions of living organisms are mediated by *enzymes;* the body activities are coordinated by *hormones*; some diseases are credited to *viruses;* and protection against many diseases is affected by *antigens.* All these different terms are specific names for body creations which in chemical composition can seemingly be classified as proteins.

Amino acids are synthesized from their constituent elements, not by our own bodies, but by plants and microbes. They are only collected, possibly simplified or transformed, and assembled into proteins by higher life forms. Animals must find their amino acid requirements in their ration, except as microbial symbiosis in their alimentary tracts supplements them. Herbivorous animals are thereby limited to those soil areas where the fertility undergirds the microbes and plants in their synthesis of the animal's amino acid requirements. It is in these fundamental body-building, life-propagating aspects of making proteins in the vegetation that the soil elements of fertility take over the control. It is only recently that proteins have begun to come into their own. They will be doing so only slowly as long as we are content to classify as proteins the whole gamut of organics of which the nitrogenous residues of ignition in sulfuric acid are multiplied by a single arithmetical factor like 6.25, or some other simple number decided upon by our majority vote. Can knowledge of such loose construction prohibit irregularities and deficiencies in the propagation and reproduction of life founded on the construction so intricate and extensive as that of the myriads of protein molecules?

Our growing knowledge of nutrition has segregated also three fatty acids as specific requisites. These are fats with 2, 3 and 4 unsaturated bonds, respectively. This chemical structure magnifies the ease and speed of either their oxidation or their hydrogenation. These reactions bring distinct chemical and physical changes converting them by the latter from oily liquids into solid fats. If their physiological essentiality should rest in this unsaturation, we need to raise the question whether the hydrogenation of our food fats is not defeating the purpose in our eating many of them. It also questions whether the numerous aerating manipulations of milking and pasteurization are not destroying the nutritional services by such unsaturated essential fats in milk. Such a hypothesis provokes itself when we remember that the calf takes its milk by means of natural nursing equipment that excludes air most completely, and when recent experiments in milking and processing in an atmosphere of nitrogen give a canned sweet milk that remains near its originally milked condition for days on the grocer's shelf.

Our knowledge of the nutrition of ourselves and of many of our animals has already listed 14 (possibly more) inorganic elements coming from the soil as requisite for the assembly of four other elements from the air and water by the synthetic processes of microbes, plants and animals providing us with food. Up to date, there are among the required compounds so synthesized in the biotic pyramid supporting us, 15 or more vitamins, 10 amino acids and 3 fatty acids. This is already a total of 46 specifically required and chemically characterized items in

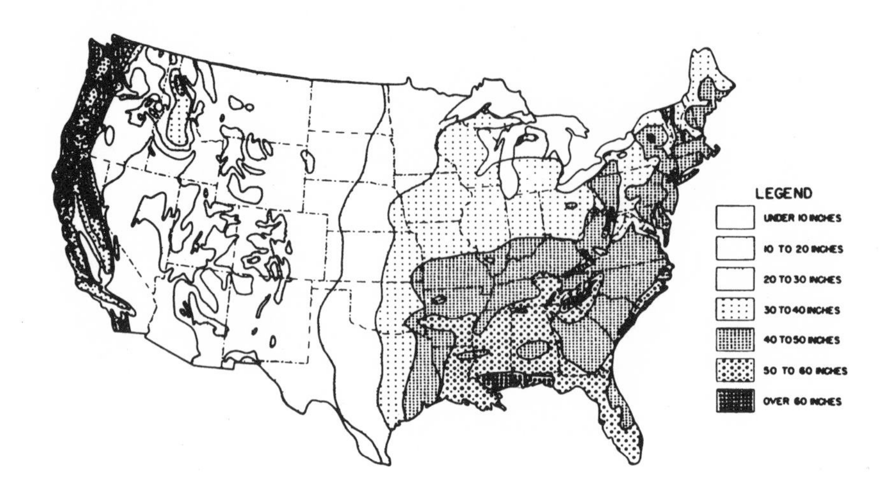

Distribution of mean annual precipitation in the United States.

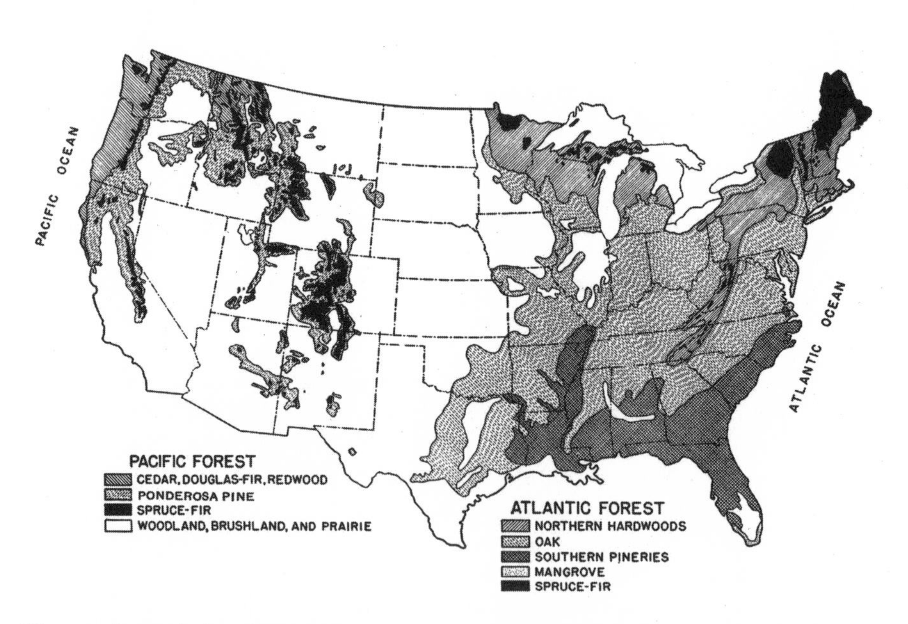

The two forest belts of the United States, the Atlantic forest and the Pacific forest, and their major subdivisions.

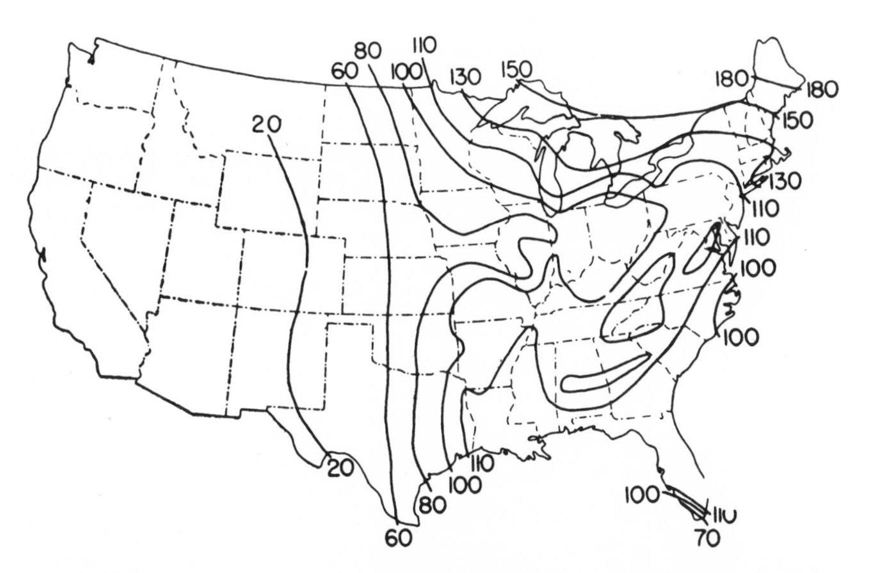

Distribution of Transeau's precipitation-evaporation ratio in the United States.

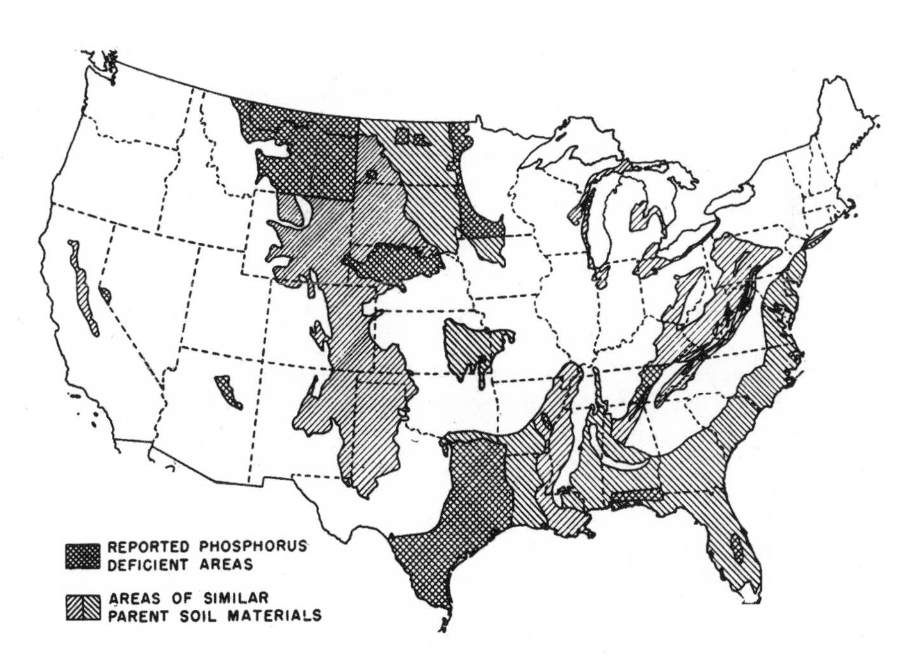

Areas in the United States subject to phosphorus deficiency.

what would seem but the overture in the program by the growing science of nutrition. Shall we be startled by our failure to keep this list (and possibly many other items still unknown) completely assembled and in proper balance at all times for complete nutrition with its resulting good health? It would seem logical to believe nutritional deficiencies the expectable rather than the unusual.

When the soil fertility, *i.e.,* the list of required inorganic elements—and possibly some organo-inorganic compounds—in the soil, is the foundation for the life pyramid of microbes, plants, animals, and man; and when the soil is the resulting product from rocks weathered by the climate, it seems most logical to look first for a natural pattern of nutrition of all life in accordance with the fertility of the soil; and then for this fertility in a pattern according as the climatic forces of the past and the present determine it. Such broader patterns of these basic forces should be helpful in elucidating and integrating the details of our nutrition which, if seperated therefrom, lose their significance or even escape us.

2. soil dynamics control nutrition

Creation of life starts with the lifeless, soil-borne inorganic elements. These serve to synthesize the meteorologically contributed elements along with themselves into the living organic compounds. It is from the soil that the provision of nourishment for all life takes its start. The rock-mineral mass of the earth and its conversion into soil determine whether the latter is sufficiently diverse and active in terms of the inorganic, or ash, elements for support of animal life. In order to fulfill the requirements for this objective, there are necessary:

1. The rock-minerals in the sands and silts containing the 14 or more inorganic elements of soil fertility indispensable in our food supply.
2. The rock-weathering processes, or the mineral breakdown by means of soil acidity to move these elements out of the fixed compounds into potentially active ions.
3. The retention by adsorption on the colloidal complex within the upper soil horizons of those active elements against loss therefrom in percolating waters, but also in forms exchangeable to the microbes and the plant roots for the hydrogen or other elements they offer in trade.
4. The exchange to the plant roots of each nutrient element in particular amount and ratio to others as permit their synthesis into the plant compounds for its growth and reproduction.
5. Their diversity in supplies and activities under continued renewals within limited soil volumes to guarantee the plants' synthetic delivery of carbohydrates, proteins, fats, vitamins, inorganics, etc., requisite for the nourishment of animals and man.

3. hydrogen in the soil mobilizes the nutrients

These dynamics of the soil make it function as a mineral reserve of the essential fertility elements in the sands and silts breaking or weathering out from there into ionic forms to be adsorbed on the clay-organo-colloidal complex, and moving from there to the microbial or root cells for their nourishment. In the opposite direction along that course, there is a movement of active hydrogen. This originates in the carbonic acid formed from the carbon dioxide of respiration by the living cells, to be adsorbed on the colloidal complex in exchange for other cations. This adsorbed hydrogen gives the colloidal complex an acid reaction. It makes this the reagent for speedier weathering of the minerals in the sand and silt reserves, or for the mobilization of their ionic contents to the colloidal complex for plant nourishment. It moves other cations into the plant roots more efficiently.

The migration of the elements from soil into the plant roots is then a highly variable behavior. It should therefore result in different amounts of—and different chemical natures of—the food compounds synthesized by the plant. Variations in products compounded by the plants occur according to the exchangeable supply in terms of degree of saturation of the clay's capacity and total amount of each nutrient element on the clay; the ratio of the amount of each element in respect to the others on the clay; the relative activity there of each ion according to its association with other particular ions; the total amount of root surface per unit of soil volume; and the metabolic performances within the plant root according as these give different colloidal attributes to the root contents and the cell membranes enclosing them.

It is the variation in the fertility and the chemical dynamics within the soil supplying the nutrients to the plants that gives the basic reasons why there are different plant species on different soils. It gives the reasons also why the same plant species on different soils should vary in the compounds it synthesizes and therefore in its delivery services of feeds and foods. It explains why on one soil—less fertile—a crop delivers mainly photosynthetic products, namely, the carbohydrates, while on another—one more fertile—the same crop delivers not only those fuel foods, but also the biosynthetic products like proteins, vitamins, organo-inorganic compounds and others. It gives these and others serving in body-building and reproduction because of more contributions from the fertility of the soil. Food in terms of more than fuel service is therefore dependent on the fertility of the soil.

4. protein potential according to climatic pattern of soil construction or destruction

The major problem in feeding ourselves and our domestic animals is one of providing proteins in sufficient quantity and of complete quality. Such higher life forms merely assemble proteins from the ingested amino acids. These are synthesized from the elements by plants and microbes. Unfortunately, we have been too readily content to measure efficiency of our management of the soil fertility by the agricultural output of bushels and tons. Also, we have apparently assumed chemical composition of plants a constant according to plant pedigree. Bulk production has been generally accepted as a satisfactory criterion.

Consequently, when the decreasing fertility of a soil brought lowered tonnage per acre of an acceptedly nutritious herbage, like alfalfa (lucerne), we substituted one whose bulk production on that same soil was high. In that substitution, there was the acceptance in the replacing crop of a higher concentration of carbohydrates and a lower one of proteins and of all the other special organic and inorganic compounds associated with, or requisite for, protein production by plants. Korean lespedeza (*Lespedeza striata*) or sweet clover (*Melilotus alba*) often replace alfalfa, for example, because the larger tonnage yields of the former over the latter on less fertile soils commonly are enticement for this substitution. This is the case especially when no more refined criterion than total nitrogen determinations is taken as the measure of their protein equivalents. It remained, however, for animal disasters, such as poor gain, defective wool, and failing reproduction to characterize the lespedeza hay, and for animal deaths from bleeding in case of the sweet clover (coumarin) to point out the need for more specific criteria in judging the services of the soil fertility in the production by "juggled" crops. In terms of nutrition, it seems wise that we should turn to the total yield of protein per acre and its completeness in terms of at least the ten essential amino acids as the criterion of our successful management of the soil for feed and food production.

5. *amino acids in forages according to trace elements applied to the soil*

More recently we have taken to measuring the total amino nitrogen (Van Slyke Method) and the concentrations of at least the ten separate essential amino acids in forages as influenced by the different nutrient elements of soil fertility.

The concentrations of these amino acids in lespedeza hay, for example, were found to vary considerably according to the different soil regions of the state of Missouri, and with the fertilizer treatments used (table 1). In alfalfa, grown on our humid soil requiring liming and other soil treatments for its successful crop use, the concentrations of these amino acids were increased by the applications of the separate trace elements, manganese and boron, and by a mixture of these with cobalt and zinc, all as supplements to the calcium (table 2). They were also increased by the use of superphosphate or potash salts in conjunction with lime. This suggests that these phosphate and potash salts may have been effective through their "impurities" consisting of trace elements. Spectrographic analysis of superphosphate revealed an extensive list of these in it to suggest that we may have been unwittingly applying trace elements extensively when superphosphate makes up such a large share of fertilizers.

Of particular interest among the effects by trace elements was the marked increase for tryptophane and lycine. These two essential amino acids are usually so low in corn (maize) protein (zein), commonly grown on similar soil, as to demand that these be supplied to the corn ration in special protein supplements even for fattening more mature animals. Tests have not yet been undertaken with trace elements as fertilizers for corn to learn whether these specific amino acid deficiencies of zein would be overcome thereby.

Table 1
Amino Acid Content of Lespedeza Hay According to Different Soil Types and Treatments (Percent Dry Weight)

Soil Type and Treatment	Valine	Leu-cine	Argin-ine	Histi-dine	Threo-nine	Trypto-phane	Lycine	Isoleu-cine	Methi-onine
Eldon -									
treated	.89	1.05	.64	.37	.63	.29	.99	2.08	.09
untreated	.91	.97	.42	.34	.56	.20	.94	1.67	.08
Lintonia -									
treated	.92	1.03	.45	.34	.62	.27	.87	1.63	.07
untreated	.78	1.01	.32	.30	.54	.18	.87	1.68	.07
Putnam -									
treated	1.02	1.28	.71	.36	.63	.24	.89	1.89	.08
untreated	.98	1.28	.56	.50	.60	.22	1.00	2.26	.08
Grundy -									
treated	1.01	1.17	.62	.36	.69	.19	.79	2.00	.07
untreated	1.13	1.46	.45	.38	.67	.19	.93	2.00	.08
Clarksville -									
treated	.85	1.02	.34	.38	.58	.25	.93	1.59	.07
untreated	.94	1.19	.36	.35	.55	.21	.87	1.38	.07

6. *climatic pattern of soil construction or destruction controls ecological patterns of plants, animals and man*

Some general observations supported by critical studies prompted the concept that the climatic pattern of the soil fertility is providing either for the biosynthesis of protein and the photosynthesis of carbohydrates giving better nutrition via a balanced ration in general, or for mainly photosynthesis of carbohydrates and nutritional deficiencies. These provisions, that is, proteins and carbohydrates, or largely carbohydrates only, come by way of the different kinds of vegetation operating on the correspondingly either high or low nutrient base of soil fertility.

The ecological pattern of the virgin vegetation contained forests for no more than wood or cellulose as the crop from the soil on either the acid, highly-developed, fertility-deficient soils, or on the rocky, undeveloped soils, all with sufficiently high and regular rainfall to maintain trees growing all season. It contained prairie grasses and legumes with production of protein as well as carbohydrates on the mid-continental soils, that are well-stocked with calcium and other fertility because of their moderate development by only moderate climatic forces. Such limited rainfalls mean intermittent summer droughts. They mean the survival only by such crops as grass tolerating periodic growth alternating with

Table 2
Amino Acid Content of Alfalfa Hay According to Soil Treatments with Trace Elements (Percentage of Dry Leaves)

Plot No.	Treatment	Valine	Leucine	Arginine	Histidine	Threonine	Tryptophane	Lycine	Isoleucine	Methionine
1.	Calcium	2.19	4.37	0.38	0.65	0.86	0.54	1.57	2.64	0.10
2.	Calcium and manganese	2.40	4.89	0.43	0.80	0.95	0.64	2.12	3.63	0.24
3.	Calcium and boron	2.13	5.55	0.41	0.72	1.07	0.85	2.13	4.09	0.17
4.	Calcium and mixture*	2.59	5.24	0.41	0.83	1.01	0.67	1.87	3.44	0.22

*Mixture of cobalt, copper, zinc, manganese and boron

dormancy periods during the growing season. This moderate soil development leaves ample fertility which makes for grass as a highly nutritious feed because of the soil fertility rather than because of this crop's pedigree. Transplanting a grass agriculture to the humid, highly-developed soil does not necessarily argue for good nutrition of the poor beasts compelled to eat it there merely because of its good reputation as feed established elsewhere.

The distribution pattern of the virgin wildlife showed the herbivorous bison herds estimated as a total of once 50 million on the mid-continental prairies. There were scattered small numbers on the humid but rarer calcareous soils now known as Kentucky's famous race horse and bluegrass area. Likewise, some of these active hulks of brawn and bone were in those limestone valleys of Pennsylvania where the high agricultural output per county breaks records today. The major post-bison area includes high-protein wheat, beef calves and beef cattle herds today. Such ecological patterns of virgin vegetation and wildlife superimposing themselves on the soil fertility pattern raise the question whether the food problem of shortage in animal protein can be solved without looking to the fertility of the soil.

The pattern of present cultivated crops has the better leguminous feed producers, rich in proteins and in the essential inorganic elements, also on soils developed under moderate rainfall and temperatures. Alfalfa, for example, requires much soil treatment for its even limited production and survival on highly weathered humid soils. Yet it grows readily and extensively on mid-western soils. This hay crop that in finely chopped form constitutes the major part of commercial livestock feeds, delivers higher nutritional service per bag when grown on those mid-continental than when on the eastern and southeastern soils of the United States.

The protein concentration in the wheat grain is low on the humid soils in the eastern part of its area, and high on semi-arid ones in the western wheat area. This fluctuates for this single plant species from a low of near 9% to a high of double that amount in no greater traverse than across the state of Kansas from east to west, or from an annual rainfall of 37 to 17 inches. With increased cropping of the wheat soils in that state, the protein concentration is falling. The figures for the national protein standards for that food grain are being revised downward. At the same time, the bushels per acre of the increasingly more carbonaceous product are going upward to give record-breaking totals of bushels very recently for the state.

In the corn crop, grown on the more highly developed soils than those given to wheat, hybrid vigor has pushed up the bushels in total and per acre. However, during the last ten years the total protein concentration in that feed grain has fallen from an average of 9.5 to 8.5% to say nothing of what might be happening to its array of specific amino acids. Quality of the same plant species varies not only with ratios of the elements in the fertility, but with the decline in their totals of supply, as these facts emphasize. Such observations call for more critical studies of the protein potentialities according to the climatic pattern as it represents soil construction or soil destruction in terms of these requisites for good nutrition.

7. *carbohydrates mainly and proteins also in crops depend on the ratios of exchangeable calcium to potassium*

A study of the inorganic composition of the crops and native plants growing on soils that are slightly developed in the western United States; moderately developed in the mid-continent; and highly developed in the east, northeast and southeast, established the significance of the calcium-potassium ratio (also calcium-phosphorus ratio) in the crop composition as a reflection of the same ratio of these fertility elements in exchangeable amounts and in relative activities on the colloidal clay complex (table 3). A wide ratio of these occurred in the forages in western United States on the recognized calcareous soils. On coming eastward, it was narrower for the crops on moderately developed soils. It was still narrower for forages and other crops grown on the highly developed, acid and calcium deficient soils on coming still farther east and southeast toward higher rainfall and temperature.

Here is well illustrated the relatively more rapid removal from the soil during climatic development of its supplies of exchangeable calcium—and some other nutrient elements apparently associated in similar physiological functions—than of the potassium. The former is generally associated more particularly with the fabrication of the proteins from the carbohydrates by the plant's life processes. The latter is generally associated with the plant's fabrication of the carbohydrates. As the soils then are not yet developed to provide exchangeable calcium generously, or are more highly and excessively developed to remove it, their exchangeable nutrients on the colloidal complex represent supplies and activities in such ratios, particularly of calcium to potassium, as to provide a fertility imbalance feeding the forage sufficient potassium for carbohydrate production and construction of the plant as a bulk-producing factory, but as little more. It fails to provide the calcium and its associated elements, presumably including the trace elements,

Table 3

Composition of Plants According to Degree of Soil Development

	Plants growing naturally on soil developed		
	Slightly (38 cases)	Moderately (31 cases)	Highly (21 cases)
Dry matter contents as			
K_2O%	2.44	2.08	1.27
CaO%	1.92	1.17	0.28
P_2O_5%	0.78	0.69	0.42
Combined %	5.14	3.94	1.97
Amounts relative to highly developed soil			
K_2O	1.9	1.6	1.0
CaO	6.8	4.1	1.0
P_2O_5	1.9	1.6	1.0
Ratio K_2O:CaO	1.2	1.8	4.5

requisite for the conversion of these energy food values into proteins and reproductive possibilities.

It is only in the central United States, then, where there was sufficient soil construction, but a minimum of soil destruction, or where the ample mineral reserves and the suite of ions adsorbed on the clay gave protein production in the virgin short prairie grass to make big herds of bison and the similar but dwindling fertility supplies in modern times to represent high protein wheat and beef cattle under our cultivation. It is there that nutrition is of a high order because of the specific soil fertility array originating from the particularly fortunate combination of glacial and windblown mineral mixtures, and moderate rainfall in the temperate zone developing fertile soils from them.

One needs only to recall the forests on the less developed and rocky soils to the west of this mid-continental area. Then there are likewise the forested areas to the extreme east with hardwoods adjoining the prairies on their east. One needs also recall that the conifers, which are much lower in the evolutionary scale, are to the northeast and the southeast. This coniferous vegetation occurs on the acid, fertility-deficient soils with clay of high exchange capacity in the northeast. It occurs on the lateritic, fertility-deficient, but non-acid soils of low exchange capacity to the southeast and the tropics. All these are areas where deficiencies in nutrition may well be expected under cultivation today when the virgin crop represented little or no protein and the wildlife was extremely sparse, if not entirely absent.

The numbers of tooth caries per inductee into the United States Navy in 1942 tell us that the ecological array of the human is part of the pattern of climatic soil construction and destruction. Arranged by longitudinal soil belts two states wide, the inductees from the mid-continental, protein-producing soils exhibited 12.08 caries per mouth. On going westward from there to soils of less construction and then to the humid west coast, the figures were 13.10 and 15.50, respectively. On going eastward from the mid-continent to soils exhibiting increasing soil destruction, the numbers of caries per mouth were 14.95 and 17.55, respectively. These facts about this exposed part of the human skeleton tell us that the human species too takes an array in these health aspects according to the climatic pattern of soils.

Extensive experimental work has been carried out using the finer colloidal clay fraction as the medium for controlled plant nutrition and growth. This means of research permits control of the combinations or ratios of exchangeable cations (some anions) as well as the total amounts offered the plant for its nutrition. These studies have demonstrated the wide diversity in the plant's synthesis and storage of sugars, starches, vitamins, proteins and amino acids according to the ratios and activities of the exchangeable ions on the colloidal complex, and seemingly to the dynamics of the plant root. The wider calcium-potassium ratio of these activities on the soil colloid favours protein production and higher concentration in the crops giving less total forage yield (table 4). Conversely, the narrower calcium-potassium ratio gives little protein production and much bulk of forage. Many other variations in plant composition have been demonstrated because of the variation in the ratios of the exchangeable or active fertility of the soil. No credence can be placed in yields of merely more bulk as a reasonable index of more complete nutritional service by the forages. Their nutritional values depend on the fertility of the soil by which they were synthesized.

Table 4

Increase in Weights of Soybean Plants but a Decrease in Total Nitrogen (Protein), Phosphorus and Calcium in the Hay Crop with Widened Potassium-Calcium Ratio in the Soil Growing Them

Exchangeable cations.M.E.			Crop weight	Nitrogen %	Nitrogen Mgm.	Magnesium %	Magnesium Mgm.	Calcium %	Calcium Mgm.	Phosphate %	Phosphate Mgm.	Potassium %	Potassium Mgm.	K%/Ca%
Mg	Ca	K												
5	10	0	14.207	2.86	407	.36	52	.74	105	.25	39	1.01	150	1.36
5	10	5	14.592	2.56	372	.36	54	.32	46	.18	26	1.90	285	5.93
5	10	10	17.807	2.19	390	.30	55	.27	48	.14	25	2.15	584	7.96

Seed content in mgm. N = 364, Mg = 16.7, Ca = 12.2, P = 39.4, K = 171

8. animal choices and feeding trials
emphasize animal nutrition via soil fertility

Observations of choices by animals at large, and under control using feeds grown on soils given different treatments, point out the animal instincts for selecting not mainly carbohydrates for laying on of fat, but proteins, inorganics, etc., for body growth and reproduction. Hogs given a choice of different corn grain in separate compartments of the self-feeder discriminated against that grown with green sweet clover plowed under just ahead of it, and in favor of that with plowed-under residues of sweet clover as a seed crop the preceding year. They were discriminating at the same time against the additions of increasing fertilizers on the soil in the former and in favor of them in the latter that made correspondingly larger yields of both sweet clover and corn in both cases. It was not the increasing yields of the crops, therefore, that invited the particular choice. Nor was it increasing fertilizers in the soil. It was a distinct animal response, via its feed, to soil dynamics involving several aspects of soil treatments.

Animal behaviors are now undergoing much closer observations. Reports of animal choices according to the soil are becoming numerous and the various explanations of the causal relations are legion. As assayers, our animals are making excellent biochemical contributions. The animal's instinctive choices of its grazing according to the soil fertility, but its ingestion of almost any plant on fertile soil, regardless of our classification of it as a weed or as a commonly accepted herbage, points out the animal's acceptance of crops according to the soil's synthetic services through them and not to the reputation founded purely on plant pedigree or scientific name. Pigs and chickens have always been running for the cow's droppings. We have only lately come to imagine that they might have long been recognizing there the "dung factor" recently considered a necessity in the poultry rations, the "animal protein factor (APF)," the "anti-anemia factor," and all the other ascriptions to what looks like the cobalt-containing vitamin B_{12}. That the pig and the chicken needing "APF" should recognize the synthetic services even as late in the series as those in the cow's alimentary tract, in bringing soil, microbe, plant and herbivorous feeder into a complex symbiotic process of feed production for them, is suggestion of the many possibly complicated integrations by which nutrition depends on the soil.

Experimental trials with ewe lambs demonstrated differences in their growth rate according as the soil growing the hays (soybean and later lespedeza) had been given no treatment, superphosphate, and lime and phosphate. During the 63-day feeding period using constant amounts of concentrates per head per day along with hay *ad libitum* but consumed at the constant rate of 2.1 pounds per head per day for all hays, there was a total growth of 8, 14, and 18 pounds per head, respectively, for the soil treatments listed. The wool of the sheep showed differences in yolk. It held up during the scouring process only where both lime and phosphate were the soil treatments. No tests were made as to possible differences in dye reactions. During the following mating season, only those ewe lambs fed on the hay with lime and phosphate on the soil came in lamb.

The larger yields of hay on lands given the two treatments permitted trials with each as feed for three male rabbits serving for artificial insemination purposes. Those males, fed hay given only phosphates, declined in their volume of semen delivered, in its concentration of spermatozoa, and in their viability. These males finally were so indifferent sexually as to refuse to breed a female in estrus. No decline in sexual ambition or activity was evident for those rabbits kept on hay grown on soil with lime and phosphate. The hays for the pens were later interchanged. In only three weeks, this latter group became indifferent and refused to mate with the female in estrus, while the originally indifferent ones became active. Three weeks were sufficient for these animals to be disturbed in their mating potentialities by the forces coming up through these herbages under control of the fertility of the soil.

Rabbits fed the same variety of hay from the major soils with and without treatments in an area as extensive as the state of Missouri demonstrated wide differences in growth, bone development, and body physiology according to the soils and their treatments. Originally as uniform as possible in breeding and selection they were widely different for the different soils after a short feeding period. We are slow to see the soil in control of the characters of the animals while clinging to a traditional faith in breeds and pedigrees as determiners of animal body form.

9. human ailments invite their consideration as deficiency diseases going back to the soil

When vitamins may be simply defined as something that will kill you if you don't eat them, we have the beginnings of the belief in ourselves possibly being killed by many things we fail to eat. Deficiencies in nutrition are gradually coming into consideration as causes of poor health. However, it is at a slow rate at which those deficiencies are being considered as originating in the soil, either directly as a shortage of indispensable inorganic elements, or indirectly as compounds fabricated through their help in microbes, plants and animals.

In order to test the hypothesis that such a baffling ailment like brucellosis in animals and man may be due to deficiencies, particularly of the trace elements, in the soil, Ira Allison, M.D., Springfield, Missouri, volunteered to cooperate and try some trace element therapy on afflicted humans as was suggested by Francis M. Pottenger, Jr., of Monrovia, California. Dosages of manganese, copper, cobalt and zinc under enteric coating have been used on more than 1,000 brucellosis patients to date. Other ailments have also been included. The changes in the blood picture and other clinical aspects coupled with the patients' reports of improvements in health have been most encouraging with hope for relief from this disturbing ailment. This hope includes amelioration of other ailments, like arthritis, allergies, anemia, eczema, angina, etc., as possibilities.

Experimental farm trials with soil treatments using trace elements for dairy cows are under way. Greater regularity of conception and heavier milk output seem to support the human evidence that some of the indispensable inorganic elements are not coming sufficiently from the soil and are giving us health troubles even if they are not killing us because we don't eat them.

10. good nutrition calls for conservation of soil fertility

Gradually our reasoning about proper nutrition is leading us back to see the fertility of the soil as possible cause of deficiencies in our nourishment. Soil science in elucidating the soil dynamics points out the limited delivery by any one soil of even the inorganic indispensables. The climatic pattern of the under—or over—development of the soil demonstrates the limited soil areas where proteins can be synthesized, and then not necessarily with complete array of essential amino acids. Trace elements, only recently recognized, seemingly play a role in the synthesis of these. Trace elements are deficiencies for animal health and for human health, too. Insufficient criteria in choosing our crops and declining soil fertility are pushing up their carbohydrate and pulling down their protein provision for us. Animals by their choices and by their health defects are demonstrating that their nutrition is a part of the soil fertility pattern. Military records of human health conditions also point back in that direction.

We are slowly coming to believe that much that we now call "disease" should be more correctly labeled nutritional deficiency. As this is more widely granted, there is a growing appreciation of the importance of the soil fertility in nutrition in general. Such recognition is making each of us more ready to take some responsibility in conserving the soil by which all of us want to be well fed and thereby healthy. It is also putting less emphasis on fighting disease and more emphasis on its prevention by better foods grown on better soils.

5.

"Your acid clay is nothing more than one that doesn't have the positive ions on it—hydrogen, calcium, potassium, magnesium, sodium and the trace elements. I've got to have 65% of that clay's capacity loaded with calcium, 15% with magnesium. I've got to have four times as much calcium as magnesium. You see why we ought to lime the soil? We ought to lime it to get it up to where it feeds the plant calcium, not to fight acidity." That was Dr. William A. Albrecht speaking into an Acres U.S.A. tape recorder. The formal paper giving expression to these findings was given at the International Society of Soil Science the day Hitler moved into Poland, 1939. It was a technical paper in which Albrecht summarized the work of about a dozen graduate students. What Albrecht was saying seems transparently obvious now. First, agronomists knew little about the real nature of fertility. Second, they knew little about the inventory of land nature had bestowed. In one Missouri county, 16% of the farmsteads had been abandoned between 1916

and 1938. In another area surveyed, 40% had been abandoned. In still another study it was revealed that 51,500 acres of Ozark land were being plowed and cropped, although not more than 17,200 acres were suitable for cropping. At the First National Conference on Land Classification, October 10-12, 1940, Albrecht hinted at the direction in-depth research was taking him and his associates. Many of the details were to find expression in The Development of Loessial Soils in Central United States as It Reflects Differences in Climate, *1942. Basically, it was found that soils differed widely as they developed from different rock minerals under variable annual rainfall conditions. This pattern changed as one moved west from the humid mid-continent to the arid regions beyond the 95th meridian, west longitude. Annual rainfall at Alamosa, Colorado, 32.5 degrees north latitude, 105 degrees west longitude, was low enough to be but 20% of the annual evaporation from free water surface. Albrecht often used Transeau's precipitation-evaporation map to illustrate the point. Constant ratios of precipitation to evaporation (as percentages) plotted as isobars to represent effective rainfall properly mapped the varied degrees of* constructive *or* destructive *soil development. These isobars delineated the ecologies of crops, livestock, land in farms, efficiency of radio reception, and defective health via problems in nutrition. Hundreds of the Albrecht papers deal with some or all of these observations.* Soil Fertility and Biotic Geography *as printed in* The Geographical Review, *1957, is concise and excellent.* Nutrition via Soil Fertility According to the Climatic Pattern, *presented in Australia in 1949, is detailed and worth study.* Declining Soil Fertility, Its National and International Implications, *presented to the Fourth Annual Convention, National Agricultural Limestone Association, Inc., 1949, also covers this phase of Albrecht's thinking. But for a matured version for presentation here, a 1954 refinement seems appropriate. Styled* Let Rocks Their Silence Break, *this paper was read before the Eleventh Annual Meeting of the American Institute of Dental Medicine at Palm Springs, California, 1954.*

Let Rocks Their Silence Break

It may well be said that our *national wealth,* our *national health,* and therefore our total *national strength* lie in our soil. If the rocks, then, from which our soils come, could speak on a par and in cooperation with the voice of our statesmen and political leaders, certainly our patriotism would be on a truly solid foundation. Since the rocks determine our soils; since the soils determine our food; and since the food of quality gives healthy bodies, it is no stretch of the imagination to see healthy bodies required to support healthy minds and the entire structure resting on fertile soils. Then, in addition, only with many individually healthy minds assembled can a democracy of strength be founded on the principle of giving independence and sacredness to the individual first and foremost—as a means of having an independence for the nation or the people collectively. "If this nation is ever destroyed, it will be from within and not from without," said Abraham Lincoln. Our internal destruction may well be viewed as taking its start when our soil resources are so highly exploited that we cannot feed ourselves generously, and when we do not find within the international union any country to do for us what we have done to others by a Marshall Plan.

It is proposed therefore to look at our country in terms of the soils to guarantee the country's future because it feeds its people by those means. These are made by the climatic forces of rainfall and temperatures, which convert the rocks into soil, and while so doing keep the assembly line of agricultural production of our foods, fibers and shelter running with output enough to guarantee our national strength. If therefore the rocks will break their silence, or speak a language we can understand, we shall see our country on a higher plane and on a larger scale. We may see a firmer foundation for our patriotism when we sing, "My Country 'Tis of Thee."

1. rocks, rainfall and our future

Since soil is a temporary rest-stop by rocks on their way to solution and to the sea, and since the climatic forces of rainfall and temperature are converting rocks and minerals into that form of potential nutrition for all that lives and grows, the climate-soil combinations by which our feeding of ourselves will be helped or hindered for the centuries ahead should seemingly be the basic guide in formulating an agricultural policy. Our boast of abilities for high agricultural production by sighting the billions of dollars worth of economically manipulated surpluses of the energy foods—while the prices of meats and other proteins are almost prohibitive—suggests such a naivety in handling national food situations in relation to the soil we use to grow our foods, that it casts doubt on our ability to arrive at any policy by using present bases commonly considered for formulating such. The boast is also more disturbing when along with the pronouncedly high agricultural production there comes the alarm about keeping the future food supply commensurate with population numbers mounting at the rate of 3% per annum. Under such a dilemma it seems time now to look, not at economics, but at the ultimately basic means of food production, which are the acres and the fertility of their soil from which source alone food must come. Our soil needs to be inventoried with an understanding of how it feeds us and how it must be managed to continue to do so. For that we must *Let Rocks Their Silence Break,* and perhaps from that voice speaking through the soil there will come some suggestion for an agricultural policy on national dimensions and on a sound foundation.

2. production persists where reserve soil minerals become the active fertility

The agriculture of the United States is by no means a uniformly significant force for food production in every section of the country. The map of the maximum intensity of the farm concentration outlines in its pattern a portion limited to the mid-continent, save for a northern extension of itself eastward as far as Ohio.

Common knowledge of this area leads most anyone to call attention to the animal production, to the grain crops, to the inviting terrain, and to a pleasant and favorable climatic combination of rainfall and temperature in that region. But it also reminds one of the occasionally dry, hot summers with drought, or even seasons with the dust bowl experiences. The western edge of the concentrated farming of the country is outlined or limited by rainfall shortages. Crop damage from this water deficiency is apt to be considered in terms of supply of this liquid mineral more than in terms of its services in converting rocks into soil, or of its value as a medium of ionization by which activity the fertility of the soil moves itself into the roots of the plants to feed them. On the eastern edge, this higher concentration of farms cuts itself down, not by the shortage of water, but by the shortage of soil fertility. This is a shortage of it in either the active form held on the clay complex, or in the reserve minerals remaining to be weathered there. With water present in the soil, and with an ample supply there in active ionic form of the chemical essentials of nutrient value, the soil serves to grow the crops that make agriculture possible. Thus the intense farm area is such because of this combination of the climate and of the soil but only as this combination represents fertility both in reserve supply and in active forms.

Extremely similar to this map of the concentration of farms in this country of ours, and outlining almost in duplicate the mid-continental area just sketched, is the area in the map of high efficiency of radio reception which marks this area as its highest. It may seem a strange coincidence that radio reception efficiency and high farming efficiency should outline the same territory, or area, in the mid-continent. But when salts in a soil represent high electrical conductivity for radio—in which the earth serves as one of the lines of transmission and the air acts as the other—and when plant production results from the ionic performances by the fertility salts possible for action only in a moist medium, it is no unusual stretch of the imagination to see electrical conduction for high radio efficiency requiring the same highly fertile soils as are demanded for high crop production. Production of crops and conduction of electric current are carried out by the same basic soil behaviors. The soil that gives high production is then in no small sense doing so by way of the electric, ionic performances demonstrated by the radio. Here then in the air and in the fields we hear the voice of the rocks as they have broken their silence, because of their decomposition and release of their chemical contents, to speak in terms of at least these two manifestations, namely, the conduction of current and the production of food.

3. climatic pattern of rock weathering gives pattern of soils in agriculture

Rock weathering represents first its disintegration into smaller particles like gravel, sand, and silt; and second, its decomposition or mineral breakdown with the different elements separated out of their compounds. Some of these elements become soluble and ionicly active. Others stay in different forms and combinations, or they may recombine. Both of these conditions represent secondary minerals, or less stability than the original. The element sodium, along with the element chlorine, goes to the sea as ordinary salt and accumulates there as an illustration of the former case. Clay as a secondary residual mineral, or limestone as a recombination mineral, illustrate the second case.

Beginning with the rock and giving it increasing rainfall with a start from the zero value, there is then more solution of the elements and more clay residue. The clay, acting as a filter, adsorbs from solution and on to itself more of the active elements. These are then exchangeable from there to the plant root. This weathering procedure represents soil construction for an increased service in plant production with increasing water as long as the rainfall is not in excess of the evaporation, or as long as there is little water going through the soil to wash out the materials made soluble. Our low annual rainfalls in western United States, and their increase on coming eastward to the 97th meridian of longitude, or thereabouts, in the mid-continent, illustrate the point in question. At the longitude illustrated—from northwest Minnesota to southernmost Texas—there is about the maximum of soil construction.

The high rainfalls on going from that point to eastern United States, which are amounts exceeding the evaporation, leach or wash the soil and remove the soluble elements to carry them on to the sea. Eastern United States, then, illustrates soil destruction so far as the fertility supply is concerned, even if there is ample and more rainfall for crop production. This annual and well distributed rainfall encourages vegetative growth. It favors farming. The decay of vegetative growth produces acids. These put hydrogen or acidity onto the clay to replace other cations of a nutrient value held there. Thus by substituting the hydrogen ion of no value at that point, we have soils that are not productive. Thus we have saline and alkaline soils in our west and we have acid soils in our east. While these, either west or east, may grow vegetative cover for the soil, such vegetation does not represent the carbohydrate-protein combination for the nourishment of animals to a profitable degree. That is more nearly well represented by the soils of the mid-continent.

Naturally, as the temperature is higher to evaporate more of the rainfall, and to make it more effective in rock weathering, we can expect such effects more in the southern half of our country than in the northern. There are differences in the effectiveness with which the same annual rainfalls operate. For the line of latitude across from San Francisco to Washington, D.C., for example, and with less than 25 inches of rainfall as a general figure in the western half, that climatic combination represents arid soils of salinity and alkalinity. There the evaporation is in excess of the precipitation. The ratio of the precipitation to the evaporation from free water surface is less than unity, or less than 100%. With more than 25 inches of rainfall at that line of latitude, or in our eastern half, the climatic combination represents humid soils. There the precipitation is in excess of evaporation, hence a serious leaching and soil acidity occur.

When the map of constant ratios of precipitations to evaporation is superimposed on the map of different annual rainfalls, the two combined serve to outline clearly the map of different soils of our country as surveyed before the causal connection between soils and climate had been so clearly established. A line running from western north Minnesota to the southern point of Texas is almost a longitudinal line serving to separate the soils on the west, which are still calcareous but are lime-deficient. Thus we speak of the *pedocals* to the west, meaning that they are soils still containing calcium for our protein crops. We speak of the *pedalfers* to the east, which do not have lime in their profile and are not producing the proteins generously in the crops.

This climatic pattern of weathering—calcium either remaining or going out of the soil profile—suggests that also other elements with fertility values are either remaining or going out in different degrees according to the climatic forces. If one travels from the arid west eastward along the line of latitude suggested, and if one samples the soils for test of their supplies of exchangeable fertility according to the increasing rainfall, it is significant that, starting with the little fertility, this increases to the maximum at annual rainfall of about 25 inches for the temperature along that latitudinal line. With the higher rainfalls, or those above this figure, the soil has a higher clay content, and therefore a higher exchange capacity. But under that higher rainfall the exchange capacity is taken over to an increasing degree by hydrogen in place of being occupied by calcium, magnesium, potassium and other cations giving plant nutrient service which hydrogen does not render there.

As a schematic diagram, we may use the mid-continental line separating the pedocals from the pedalfers as the high point of a fertility curve, which drops from there to the lower or base levels, both on going toward our western arid region and on going to our eastern highly humid area. In this diagrammatic scheme we have the basic principles of rock weathering under the climatic forces of our country as these represent protein potential in food production of a higher order in the mid-continent, but a low protein potential in that major food factor as one goes to the arid west or the humid east. With such a picture of differing degrees of development of our soil and a map representing an inventory of the country's soil fertility assets before us, should we not have in these a basis from which to set our policy about future conservation or future exploitation of the resources of the soil by which we as a nation shall either survive or perish, respectively?

4. learn how nature feeds our crops

Much that has fertilizer value has been applied to our soils. We have increased crop yields significantly by that practice. But the millions of tons of limestone, rock phosphate, and other fertilizers used, represent less than one-third of the fertility flowing through the soil's assembly line for agricultural creation. The continued natural rock breakdown within the soil to make clay and to keep that colloidal complex stocked with exchangeable nutrients for the plants has been nature's way of contributing more than two-thirds of the materials which keep the agricultural assembly line turning out our crops and our nutrition. While we may believe from the millions of tons of fertilizer used that we have really been feeding the crops, yet that larger share coming naturally from the soil represents what the crops are rustling up for themselves. It is quite evident then that we should study nature and learn how she feeds our crops for us, since the greater share of agricultural production is still her contribution by methods not yet very familiar to us. It would seem a poor agricultural policy built on the belief that we feed the crops and that we direct agricultural production at the creative level of the soil under the climatic forces.

When our exploitation of the soil has brought us to producing mainly carbohydrate crops, while nature has been compelled to produce the protein crops required to keep life surviving, it would seem a more sound approach to an agricultural policy if we should formulate it by considering the more natural ways through which production is brought about and life is maintained—rather than those artificial ones, by which we may be deceived when we believe we are managing it so efficiently even to the point of economic surpluses. Here is an opportunity where the rocks might break their silence and tell us how in their traverse from the mountains to the sea, with their temporary rest-stops as soil, they have fed the microbes, the plants, the animals, and ourselves over the centuries. Can we be a strong nation and be content with any other agricultural policy than one concerned with the maintenance of fertile soils for food by which we ourselves are maintained?

5. *roots may be growing or only going*

The distribution of plant roots through the soil is usually considered a matter depending on the crop or what you sow. It is a character believed even transmissible from generation to generation of crops, when plants like alfalfa are called *deep rooting* plants. Another legume, lespedeza, is classified as a *shallow rooting* plant. One needs only to try growing alfalfa on less fertile soil where lespedeza succeeds and dominates to find that alfalfa surviving there has changed its character to become a *shallow rooting* crop like the lespedeza. Unfortunately, on a soil offering no more fertility than the limited amount that supports the crops like lespedeza—which does little in creating feed value—the alfalfa crop cannot survive very long. Its high protein-producing processes demand large amounts of fertility not found in a shallow surface soil. These deficiences for the alfalfa make this crop become then *shallow rooted* in spite of what might have been considered attached to its pedigree.

Alfalfa is native to less highly developed soil, *i.e.,* those less leached. They are also less often and less regularly watered by rain. Consequently, its roots go deep. Thereby the crop survives because only from much soil depth can it get ample water and plenty of fertility to support the complex physiological functions by which it grows and makes the nutritional values we treasure so highly. It requires much mineral or inorganic fertility. Consequently, it grows well if it goes down where the rock breakdown still offers generously in those respects. It grows naturally in the western prairie and plains area where the rocks speak up naturally as plant nutrition more than commercial fertilizers do so artificially.

The plant roots growing through the soil as very fine ones search out and seem to find the soil centers of much fertility. In those fertility centers, then, the roots suddenly become clusters of large ones. Roots may enter a sewer, for example, and grow to a cluster which clogs the drain. Observations of the roots in the soils and efforts to separate the soil water and the soil fertility as factors in plant growth, suggest that the size of the plant root reflects the fertility supporting it. The locations of differing masses of the same root within the profile at different depths reflect the horizon at which the soil water and the soil air are well balanced for root growth. If the supplies of both fertility and moisture are regularly maintained in a shallow, well-aerated surface soil, then that limited layer grows the crop as a shallow rooted one. If that layer is not maintained in moisture, then the roots must move away from even the air, but also from the fertility left there, and must go to depths for more moisture, fertile or infertile as such may be. Crops so deep rooted, primarily for moisture, will give yields according as the deeper layers are fertile also. Thus it is evident that it is the soil fertility that makes the plants either deep rooted or shallow rooted and not their pedigree.

Plant nutrition by root contact with the soil occurs only in the presence of both moisture and fertility. The latter needs be in both forms of the adsorbed, exchangeable ions on the soil colloid and in the compounds of the mineral reserves as found in the silt fraction. Without the latter as reserves in the rocks to restock the soil when it is said to be resting, the productivity of such soil is soon exploited. The roots are growing bodies reflecting the soil centers of nutrition. They are not merely going under orders from the tops of the plants.

6. *if water doesn't bring nourishment to plants, they must take it from the soil and rocks by trading acid for it*

The methods by which the plants nourish themselves were first demonstrated by putting the plant roots into very dilute aqueous solutions of salts of the essential elements found in the plant's ash. Plants can be grown by this hydroponic procedure. But in nature one fails to find enough fertility in the soil solution to do much more than start a crop, much less grow it to maturity. Instead we find that in the soil the nutrients are held in quantity as adsorbed and exchangeable ions on the colloidal clay-organic complex. The ions are held firmly on this against their being readily taken by the root. They cannot be washed therefrom by pure water. But if some carbonic acid is put through the soil—no more than that put into pure water by breathing through it—then the acidity, or hydrogen ion, will exchange to the clay for the positively charged nutrient ions held there.

Thus as a plant feeding process, we envision the respiring root giving off carbonic acid. This waste product of respiratory origin gives active hydrogen to exchange to the clay for what positively charged ions are there of nutritional service. If there is but little clay in proportion to the plant roots, its store of fertility is soon given up to make the clay very acid. If we should apply water soluble fertilizers, like salts of calcium or magnesium or of potassium, for example, these applied elements are caught up and taken out of solution by the plant and the hydrogen is given by the soil to the solution. They are held there by similar adsorption phenomena and can likewise be quickly exchanged for hydrogen on the root. Thus the clay, when separated from the rock of its original origin, or from mineral fragments of original material scattered through the soil, is soon exhausted of its fertility or of its power to produce.

By many of our soils which have not been extremely high in clay content, their continued production under our cultivation of them has not been due to their larger store of exchangeable nutrients originally held in the virgin clay. Instead, it has been due to the reserve minerals left there still unweathered, being decomposed by the more acid clay, made so each season by cropping. Thus they give up from the inactive mineral contents those ions which can be so weathered out, can be adsorbed on the clay, and represent a stock of fertility being replenished during the period considered as that when "the soil is resting."

The original rocks left in the soil have been the "sustaining fertility" to the extent of more than two-thirds of the output commonly called "agricultural production." These rocks have not yet broken their silence nor spoken boldly enough for us to appreciate either their past importance or their potential for sustaining two-thirds of our future production. We have been too intent on what is only "starter fertilizer" for crops, even if it has represented about one-third of the total crops we produced.

In reality then, the assembly line of agricultural production of each year starts from the rocks and the minerals mixed as silt and sand with the clay of the soil. Only as these rock fragments have been weathered by the acid taking its biochemical origin from the respiring plant roots, or from the microbes living on the decaying organic matter, have the nutrients passed on to the plant roots, and has crop growth been possible. The magmatic rocks of the original earth's crust are still making more soils as they are on their way to solution and to the sea. Only as that process within the soil is highly dynamic now, or or has been very active in the past and stored its resulting fertility in the soil and its organic matter, can we expect the soil to nourish the plants and thereby our animals and ourselves.

In the chemical dynamics serving in plant nutrition, the clay is a *jobber* to which the roots can move during their growth period and from which they can get a large supply of nourishment quickly. The more stable, or the more permanent organic matter in colloidal forms behaving like the clays, serves likewise. But these jobber performers in crop growth must depend on the original rocks, on the decay of crop residues, and on the fertility treatments given the soil, to keep themselves well supplied with what they can hold or pass on but what they themselves cannot create.

7. plant diets for their making of mainly carbohydrates or these plus proteins

Accordingly, then, as the above listed suppliers to the clay give different combinations of calcium, magnesium, potassium, manganese, copper, zinc, etc., to it, so will the clay offer different ratios of these as the nutrition for the plant. Whether this is a plant diet, properly balanced for its synthesis of carbohydrates accompanied by ample proteins synthesized simultaneously, depends on the ratios of these ions exchangeable from the clay-organic complex or actively included in the exchange capacity of the soil. Thus the clay as a jobber passes on to the plant root different combinations of ions according as these are passed to the clay, and as their behaviors in company with each other move them under their combined activities into the plants. These combinations may be plant nutrition that makes mainly bulk of vegetative mass or as it helps the plant to synthesize essentials of more complete nutritional help to animals and to man.

To date, the protein production by crops has been high where much of the original rock is left in the soil for it to speak in terms of nourishment for high-protein legumes. We use the limestone rock, the phosphate rock, and even granite to restore protein-producing crops on soils where forest trees, or only wood, were the virgin crops. We have called these the acid soils, and have deplored their low food crop potential. We have only to realize that by means of many kinds of rock fertilizers—and other nutrient supplying minerals to build back the sustaining fertility of the soil—will the natural soil acids process those raw rocks to make the soils more productive again. Our technological view of acidity, or rather our strictly chemical view of it (rather than a biological view) has delayed the day of using this very condition to treat the raw rock substances that would rebuild the fertility according to mother nature's way of doing such.

Observations on the migration of the ring-neck pheasant recently from the northwest into the cornbelt suggests this bird's spread over that territory in connection with the treatments of the soil with agricultural limestone. While the high calcium diet is a requisite for this egg layer, it is the belief that it is not the high calcium content of the feed so much that favored the migration. Rather it seems that the larger particles of limestone serving as calcareous grit in the bird's crop are the factors giving this bird this change ecologically. If this reasoning is correct, it is the voice of the coarse lime rock that speaks to call this game bird into the cornbelt.

8. organic matter and rock together—organic plus the inorganic—make soil fertility more sustaining

Also we dare not overlook the role of the organic part of the soil in growing our crops. This acts as a biological fertilizing process in contrast to the technological fertilizer treatment in crop growing. The rocks within the soil, and likewise the fertilizer salts, speak differently when in company with generous quantities of organic matter in the soil. The clay, which may be handling the inorganic elements within the soil for plant nutrition, according to the different ratios and activities resulting in the plant's output of either mainly carbohydrates, or also some proteins according to a certain combination of the inorganic substances of the soil as revealed through the help of the soil tests, handles the situation very differently when the inorganic ions and molecules are put in company with similarly adsorbed organic ones. The small amount of soil organic matter becomes decidedly significant in plant nutrition when it modifies also the inorganic performances.

Of even the inorganic half, we have emphasized the cations and neglected the anionic part. So far we have no clear concept of how the bicarbonates, nitrates, phosphates, sulfates, etc., really are held in, and delivered by the soil to the advancing plant root. Also when we see the clay as a source of the cations, we do not see that part completely when the cations of the trace elements defy our learning of their behaviors there because the amounts of them biologically significant are not even within the pale of our chemical measurement of them. Thus a significant share of even the inorganic half of plant nutrition is still neglected.

We are still deaf to the voice of this portion that originates in the rocks. We have not yet heard it speak very plainly in terms of nutritional services to the plants and to ourselves when trace elements are still in question by many.

9. the organic cult vs. the chemical camp

One dare not swing all emphasis to the *inorganic* contribution to crop production by the rocks through the soil. Nor dare one swing it wholly to the returned *organic* contribution when this was originally grown on that same soil. Does the return of organic matter to its place of origin eliminate a deficiency of any inorganic essential there before that mature crop was grown? Do carbon, hydrogen, oxygen or nitrogen, all of atmospheric origin going into the soil, increase the stock there of any inorganic elements? Plant nutrition supported by organic compounds taken up as such by the plants may be means of synthesis of the more complex compounds. We expect more complexity because of the delivery of the organic starter compound into which more complexity can readily be built by that help than if the plant were required to synthesize that starter compound first. But even that extra chemical complexity added, or chemically attached, to the starter contribution may call for extra inorganic compounds of rock-mineral origin.

More organic complexity in synthetic services by the plants demands more variety in the inorganic elements contributed, including the "trace elements." Service by these trace elements does not represent inclusion of the inorganic elements in the final organic synthetics for which the inorganics may be only tools via enzymes, hormones, etc.

As a case in illustration, the microbial syntheses within the cow's paunch and the elaboration, by the microscopic life forms, of protein equivalents, which lessen the required protein feed supplements, may well be cited. The organic chemical, urea, is fed as a starter compound with molasses and other highly carbohydrate substances. The urea, quite different from other nitrogen fertilizer compounds, represents the amino nitrogen (NH_2) attached to a carbon.

$$NH_2 - \underset{\underset{O}{\|}}{C} - NH_2$$

This is the structure of the nitrogen in the amino acids of protein. This starter compound serves better, apparently, than other nitrogen compounds devoid of this similarity to the amino structure. But even that service by the urea is improved, according to the studies by the Missouri Agricultural Experiment Station, when the inorganic elements of the ash of alfalfa are fed along with urea, molasses, etc. But it is more significant to note that the ash increases the efficiency of these processes of protein synthesis by the paunch microbes according as it comes from alfalfa grown on soil more completely built up by fertilizers for growing this high protein crop better.

These observations suggest that even the synthesis of the protein equivalents by microbes, or single cell synthesizers of these complexities, calls for a combination of helps, both organic and inorganic. More help comes from the latter when these ash elements of rock origin in the ultimate, are selected out of the soil and delivered as inorganic elements by a crop which itself is a producer of particular protein complexes which we have long considered of high nutritional values for animals and even lately for man.

10. glaciers are rock grinders for soil making and our survival

We can now view agricultural production over historical time periods in terms of the sustaining soil fertility. We see production maintained by the original rock and mineral fragments as reserve fertility being weathered by the soil acidity of root and microbial origin in the carbonic and other acids produced. We are now in a better position to see some geographic regions that were quickly exhausted because they have not recently had significant supplies of reserve minerals left in the soil. This latter will be true of areas highly developed under the intensive climatic weathering factors of high rainfalls and temperatures. Soils under growth, particularly coniferous ones, are suggestive illustrations. It was on those soils put under agricultural use where the starter fertilizers first made their contributions. It

Effects of Ash of Alfalfa from Three Types of Soil on the Digestion of Cellulose by Sheep

	% Cellulose Digested	% Protein Digested	Bacterial count x 10^9	V.F.A.* mgm%**
Basal ration	42.5	37.1	21	3.5
Basal ration + ash - alfalfa Sarpy fine sandy loam	46.4	39.2	29	5.7
Basal ration + ash - alfalfa Fertilized Knox silt loam	44.3	—	26	4.1
Basal ration + ash - alfalfa Limed Hagerstown silt loam	42.5	37.2	22	3.9

*V.F.A. = Volatile fatty acids
**Mgm% = Mgms/100 gms. or cc.

was on such soils where they became a regular necessity for significant crop production.

We can also see some soil regions under cultivation for a long time giving little suggestion of exhaustion of their fertility. These soils represent less development of themselves under only moderate climatic forces of rainfall and temperature. The Mediterranean area, with its particularly moderate climate characterized by the name "Mediterranean Climate," is a good illustration of soils that have been slowly and regularly renewed at the lower part of their profile by rock weathering at rates duplicating those at which they were depleted at the upper part of it. The glacially deposited soils with no extremely long geological history of development to their credit, and with their contents of heterogeneous rock and mineral mixtures, are similar cases of extended history of production under agricultural use. We are coming slowly to recognize the rocks not only as the origin of the soil, but also as the source within it from which the soil renews its creative strength for each successive crop.

11. man concentrates around rock-grinding centers and freshly pulverized rock deposits

In the ecological pattern of man scattered over the various parts of the world, it is interesting to note that the regions of the rock-grinding processes by glaciers and freshly pulverized minerals in stream deposits are literally the oases for food around which humans collect most densely. It is around these where probably the human species will survive the longest. The history in the valley of the Nile River, with its ancient tombs of Egypt on the high and dry burial grounds overlooking the periodically flooded lowlands of that stream, tells the story of these as oases of nutrition from lands under cultivation for centuries because of pulverized rock washed out of the mid-African Islands and delivered to those flood plains. But now that the dam built across that river at Assuan, near the upper end of the Egyptian area of flood plain, intercepts much of the mineral as sediment under water impounded above the dam, the production potential for that country from the flood plain representing its very lifeline has been seriously reduced. Rock grinding, whether by precipitous streams or by them slowed down to glacial movement in regions of perpetual frost, exhibits itself as the means by which the soils continue to grow foods over extended time periods. Regions of highly weathered soils, stimulated to crop production by no more than starter fertilizers, have a short agricultural history.

Other illustrations of rock grindings supporting human survival deserve mention. The Yellow River in China, like the Nile where the cities are along the flood plains while the uplands are barren, has also tremendous numbers of people collected along itself. The country above it has been denuded even to the point of digging up tree roots for fuel and bringing on the consequent serious erosion. As a result of that situation, the Yellow River floods seriously. However, while so disastrous in human drownings, it is conversely beneficial because the yellow, unweathered silts it carries out of the arid highlands as new hope for another crop to the starving over-population still remaining. Freshly pulverized rocks and minerals distributed over the flood plains near the mouth of that stream are the lifeline to which those people must cling for survival. Older countries and civilizations serve to illustrate this aspect of the ecological pattern of man in relation, not only to the soil as acres but as the material it handles in starting the assembly line of agricultural production, namely, the raw materials of the rocks themselves.

In India, the Ganges River has become sacred largely because it is literally the mother "on the breast of which many of the people are fed." It has been their life support, not because it is water as a liquid, but because its waters originate in the glaciers of the Himalaya Mountains where high annual precipitation makes the glaciers for grinding the rock minerals. These are carried down to the flood plains of that river where they speak out as a sustaining fertility of the soil and thereby as the sustenance of the people. In the other sector there, or what was once a part of India, that is Pakistan under much lower annual rainfall, the Himalayas of Kashmir pour their glacially pulverized rock into the Punjab, that is, The Five Rivers, of which each with its blessing in the form of water for irrigation is also blessing in the form of fertility by which food is produced. Shall we be surprised when, in the minds of these humans existing at bare survival levels, such blessings in the form of the pulverized rock delivered regularly by the rivers should give those waters sacredness?

The glaciers of the Alps in Switzerland, where precipitation is generous, make the Po River, rising from the southern side of those mountains, the lifeline of northern Italy. That river represents agricultural production far above that coming from the rocky backbone of the rest of that country represented by the less elevated and drier Appennine Mountains. From the northern side of the Alps the pulverized rocks of glacial origin are making the streams of Switzerland milky and serving to give the Rhine River its load of portable silt. This is the sustaining fertility, not only for the Germans along the flood plains of that international line of travel, but it is also the lifeline for the Netherlands, or the Lowlands, where those rock grindings have been so jealously treasured even to the point of rescue from under the sea in extensive reclaimed areas.

12. may our rocks break their silence

In the United States, our Rocky Mountains are high in elevation but not so in precipitation. As a consequence, they are not extensively glacier covered. But their disintegrated rocks as unweathered minerals are carried in the limited seasonal rainfalls and melted snows to the eastward by the many rivers running in that direction. The gradient of 4 feet per mile in circuitous flow, representing nearly 7 feet per mile as the crow flies, represents large carrying and grinding powers for the rocks moved from those mountains. It gives also the discharge of the tremendous fertility load into the broad flood plains of the Missouri River, extending from beyond north of eastern Nebraska down into the mid-state of Missouri, of which the gradient for that part of the stream is less than a foot per mile. Since this fertility load comes in the spring flood, and since the stream is of no great size for the rest of the year, the dry, broad flood plains in the mid-continent are the source of the silt to be picked up by the southwesterly winds and to build the loessial soils of north Missouri, Iowa, etc., with this annual fertility renewal amounting to nearly a thousand pounds per acre. Processed in our acid soils as these freshly delivered, little-weathered rock flours are regularly, they have been the sustaining fertility that goes unappreciated. They certainly are not recognized in any engineering planning that would put large dams across the Missouri River serving as more of a lifeline in the past and representing such for the future than the people appreciate. Let us hope that the rocks moved in by the Missouri River may break their silence before too many dams are constructed to sever the lifeline that must even be strengthened for our future food security.

Perhaps also in our economic planning the rocks, minerals and soils may be speaking for a sounder agricultural policy. Perhaps if that is built on food more than on finances, it will transcend the realm of economics and escape the hazards of financial manipulations, inflations and similar economic irregularities.

"We have for the last 25 years been coining (printing) money and loaning it to foreign governments to buy our surpluses. Thus we have shipped the fertility of our soil and the riches of our mines and wells to foreign lands. It is as though we were petitioning to become Crown colonies of Europe once again. Any agricultural prosperity built on the sale of agricultural products abroad is unsound. The siphoning off of wealth that we create into implements of war is equally destructive of our culture and our civilization because it undermines the whole structure economically." So said Dr. Jonathan Forman.

Perhaps when we give ear more attentively to the voice of the soil reporting the ecological array of the different life forms (microbes, plants, animals, and man) over the earth, we shall recognize the soil fertility—according to the rocks and resulting soil in their climatic settings—playing the major role through food quality rather than quantity, surpluses, prices, etc., in determining those respective arrays. Man still refuses to see himself under control by the soil. Then when we take national inventory of the reserves of fertility in the rocks and minerals mobilized so slowly into agricultural production in contrast to the speed of exploitation of virgin and remaining soil fertility, perhaps thrift and conservatism with reference to the mineral resources of our soils will be the first and foremost feature of our agricultural policy. By that policy we shall build the best national defense, if there is any basis for contending that if wars must come, food will win them and write the peace. May rocks their silence break and speak nationally through a better knowledge of soil for food as the basis of national health and thereby a national strength for the prevention of war and for the simplest road to peace. Our future national strength must rest in our soils.

6.

Most of Dr. William A. Albrecht's heavy technical papers had become a part of the scientific literature by the end of the 1940s. Even before then, the great scientist had turned his energies to the task of communicating with farmers. It was the farmer, after all, who made it happen. Spoken words from the podium reached only a few. Articles in the public prints reached many more, and for this reason Albrecht often penned reports for farm papers, local dailies, and for consumer magazines with reach and numbers. By the end of World War II, Albrecht's theme had become clear. The Creator had not distributed fertile soils evenly over the land areas of this world. Most of the passengers on spaceship earth were crowded into seven breadbaskets—half in Asia and the South Pacific, one-fourth in Europe and Africa, and one-sixth in North and South America. Man's manipulation of fertility requirements for soil systems was primitive and often at variance with nature's requirements. In short, men were what they ate, and their hopes and aspirations were unalterably linked to the soil on which what they ate was grown. This item from the South Haven, Michigan Daily Tribune, *March 22, 1945, illustrates the kind of press work Albrecht was doing during and after the great war.*

War Is Result of Global Struggle For Soil Fertility

Soil fertility determines whether plants are foods of only fuel and fattening values, or of body service in growth and reproduction. Because the soil comes in for only a small percentage of our bodies, we are not generally aware of the fact that this 5% can predetermine the fabrication of the other 95% into something more than mere fuel.

Realization is now dawning that a global war is premised on a global struggle for soil fertility as food. Historic events in connection with the war have been too readily interpreted in terms of armies and politics and not premised on mobilized soil fertility. Gafsa, merely a city in North Africa, was rejuvenation for phosphorus-starved German soils. Nauru, a little island speck in the Pacific, is a similar nutritional savior to the Japanese. Hitler's move eastward was a hope looking to the Russian fertility reserves. The hoverings of his battleship Graf Spec, around Montevideo, and his persistance in Argentina were designs on that last of the world's rich store of less exploited soil fertility to be had in the form of corn, wheat, and beef much more than they were maneuverings for political or naval advantage. Some of these historic martial events serve to remind us that "an emtpy stomach knows no laws" and that man is in no unreal sense, an animal that becomes a social and political being only after he has consumed some of the products of the soil.

1. differences give us an east and west

In view of our youthfulness as an extensive country, our different geographic areas have registered themselves mainly as differences in body comfort, whether hot or cold, wet or dry. We have not yet marked out our country into the smaller patchwork districts with distinctive local colorings as the Old World has in the opinion of visitors from the New World. Limitations in travel, difficulties in food delivery, and all the other restrictions now making us more local, will soon emphasize differences and deficiencies according to the soils by which we live.

Geographic divisions to give us an east and west, and a north and a south, for the eastern half of the country, are commonly interpreted as separations, according to differences in modes of livelihood, so-called customs, or political affiliations. Differences in rainfall and temperature are readily acknowledged. But that these weather the basic rock to make the soils so different that they control differences in vegetation, animals and humans, by control of their nutrition is not so readily granted. That "we are as we eat" and that we eat according to the soil fertility, are truths that will not so generally and readily be accepted. Acceptances are seemingly to come not by deduction but rather through disaster.

The heavier rainfall and forest vegetation of the eastern United States mark off the soils that have been leached of much fertility. Higher temperatures in the southern areas have made more severe the fertility-reducing effects of the rainfall. Consequently, vegetation there is not such an effective synthesizer of proteins. Neither is it a significant provider of calcium, phosphorus, magnesium, or the other soil-given, fetus-building nutrients. Annual production as tonnage per acre is large, particularly in contrast to the sparsity of that on the western prairies. The east's production is highly carbonaceous, however, as the forests, the cotton, and the sugar cane testify. The carbonaceous nature is contributed by air, water and sunlight more than by the soil. Fuel and fattening values are more prominent than are aids to growth and reproduction.

2. fertility pattern of europe

The more concentrated populations in the United States are in the east and on the soils of lower fertility.

In Europe, the situation is similar. It is western Europe that represents the concentrated populations on soils of lower fertility under heavier rainfall. Peoples there reached over into the pioneer United States for fertility by trading for it the goods "made in Germany." More recently, the hard wheat belt on the Russian chernozem soils has been the fertility goal under the Hitlerian move eastward. Soil fertility is thus a cause of no small import in the world wars.

Life behaviors are more closely linked with soils as the basis of nutrition than is commonly recognized. The depletion of soil calcium through leaching and cropping, and the almost universal deficiency of soil phosphorus, connect readily with animals when bones are about the complete body depositories for these two elements.

The distribution of wild animals, the present pattern of distribution of domestic animals, and the concentrations of animal diseases, can be visualized as superimpositions on the soil fertility pattern as it furnishes nutrition. We have been prone to believe those patterns of animal behaviors wholly according to climate. We have forgotten that the eastern forest areas gave the Pilgrims limited game among which a few turkeys were sufficient to establish a national tradition of Thanksgiving. It was on the fertile prairies of the midwest, however, that bison were so numerous that only their pelts were commonly taken.

3. effect on cattle reproduction

Distribution of animals today reveals a similar pattern, but more by freedom from "disease"—more properly freedom from malnutrition—and by greater regularity and fecundity in reproduction. It is on the lime-rich, unleached, semi-humid soils that animals reproduce well. It is there that the concentrations of diseases are lower and some diseases are rare. There beef cattle are multiplied and grown to be shipped to the humid soils where they are fattened. Similar cattle shipments from one fertility level to another are common in the Argentine.

In going from midwestern United States eastward to the less fertile soil, we find that animal troubles increase and become a serious handicap to meat and milk production. The condition is no less serious as one goes south or southeastward. The distribution patterns of milk fever, of acetonemia, and of other reproductive troubles, that so greatly damage the domestic animals industry, suggest themselves as closely connected with the soil fertility pattern that locates the proteinaceous, mineral-rich forages of higher feeding values in the prairie areas but leaves the more carbonaceous and more deficient feeds for the east and southeast with their forest areas. Troubles in the milk sheds of eastern and southern cities are more of a challenge for the agronomists and soil scientists than for the veterinarians.

The influence of added fertilizers registers itself pronouncedly in the entire physiology of the animal. This fact was indicated not only by differences in the weight and quality of the wool, but in the bones and more pronouncedly in semen production and reproduction in general. Rabbit bones varied widely in breaking strength, density, thickness, hardness, and other qualities besides mass and volume. Male rabbits used for artificial insemination became sterile after a few weeks on lespedeza hay grown without soil treatment, while those eating hay from limed soil remained fertile. That the physiology of the animal, seemingly so far removed from the slight change in chemical condition of the soil, registered the soil treatment, is shown by the resulting interchange of the sterility and fertility of the lots with the interchange of the hays during the second feeding period. This factor of animal fertility alone is an economic liability on less fertile soil, but is a great economic asset on the soils that are more fertile either naturally or made so by soil treatments.

4. animal instincts are helpful

The instincts of animals are compelling us to recognize soil differences. Not only do the dumb beasts select herbages according as they are more carbonaceous or proteinaceous, but they select from the same kind of grain the offerings according to the different fertilizers with which the soil was treated. Animal troubles engendered by the use of feeds in mixtures only stand out in decided contrast. Hogs select different corn grains from separate feeder compartments with disregard to different hybrids but with particular and consistent choice of soil treatments. Rats have indicated their discrimination by cutting into the bags of corn that were chosen by the hogs and left uncut those bags not taken by the hogs. Surely the animal appetite, that calls the soil fertility so correctly, can be of service in guiding animal production more wisely by means of soil treatments.

Dr. Curt Richter of the Johns Hopkins hospital has pointed to a physiological basis for such fine distinctions by rats, as an example. Deprived of insulin delivery within their system, they ceased to take sugar. But dosed with insulin they increased consumption of sugar in proportion to the insulin given. Fat was refused in the diet similarly in accordance with the incapacity of body to digest it. Animal instincts are inviting our attention back to the soil just as differences in animal physiology are giving a national pattern of differences in crop production, animal production, and nutritional troubles too easily labeled as "diseases" and thus accepted as inevitable when they ought to have remedy by attention to the soil. The soils determine how well we fill the bread basket and the meat basket.

7.

Dr. Albrecht found time for students—first, last and always. With unremitting zeal, his contributions emphasized the fundamental necessity of feeding plants, animals and human-kind through ministrations to the soil itself, correcting the deficiencies of diet at their point of origin—in soils that had been assayed and found wanting. This little statement came sub-titled, A Personal Message From Dr. William Albrecht, Chairman of the University Soils Department. It was published in College Farmer, February 1949.

The Fundamentals of Soil Survival

The study of soils is a dual opportunity. First, it is to become familiar with the science of the great natural resource by which we live; and second, to learn the practices by which we can manage the soil so that it will feed us now and yet feed those who come after us.

Like the study of all other natural things, that of soils soon becomes an organized body of knowledge or a science for those students who enroll in this subject in the curriculum of the College of Agriculture.

The wise management of the soil, as one of the fundamental practices of agricultural production calls for the use of soil while at the same time we are conserving it. It is necessary to study the other sciences by the combination of which we get our soil science. We need to emphasize the understanding of the soil as this gives better understanding of the more secure foundation of agriculture itself.

1. agriculturists create food

The service of agriculture is that of creating our food. The assembly line of this manufacturing effort starts in the soil. It is from there that the lifeless, inorganic elements move into the plant roots to become the living organic compounds.

It is from those interactions between the clay molecule, supported by its reserve nutrient supply of the minerals, and the plant roots, connected through the plant tops with sunshine energy, that crop production is possible. It is the understanding of these fundamentals by which the science of the soil gives us a recognition of the broader scope of agricultural production as a whole.

2. soils—our challenge

The study of soils is therefore an education as well as a training. It is a cultural subject to give one comprehension and appreciation of a great body of facts organized into a science. Soils is a practical subject to enable one to support himself in the work of food production.

It is a challenge to any student interested in learning and disciplining himself. It puts at his disposal the teachings of our College and the experiences of the Experiment Station. To each it makes available much that has gone before for some time and what has only recently been discovered, all for the betterment of the business by which agricultural folks live.

3. research proves theories

The laboratories for teaching and for research; the fields under long-time soil treatments with the crops testifying accordingly; and the accumulated wisdom of those persons who have built it into the records of the College, are all available to those students who truly want to study and to improve themselves.

The staff in soil teaching and research has been enlarged recently through both the young and the more experienced men. New outlying fields are being put under experimental control. Increased sums of money are being spent for research, for teaching and for demonstration of the facts about soil.

4. people are soil conscious

These facts are demonstrating the need to look to our soil. Industry is becoming soil conscious and employing men to help them make their business relate itself to soil conservation in its broader aspects.

Many industries and businesses are looking more and more for men who know about soil and who can guide them with a view of their future in relation to our wise use of the soil and, therefore, our greater national security.

Opportunities are increasing for the services of men well educated in the science and trained in the proper use of our soil. Any young man planning some educational helps for himself in agriculture as his future may well look to a more intense study of the subject of soils.

• • •

The present problems in agriculture seem to have arisen because of the disregard of the simple fact that, in the main, agriculture is normally the art of cooperating with the natural behaviors of the many life forms lower than man in the course of evolution. By the natural creation of these simple life forms, and their products serving in his support, man and his survival are a consequence and not the creative power, nor the cause. We have erroneously assumed that agriculture with all its living forms of plants and animals may be managed with assembly line speed, and economic controls, from nature's raw materials to sales of finished products according to man's economic, industrial and technological planning.

• • •

In the economist's concepts, apparently the economic finality of agriculture consists of one that includes only buying and selling.

• • •

Slaughter is, too often, an economic venture in disregard of the animal's poor health under an inspection geared to the clinical cases only while it is unmindful of the subclinical and of incipient extinction of this domestic species.

• • •

No wild animal chooses to be fattened. While there is the increased enshrouding of every capillary of the blood vessels and every cell with a thickening layer of fat, the cells normally fed by the diffusion of the nutritives from the capillaries to them will become more starved. Their excretionary products will accumulate, since fat hinders the two-way ionic and molecular exchanges between the capillaries and the cells to give hidden hungers and excessive accumulation of metabolic wastes.

• • •

We are slowly accepting the postulate that selection and propagation of species, mainly for economic advantage, may have accentuated successive losses in the generations moving the life stream forward.

• • •

Wild animals chose their own medicine according as the soil grows it, and thereby exemplify better health and survival on their own than our domestic ones do under our management.

BACKGROUND

Insoluble Yet Available

Insoluble Yet Available *is a keystone entry among the Albrecht papers. As a mature expression of his lifetime of work, it poses questions, answers them, and then explains the answers. It stands as sublime irony that modern fertilization has so little to do with fertility. Dr. William A. Albrecht sensed this as he entered the lion's den to plead for an end to "diminishing returns" agriculture. In the concluding paragraphs of this paper, Albrecht mentions Dr. A. W. Blair of New Jersey, and he talks about the pioneer scientists who "did not visualize productive permanence in soils treated with water soluble salts."* Insoluble Yet Available *calls to question the very concept now written into law, the idea of soluble fertilizers. It questions attempts to legislate biology because such attempts have the effect of putting to death the inquiring mind on which scientific advancement depends. Those who have read these papers so far will be able to understand the complex concepts presented here.* Insoluble Yet Available *was first presented as a talk before a meeting of the Arkansas Plant Food Conference. It was later reprinted in* Farm and Garden, *a British publication, in February 1962.*

Insoluble Yet Available

The creative act by a soil when it grows our crop plants is still much of a mystery and miracle. We do not yet comprehend fully the *modus operandi* of the plant roots connecting with the soil and getting nourishment from that source.

Even though soil comes from the ancient rock, yet plants contain only about 5% of themselves as the elements of rock origin. The mystery of how the rain-drenched soil can resist being dissolved away and yet pass nourishment through the root to the plant may soon yield to the scientific efforts of clarification and fuller understanding. Such clearer vision of the soil's services in growing crops can do much to elevate the appreciation of the soil, to encourage more judicious management of it, and to bring wider conservation of the only creative power by which all forms of life must be fed.

The soil's behaviors, like those of many other things in nature, do not conform completely to the laws we learn in laboratory chemistry and by which we have been explaining the natural nutrition of plants. In the laboratory, the solubility of a substance in water, in alcohol, or in certain reagents is commonly the foremost criterion for its description and classification. We are mentally disturbed, then, when we find that some natural substance or commercial product, commonly soluble in the laboratory, behaves as an insoluble one, or vice versa, when put into the soil.

We experience similar confusion when, for example, some *insoluble* substances on coming in contact with the mucous membrane of our body, like that of the lungs on inhalation, is *soluble* and harmless there. We are more disturbed when some so-called *soluble* substances are as disturbing (and fatal) to the lungs as the insoluble silicates of rock dusts are in the case of the disease known as silicosis. Land plaster, consisting of the mineral, gypsum, or of the chemical compound, calcium sulphate, is considered soluble in the laboratory. Yet its inhalation brings a breakdown of the lung tissues which on x-ray examination suggests silicosis or even calcification by its accumulations there.

1. sulfur—case in point

The mental confusion grows still larger when we consider the element sulfur an insoluble one but find that the inhalation of flowers of sulfur for a long time is a case of absorption and one without scar of the lung membranes. According to these two cases, what is classified as a chemically *soluble* substance in the laboratory may be an *insoluble* one by contact with this kind of living tissue, and vice versa.

In a corresponding confusion, the plant root as a living tissue in contact with the soil has not been understood in its absorbing activities because the properties and behaviors of substances in the soil have been assumed to be the same in relation to that living form of nature as they are when tested in relation to chemical reagents, or in solubility tests of them, in the laboratory. In transplanting our knowledge, gained from studies in the laboratory, as only a vision (not tested by plant growth) of what might be the situation or phenomena in the soil, the visionary concepts are not necessarily in accord with the natural facts.

Chemical inspection of fertilizers and regulatory controls of their sales became a practice according to the early beliefs that water solubility of them was an index of their absorption by the plant roots. Yet, after their application, some water soluble fertilizers are washed out and lost from the soil while others become highly insoluble and of very limited uptake by plants regardless of that property of solubility in water. But fortunately we are moving toward a fuller comprehension of the facts of nature according as the accumulating research—under its higher refinement through technological aids—is simultaneously refining our concepts and visions of plant nutrition. Fuller understanding is coming about because we see in more detail, and we comprehend more relations of causes and effects between the soil and the plant.

2. roots find nutrients still available in soils highly leached by rainfall

Just what the plant root does while going through the soil and gathering the many ash or non-combustible elements (also some combustible organic molecules) of nourishment from the supposedly insoluble, pulverized rocks that are weathering to give out their minerals under the forces of the climate, has been another one of those confusions in our thinking. That came about also because it was founded on the criterion of solubilities and insolubilities according to laboratory chemistry. When the nutrient elements considered *insoluble in the lab-*

oratory are *available to*, or taken into, *the plant,* we are content to accept it as mystery. It must then be more of a mystery when the rain falling in excess of evaporation goes down through the soils as drainage water to carry the solubles out and down into wells, into underground streams, and into the sea to be added to the accumulation of salts there. Yet it is on those washed-out soils that the plant root gets *available* nutrient elements which control the growth processes elaborating carbon dioxide from the air along with water from the soil into sugars, starches and other carbohydrates by means of sunshine as energy source.

It is that water going down through the soil that has been carrying away to the sea—as a beneficial service—both the sodium and the chlorine of which neither dares be in the soil as a salt in more than fractions of a percent, if food crops are to be grown. It is the rainfall that has been weathering rocks into parts as solution and parts as suspended clay, and by that rock destruction has been giving us the soil from which not only plants but all living bodies are created. The silicon, or the larger share of most any rock, forms the clays. They do not rush off to the sea. Rather they, as a non-nutrient, hold back while a share of the active elements in solution may move out with the water. But again, the larger share of the solubles, including nutrients, is quickly inactivated by being adsorbed on the clay. That process duplicates the manner in which lime or calcium in solution in hard water is adsorbed on the colloidal compound in the household water softener.

Nature exhibits these effects of adsorbing the soluble elements extensively whenever any kind of substance is so finely divided that it behaves according to its total surface rather than according to its mass or total weight, or when such a finely divided substance stays in suspension to be called a *colloid.* Clays are classified under that chemical category. It is the clays of our soils, then, by their adsorption on themselves of calcium, magnesium, potassium, iron, aluminum and others, that hold the nutrient elements so that they are not soluble to be carried away in the drainage water but yet will move into the plant root to serve as plant nutrition. Colloidal chemistry, or the surface chemistry of suspensions, came along after the chemistry of solutions was well established, had explained much, and was supposedly well comprehended.

3. the mills of god must grind insoluble rocks as source of available plant nutrients

We visualize the weathering of rocks to give us our soils by the processes in which the elements of positive electrical charge, initially combined with the silicon and oxygen of negative charge as minerals, are broken apart to become chemically active separates. That condition permits their recombinations into other or secondary minerals and compounds. The silica (silicon and oxygen) combines with water into a large molecular grouping which is not very active, or not significantly ionized, like the smaller positively charged elements are. [*Silicon's behaviors represent slow-speed, inorganic chemistry much as in the case for most all organic substances behaving mainly as large molecules rather than as ions.*] The silica, then, forms the clay as its possible combinations with iron and aluminum to remain as a gelatinous covering on the weathering mineral crystals. Consequently, from the start of the weathering of a mineral fragment, the potassium, the calcium, the magnesium and the others coming out of it as active separates are enshrouded by the clay envelope on which they may be adsorbed and where they are not really free and active ions to be carried away in solution in the drainage water. Also, they are not soluble in water in the true sense of that terminology. *They are insoluble but yet available.*

The elements adsorbed on the clay colloid behave as a group, or, figuratively speaking, as if lined up in company front within the clay envelope. From there each may move out if some ion (or ions) of corresponding electrical charge approches from the outside and replaces it with the particular expenditure of energy required for the exchange. This involves what may well be called fair trade, or exchanges on the basis of at least nearly equal properties. Those include chemical valence, energies of replacement, and size of ions or molecules concerned.

That is the explanation in about the simplest terms, but not in complete detail, of the physico-chemical phenomena of the soil when we call it *adsorption and exchange of cation.* It represents the prominent soil phenomena so far as the positively charged nutrient elements like calcium, magnesium, potassium and others in less amounts are adsorbed on the inorganic clay colloid to be *insoluble to percolating water but yet available to the plant roots.*

Since the organic matter of the soil, especially the more stable fraction resulting from microbial digestions of it and considered the humus, also has adsorptive and exchanging properties duplicating those of the clay, we can expect the nutrient cations to react with the organic matter fraction of the soil according to the same principles. Fortunately, the humus was a higher adsorption and exchange capacity per unit weight by several times that of the clay, so that by building extra organic matter into the soil there results a higher adsorptive and exchange capacity of it.

This soils phenomenon of adsorption and exchange was initially spoken of as base exchange, since calcium, magnesium, potassium and sodium are considered basic in reaction, contrasted to acidic, which is the reaction of the cation hydrogen. We use the bases to neutralize the active, sour-tasting hydrogen ion, or to make it inactive and not ionic. But since hydrogen, like these other elements cited as bases, is also a cation of positive charge, and will also be adsorbed along with them and exchanged for any of these others on the clay and humus, all in like manner, it is more nearly correct to speak of *cation adsorption and exchange* by these two colloidal fractions of the soil (clay and humus) than to speak of only *base exchange.*

4. adsorption and exchange permit quantitative assessment of available cations

In speaking about cation exchange in terms of quantities concerned, the amounts of cations adsorbed are so small, relative to the clay or humus acting as adsorbers of them, that we speak of milligrams of cations adsorbed per 100 grams of clay or humus (parts per hundred thousand). But since hydrogen is the common chemical unit of equivalence of all elements, as for example, one gram of hydrogen is the equivalent of 20 grams of calcium, or 12 grams of magnesium, and of 39 grams of potassium, we speak of the *cation exchange capacity* (CEC) *per 100 grams of clay* in terms of a certain number of *milligrams equivalents* (ME) of hydrogen.

Thus the colloidal clay fraction of a soil type like the Putnam silt loam is said to have a cation exchange capacity (CEC) of 65 milligram equivalents (ME), (CEC) = 65 (ME). Accordingly, 100 grams of that colloidal clay (according to accurate measurements) could absorb a total of 65 milligrams of hydrogen alone; or 65 x 20 = 1,300 mgms. of only calcium; or 65 x 12 = 180 mgms. of magnesium; or 65 x 39 = 2,535 mgms. of potassium only. By that arrangement in equivalents of the element hydrogen, one can visualize the total exchange capacity satisfied by combinations of extensively varied amounts of each of those three cations cited, to say nothing of including some hydrogen, or of adding also all the positively charged trace elements in their extremely small quantities. By that situation each element can represent a wide range of its quantities according as those are different percentages of the total cation exchange capacity (CEC) including a whole suite of cations.

5. *adsorption and exchange gave techniques for measured variation in plant's chemical composition according to varied soil fertility*

By use of the colloidal clay of the Putnam silt loam with controlled quantities of the nutrient cations adsorbed on it as percentages of saturation, that is, in specified ratios to each other, it was discovered that such variation produced different growth effects, and different chemical compositions (carbohydrates *vs.* proteins), of a legume crop like the soybean. Thus, the varied suite of cations adsorbed may represent a balance, or an imbalance, in the nutrition of the plant. As a suggested balance, for example, the adsorbed calcium may well be 75% of the CEC, the magnesium, 7.5 to 10% of it, and the adsorbed potassium, about 3%, while the remainder of the CEC is taken by hydrogen (to give some acidity) along with a remaining small part taken by the many unmeasured cations of both known and unknown trace elements. Thus we can now *test soils for their available nutrient cations* (not only those soluble in water) and, accordingly, can speak of the soil offering a *plant ration of insoluble but yet available cations held by the soil* according to which ration or plant's diet the varied growth of the plant responds and a varied final chemical composition results in terms of its nutritional values as feed or food.

6. molecular nature of clay mineral modifies adsorption-exchange activities

Because a given amount of a cation is adsorbed on the clay, it does not follow that an exactly similar amount will always be exchanged readily. In the case of potassium, as in illustration, of a given amount adsorbed on soil colloids only a part may be exchangeable (by laboratory test) while the balance is considered as *fixed* and not exchangeable or soluble. Just what structural property of the clay molecule brings about this particular behavior of the potassium is not entirely clear in vision or concept. Other elements, too, are not completely exchangeable. Hence, the forces holding the adsorbed cations and the energies required for exchange are not fully understood. Nevertheless, the colloidal aspects of the clays (made up as they are of different clay minerals) have taken on helpful quantitative meanings by the terms *adsorption, exchange, cation exchange capacity, percentage saturation of cation exchange capacity,* etc.

Those are the behaviors and quantitative assessments of the colloidal properties of the major active fraction of the soil. These contribute clarification of the previous confusion about soluble substances going into the soil to become insoluble against the loss therefrom to the percolating waters. Knowledge of those soil properties has done much to explain how such *insoluble elements* (adsorbed on clay or humus) *are nevertheless available* for exchange to the plant root, or to water, according as either offers to exchange some other cation in the equivalence of the requisite properties. It explains more clearly than ever before how the root can come along with hydrogen, which is the acidic cation enshrouding the root because that plant part respires carbon dioxide to make carbonic acid,and can exchange such from that source for nutrient cations which feed the plant while making the soil more acidic. The uniqueness in that process of nature lies in the fact that the waste product of the root's respiration is the basis of its power to take nutrients off the clay or to make *the insoluble become the available nourishment.*

7. we have not improved on nature's laws by use of water-soluble fertilizers

In contrast to the confusion in our attempts to manage plant nutrition more efficiently than nature by our use of water soluble (salt) fertilizers in extra quantities, nature has long been using a simple practice of mobilizing the insoluble elements from the rock minerals into the adsorbed and insoluble condition on the clay and humus. By the same principle, nature moves them from there into the plant, by use of the root's own waste from respiration, namely, carbon dioxide. In the presence of water, that waste has given carbonic acid—nature's most universal acid—and its active hydrogen to be exchanged to the clay and traded there as an active *non-nutrient cation for the nutrient cations* and their movements into the root as plant nutrition. If we will comprehend and follow nature's practices and principles rather than our own confused thinking, there is still hope that we shall learn of, and have higher appreciation for, nature's successes in crop production for healthy plant survival. Perhaps we can see that nature practices scrupulous conservation, when, in fact, making *available* to plants—more generously than we realize—the essential fertility elements which we believe unavailable because we classify them as insoluble in terms of chemical inspections for regulatory purposes.

8. organic plant nutrients through living soils

In our attempts to comprehend how the insoluble rock can make soil which, under proper moisture, will nourish plants, we have too long believed that the nutrient inorganic elements from the soil must be water soluble. That has seemed necessary, at least for the first step of that service, namely the entrance of them as ions into the plant root, or for their movement through the outer wall or membrane of it.

The states' inspections, and their license of commercial fertilizers, use water solubility and the solubility in citrate solutions, as the criteria of the fertilizers' services. The inspection assesses fees on the tonnages of fertilizers distributed according to those solubility requirements for the nitrogen and potassium units and for those of phosphorus, respectively.

9. nutrients are not washed into the plant by the transpiration stream: they enter under their own power

In that contention that solubilities of high order are required for entrance into the plant root, we are apt to believe also that such entrance is connected with the large amount of water moving from the soil into the root, passing through the plant, and evaporating to the atmosphere from the leaf surface. More water is moved through and transpired by the plant according as the evaporation rate from the leaves increases with the rise of the daily temperature, the wind, or air movement over the leaf surface, the lower humidity of the atmosphere, and the larger supply of water in the soil. But because there is a decided flow of water from the soil through the plant for evaporation to the atmosphere, that is not proof that the fertility elements are necessarily moving along that same course because of that current of water as transpiration. Calcium, magnesium, nitrogen, phosphorus, potassium, and all the other essentials are not swept into the plant because they are applied to the soil in water soluble forms of fertilizers and flooded in, as it were.

There are natural facts, some readily demonstrated in the laboratory, which refute such erroneous beliefs that the water solubility within the soil is a requisite for fertilizer availability and flow with the water into the growing crop. As the first fact, plants will grow and their nutrients will move normally from the soil into the roots without the evaporation of water from the leaves. A potted plant, enclosed in a water saturated atmosphere with carbon dioxide under a glass bell jar in the light, will grow normally. This fact tells us that while the transpiration stream is halted because the saturated atmosphere will not take any water of evaporation, the fertility elements are, nevertheless, flowing into the plant from the soil.

In research at the Missouri Station, some soybean plants were grown on soils of such low saturation of the clay by calcium, that the totals of nitrogen, phosphorus and potassium in the total crop of tops and roots were less than those of the planted seed. Such facts tell us that the fertility elements may flow out of the root, or in the reverse direction of the flow of the transpiration stream of water.

That same reverse flow of fertility can be demonstrated under the conditions used for the potted plant within the bell jar, or when there is no flow of transpiration. Such facts inform us that even in the absence of water movement within the plants, the nutrients will move either into, or out of, the plant, entirely independently of either the static or the flowing condition of transpiration water. Forces, other than the water flowing into the plant root, must move the fertility elements serving in connection with plant nutrition.

Still as another situation, the desert plants have shown according to research reports by Dr. Went, now Director of the Missouri Botanical Gardens, that nutrients go into the roots for nourishment of the plants when in the daytime the water is transpired to move from the soil to the atmosphere. Then, also, they go into the roots when at night time the atmospheric moisture of condensation moves from the plant back to the soil sufficiently for plant survival through such diurnal reversals in movement of the limited moisture supply.

These facts deny, categorically, any necessity of water solubility of nutrients for their flow into, or within, the plant for any delivery services of them by the transpiration. They tell us that the fertility, which is feeding—not watering—the crop plants, behaves according to certain laws of physico-chemical relations within the soil and plant, while the water movement behaves according to the meteorological conditions and the climatic situations controlling the conversion of water from the liquid to the gaseous form and vice versa.

Water solubility of plant nutrients in the soil is not the rule of nature for their services to plants. Rather, they are naturally insoluble there, by which condition they remain there against loss through leaching out of the soil. By virtue of that condition they are still there when the growing root comes along. But that fact does not deny their being available through other mechanisms than aqueous solution.

10. natural plant growth emphasizes the insolubilities of both the organic and the inorganic nutrients

When any plant species established itself naturally in its particularly well-suited climatic setting, the crop is at its best in growth, in self-protection against pests and diseases, and in reproduction of its kind. But those conditions occur only after many annual crops have grown and died in place to build up a significant supply of its own organic matter on the surface of the soil. As an example, we speak of the forest floor consisting of the accumulated leaves, needles, etc., as a spongy mass of the forest. We speak also of the grassy mat of the prairies. Then, there is the accumulated organic matter within the upper horizon of the soil profile because of which we call it the *surface soil.* Then with more emphasis on the extra organic matter as the very top of the profile we say the *prairie sod.* These are the situations when the crops are naturally at their climax, or at their best.

These organic accumulations are some of the means by which nature retains the previously accumulated inorganic combinations until their microbial decompositions set them free as ash elements again. Microbial respiration, or nature's slow process of burning out the carbon to escape as gas, gives the energy by which there is the *living soil.* The organic, not the inorganic, part *puts life into the soil.* During its decay, some organic matter moves downward in its colloidal forms within the soil profile. It may precipitate there in combination with iron, manganese, and other elements as insoluble layers.

All this emphasizes the insolubilities of both the organic and the inorganic substances in connection with nature's processes for accumulations—not removal—of the organic matter both on, and within, the profile. It is those accumulations of *insolubles,* to become *availables* later, by which nature grows crops at their best in terms of freedom from pests and diseases, of dense stands, and of big yields. The living soil under the climax crop represents much of the recently contributed, highly carbonaceous substance on which an enormous population of microbial life can feed for its energy which was absorbed by and stored within the previous plant generations as sunshine energy. Consequently, a virgin, living soil represents ample energy food for its microbial populations, but a decided shortage of readily available foods for growth, like nitrogen, phosphorus, sulfur, calcium, magnesium, potassium and others. Therefore, if any one of these elements comes out of the decay activities which convert it into available forms, that element will be quickly taken by the microbes as foods for their growth, through its balancing of the excess of those serving only as food of energy source.

Such are the processes by which nature keeps the fertility of the soil insoluble. She conserves and maintains it most scrupulously because it is building a rapidly multiplying microbial population in the soil active in that performance. That conservation is so wisely managed through the biochemical activities of the soil organic matter by which the plant and microbial nutrients are kept insoluble in water but yet available for plant growth.

Nature's management, as just outlined for her accumulating organic matter of the soil, duplicates, in principle, what the straws, and other plant residues and wastes, do as bedding under stabled livestock to conserve the soluble fertility in the animal urine and feces. The bedding is a microbial food representing excess energy. It is what has been called *go* food. But the urinary salts, especially the nitrogen, are their food for cell growth. These are *grow* food. When the latter as solubles are added to the bedding, the microbes grow rapidly from this growth food as supplement to the energy food, or carbon, of the bedding. While about one part of carbon is built into the living cells about two parts are burned by them for energy to escape to the atmosphere as respired carbon dioxide gas.

Thus with this reduction in the soil's supply of energy substance through carbon escape, any growth substance like nitrogen, etc., is carefully conserved within the soil's living cells which consume it the moment any bit of it becomes available as the remains of one generation to be the food for the next one. This process continually conserves the living potential, the growth food or the creative power of the soil. The originally wide ratio of the carbon (carbohydrate) to the nitrogen (proteins) in the straws as plant residues from which the protein-rich parts, like seeds, were harvested, becomes narrower and narrower to approach that of the higher protein concentration of only the living microbial cells. Those must then consume each other when the energy supply in fresh organic matter like the bedding is all consumed. Such are the biochemical results of the heating processes occurring in urine wetted bedding in barn waste on its way to become what is truly manure in its fertilizer form, and occurring also in any highly woody, organic wastes when composted with some nitrogen and other salts of commercial fertilizers to bring on the process of decay.

But very soon those cycles of re-use and careful retention of the growth foods will have run their course and the surplus energy foods will have been burned out. Then the manure ceases to heat. It will no longer be keeping the soluble growth elements as cellular, excretory wastes combining into insoluble living substances. At that moment, the carbon-nitrogen ratio of the manure (or the soil) has become so narrow that dead microbes are consumed by the surviving others for energy foods as well as for growth foods. After that moment the growth elements made soluble will not be made insoluble so quickly, since there is no more energy food in reserve as fresh organic matter. It is then that those solubles may serve as the plant's nutrition or, like in the unsheltered manure heap, they may be leached away by rainfall. It is after the heating stage is over that manure serves as fertilizer to feed the growing crop, when during the heating process it was serving to feed the microbes, for which plant roots are no competitors within the soil.

11. the rise then fall of natural conservation of fertility give the successions of different climax crops

Under nature's management of the soil, it is the period of pre-climax crops which represents conservation first of the inorganic fertility and later of both the inorganic and the organic. Man's taking over of the climax or virgin crops and his putting the soil under cultivation prohibits natural conservation. It is replaced by exploitation or by expenditure of the many fertility values which nature had assembled and preserved as insolubles but availables for microbes and plants because of her maintenance of much carbonaceous organic matter of a truly living soil. Crop values as nutrition of warm-blooded bodies must decline accordingly as the living soil moves slowly toward its own death.

After a natural crop, like grass on the plains or prairies with its natural animal herds, has reached its climax because not much more inorganic fertility can be added to the annual cycles of fertility turnover because the roots do not reach it, or after annual rainfall has increased to leach more of it away or erosion removes it, then the climax crop is slowly replaced by another which will later reach its climax. *But that succeeding crop represents a lower level of physiological processes especially lower than those we see in protein-rich plants to feed warm-blooded bodies. Hence those crops represent less nutritional support for animals and man. In that decline in the succession of crops, the forests, especially the coniferous, can be one of the last of the climax situations of the declining fertility support by the soils.*

12. contents of plants are insoluble and highly organic yet biochemically active

Just as the living soil represents little that is water soluble, or salts in solution, likewise there is little that is water soluble within the plants and within any living tissues growing on the soil, save that some potassium is in solution surrounding the plant cells. Occasionally, some potassium is washed out of the shocks of harvested wheat to be indicated by the encircling better growths of the stubble crop of red clover. It may be similarly washed out of the shocks of fodder-cut corn marked by better fall-seeded wheat around those shocks left for a considerable time during a rainy post-harvest season.

Crop production does not represent chemical processes dependent on the creative fertility elements of soil origin in water soluble states. Those biochemical processes of living matter do not depend on compounds with a high degree of ionization, or of their separation into their component, electrically-active elements for ready conduction of current and speedy recombinations into other compounds. Inside of the single cell and inside of the specific collections of them as particular tissues and living bodies, the very opposite conditions prevail. Proteins, the only compounds which are living, may be broken down into what is called *amino acids,* but they are not acidic in reaction when combined to form the living protein tissue.

While our concepts of inorganic and mineral chemistry emphasize the elements as active ions, those creative atoms are not in that ionic condition to any degree of significance either in the original rock minerals that go to make soil, or in the life forms of microbes, plants, animals and man that are grown from that soil. Those *elements* need to be viewed *in their compounds,* in their *larger molecules* and as *adsorbed ions* to give the lowest possible ionic activity when held by colloids requiring energies, or work, to exchange them and make them more ionic, or separately active. Living soils are stocked with organic compounds, which control the chemical activities of the essential inorganic elements emphasized for their water solubilities but are unknown, or forgotten, for their unique combinations into un-ionized and insoluble parts of organic reactions, and which supply nutrient directly in the form of those inorganic compounds.

13. science learns facts of nature, then interprets her secrets of long standing

Scientific studies are uncovering what seems like a natural protection, via the soil organic matter and the living soil's dense microbial population, against excessive salts in the soil. Nature uses the inorganic elements mainly as inclusions within larger organo-complexes. This has demonstrated itself as a means of moving more inorganic fertility of the less soluble nature from the soil into the plant and about within it. This emphasizes *the availability of the insoluble.* Now that we have discovered, in the laboratory, the process which we call chelation whereby an inorganic ion is stably fixed within a larger organic molecule so that both are taken into the plants for their better nutritional service there, we comprehend what has been nature's way by which the insoluble fertility of the soil has become the available to the microbes and the plants during the past ages. But only slowly are we putting science under nature's secrets by which she grows better crops by returning more organic matter back to the soil.

Chelation is the term that interprets what observations in agricultural practice have often suggested. It is a widely recognized fact, for example, that both the soluble and the insoluble phosphates, and particularly the former, are more effective for improving the crop when first mixed with barnyard manure. Also, the simultaneous applications to the soil of ammonium salts and soluble phosphates will mobilize more phosphorus into the plants for better crop growth. Sodium nitrate applied in combination with the latter does not. Calcium nitrate in place of the sodium even reduces the mobilized phosphates, as compared with what occurs when soluble phosphates are applied alone.

The element nitrogen, the common symbol of our only living compounds, *i.e.*, the proteins, is prominent in bringing about the chelation of calcium, magnesium, iron, cobalt and a list of other elements, as examples. It is the nitrogen, in particular, that serves to connect the inorganic elements more stably into the large organic unit of the final complex that results from chelation. It suggests the proteins, and other forms much like them, as the major means of chelation which robs the inorganic elements of their common property and chemical activity which we usually emphasize about them, namely, their solubility and their ionization, respectively.

This concept of chelation should not stretch our vision beyond its elastic limits when chlorophyll, the green coloring matter in every leaf, is an age-old illustration of nature's use of chelated magnesium. In its chemical composition this photosynthetic agent represents about one part of the inorganic in about 40 of total organic for building sugar from water and carbon dioxide under the sunshine's energy. Magnesium is similarly chelated in a long list of other enzymes of both plants and animals. Hemoglobin in our blood is a case of chelated iron with about one part of it in 50 parts of organic matter. This is the means for taking up oxygen from the air in the lungs to be carried by the bloodstream and given up to the tissues while the iron is not ionic.

Then cobalt is also chelated into vitamin B_{12} in a similarly wide ratio of the inorganic to the organic part of the complex. The importance of this chelation compound was discovered by chickens, taking to the cow's droppings, long before we as chemists had any vision of it. Natural chelation under the dynamics of organic matter and microbes, may be expected to illustrate itself more widely now that much research work is studying it, and to increase our appreciation of this one of nature's secrets, now that the chemistry has given foundation and pattern for our visions of the chemical aspects a bit more fully.

14. commercial chelators give the means for determining the effectiveness of naturally insoluble yet available nutrients in soil for plants

The commercially available chelator, ethylene-diamine tetra-acetic acid (EDTA) was put into soil of no iron content along with one-half of the plant's roots, while the other half was in similar soil given iron as well as ample phosphates to make both less available according to their solubilities. Yet this commercial chelator was taken by the plant roots from the one-half of the soil. It served to mobilize the iron from the other half into the plant roots and to correct the chlorosis of the plants which occurred under similar soil conditions omitting the applications of this special chelator. More significant, however, was the additional demonstration which added water leachings from a highly organic soil as a substitute for the manufactured chelator, EDTA, only to find that this natural substitute served in iron mobilization for the cure of chlorosis just as the EDTA did.

Such facts help us to visualize the services by the soil microbes in their absorption of the salt shock when ammonium phosphate is added to the compost heap; when its soluble inorganic nutrient elements are taken into insoluble, much larger organic complexes; and when, thereby, both parts are made more available as nutrition entering the roots of the crop plants. We are gradually visualizing that, *within the soil, the organic fertility as chelating forces is dominant over the inorganic in about the same ratio as the combustible organic part of the plants is dominant over the incombustible, the inorganic, part there.*

15. phosphorus in organic matter is more efficient than its other forms in the soil

That such ratios prevail, relative to the advantages of making the insoluble become the available via the soil organic matter, was recently demonstrated by some research using barley as green manure for feeding radioactive phosphorus into a crop of soybeans, according to the unpublished data from the research studies by Vernon Renner of the Missouri Agricultural Experiment Station.

Barley plants were grown on sand cultures with a controlled nutrient medium containing radioactive phosphorus. They were harvested, dried, pulverized, sampled for chemical analyses, and then that pulverized organic matter mixed thoroughly into the soil in the ratio of one part of the former (organic matter) to 500 of the latter soil. This represented the common field rate of two tons of this dried green manure per two million pounds of soil per acre per plowed depth.

The soybeans were harvested after a growth period of 60 days and chemical analyses were made to determine their contents of total phosphorus and of the radioactive phosphorus. This would, via radioactivity, determine the phosphorus contributed to the soybean roots by the green manure applied to the soil. The total phosphorus, minus the radioactive part, would determine the phosphorus coming from the soil, which was a mass 500 times as large as the applied organic matter in the barley as the green manure.

This separation of the phosphorus given by the organic matter of the soil from that given by the larger, originally more inorganic part of the soil, showed that one part of the phosphorus taken into the soybeans was radioactive, therefore, taken from the green manure turned under. Five parts of the phosphorus taken into those soybean plants were not radioactive, hence were taken from the original soil.

Thus, it is established very clearly, that when the phosphorus coming from the barley as organic matter which was only one part while the soil was correspondingly 500 parts, yet phosphorus from those two sources, respectively went into the plant in the ratio of one to five; the phosphorus from the soil organic matter was just 100 times as effective in feeding the insoluble but yet available phosphorus to the soybean plants as the soil was when its soil test by an extracting chemical reagent reported the soil "high" in its available phosphorus.

16. knowledge about, and appreciation of, the soil as creation's starting point come slowly and late

Some of our pioneer agronomists, as able chemists and scholars, may have had a vision of nature's unique phenomena of the chelation of inorganic fertility of the soil by the organic matter, and microbial processes connected with it, in what they considered *the living and creative soil* when, as an example, nearly half a century ago Professor A. W. Blair of New Jersey said, "*It is well known, for example, that by judicious use of lime and vegetable matter on the soil, reserve of locked-up mineral plant food may be made available.*" Others spoke about maintaining the soil fertility and a permanent agriculture by returning organic matter to the soil in combinations with natural rock fertilizers. Those pioneers did not visualize productive permanence in soils treated with water soluble salts.

Now that we appreciate the rapid decline in the productivity of many soils and our inability to call from the chemical shelves every classified chemical, inorganic elements required to hold up crop yields in quantity, much less in nutritional quality, we may realize the weakness in retaining water solubility as the criterion for materials applied to the soil for agricultural crop nourishment. We are slowly seeing the need to nourish also the microbial crop which makes the living soil, and is a favorable help in chelation and thereby in our getting better quality of crops as food and feed. We are learning that, through the organic matter of the soil and its microbial processes, much that is the insoluble yet available fertility is an effective natural creator of excellent crops, and those with a high degree of regularity.

LESSON NO. 1

Fertility

8.

A few years before Dr. William A. Albrecht passed from the scene, Acres U.S.A. taped an interview that in effct summarized the hundreds of scientific articles the great scientist had accounted for. "On the basis of your research, should fertilizers be soluble?" became the routine opener. "No," said Albrecht. "Fertilizers are made soluble, but it's a damn fool idea. They should be insoluble but available. Most of our botany is solution botany. When we farm as solution botany, the first rain takes out the nutrients. There's a big difference between the laboratory and the farm." Albrecht knew whereof he spoke. He in fact worked out his vision of the Creator's balance in the laboratory. Because of his innate interest in medicine, he figured there had to be an optimum fertility load for a soil system, just as there had to be health-giving nutrition levels for human beings. In The Case for Eco-Agriculture, this chapter of Albrecht's career is covered with brevity. These extract lines are quotes for all practical purposes. . .Since the soil colloid was a catchpen for cation nutrients, Albrecht separated the finest part of the clay out of Missouri soil by running it in a centrifuge at 32,000 r.p.m. Albrecht and his students put the processed material into an electrical field and made it acidic. They took all the cations off so that the end product was an acid clay. This way the researchers could titrate back given nutrients, mix them, and balance them. Albrecht found that for best crop production, the soil colloid has to be loaded with approximately 65% calcium, only 15% magnesium, and that the potassium cation must be in the 2 to 4% range. Other base elements had to account for 5%. These, Albrecht held, were the saturation percentage figures when nature functioned at her finest balance and was capable of producing healthy crops.

Albrecht's career became one of proving out ecological agriculture, and demonstrating how declining fertility levels brought on insect attack for the crops, and ill health for animals and man consuming them. He was well aware that even with growth stimulants and super-mining techniques, the economist's curve showed production in the U.S. had topped out, and was on a downhill slide. Ever increased use of NPK and toxic chemistry was being accompanied by diminished returns, with quality at the vanishing point.

One of Albrecht's great friends was E. R. Kuck of Brookside Dairy Farms, New Knoxville, Ohio. In fact it was a case of animals telling Kuck about vanished nutrition that launched Brookside Laboratories in the first place. After plastering the walls of a new calf barn, Kuck noticed that the animals literally ate the plaster off the walls of their stalls. Calves were scouring at the time. Almost immediately, scouring stopped, and Kuck—consulting with Albrecht—determined that the hungry animals were in fact after the calcium carbonate and the magnesium carbonate in the plaster material. These nutrients had been mined out of the soil of the dairy operation and never replaced.

The sheer desperation of those animals emerged time and again in Albrecht's writings. "Our dietary essential minerals are taken as organo-inorganic compounds. We are not mineral eaters. Neither are the animals. When any of them take to mineral box, isn't it an act of desperation?" On another occasion, Albrecht came right to the point. "Cows eat soil or chew bones when ill with acetonemia, pregnancy troubles, or deficiency ailments. Hogs root only in the immediate post-winter period after confinement to our provision for them and their behavior suggests past deficiencies to be quickly remedied in desperate digging of the earth."

Other observations came to the fore at Brookside. Soil and leaf tests revealed a wide variance in crops produced on treated vs. untreated acres. In December 1946, Kuck issued a report, Better Crops with Plant Food, published by the American Potash Institute. The article was picked up by the farm press and given wide distribution. Feedback arrived almost immediately—some 1,000 letters from farmers telling of their own problems and asking for help. There were also 16 letters from academic folk condemning Kuck in vitriolic terms, and denouncing the significance of magnesium in animal nutrition. Moreover, the college people seemed to think that no farmer had the right to make such observations and tell it the way it was without the imprimatur of credentialed people.

That the outdoor biology known as living soil had a game plan has not been fully communicated to farmers even now. It is the conceptualization of laboratory scientists that nutrients for plant life must become ions, cations if they are charged positively, anions if they are negatively charged acid elements. They enter plant roots the way electricity flows, so to speak, in chemical equivalents. This has been compared to the action one gets when positive and negative poles of a battery are connected. Interaction of cations and anions dovetail with the sun's energy, carbon, hydrogen and oxygen to sustain plant life.

The first order of business for the soil colloid, then, is to hold nutrients—nutrients that can be traded off as the roots of plants demand them. The soil laboratory measures the holding energy in the clay and the humus.

Almost all laboratories report this energy as cation exchange capacity, and they do this in terms of milliequivalents, or ME. Some few farmers like to think of all this the way an electrician thinks of his payload, in terms of volts and amperes. The only difference is that the soil laboratory measures colloidal energy in terms of milliequivalents of a total exchange capacity, since soil colloids—composed of clay and humus—are negatively charged particles, or cations. Because anions are not attracted by negative soil colloids, they remain free to move in the soil solution, or water.

A milliequivalent is the exchange capacity of 100 grams of oven-dry soil involved with 1/1,000 of a gram of hydrogen (hence the prefix, milli). *Each cation element has its own weight. Calcium carries 20 on the Mendeleyeff Periodic Chart of the Elements. Potassium is 39. Magnesium is 12.*

Fossil fuel fertilizer people knew all this even before NPK fertilizers were turned into foo-foo dust, good for what ails a soil system according to low or high analysis. In fact, the old National Plant Food Institute (now The Fertilizer Institute) once circulated a rather sound presentation styled The Living Soil. *In it was revealed the intelligence that "milliequivalents can be converted to pounds per acre if you know the equivalent weight of the element involved. Multiply the equivalent weight by 20 [the computed factor] to get pounds per acre for one milliequivalent per 100 grams of soil." The chart weight of calcium is 20. Therefore, 20 times 20 equals 400, or 400 pounds of calcium per acre.*

Cation exchange capacity of a soil depends on the type of clay and the amount of humus. These can vary widely. There are CEC figures as low as 1 and as high as 80. The clays themselves vary—kaolinite in the south often measuring 10 to 20, montmorillonite in the west measuring 40 to 80. CEC values for organic matter are higher yet—100 to 200 ME being common. Obviously, pure sand or gravel would have a CEC of near zero. But the moment sand or gravel starts accumulating colloidal clay and humus, its CEC can go up dramatically.

Eco-farmers have long recognized that what you do to a soil system depends on the condition of that soil system in the first place. To treat a 6 ME soil the same as a 20 ME soil amounts to the same thing as powering a go-cart with an airplane engine.

Since it would take 400 pounds of calcium to fully load the top 7 inches of an acre of 1 ME soil, it would take 4,000 pounds per acre for a 10 ME tested soil. Needless to say, no farmer would want to go whole hog on one nutrient, and neglect the rest. Such an outlook would come terribly close to adopting a simplistic NPK mentality.

The function of the soil audit, then, is the measure the cation exchange capacity and to determine the nutrient status a crop might rely on. Positively charged elements, of course, are potassium, magnesium, zinc, copper, sodium, etc.

These many principles became embodied in the soil audit and inventory report Albrecht helped E. R. Kuck design for Brookside Laboratories. One such report rendered for Albrecht's property in Illinois appears on the next page.

That health problems started with the soil had long been an Albrecht contention. And now, with strong allies, Albrecht was able to balance soil on both the test plot and the farm.

On Albrecht's farm the exchangeable cations were tabulated ever so carefully. Once the CEC had been determined, the scientist developed desired values, added or subtracted values found, and prescribed the appropriate treatment—calcium, magnesium, potassium. The anion nutrients values were also determined and repaired. Albrecht could grow clean rows and infected rows, depending on how he altered soil fertility. Balance was the one thing a farmer couldn't overdo, Albrecht observed. This made low exchange capacity soil a tough customer, with fine-tuning a delicate matter. Dealing with paper-thin holding power in a soil system was like doing jeweler's work with an anvil hammer. This shortfall, Albrecht believed, wasn't helped much by prevailing folklore.

Somehow folklore became law in federal cost-sharing programs. The bureau people were always trying to erase soil acidity with lime programs, not because they understood balance, but because they loved simple answers for complex problems.

By the early 1930s, soil mining techniques in the United States had come to a climax point. In the west this meant a hot-dry cycle dust bowl. In the east, it meant faltering production. The first give-away lime program surfaced in 1933 under the

Account Dr. William Albrecht City ______ State ______

Service Representative ______

		1	2	3
House, Plot or Field No.		1	2	3
Soils Sample No.		Untilled	FIELD IN USE	Compost
Total Exchange Capacity (ME)		24.00	30.00	37.50
pH of Soil Sample		7.5	7.9	7.8
Organic Matter, Percent		2.40	2.30	8.30
ANIONS — NITROGEN:	lbs/acre	62	59	111
SULPHATES:	lbs/acre	1714	2171	3345
PHOSPHATES:	Desired Value	253	275	315
as (P_2O_5)	Value Found	5241 P_2	6173 P_2	4034 P_2
lbs/acre	Deficit	113	96	646
EXCHANGEABLE CATIONS — CALCIUM:	Desired Value	6048	7200	9000
lbs/acre	Value Found	6240	8224	6768
	Deficit			2232
MAGNESIUM:	Desired Value	980	1440	1800
lbs/acre	Value Found	1040	1480	2880
	Deficit			
POTASSIUM:	Desired Value	749	796	850
lbs/acre	Value Found	2336	1648	4800
	Deficit			
SODIUM:	lbs/acre	84	62	470
BASE SATURATION PERCENT				
Calcium (60 to 70%)	80% (Calcium + Magnesium)	64.97	68.60	45.34
Magnesium (10 to 20%)		17.83	20.32	31.75
Potassium (2 to 5%)		12.45	7.04	16.48
Sodium (.5 to 3%)		.75	.43	2.73
Other Bases (Variable)		4.00	3.61	3.70
EXCHANGEABLE HYDROGEN (10 to 15%)		–	–	–
Salt Concentration (p.p.m.)		1530	600	6600
Chlorides (p.p.m.)		100	None	1000
Boron (p.p.m.)		17.2	18.5	9.6
Iron (p.p.m.)		9.9	6.6	9.9
Manganese (p.p.m.)		59	25.3	45.7
Copper (p.p.m.)		28.2	28.2	176
Zinc (p.p.m.)		74	61	61

Remarks and/or suggestions:

AAA. Possibly half the soils in the U.S. were helped by liming. "Lime and lime some more" became a catchphrase. "You can't overdo it." As with most programs, this one was continued to comply with political strategy. Two decades after the liming program had come into being, Albrecht appeared before a Senate Subcommittee to explain the name of the liming game.

Statement Presented by Dr. William A. Albrecht, Chairman of the Department of Soils at the University of Missouri College of Agriculture Before the Senate Subcommittee on Agriculture, April 29, 1954

According to the 1954 *National Bulletin for the Agricultural Conservation Program* on page 6, there is the following directive, "For those practices which authorize federal cost-sharing for minimum required applications of liming materials and commercial fertilizers, the minimum required application on which cost-sharing is authorized shall in each case be determined on the basis of a current soil test."

At this moment the evolution of the techniques of soil testing has not progressed far enough beyond what is apt to be mainly laboratory gadgetry. It can scarcely yet replace the diagnosis of, and judgment on, each specific case. In my humble opinion, therefore, I doubt the wisdom of a mandatory regulation on the national level, for any soil treatments with applied materials, whether limestone, rock phosphate, soluble fertilizer, etc.

The simple manipulation of soil tests and the prescription of soil treatments therefrom will often not serve as a reliable guide for modifying the soil effectively for better crops. This situation can be most simply illustrated in our common testing for the degree of soil acidity, reported in terms of pH. This acidity is an irregularity in plant nutrition, not because of the specific degree of acidity presence (however accurately measured), but because of the absence of several fertility elements lost as the acidity developed.

1. variable with geographic location

Acidity is a common soil condition in many parts of the temperate zone. It occurs where the rainfall gives water enough to go down through and to wash out much of the fertility. In general, if the rainfall is high enough to provide plenty of water during the crop-growing season, there will also be enough water, with its carbonic acid, to leach the soil or to give it hydrogen in place of much of its supply of plant nutrients, and thereby make it acid.

Acid soils differ in their degrees of acidity, expressed as negative logarithms or what are known as pH values. Accordingly, then, a higher degree of acidity is represented by a lower pH value, or figure. This merely means that as more hydrogen has come in, less fertility or other positively charged nutrients are left there. Timbered soils of the eastern United States are acid; also those of the eastern edge of the prairie are sour. Acidity is a natural condition where soils have had rainfall enough so they have been growing much vegetation. Such soils have therefore been subjected to a leaching force taking the fertility downward, and to the competitive force of the vegetation with its roots taking the nutrients upward. In the latter case, those nutrients are built into organic combinations of them *above the soils of the forests and within the surface soils of the prairies.* Consequently, acid soils have distinct surface soil and subsoil horizons in their profiles. They are naturally low in fertility, especially in the subsoil, and have been growing mainly carbonaceous or woody vegetation. It is no small soil's problem to have such soils grow protein-supplying, mineral-rich, highly nutritious crops.

Natural soil acidity is in reality, then, mainly a shortage of fertility in terms of many plant nutrients reflected in the physiological simplicity of the vegetation. This is the situation because the soil has been under cropping and leaching for ages. This was true before we took over to intensify these fertility-depleting effects. This, then, is the condition of the soil that prompts the common question, "How can we grow mineral-rich, fertility-consumng forages, like the legumes, of good feed value for such high-powered, protein-producing animals as cows?"

2. *a little science led us astray*

It was the growing agricultural science of the early decades of the 20th century that brought liming of the soil back as a more general agricultural practice. We cannot say that liming was an art carried over from colonial days. It had been pushed out when fertilizers came into use. Liming the soil has become an extensive practice under the encouragement of an embryo soil-testing service. That service was guided by the belief that the applications of limestone, which is a carbonate of calcium; of hydrated lime, which is an alkaline calcium hydroxide; or of quicklime, which is a caustic oxide of calcium, are all beneficial for crop growth because each of these is ammunition in the fight against soil acidity, or against the high concentration of hydrogen in the soil, or the soil with a low pH figure, as the degree of acidity is now regularly expressed.*

This struggle to drive the active hydrogen ion, or acidity, out of the soil was

**The pH figure is the logarithm of the concentration of the active hydrogen or ions, not molecules. Distilled water has one millionth of a gram (.000001 or 10 to the minus 7 gm.) of active hydrogen per liter, said to be a pH of 7.0. Such water tastes flat. With a bit more*

aided by the technological advancements giving us instruments and equipment that measured the hydrogen ion to a finer degree than known before. *The ease and speed with which soil acidity could be detected and measured encouraged the widespread testing of soils.* This activity discovered soil acidity almost everywhere in connection with extensive agriculture. Through the help of the pH-measuring gadgets we were impressed by the apparent universality of soil acidity. Only a few humid soils were not seriously stocked with acid.

We discovered that for acid soils, in general, the productivity was lower as the degree of acidity was higher, or as the pH value for the soil was a lower figure. From such a discovery we might expect to conclude—even though it was later found to be the wrong conclusion—that the *presence* of the large concentration of hydrogen ions in the soil, or a certain low pH value, was the *cause* of the poor crops. This conclusion would be expected also from the bigger troubles in growing the proteinaceous, mineral-rich legumes of higher feeding values and of more physiological complexity through which these values alone are possible.

The extensive use of limestone in the corn belt has now multiplied itself into the millions of tons of these natural rock fragments that are annually mixed through the soil. This increased use was prompted by the belief that limestone is beneficial because its carbonate removes the acidity of the soil, and that soil is most productive if it is neutral, or when it has no active hydrogen ions in it; that is, when the soil has a pH of 7.0, or near that value. Under these beliefs (now known to be poorly founded), we have become belligerent foes of soil acidity. Limestone has become the ammunition for fighting this supposed enemy hidden in the soil. With national financial aid, we have been prone to believe that in putting limestone on the soil we can follow the old adage which says, "If a little is good, more will be better." We are just now coming around to a fuller understanding of how nature grew crops on the acid soils before we did, and that crops are not limited to growth within certain pH values when they are well nourished by ample, well-balanced fertility.

3. science now shows limestone feeds crops— doesn't fight soil acidity

Only recently have we recognized the fallacious reasoning behind the conclusion contending that it must be the presence of acidity in the soil that brings crop failure when liming lessens both the degree of and the total of the soil acidity while making better crops at the same time. While the convenience of soil-testing gadgets for refined points of pH was encouraging this erroneous belief about soil acidity as an enemy, it was the diligent study of the physiology of plants, of the colloidal behavior of the clays growing them, and of the chemical analyses of them all that finally pointed out the errors of such hasty conclusions. It indicated that

acidity or ionized hydrogen (.00001 or 10 to the minus 6 gm.), or more acidity at a pH of 6.0, it tastes better. Thus, as the pH value is smaller, the degree of acidity is higher because there are more active hydrogen ions per unit volume as the hydrogen ion concentration is higher.

the presence of soil acidity is not detrimental, but that the absence of fertility, represented by the acidity, is the real trouble. On the contrary, some acidity can be, and is, beneficial.

We now know, of course, that in applying the limestone, which is calcium and magnesium carbonate, there is possibly some reduction of the degree and the total acidity by the carbonate portion. At the same time, there is applied also some calcium and some magnesium—nutrients highly deficient in the leached soils—to nourish the calcium-starved and magnesium-starved crops. These nutritional services come about both directly and indirectly. We have finally learned that it is this better nourishment of the crops, rather than any change in the degree of acidity, or any raising of the pH value, of the soil that gives us the bigger and better crops. Unwittingly we have been fertilizing the crops with calcium simultaneously while fighting soil acidity with the carbonate, the hydroxide, or the oxide of lime.

Regardless of our ignorance of how lime functions, we have unknowingly benefited by using it. However, an erroneous understanding of what happens to crops and to the soil when we apply lime, cannot successfully lead us very far into the future. We dare not depend forever on accidents for our good fortune. We cannot continue to grow nutritious feeds under the mistaken belief that we do so merely by changing the pH, that is, the degree of acidity, or by the removal of the soil acidity through the use of plenty of any kind of carbonates on our humid soils. Wise management of the soil to grow nutritious feeds can scarcely be well founded on facts so few and so simple.

4. simple tests show lime is beneficial through calcium

Should you decide to demonstrate for yourself the truth of what has been said above, you might apply some soda-lime, or sodium carbonate, to acid soil. This will increase the pH of the soil. It will reduce its total acidity. But while this soil treatment will rout the enemy, i.e., soil acidity, and raise the pH toward 7.0, it will still not give successful crops. Merely removing the acidity by a carbonate (of sodium—a non-nutrient—rather than of calcium, in this case) does not guarantee the successful growth of the crop.

As proof that it may be calcium as plant nourishment that is the helpful factor in liming a soil, one can repeat Benjamin Franklin's demonstration and apply calcium sulphate, that is, gypsum, to the soil. One might even apply some "Dow Flake," a calcium chloride. Either of these calcium-carrying compounds will make the soil more acid; either will *lower* the pH decidedly. In spite of this fact and because they add calcium, the gypsum and the "Dow Flake" will improve the crops on the initially acid soil either left so, or made more acid. We are now resurrecting the ancient art used by Benjamin Franklin, for whom liming the soil was a matter of fertilizing it with calcium sulphate, and not one of fighting soil acidity with calcium carbonate.

5. soils made neutral are not necessarily made productive

While we were fighting soil acidity, we failed to notice that most of the populations of the world are concentrated on acid soils. They are not in the humid tropics, where the soils are not acid or where the clay doesn't adsorb much hydrogen or even much of any nutrient cation. Nor are they on the arid soils that are alkaline (high pH values)— a reaction opposite to the acid (low pH values). Soils that are not acid are not necessarily the supporters of many peoples. Yet in fighting soil acidity we labor under the belief that if a soil were limed to the point of driving out all the acidity, such a soil should be highly productive.

We now know that even while a soil may be holding considerable acidity or hydrogen, it may be holding also considerable calcium or lime. To a much smaller extent of its exchange capacity, it is also holding nutrients other than calcium. Among these are magnesium, potassium, manganese, and others. But these in total are held in much less quantity and by less force than are either the calcium or the hydrogen—the former a nutrient and the latter a non-nutrient cation, or a positively charged ion. Should we put on lime or calcium enough to drive all the acidity out of the soil, that is, to make it neutral or to bring it to a pH of 7.0, by putting calcium in place of the hydrogen, all the other nutrients would be more readily driven out than would this acid-giving element.

Liming the soil heavily, then, does not necessarily drive out only the acidity, i.e., the hydrogen cations. Instead, it would also drive out all other fertility cations except calcium. It might load the soil with calcium so completely that it could offer only calcium as plant nourishment. Plants would then starve for other nutrients even though on a neutral soil. Plants on such a non-acid but calcium-saturated soil would be starving for all the same nutrients, except calcium, as they do on the acid soils. Making soils neutral by saturating them with calcium does not, therefore, make them productive. This is the situation of some of the neutral (pH 7.0 and higher) semi-arid soils of our western states. In our struggle against soil acidity we need to remember that neutral soils are not the productive soils. Instead, productive soils are the acid yet fertile ones that feed us and nourish the major portion of the other peoples of the world.

6. some soils need their pH changed much, others but little

By considering the increasing degrees of soil acidity simply as increasing deficiencies of fertility, we find in nature, in general, that as the degree of acidity is higher (pH figure is lower), the adsorption or exchange capacity of the soil* is saturated to a higher degree (larger percentage) by the positive ion, or cation, hydrogen. With this higher saturation by hydrogen, there are more hydrogen ions per unit of exchange capacity active or not held inactive by the soil; consequently

**The exchange capacity is expressed as milliequivalents (ME) per 100 grams of soil. One ME means 1 milligram (.001 gms.) of hydrogen or its equivalent. One ME per 100 gms. equals 1 gm. in 100,000 gms. or 10 in a million, or 10 pounds in a million or 20 pounds in two million of soil (acre 6⅔ inches deep).*

the degree of acidity is higher. The acid is stronger. There are more hydrogen ions to make contact with the measuring electrode, and the pH value is therefore lower.

The same holds true for the degree of saturation of the soil's exchange capacity by calcium, or magnesium, or potassium. As the degree of saturation by any one of these nutrient cations goes higher, more of it is active in making contact with the plant root and in getting into the growth activities of the plant. These cations are nutritional helps coming from the same source as the hydrogen cation, namely, the soil colloid. The hydrogen coming from there is not. *Because we have had the gadgets to measure the hydrogen, we have emphasized the pH, or the presence of a certain degree of acidity.* We have not emphasized the absence of all the many fertility cations resulting because the hydrogen has replaced them on the negatively charged exchange complex or colloid of the soil. We have not had simple gadgets to measure them.

If, then, we should have a sandy soil, for example (low in clay content and thereby low in exchange capacity) with a low pH value or a seriously high degree of acidity, or hydrogen saturation, this would conversely represent a low degree of saturation of its exchange capacity by calcium, magnesium, potassium and other nutrient cations. Accordingly, then, the addition of but a small amount (two tons per acre) of limestone (calcium and magnesium carbonate) would move the pH value up decidedly or shift the degree of acidity to neutral.

This is similar to changing the degree of heat of a cup of scalding hot coffee by putting an ice cube in it. Here a little ice lowers the degree of heat very much where there was a little total heat. With the sandy soil's low exchange capacity there could not be much total hydrogen, even if the degree of activity by it is high; consequently, the pH is changed decidedly toward neutral and the acidity is completely removed. The exchange capacity is loaded very highly, in turn, by the calcium or magnesium of the limestone. Then, accordingly, these two nutrients become highly active in plant nutrition to make legume crops succeed well where on this sandy soil they may have previously failed.

By doing no more than using the gadgets, in this case, to measure the change in pH (or the degree of acididty), resulting from liming the soil, one would conclude readily that the crop grew better because the pH was changed or the acidity was neutralized. One would not concern himself very commonly about the increased amounts of calcium, magnesium, and other nutrient ions which became so much more active to nourish the plant better. We have no simple gadgets to measure these effects; hence we attribute the crop improvements to the wrong causes. Grazing animals probably make no such mistakes in their choices of forages, judged according to nutritional values in terms of calcium and magnesium, rather than in terms of pH of the soil. If, on the other hand, we should have a heavy clay soil (high in exchange capacity) with a seriously low pH value, or a high degree of acidity, and conversely, of a low degree of saturation of its exchange capacity by calcium, magnesium, potassium and other nutrient cations, the addition of the same amount of limestone as was applied to the sandy soil would not change the pH value or the degree of acidity significantly.

This would be similar to dropping an ice cube of carmine solution into a bathtub of scalding hot water. The cooling effect would not be recognized, but the coloring effect would. Because the larger exchange capacity and larger total amount of hydrogen would keep almost the same amount of it active in spite of the relatively small amount of limestone, there would be no measurable change in the pH of the soil as it is commonly sampled and tested. Yet the calcium (magnesium) carbonate would react with the soil to be adsorbed by it on the soil's colloidal complex. It would exchange from there to the plant roots to improve the legume crop, even if much active hydrogen left in the soil maintained the pH of the soil near the initial value. There would be focal points of calcium (magnesium) in the soil (significantly so if the limestone was drilled) to exchange these nutrient cations more actively to the plant roots than before the soil was limed. Hence the two tons of limestone on the clay soil, initially considered of seriously low pH value, may not have changed the pH, although it established the legume where it failed previously.

Thus clover may be established on the sandy soil by two tons of limestone where the pH was changed to a decidedly higher value. Also, clover may be established on the clay soil by two tons of limestone where the same pH value was not changed significantly. Plant behaviors tell us that the changes in the pH values are not contributions to the improvement of the crop growth. On the contrary, it is the liming as a remedy of the fertility deficiency in its application of calcium (magnesium) for the crop, and not the change of the pH or degree of acidity by the carbonate that is the responsible factor in crop betterment. Different soils may differ widely, then, in the extent to which their pH values (or degrees of acidity) are changed by soil treatment for growing better crops, when it is the deficiency of fertility—and not the degree of acidity—that is the cause of the trouble on so-called "acid" soils.

7. *experience is a thorough teacher*

It was just such a case of confusion about what pH really means and how liming the soil serves plants as cited above, which put a county agent of Missouri into an embarrassing predicament many years ago. This occurred when the campaigns for *"Lime for Clover and Prosperity"* and *"Lime for Legumes and Livestock"* were at their height. A farmers' meeting under his leadership was held one autumn day on a farm where the soil was about to be prepared for wheat serving as a nurse crop for red clover. The crowd assembled at the first field on the river hill, a windblown bluff of very fine sandy loam. After testing the soil and finding it seriously acid (of low pH figure), it was agreed that, according to this pH report on the soil, two tons of limestone per acre were needed to grow red clover.

The farmer crowd then moved down into the river bottom to the second field, made up of a heavy clay soil This soil, under acidity test, revealed the same degree of this trouble. The same amount of limestone per acre, two tons, was deemed necessary as for the field of fine sandy loam.

After both fields had been plowed, the two tons of limestone per acre were applied. The soils were disced, the wheat was seeded and followed by the clover seeding the following early spring. The clover stands in the wheat stubble the next autumn were excellent in both fields.

This was the reason, and considered a good setting, for another farmer meeting on liming and soil acidity. This one started again with the fine sandy loam soil, and the test of its pH to reveal the absence of acidity and the good clover there, supposedly because the pH value of the soil had been changed to that of neutrality. But in the second field, with its clay soil, the degree of acidity under re-test was the same as before the limestone had been applied. Yet the crowd observing this test was standing in an excellent stubble crop of red clover. The county agent was in a predicament. Here the liming treatment of the soil established clover in the second field without changing the pH of the soil, when on the first field he had just pointed out that liming the soil established clover because, as he erroneously explained, it had changed the degree of acidity, or raised the pH of the soil. The pH of these two soils was not shifted to the same extent, nor was the acidity changed by the same degree; yet the clover shifts were the same, namely, from failure to good stands.

8. plants are sensitive to degrees of fertility deficiency, not to degrees of acidity

That plants are not "sensitive to, or limited by, a particular pH value of the soil" was demonstrated by experiments at the Missouri Agricultural Experiment Station some years ago. The clay fraction was taken out of the Putnam silt loam, and electrodialized to make it completely acid (replacing the nutrients by hydrogen) which gave it a pH value of 3.6. Six lots of this soil were set apart. Each was titrated with limewater to reduce its acidity to a certain particular degree, or specific pH figure. The lots represented the following series of pH values, namely, 4.0, 4.5, 5.0, 5.5, 6.0 and 6.5.

Enough clay was taken from each lot of this series of clays to represent .05 milligram equivalents (ME) of exchangeable calcium per plant for a total of 50 plants, and put into pans of equal amounts of quartz sand. In a second series, enough clay from each lot was taken to provide .10 milligram equivalents (ME) of calcium per plant in each of the six pans of sand, or doubling the amounts corresponding to the different pH values in the first series. A third series of these same different pH values of the soil was set up similarly, except that enough clay was used to provide .20 milligram equivalents of calcium per plant, or four times as much calcium in each pan as in the first series. Thus the three series were triplicates in pH values of the soils, but there was more clay, more adsorptive-exchange capacity, and more exchangeable calcium by two and four times in going to the second and third series from the first.

Observations of the soybean crops grown on these pans suggested different so-called "sensitivities to pH values" by this crop. By observing the first series containing the least amount of clay in the sand (the most sandy soil), and thereby the least amount of calcium for the crop (.05 ME per plant), one would have concluded that soybeans are sensitive to a pH of 5.5 but are not so to a pH of 6.0. Had one observed only the second series, the corresponding figures in one's contention would have been pH 5.0 and 5.5. But had only the third series been open for observation, one would have put the "sensitivity" value at pH 4.5 and not at pH 5.0.

Here, then, a lower pH value (or a higher degree of acidity) by as much as ten times was the difference in sensitivity brought about by offering four times as much exchangeable calcium as nutrition for the plants. This was brought about merely by giving more clay, a heavier texture, to the soil at the same degree of acidity in the series, or at the same pH. More exchange capacity offset the significance erroneously ascribed to the pH.

Still more significant was the change in the pH values of the clay-sand soils as the result of growing the soybean plants on them. The three more highly acid soils of pH values, 4.0, 4.5 and 5.0 in the three series *had all become less acid.* The growth of the crops had made their pH values in these nine cases move upward, or shift toward neutrality. The three less acid soils of pH values 5.5, 6.0 and 6.5 in the three series *had all become more acid* in consequence of growing the crop; or this crop growth had moved the pH values downward in these nine soils, away from neutrality. These shifts in pH values were as much as 1.5 pH, where the original pH values were 6.5. The soils in these instances were made 32 times more acid by only the partial or limited growth of the crop. Surely, then, when the growth of the crop, or the activities of the plants by way of the roots' contact with the soils make these soils 32 times more acid, one would scarcely say that it is the pH of the soil to which the crop is sensitive, or that a crop will grow only when a certain restricted pH or degree of acidity of the soil prevails. Plants are very sensitive to minute degrees of fertility deficiencies, but certainly not to degrees of acidity of very wide range.

9. wise soil management aims at plant nutrition rather than acidity removal

Liming the soil, then, as a matter of putting active calcium and active magnesium into the acid soil, or even into one that is not necessarily acid. These two elements need to be in certain ratios of their respective degrees of saturation of the exchange capacity of the soil, if that soil is to grow legumes or protein-rich crops. For calcium, this may well be 75% while for magnesium it may well be near 10%. For potassium the percentage saturation of the exchange capacity of the soil occupied may be from 2 to near 5. Just what percentage the other cations, especially the trace elements, should each occupy has not yet been specifically suggested. We need, then, some gadgets to measure the activities of the calcium, the magnesium, the potassium and other nutrients. *Instead of becoming so serious about the pH of the soil, we need to become much more serious about pCa, pMg, pK,* etc., since these are activities of the nutrient ions and would help us get a picture of the dynamics by which these move toward the plant root for entrance there and nutrition of the plants.

Since the hydrogen ion, a non-nutrient, is positively charged, as are the nutrients calcium, magnesium and potassium, it is significant to consider hydrogen along with these as the combination representing almost the total exchange capacity of the soil. From this total capacity the ratios of the percentage saturation of that capacity by the nutrients (and also by hydrogen, a non-nutrient) may be calculated and adjusted by fertility treatments for most efficient plant nutrition.

The pH, then, serves to suggest the degree to which the total potential stock of nutrients in the soil has been replaced by hydrogen, a non-nutrient; but it gives no suggestion as to which nutrients are grossly deficient or to what degree the nutrients are imbalanced. It is not an indicator of what kinds or amounts of fertility are required to make the soil productuve, and has therefore been a hazard to keeping soils productive and in nutritional balance. Undue emphasis on, and attention to, the pH of the soil to the extent of disregard of the soil's fertility saturation for plant nutrition suggests itself as a case where "a little knowledge can be a dangerous thing."

10. the farmer must conserve the soil

Similar illustrations could be cited for soil treatments with fertilizers, including nitrogen, phosphorus, potassium, magnesium and other elements. Suffice it to say that as one cannot diagnose human illness and prescribe on a mass scale, so it is similarly hazardous to believe that soil illness can be handled likewise. Soil fertility conservation must come about under the interest, knowledge and effort of the owner of that soil. The soil as his patient will be healthy because of his attention to it as the nurse who follows the diagnosis of mature judgment in wise soil management, rather than suggestions from simple gadgetry.

In my humble opinion, this Committee may well consider permitting the decisions as to the operation of the program of agricultural conservation assistance to be made by the farmer without mandatory restrictions which have appeared in the 1954 Agricultural Conservation Program for the first time since this program began in 1936.

9.

Release and adsorption, availability and taking into storage—these things commanded much of Dr. Albrecht's attention when he wrote so that all could understand. Albrecht was at his best when he dealt with the convention that "availability" of plant nutrients is synonomous with water solubility. This short item appeared in the Journal of the Soil Association, London, 1955, as an abstract of an earlier version in Natural Food and Farming.

More Thoughts on "Availability"

Because we turned away from much of the art of agriculture, we have serious confusion. Let us comprehend the fact that soil fertility, properly coupled with plant nutrition, is a form of creation—a form of outdoor biology—and is not a matter merely of scientific technology. We seem to have lost sight of the fact that the creative business of agriculture has always started in the soil.

In terms of wise fertilizer use, the most shocking confusion prevails when we talk about soluble fertilizers—considering water as the agency for solution—and then we make laws requiring that fertilizers must be water-soluble and thereby so-called "available."

In fact and in nature, these soluble fertilizers are never taken out of the soil, because the plant takes them into itself along with the water that it takes from the soil. The use of the major amount of water by the plant is that keeping the respiring leaf tissues moist for the exchange of gases—carbon dioxide and oxygen. That escape of water from the leaf is called "transpiration."

The transpiration stream of water from the soil, through the plant, and into the atmosphere, is independent of the nutrient stream from the soil into the roots. The water uptake by the roots is the result of atmospheric conditions favoring evaporation from the leaves, with a set of dynamics which are more than a match against the forces holding the water on the surfaces within the soil.

Nutrient intake by crops is a function of three colloids, or possibly four, in contact. First of all, there are the nutrients on the clay colloid, or on the organic colloid of the soil. The soil colloid is in contact with the root membrane, which is another colloid. That root membrane is in contact with the contents of a cell on the inside—namely, the protoplasm (or cytoplasm). In turn, that cell is in contact with another cell. In that you have the combination of the three or four colloids in contact. The movement of the nutrient ions from the clay into the root membrane and into the cells follows the chemical laws controlling their traverse there because of the differences in activities, adsorption capacities, interfering ions and other factors along that line.

The idea that the "drought" is responsible for the failure of plant nutrition still persists. But what is commonly called "drought" isn't trouble in terms of water only. It is apt to be due to the fact that the upper layer of the soil—where the fertility is—dries, and the roots must go down through a tight clay layer, which has almost no fertility. Then, because of crop failure in the absence of plant nutrition in that soil layer of stored water, we try to blame the drought or the bad weather. During drought we don't use water to the best of our ability. We neglect to remind ourselves that the plant is about 95% air, water and sunshine—and only about 5% fertility. We are too indifferent to that fact to consider carefully how we can use that 5% as the requirement to produce the other 95% of plant growth.

10.

Why do pine needles decay so slowly? How do calcium and phosphorus fit into the microbial diet? Why is degree of acidity, like temperature, a condition, not a cause? Here, Dr. Albrecht both answers the questions and explains the answers.

Calcium emerges as the prince of nutrients, and this paper must rate attention as one of Albrecht's finest presentations. Calcium *was first published in* The Land, *December 1943. At that time,* The Land *was the publication for Friends of the Land, a sophisticated group of far-seeing Americans whose laureate was Louis Bromfield. For many years* Calcium *was kept in circulation by the Lee Foundation for Nutritional Research. In the early 1960s many of the Lee Foundation supply of reprints were restricted under Food and Drug Administration supervision as a condition for a "truce" between the regulatory agency and the organization using products from a free press for consumer education.*

Calcium

Calcium is at the head of the list of the strictly soil-borne elements required in the nourishment of life. It is demanded by animal and human bodies in larger percentages of total diet than any other element. Its own properties, as for example its relative solubility in some forms; its pronounced insolubility in others; its ease of displacement from rock and soils by many elements less essential, and the multitudinous compounds it forms; all make it the mobile one of the earth's nutrient ions.

These properties are responsible for its threatening absence from our surface soils that are bathed in the pure water of rainfall, and for its presence in the water at greater soil depths in the distressing amounts that make it appear as stone in the tea kettle or as post-bathing rings in the bathtub.

These same properties, that seemingly impose shortages and hardships have given cubic miles upon cubic miles of limestone in geologic sea deposits to be later uplifted as land areas widely distributed in close proximity to the soils now suffering shortages of calcium needed for plant and animal nutrition.

While calcium is moving by aqueous aid in this cycle from the surface soil of our land to the sea and back to our soils again, this very nomadic habit makes possible its services in nourishing life. Like other natural performances, it does work while running down hill. If maximum benefits to life are to accrue while this natural cycle continues, we must understand it and help to fit life into it.

An understanding of calcium and its role in the nutrition of life is the start in getting acquainted with the first on the list of all the soil-given elements. Its behavior bids fair to be profitably elucidated through the help of our observations of animals, animal assay methods, and other bio-chemical behaviors. When all the other soil-given elements are similarly studied, they will no longer remain as micronutrients beyond our general understanding as they must when the research light has no more candle-power than that of simple chemical analysis. Calcium may well be the "test case" or "pilot plant" experience to guide our thinking and understanding of all the other nutrient elements of the soil for nutrition of microbes of plants, and of animals.

1. the science of analysis

Chemistry has long been the science of analysis. Nature has presented herself as something to be examined, to be taken apart, and to have its parts measured, named, and classified. Functional significance of each part was assigned as fast as experimental procedure could study each as a single variable while all others were held constant. Only recently has chemistry become the science of synthesis. Its synthetic efforts are now giving us dyes in manifold colors and fibers of rayon and nylon for fabrics that fairly rival the rainbow itself. Nutritional minerals and medicinal compounds as complex as the vitamins themselves are now products of the chemist's skill.

Nevertheless, nutritional studies still move forward mainly on the pattern of analytical procedures. Many are the parts and the factors in nutrition that remain unknown. We are still wondering how many golden eggs can be laid by that great goose known as nature. The list of carbohydrates, proteins, fats, minerals, and vitamins has had increasing numbers of compounds within each of these to be given particular emphasis. A list of a dozen or more minerals coming from the soil has given each importance far beyond the magnitude each of them represents as a percentage of the body composition or of our daily diet. The vitamins of recent recognition as essentials on the dietary list have already increased in number until a total of about 50 is certainly going to drive many people to the drugstore. Three specific fatty acids are now listed, and some 30 amino acids must be ingested if nutrition is to be without some health troubles.

Synthesis has not yet been much used as a technique to help in our understanding of biological behaviors. We have not yet formulated the ideal toward which we are striving because normal bodies and good or perfect health are yet widely unattained. The analytical procedures and single-element controls are still in vogue, unsatisfactory though they may be. The isolation of one essential compound and the demonstration of its essentiality by abnormalities its absence invokes, is still the main procedure in nutritional studies. Plant physiology, likewise, demonstrates the plant troubles when, for example, the calcium supply is varied, or when phosphorus is reduced, or either is completely withheld. All the separate items on the essential list have had their individual effects demonstrated, and we are mapping the world in terms of their individual absence. Little has been done, however, to vary two or three elements at the same time. The number of combinations would run the experimental trials into legion, and consequently such experiments have not yet been undertaken extensively.

But such multiple variations are the situations in nature where all the soil-given nutrients, for example, may be varying during the life or growth cycle of a single organism. It is impossible, therefore, in natural performances to segregate the effects of separate elements. They can be evaluated only as a summation in terms of the final plant or animal. It is for this reason that we must resort to the bioassay method. It becomes necessary to use the animals themselves to obtain more gross results of value in terms of our own life before all of the intricate individual processes can be learned and life itself synthesized thereby.

Nutritional thinking, however, is moving forward rapidly. It is not limited to compounds like the carbohydrates or proteins and the chemical reactions they undergo. It is giving detailed attention to the catalysts that speed these reactions, if vitamins can be considered in this category. Body catalysts for improved mineral management, like thyroxin for example from the thyroid gland and the activities of the parathyroid in control of the calcium and phosphorus in storage and in circulation, they are bringing into the limelight the importance of supplies of these elements as well as their catalysts.

Calcium behavior in nutrition is no exception to this concept of complex interrelation when its supplies in the bones, in the blood stream, and in the alimentary tract may be moved through this series in either direction according to certain relations of its amounts to the supplies of the catalyst, vitamin D, exercising control. Then when there are a dozen soil-given elements, each with its variable supplies and possible catalyzed behaviors to be synchronized, the possibilities for shortages or deficiencies multiply themselves quickly. Attention to calcium can only be in terms of its deficiencies as gross manifestations, when all of its many functions are not yet catalogued.

The soil is formed from the rocks and the minerals by the climatic forces of rainfall and temperature. The presence or absence of calcium in the soil has long been the soil scientist's index of the degree to which the soil has been developed or to which the rocks have moved towards solution. As rocks are broken down to form soil by increasing but not large amounts of rainfall, there is an increase in the soil's content of active calcium. Then as the larger amounts of rainfall go higher and temperature increases also there is calcium depletion. Life forms, whether of the lower, like the microbes, or of the higher, like plants and animals, all are part of this calcium picture. The distribution of the different plants and of the different animals as well as their densities of population take their ecological pattern very much according to the calcium supply. The United States divide themselves readily into the east and the west, according to the lime content of the soil. The dividing line across central United States puts lime-rich soils to the west and the calcium-deficient soils to the east. This is also according to the lesser amounts of rainfall to the west and to the higher rainfalls and temperatures to the east, these differences having been so related as to weather the soils just enough to leave those in the west with calcium, and to carry the weathering to the point of the removal of the calcium in the east. Higher temperatures and rainfall as in the southeast, not only remove the calcium but change the clay complex so that it has little holding or exchanging capacity for any of the soil mineral elements.

In these facts there is the basic reason for calcium deficiency and many other deficiencies in the humid tropics. Here is the basic reason for the confinement of the population of the wet tropics mainly to the seashores where fish return the flow of soil fertility in part from the sea back to the land. Such facts account for the sparsity of population in the humid tropics and yet we marvel at the tremendous vegetative growth of jungle densities. We forget that its contribution for human use is mainly wood or fruits, which if not actually poisonous have little food value and at best only drug value as the coffee, the cinchona, and the alkaloids. It is this larger soil picture with its highlights of calcium presence and its shadows of calcium absence that makes the pattern to which all life, whether microbe, plant, animal or man must conform.

Microbes as the agencies of decay testify to the level of the nutritional conditions by their rates of destruction of the debris which they rot or on which they feed. Pine needles decay slowly because they are grown on a calcium-deficient soil and are consequently deficient in this element essential in the diet for microbes, and deficient in all the nutritive values associated with calcium in plants. Timothy hay and timothy sod decay slowly. Clover hay and clover sod rot quickly. "Clover and alfalfa hays," says the farmer, "are hard to make because they spoil so quickly." This is merely saying that such hays, as products of soils rich in calcium and therefore themselves rich in this element, allow the bacteria to multiply faster, or nourish themselves better. Consequently, they consume clover and alfalfa hays more rapidly than they consume timothy or pine needles. Cattle choices agree with the microbial choices.

Rapid decay of certain substances points to these as balanced diets for microbes and is an index of chemical composition and nutritional value for higher life forms that we too often fail to appreciate. We have been thinking of the disappearance of the debris as it rots and have not been measuring the growth of the microbial crop. Microbes, as a kind of guinea pig, for quick evaluation of the dietary contributions of the substances on which they feed, offer a neglected scientific technique for judging much that might be considered human food. Insects can serve likewise. If neither microbes nor bugs care for certain substances should these be considered as of food value for higher life forms? Whole wheat flour "gets buggy" so much more readily than white flour and by just that much is it a more wholesome food.

Calcium for microbes in the soil's service as a plant food factory has only recently become appreciated. Legumes cooperate with nodule bacteria for the appropriation of nitrogen from the air in many soils only when calcium is supplied as lime. Not only the plant, but the legume microbe, too, makes high demands for calcium. The microbe separated from the plant must be given liberal supplies of calcium if this cooperative struggle for nitrogen or protein is to be successful.

3. microbial decay process

Microbial decay processes within the soil by which nitrogen as ammonia is converted into nitrate also depends on the calcium supplied. Unless the clay of the soil, for example, has calcium present in liberal amounts, this conversion of nitrogen does not proceed rapidly. The function of calcium, as it makes the phosphorus of the soil more effective, was suggested by microbial behaviors. With calcium and phosphorus absorbed on a clay medium, the growth of certain microbes made the medium acid while other, but closely similar, microbes made it alkaline. This difference occurred because both calcium and phosphorus are brought off the clay and into solution with the result that intermittently one or the other of these dominates over the other; phosphorus dominating to make the medium acid, calcium domination to make it alkaline. Microbes apparently separate both calcium and phosphorus at the same time from the absorption forces of the clay but consume one or the other differently after this separation to bring about the acidity or the alkalinity. Here are calcium and phosphorus in the microbial diet, and here they are closely associated in their nutritional services just as they are found associated in plants, and just as they function together in animals mainly as the compound of the different calcium phosphates.

Microbial nutrition suggests itself as indicator of soil fertility and therefore of plant and animal nutrition. Microbes, as they grow rapidly and rot vegetation quickly, or conversely as they grow slowly and rot it slowly, indicate the soil nutrient supply by revealing the composition of the products grown by that soil. Pine needles decay slowly. Oak leaves decay slowly, but elm, linden and other soft wood leaves decay rapidly. The Swiss farmer selects leaves from the portion of the forest with the soft wood trees for bedding litter for his cows and goats because these leaves rot more completely in the manure than oak leaves do. The service of the leaves in decay when mixed with animal excrement and in the return of their nutrients to nourish the grass are judged by the Swiss farmer through this microbial indicator. The rate of decay can be taken as a universal indicator of the nutrient balance for microbes and therefore as balance for higher life forms.

The organic matter produced on a soil and going back into the soil reflects by its rate of decay the plant composition and therefore the soil fertility producing it. The size of the microbial crop as reflected by its activity and like any other vegetation is determined by the nutrients being mobilized in the soil. If calcium is deficient there, then the organic matter grows a less proteinaceous composition or is mainly of carbonaceous content. Such vegetation is a poor microbial diet. It reflects this fact when it accumulates or remains for a longer time while the proteinaceous, or more calcium-rich decays more rapidly.

The microbes, as lower plant forms, give us the ecological pattern of higher plants serving to nourish higher animal forms. They point out, in general, that the vegetation produced on soils amply supplied with calcium is mineral-rich and proteinaceous to serve the microbes well in their nutrition. On the soils deficient in calcium, the vegetation is carbonaceous, protein-deficient, mineral-deficient, and lacking in many organic and mineral complexes requisite not only for microbes but for the higher life forms as well.

Microbes give us this larger ecological picture in agreement with the soil map of the United States. Prairie soils or calcareous soils, with their proteinaceous and mineral-rich vegetation are in the west, and forest soils and carbonaceous vegetation are in the east. Calcium is the index factor associated with these differences. As a very helpful factor, it needs to be given recognition and attention in connection with the larger picture of crops and foods of correspondingly variable nutritional values produced on these different soils.

4. delayed appreciation

The delayed appreciation of the significance of calcium in plant nutrition may be laid at the doorstep of a confused thinking about liming and soil acidity. The absence of lime in many soils of the non-temperate zone has long been known. Lime in different forms such as chalk, marl, gypsum or land plaster, has been a soil treatment for centuries. Lime was used in Rome in times B.C., and the Romans used it in England in the first century A.D. Chalking the land is an old practice in the British Isles. The calcareous deposits like "The White Cliffs of Dover" were appreciated in soil improvement for centuries before they were commemorated in song. Liming the soil is a very ancient art, but a very recent science of agriculture. It was when Leibig, Lawes and Gilbert and other scientists began to focus attention on the soil as source of chemical elements for plant nutrition that nitrogen, soluble phosphate, and potassium became our first fertilizers. It was then that the element calcium and the practice of liming were put into the background. Unfortunately for the wider appreciation of calcium, this element in the form of gypsum was regularly a large part of the acid phosphate that was applied extensively in fertilizer to deliver phosphorus. Strange as it may seem, superphosphate fertilizer carries more calcium than it does phosphorus, and consequently calcium has been used so anonymously or incidentally that its services have not been appreciated. Fertilizers have held our thought. Calcium was an unnoticed concomitant. It has been doing much for which the other parts of the fertilizers were getting credit. Appreciation of the true significance of calcium in plant nutrition was therefore long delayed.

More recently soil acidity has held attention. This again has kept calcium out of the picture. Credit for the service of liming has been going to the carbonates with which calcium is associated in limestone. It was a case of the common fallacy in reasoning, namely the ascribing of causal significance to contemporaneous behaviors. Here is the line of reasoning: "Limestone put on the soil lessens the acidity, and limestone put on the soil grows clover. Therefore the change in acidity must be the cause of the growing clover." At the same time, there was disregarded the other possible deduction, namely: "Limestone put on the soil applies the plant nutrient calcium. Therefore the applied calcium must be the cause of the growth of clover."

The labeling of calcium as fertilizer element of first importance was delayed because scientists, like other boys, enjoyed playing with their toys. The advent of electrical instruments for measuring the hydrogen ion concentration gave tools and inducement to measure soil acidity everywhere. The pH values were determined on slight provocations and causal significance widely ascribed to them, when as a matter of fact the degree of acidity like temperature is a condition and not a cause of many soil chemical reactions. Because this blind alley of soil acidity was accepted as a thoroughfare so long and because no simple instrument for measuring calcium ionization was available, it has taken extensive plant studies to demonstrate the hidden calcium hungers in plants responsible in turn for hidden but more extensive hungers in animals. Fortunately, a truce has recently been declared in the fight on soil acidity. What was once considered a malady is now considered a beneficial condition of the soil. Instead of a bane, soil acidity is a blessing in that many plant nutrients applied to such soil are made more serviceable by its presence, and soil acidity is an index of how seriously our attention must go to the declining soil fertility.

Now we face new concepts of the mechanisms of plant nutrition. By means of studies using only the colloidal, or finer, clay fraction of the soil, it was learned that this soil portion is really an acid. It is also highly buffered or takes on hydrogen, calcium, magnesium, and any other cations in relatively large quantities to put them out of solution and out of extensive ionic activities. It demonstrated that because of its insolubility, it can hide away many plant nutrients so that pure water will not remove them, yet salt solutions will exchange with them. This absorption and exchange activity of clay is the basic principle that serves in plant nutrition. This concept comes as a by-product of the studies of calcium in relation to soil acidity.

Imagine that a soil consists of some calcium-bearing minerals of silt size mixed with acid clays. The calcium-bearing mineral interacts with the hydrogen of the acid clay. The hydrogen goes to the mineral in exchange for the calcium going to the clay. Imagine further that the plant root enters into this clay and mineral mixture. It does so more readily because of the presence of the clay. It excretes carbon dioxide (possibly other compounds) into this moist mixture to give carbonic acid with its ionized hydrogen to carry on between the root acid and the clay particle and the mineral. The hydrogen from the root exchanges with the calcium absorbed on the clay in close contact.

Thus plant nutrition is a trading business between root and mineral with the clay serving as the jobber, or the "go-between." The clay takes the hydrogen offered by the root, trades it to the silt minerals for the calcium and then passes the calcium to the root. Thus nutrients, like calcium, and other positive ions as well, pass from the minerals to the clay and to the root, while hydrogen or acidity, is passing in the opposite direction to weather out of the soil its nutrients mineral reserve and leave finally the acid clay mixed with unweatherable quartz sand. Acid soils are, then, merely the indication of nutrient depletion.

5. moving other nutrients into the plant

Calcium plays more than the role of moving only itself into the plant. This element is serving, apparently, in the mechanism of moving other nutrients into the plant (and possibly excluding some non-nutrients). Careful studies of plants growing on colloidal clay have shown that no growth is possible unless calcium is moved into the crop. As the supply of calcium becomes lower, the crop may be growing but losing back to the soil some of the potassium, some of the nitrogen, some of the phosphorus, and even some of the magnesium planted in the seed while taking none of these nutrients from the soil. Unless the calcium is serving its function in the plant, the crop may be growing and contain in both the top and the root less nitrogen, potassium, and phosphorus than was in the planted seed.

Here is an unappreciated service by calcium. Calcium is associated with more delivery of phosphorus to the crop when a phosphorus application on limed land, for example, puts three times as much of the phosphorus into the crop as when phosphate fertilizer is put on without lime. It is associated with the more effective movement of nitrogen from the soil into the crop. It plays some role, possibly in the mechanism of the root membrane through which phosphorus goes more efficiently perhaps as a calcium phosphate than as any other form. Nitrogen may go through more effectively as a nitrate. Here are some services by calcium of which the details of mechanism will be fully elucidated only by future researches.

Calcium plays what might be termed the *leadership* role amongst the nutrient ions not only as to their entrance into the plants but also as to their combination into the proteinaceous compounds around which cell multiplication and life itself center. As the protein concentration of forages rises, there is also an increase in the calcium concentration. Also there is accumulation of evidence that with the increase in protein there goes an increase in vitamins. Legumes, the more nutritious of the forages, have long been known for their demand for calcium and high content of protein. They are also high in other minerals, so that calcium in the plants seems to synthesize the soil-borne nutrients into the organic combinations though it does not itself appear as part of the final products. Potassium, quite unlike calcium, is more directly effective in the compounding of air and water into carbohydrates, and like calcium does not itself appear in them. Potassium is effective in making bulk, or tonnage, of forage. Calcium is effective in bringing higher concentration of proteins, and other nutrients essential, within that bulk. According as the active calcium dominates the supplies of nutrients in the soil, so proteinaceousness, and with it a high content of growth minerals, characterize the vegetation produced on the soil. As potassium dominates, there is plenty of plant bulk but its composition is highly carbonaceous or it is dominantly woody.

Here is a general principle that is helpful in understanding the ecological array of vegetation. According to it, the vegetation is highly proteinaceous and mineral-rich on our prairies in the soil regions of lower rainfall or those soils retaining a high mineral content with calcium prominent. Contrariwise, vegetation is mainly wood, or like the forest, on the more leached soils with low mineral content but with potassium naturally dominant.

This ecological picture served as a stimulus for some soil studies of the chemical activities of the potassium and the calcium when present on the clay in different ratios. Professor C. E. Marshall of the University of Missouri has designed electrodes and membranes for measuring the ionic activity of calcium and potassium of the soil in the same way as hydrogen ion activities are measured. His data of what might be called pK, and pCa in the same manner as we speak of pH, demonstrate clearly that the ionic activities in a mixture of elements are not independent of each other as is true in mixtures of gases. Rather they are complimentary in some combinations, or opposing in others. Considering calcium and potassium in combination, the latter gains ascendency in relative activities as the ratios between the calcium and the potassium become narrower. Thus as calcium is more nearly weathered out of the soil, potassium becomes relatively more active in moving into the plant. Here is the physico-chemical soil situation that provokes the protein-carbohydrate relation which in turn represents the "grow" foods versus "go" food situation so prominently basic in our hidden hungers and the disturbed animal nutrition.

The soil as it is nourishment to make one kind of plant or another kind, that is, whether the plants are truly nourishment for animals or are only so much internal packing material, is the real basis for and real help in understanding the animal distribution whether wild or domestic, whether lower or higher. Cattle growing with ease and success is common in Texas, but meets increasing difficulties as one goes eastward. Donkeys are "sweethearts of the desert" where their sure-footedness—and sturdy but fine bones—among rock soils may well be associated with highly calcareous feeds grown in more arid regions. Crossbred with the horse, the resulting hybrid, or mule, is at home farther east in higher rainfalls and on more leached soil areas. But even then, he is found most commonly on the limestone soils of Missouri, Arkansas, Kentucky and Tennessee. Grown to maturity in these areas and then shipped to the cotton south, the mule survives to render labor because with no hope for posterity his calcium supply transported within him in his bones is not depleted by reproductive demands. Picture further the sheep and goat according to their concentration on different geographic areas, and with them to go the increase of so-called "troubles and bad luck" in raising them as they are in the humid, more acid soil regions. The soil fertility, so prominent as the foundation of animal reproductive performances, has not been appreciated. We need to see our most nutritious foods in those animal products connected with the reproduction of the animal, namely eggs and milk. Reproduction goes forward only on a plane of liberal supplies of soil fertility and is therefore the safety factor in our living and can be a safety factor in our thinking.

This picture of animal ecology and its nutritional reasons based on the soil does not present itself without calcium playing a prominent part in the causal forces. Soil treatments that supply calcium in the humid regions are readily detected by the animal if it is given an opportunity to manifest choice of grazing the herbages on differently treated areas. Domestication has dulled the instinct of wise food choice in some animals, as for example, greediness of the high-producing milk cow brings bloat on herself. This is less common among cows not so highly domesticated toward intensive milk production. Nevertheless, there is still enough appetite instinct left in the cow when she refuses to eat the grass spot where urine was dropped to bring about unbalanced plant composition by an excessive nitrogen application. Her refusal to eat sweet clover, and her preference for bluegrass over white clover, are all evident that the foster mother of the human race is her own nutritionist and knows her carbohydrate-protein ratios for a well-balanced diet. She is demonstrating daily her appreciation of the whole series of effects and causes in variable plant composition as they go back through vegetation to the soil to demonstrate relations between calcium and potassium, and other nutrient ions of the soil. We need to observe our animals and learn from them. When we accuse the mule of stubbornness in his refusal to eat or drink, the reflection may not be on this dumb beast as much as on his master, too stubborn to learn from nature.

6. biological assay

By means of the biological assay, our animals are telling us that they can be fed more efficiently by wise fertilizer treatments of the soil. Soil treatment is not merely a case of the plants' service as mineral haulers to carry calcium from the field to the animal feedbox, but rather, lime, for example, is applied to make the plant factory much more efficient in gathering its various nutrients from the soil and still more efficient in synthesizing these with carbon, hydrogen, and oxygen from air and water into the extensive list of complexes and compounds whose service to growth and good health we are slowly unravelling. Soil management is more than a practice guided by economics; it is a responsibility of nourishing microbes, plants, animals and humans to their best growth and health.

Animal growth studies testify to the importance of liming for its help in better animal nutrition. Sheep studies and rabbit feeding trials using Missouri soils with various forages point clearly to the more efficient conversion of roughages into meat when lime was used as a soil treatment. Increases as high as 50% in animal growth from the same amoung of feed consumed are efficiencies that surely cannot be disregarded when food is to win the war and protein is the particular food deficiency. In terms of production with reduced labor, the better feeding of animals by means of soil treatments surely must not be neglected when the same acre with a constant labor input can deliver so much more as food products in the form of meat.

Improved products from animals come in for greater efficiency also by way of lime and fertilizer used on the soil. The wool of the sheep was improved when fed hay grown on soil given lime and phosphate over that grown where only phosphate was used. Fatty secretions were visibly different but quality differences in the fiber were revealed on scouring the wool. Milk, another animal secretion, so commonly considered of constant nutrient value has permitted rickets in the calf, irrespective of ample green feed and ample sunshine for both mother and calf on soils deficient in lime.

Animals and their products have been a safety factor in man's diet in that animals are additional helps in collecting from wider range all possible helps toward the food man needs. Historic man's survivals have possibly been more largely the result of his herds and flocks than we are wont to believe. But even with the help of animals the soils may still be so deficient that animal products fail to give the full service commonly credited to them. As we push meat and animal products out of their more common place in our diet, and as we go more nearly to strictly vegetarianism, the attention to the soil is all the more important. Man dare scarcely circumvent the animal in the biotic pyramid suggested by Aldo Leopold that includes soil, microbes, plants, animals and man in that order from the bottom upward. He may claim vegetarianism, in that he survives without consuming meat, but seldom does he exlude milk, cheese and fish completely and become strictly vegetarian. It is true that nations, highly vegetarian like the Chinese and Japanese, consume mainly rice. It is granted that they are highly vegetarian but not to the exclusion of the minimum of some chicken broth or some fish. The fish required for survival may be only the head of this animal as reported by J. B. Powell, the journalist of Japanese prison experience, whose refusal to eat even this amount of animal products, cost him his feet, lost by gangrene, when the companion prisoners as fish head eaters were not so unfortunate.

Current attention to the calcium and other fertility elements in the soil promises better nutrition and health. Although proper nutritional requirements are still very much the result of chemico-analytical thinking, we are still discovering more essential parts in the proper diet. We are not yet strictly synthesizing purely chemical diets. A significant share of a good diet still consists of the so-called "natural" food so that nature is still, for many of us, the best dietician when we prefer to stake our future health on omnivorousness and plenty of natural foods from fertile soils. Synthetic diets will meet the supreme test, not when they are merely able to make animals grow or to keep them alive, but rather only when they can carry animals through their regular reproductive cycles for several generations with numerous and healthy offspring. The natural growth processes, initiated and guided in the main by the chemical nutrients coming from the soil, are still the main basis of nutrition. Calcium as the foremost element on the list of the nutrients demanded from the soil has given us a pattern of the importance of its role in nutrition. It points with suggestions of importance to the other nine or more elements whose nutritional significance we do not yet understand as well. Observations and scientific studies under the present appreciation of our soil as the basis of health will soon increase our knowledge of nutrition when we recognize the larger principles as they apply to all the life forms including microbes, plants and animals no less than man. We shall rapidly come to recognize that our national health lies in our soil and our future security in soil conservation guided by nutrition's universal laws.

11.

"We are not yet strictly synthesizing purely chemical diets," warned Dr. Albrecht in the paper styled Calcium, circa 1943. Albrecht had no way of knowing that Nazi Germany was, indeed, attempting the feat nevertheless. I. G. Farben officials fed prisoners of Germany on 2,000 calories of totally synthetic meals during World War II—synthetic vitamins, synthetic fats, synthetic oils, etc. Prisoners lost weight and were unable to function in a few months. Finally Himmler took on the color of a humanitarian to save the prisoner people from the "knowledge" of the scientists, who were covering their errors with unspeakable medical tests. When told of the Nuremburg documentation of this report, Albrecht reaffirmed a conclusion Andre Voisin stated in Soil, Grass and Cancer *thus:*

•

what one knows about nature

what one does not know about nature

At that time Albrecht read aloud this quotation from Schopenhauer to the editor of Acres U.S.A.

"The value of what one knows is doubled if one confesses to not knowing what one does not know. What one knows is then raised beyond the suspicion to which it is exposed when one claims to know what one does not know."

"Include this in your collection when you get to it," Albrecht said, extending Calcium Membranes in Plants, Animals and Man. *This paper appeared in* The Journal of Applied Nutrition *in 1968, and represents the kind of refinement his last years accounted for. It is included after* Calcium *because this is where it belongs in terms of thought, if not chronology.*

Calcium Membranes in Plants, Animals and Man

When Professor Curtis F. Marbut made his survey and classification—published with maps after 35 years of work—of the soils of the United States, his first and foremost, visible criterion of soil productivity and general differentiation was the variable concentration of the nutrient element, calcium, in the horizons of the soil profile. Accordingly, those differences in profiles divided the arid, saline and alkaline soils of our west from the humid, acid, lime- or calcium-deficient soils of our east, roughly, by an almost longitudinal line (97 degrees west) running from the northwestern corner of Minnesota to the southernmost tip of Texas.

That line bisected closely, the highly productive prairie or grassland soil area, distinguishing our mid-continent's agricultural productivity. It also separated out that area of healthier teeth and more acceptable draftees for the Navy and the Army, as well as calling attention to agricultural production of our short-lived, quickly-fattened pig crops of the eastern mid-continent in contrast to the highly proteinaceous longer-lived, naturally grown beef crops from the western prairies and plains.

Marbut's report did not bring the soil-borne, nutrient element, calcium, into its later recognized, physiological importance, particularly in controlling the living membranes in their separatory, bio-chemical performances, like those of plant root-hairs managing the ingo from the soil of the balanced plant-nutrition: the controlled nerve environment sometimes upset in tetany: and the intestinal and blood-vesseled walls as they are disturbed by excessive magnesium for purges, or in case of temporary albuminuria, and possibly other absorptive and excretionary biotic functions.

1. our mineral nutrients come via the soil—plant—animal—man route

The biotic pyramid with man at its apex, and other vertebrates next below him, is big reason why those bodies demand higher concentrations of calcium than of any other element. Calcium, as the major soil-borne one, must be brought through uptake by the plant roots' exchange for it of the positively-charged hydrogen, which is made cationically active as an ion in the soil moisture by the carbon dioxide excreted as the root's respiratory waste. That acid hydrogen moves from the root readily to exchange for calcium (and similar other cations) held in adsorbed, or available, form on the colloidal clay-humus complex. More slowly, the adsorbed hydrogen remaining to give the fertility exhaustion of acid-clays of soils by cropping or leaching, breaks down limestone particles to mobilize mineral calcium (similarly also magnesium and potassium reserves) out of rock fragments, to initiate the transport of nutrient, cationic elements from the soil (S) to the plant (P) to the animal (A) and finally to man (M) by the S-P-A-M route, as the soil's creative process via climatic forces in particular geoclimatic settings.

Considering not only man's nutritional support, that upward distance toward higher biological complexity represents a serious hazard. There is the danger of calcium deficiency inviting the ailments of rickets, the cow's parturitional irregularities of milk-fever, acetonemia, or the animal's sudden nerve break-down by tetany, and the many physiological upsets of humans, like bursitis, arthritis and other long-baffling, functional failures associated with erratic calcium metabolism.

While the plowed acre of surface soil (weighing two million pounds) might contain nearly 70,000 pounds of calcium, the average soil is liberally supplied plant nutrition-wise if its adsorbed, exchangeable, hence available, supply for the plant roots amounts to but one-tenth of that, say 6,000 pounds per plowed acre layer. From that surface horizon, the vegetation takes, as a very general mean, about 12-15 pounds per ton of its own dry matter (less than one pound per hundred), when mineral-rich and protein-rich legumes, like red clover and alfalfa, require the higher figure of 40 pounds per ton (2 pounds per hundred). Out of the vegetation (assuming vegetarianism) the healthy, adult human body (weight 70 kgm.) must have 2.6 pounds of calcium (1.6 pounds per 100) come along the S-P-A-M route to realize that when much of the humid soil area is weathered into the so-called "acid" condition, the dangerous shortage of calcium is to be expected.

Also, in even the arid soils, which are neutral, saline or alkaline, their salinity, due to excesses of magnesium, potassium and sodium in adsorbed and soluble forms brings on an imbalance within this quadruple cationic array of the last three against the one of calcium. These three separate excessive saturations, when normally magnesium is 7-15%, potassium is but 2-5%, and sodium is 0.5-2.5%, disrupt calcium movement into plants when its normal saturation of the colloid capacity must be up to 60%, and more than the sum total of such for the three disturbing cations. Accordingly, on arid soils, now coming to be required for production by the pressure of increasing population, calcium has recently been demonstrated as taken more generously increasing by even non-legumes when it is applied in contact with the seed. This brings better crops in the face of what appears to be excess pounds of calcium already in the soil horizons. This treatment by intimate seed-contact has increased the yields of both non-legumes and legumes to say nothing of improving simultaneously the nutritional quality of both kinds of crops.

2. plants require calcium early—and hold it firmly—in their roots

Ash analyses aplenty have demonstrated that high-protein crops require relatively higher concentrations of calcium in their plant tissues. This is true, especially of the legumes. The microbes, giving root nodules of legumes their power of using, or fixing, atmospheric nitrogen, also require high calcium, whether cultured in cohabitation with plants, or separated therefrom and grown on specially compounded laboratory media for use as inoculation of seeds on new soils requiring the introduction of the nodule-producing Rhizobium species.

Both non-legumes and the legumes respond, by their extra growth, to calcium availability from the soil early in their life. That this is true for non-legumes can be demonstrated easily by dusting ordinary beet seeds—normally seed clusters—not single seeds, or field planting of sugar beets with calcium as gypsum or finely pulverized calcium carbonate (agricultural limestone) in contrast to similar seeding rates without this fertility treatment, to find the increased stand counts resulting from increased germination where calcium was offered at the outset.

Starting soybean seeds as one set in a mixture of quartz sand plus limestone and as another set in quartz sand only for a period of ten days and then after the ten-day period of growth transplanting each set separately to a constant type of soil of relatively low degree of calcium saturation (less than 60% of total exchange capacity), it is significant to note that (1) on transplanting, the seedlings carried 17.07 milligrams per 100 seedlings from no calcium contact during their starting ten days of growth, but carried 30.14 milligrams of calcium (an increase of 76% in the ten days of calcium contact); (2) their nodule production ranged from 3.4 to 15.0 without calcium as starter but from 36.6 to 38.9 when given the calcium early; (3) their weights were alike, 4.77 grams from both early treatments but their heights were 7.0 and 12.0 centimeters with the latter's contact with calcium. Sampling at every ten days of growth after transplanting revealed (a) taller plants by as much as 30%, (b) heavier plants by 25% by 20 days; and more total nitrogen in the plants from 8.5 to 30.7% for the five sampling intervals over a total of 50 days.

Evidently, the favorable growth effects by the calcium uptake by the plants during but the first ten days of their growth period were not prohibited from their continued expression of better plants by even the soil of low calcium saturation into which the 10-day old seedlings were transplanted.

That simple fact of the plant roots holding firmly their earlier calcium contents when growing later in soils of low calcium offerings was again demonstrated more fully by use of the isolated, highly purified colloidal clay mixed with quartz sand as growth medium for soybeans at controlled pH values and specific totals of exchangeable calcium. These plants demonstrated no loss of seed-borne calcium back to the soil. Simultaneously other cases of both anions and some cations moved from the plants in that reverse direction from plant back to soil, to increase the clay's pH values. But the acid colloidal clays of very low pH values removed potassium seriously from the plants back to the clay along with other substances to give bean vegetation of significant forage yields, but that with less of nitrogen, and of phosphorus, as well as of potassium in the total crop (tops and roots) for only these three elements tested, than were initially in the planted seed.

In these facts about the ingo and the outgo of nutrients between plant roots and the soil as the balances and imbalances of the latter affect that exchange, there is seemingly the evidence that the plant roots mobilize early-taken calcium with themselves so that this element is carried forward with the extending root growth in sufficient quantities into otherwise calcium-deficient, deeper soil horizons in the profile to support normal nodulation and nitrogen fixation. Both of these had failed when no calcium was offered in connection with the seeding on such hostile soils to legume crops.

3. nature uses similar, advantageous, bio-physico-chemical principles in warmblooded bodies as well as plants

In accordance with the reported preceding natural demonstrations, dare we not envision the importance of calcium in the cell wall of the root hair (often called a "semi-permeable" membrane) as that unique bit of plant structure controls the intake and the outgo to accomplish "naturally controlled plant nutrition for healthy growth" of vegetation?

In studies of controlled plant nutrition by means of the colloidal clay technique and plants showing profuse nodulation and ample fixation of atmospheric nitrogen as criteria of healthy plant survival, there were demonstrated the shifts in equilibria by the plant's lowering of the clay's pH when that represented both high degrees of saturation by, and high total of, calcium in the soil to favor plant growth; and, vice versa, there were produced shifts by raising the clay's pH when it represented both low pH by, and a low total supply of, calcium of the clay to limit plant growth.

Such a vision of calcium's controlling role is prompted by the reported research of R. D. Preston (Great Britain) which has shown by the electron microscope that calcium is a cross-tie structural part of cellulose membranes, including the cell walls functioning probably in giving variable permeabilities for different ionic passages by which biotic materials are separated and controlled. It is, then, expectable that membranes like the cell wall of the root hair would require a clay-contact environment of cationic balance and totals, as colloidal saturations, duplicating closely what is the balanced nutrition of the plant required for its nodulation and nitrogen fixation, or for its normal protein synthesis during regular healthy survival.

For the more nearly balanced, nutrient, cationic array as percentages saturation of a soil's exchange capacity for legumes the following values have been found valuable guides in general; namely, calcium 60-75%, magnesium 7-15% (10-20%), potassium 2-5%, and sodium 0.5 to 3.0%. For the non-nutrient cation, hydrogen, serving as activator or a mobilizer of the other cations off the adsorbing clay-humus colloid because of the hydrogen's high energy of adsorption, the figure 12% has been considered. This set of figures shows immediately that under lower percentages saturation, but any very small excesses of sodium, potassium, and even magnesium become more disturbing as imbalance for those cations functioning at only higher saturation like calcium. According as the last three become excessive, they prohibit adsorptive space for hydrogen as an activator or mobilizer of the adsorbed nutrients, and likewise for calcium in balance which requires from 60 to 75% of the soil's adsorptive potential. Desert soils can then be cited as mainly a case of calcium deficiency, not only in its role as a constituent of living tissues, but also in its functions within membranes controlling movements of other nutrients or failing as such. That functional failure has suggested itself most forcefully when many tests of desert soils showed excessive saturations of magnesium, or potassium or sodium, or of all three while calcium was often shown as of insufficient degree of saturation, and of better plant growth resulted by fertilizing the crop with calcium applied by contact at seeding.

4. cationic imbalances disturb membrane functions in animals and man

Now that imbalances amongst the cations in the soil can be understood as disturbing to plant nutrition, we can appreciate the increasing emphasis on warm-blooded body disturbances, premised on insufficient calcium and its imbalance with respect to the other three cations, especially in cases when no single specific element can be specified as cause.

Tetany, long recognized as a baffling trouble of ruminant animals, particularly dairy cattle and sheep, during shift from dry feed under housing to luscious growth of young herbage, is one example. The convulsions preceding the animal's death, suggest nervous irregularities and point to disturbed magnesium-calcium interrelationships when these two soil-borne, plant nutrients exercise either contrary or similar effects on neuromuscular transmissions.

For the animal, this is also a cation imbalance of the feed's chemical composition which is seemingly responsible with suggestions that the imbalance in the soil's exchangeable cationic supply has moved itself up higher in the biotic pyramid by one or two living strata; namely, soil to plant to animal, the S-P-A route. The interrelations of such small amounts as may bring nerve excitation are still too intricate for us to separate out two factors like the amounts of calcium and magnesium as balance or imbalance in the fluid which bathes the nerves. "The increase in acetylcholine and the extended prolongations of its effects can give rise to neuromuscular upsets."

Living substances, especially proteins, do not have, nor tolerate, a high salt content or ionic contact, save as any salt may be separated from even the cell's protein by confinement within the cell's vacuole; by extra-cellular restrictions; or by movement within the excretory system enroute to elimination. Living tissues and processes are highly "buffered," by being organo-molecular and not "salty" or "ionic." It is by the "ionic" states of calcium, magnesium and similar elements that the troubles called "imbalances" come about.

In the human body, the blood is a colloidal suspension, scarcely ionic as amphoteric state, with limited adsorbing powers. Its calcium is almost un-ionized but plentiful as "serum calcium." Its iodine, similarly not significantly ionic, is considered as "blood plasma iodine, B.P.I." The "balance" of cations in the human body points toward the dangers from imbalances, no less serious than such for the plant's root hair membrane's calcium which is shifted by contact with imbalanced cationic saturations of the clay-humus colloid bathing the root.

For theoretical considerations as body illustrations there may be cited (a) the modification of the vessels of the kidney excreting high magnesium in the urine as blood flows through, and (b) the shifts in the calcium-membrane to magnesium-membrane in the intestinal wall—similarly to those in the cell walls of the plant's root hair—when magnesium sulfate (Epsom salts) has been absorbed into the bloodstream to b excreted after having changed the intestine's functions to make it purge itself by the above salt dosage of either the sulfate or hydroxide as "milk of magnesia."

Shall we not envision the excessive ionic magnesium put into the intestine, while being partly absorbed into the bloodstream but simultaneously displacing the calcium in the cellular wall or calcium-membranous structure, as making the wall the equivalent of a magnesium-membrane and disrupting the power there of controlling intake and outgo? Should we not expect, then, that membrane's allowing the larger amounts of water and other liquids and substances dumped in from the bloodstream to be the flushing agency? Then should we not expect the duration of the purge to be no longer than the time required from the bloodstream to absorb the magnesium from the cells of the intestinal wall and replace it with calcium to restore the membrane's normal balance of cations and physico-chemical control against excessive losses of liquids and other essential matters from the blood to the intestinal canal?

The use of the "Combistix" tests for (a) urinary protein, (b) glucose sugar, and (c) pH simultaneously of a diabetic after "milk of magnesia" purge, demonstrated the kidney upset when the test for proteinuria showed up as excessive and was incidently observed in connection with the test for glucose via hyperglycemia. Thus, the absorbed magnesium later thrown off after the magnesium purge, represents a cationic imbalance by excessive magnesium in the bloodstream upsetting the normal controls of excretion by the membranous walls of the kidney shown by their allowing the protein overflow, or the albumenuria, as the result.

5. *summary*

In our increased concern about human nutrition, not only for more mouths but also for healthy bodies, we need to think along the S-P-A-M route of nature's design for creation, or "from the ground up" first, and then for economic advantages second. When we dose our soils so heavily with salts or highly concentrated, incomplete fertilizers, should we not be reminded that for a normal "living soil," as nature had it before man's soil management took over, we are too destructive now for those soils to last economically even beyond a second generation. Our fertilizers compounded for advantages of cheaper freight costs and lowered labor loads in their use have lost sight of the requisites of balanced plant nutrition via the soil's adsorptive and exchange capacities in relation to its clay and humus contents. Those as saturations are suffering from imbalances of the four cationic nutrients, namely, calcium, magnesium, potassium and sodium, with calcium required more highly as soil saturations than the other three combined but yet pushed into a calcium deficiency state by the others when it is the first requisite in our body ash.

12.

The topic of limestone accounted for dozens of papers by Dr. William A. Albrecht. Some were written early in his career. Others were written or recycled as late as the early 1970s. Agricultural Limestone for Better Quality of Foods—What Fineness for Agricultural Limestone *states the case for the calcium nutrient so that all can understand. It was printed in* Pit and Quarry *in May and July 1946.*

Agricultural Limestone for Better Quality of Foods—What Fineness for Agricultural Limestone?

Those who have been grinding limestone for the purpose of treating soils and growing better crops have been raising the question, "Just what shall the fineness of this material be?" One need not be a chemist to know that the finer it is pulverized the more rapidly it will react with the soil. Then, too, one need not have much experience with rock crushing or a grinding business of any kind to recognize the economic aspects and to remind oneself that pulverizing any rock to fine powder becomes an expensive operation. Producing agricultural limestone centers around the following two short questions: (1) How much speed of reaction is needed by the limestone in the soil? and (2) What fineness of stone will properly give it? After these questions are answered, the costs of crushing can be quickly calculated by anyone with a little stone-crushing experience.

But when we have been considering commercial fertilizers in terms of their "availability" only as they are soluble in water and in very dilute organic acids in laboratory tests, and when we have set up regulations to make commercial fertilizers conform to that concept, we are easily inclined to feel that possibly limestone for soil treatment should be highly pulverized. If so finely ground, it too would react very rapidly with dilute acid on tests in the laboratory. Limestone pulverized finely enough to pass through a 100-mesh sieve has been found equivalent in many respects to the behaviors of the water-soluble lime hydrate and of the quick-lime. Consequently, by reasoning on the basis of this solubility criterion we are apt to cling to 100-mesh fineness as an ideal.

1. limestone has become an essential for the soil, by way of its coarser products

But then, the extensive farmer experience with mill-run, 10-mesh, or even quarter-mesh sizes of limestone as soil treatments has demonstrated good results in crop improvements so generally that these coarser materials represent the preferred ones in the light of what stones have been most generally obtainable. Results from the use of even these coarser ones have labeled limestone as an essential soil treatment. Such facts matched up against chemical concepts of the necessity for exceptional solubilities pose a real problem in deciding just how finely ground our agricultural limestone should be.

The history of the use of limestone had its origin in the applications of quick-lime and hydrated lime as the earliest forms of calcium carriers. Both of these are water soluble. It is true that they may become air-slaked, or be converted into the calcium carbonate forms like the limestone. But, even if used in that form, they are so finely pulverized as a consequence of having been burned that they are rapidly reactive. But now that limestone, and not the burned lime, is the main form used; now that its use annually is running into millions of tons within a single state; and now that is application on the soil as a practice in soil conservation is enjoying support in payments from the federal treasury, there may soon come regulations regarding fineness of grinding according to tests on standard screens. What these regulations as limits in size, or mesh should be can be decided only on the basis of experimental evidence, with farmer experience included, and not on arbitrariness without foundation in facts.

2. *concepts of the chemical laboratory transplanted to the soil may be misfits*

In settling this matter, there may be an error in transplanting concepts from the chemical laboratory to the soil. Rapid solubility through extreme fineness of grinding may sound well as speedy chemical reactions, but is this a desirable aspect for the mineral rock materials that compose the soil? The soil is merely a temporary rest stop by rocks and minerals on their way to solution and the sea. Do not plants naturally start growing long before the rocks have traveled very far along this journey? Are not most plants feeding very effectively when the soil is much nearer to the condition of the rock than it is to the complete solution?

We may be making an error in thinking that the rapid solution aspect of limestone is essential, much as we made another error in thinking about limestone when we believed it serving to make crops grow better merely by reducing the degree of acidity of the soil. Some simple laboratory tools and tests by which the degree of soil acidity could be measured, prompted the widespread convinction that when liming a soil improved the crops on it and lessened the degree of soil acidity at the same time, it was the reduction of the acidity that was the cause of the better crops. That this line of reasoning was an error has been clearly shown by extensive studies of the physiology of leguminous plants on acid and limed soils. We now know that the liming of a soil is beneficial because of the calcium—and possibly the magnesium—it supplies for services as a nutrient rather than because of the carbonate it delivers for reducing the concentration of soil acidity. There is the possibility that our belief that limestone ought to be highly soluble may be resting on clouded thinking as was the case in ascribing benefits for crops from limestone to its effects in neutralizing soil acidity.

3. *experiments point to propriety in mill-run, 10-mesh fineness*

Experiments were set up at the Missouri Agricultural Experiment Station to study this question of fineness of grinding of the limestone as this determines the services in the improvement of the growth of the crop. Highly-pulverized stone, all of 100-mesh fineness, was compared with mill-run, 10-mesh stone. The latter had nearly 25% passing through the 100-mesh sieve with the remaining 75% well distributed over the other regular test screens of finer openings. The entire mill-run sample passed through the screen of 10-mesh.

These stones served as soil treatments in a two-year crop rotation of corn and oats. The latter served as nurse crop for sweet clover. When the sweet clover and oats were seeded, limestones of each of these two finenesses were drilled at different rates in order to help in starting the sweet clover. The soil was known to be acid in reaction and deficient in calcium for this legume crop. Calcium chloride and calcium sulfate were also used in rates of calcium delivery comparable to that by the stone. These last two do not reduce the acidity of the soil. In fact their chemical combinations as sulphate and chloride should represent increases in this condition because of the sulphuric and hydrochloric acids represented in these compounds. In addition to drilling these soil treatments of calcium regularly each year with the oats seeding in the crop rotation, the highly pulverized stone and the mill-run, 10-mesh stones were also used as heavy initial applications of two tons per acre at the outset of these tests.

The results of these trials were measured as increased yields of the corn crop because of the larger amount of sweet clover grown and turned under as a green manure ahead of the corn crop. Two series of plots provided a corn crop every year. Eight crops have been grown. Their yields report the fallacy of the theory that greater solubility of limestone through finer grinding should make this more efficient in growing more sweet clover and thereby larger yields of corn following.

As an average of eight corn crops, the drilling of the 10-mesh, mill-run stone gave eight bushels more corn grain than similar use of 100-mesh stone. This trial with its low rate per acre of using coarser stone was superior to pulverized stone drilled at even higher rates. The sulphate and chloride were by no means the equal of the mill-run, 10-mesh, though slightly better than the 100-mesh in these trials. Where the stones were initially put on as a two-ton application, the results show that the 10-mesh was superior to the 100-mesh, used at the same rate, in bringing on better crops of corn through sweet clover ahead of them.

4. less speed and less solubility suggest time as a necessary fourth dimension.

Such facts suggest a definite principle in soil reactions with lesser solubilities and lower rates of reaction much more effective than are the behaviors of high solubilities and speedy reactions. Soil reactions emphasize the greater time element, or what may well be considered a "fourth dimension," so common in the geological concepts involved.

Where the pulverized limestone, or the chemical salts of chloride and sulphate were drilled, the sweet clover crop got off to a good start in the first year of its biennial life, but it failed to carry well over into its normal second year. But where the mill-run, 10-mesh limestone was drilled, and even at very low rates per acre, the sweet clover grew well. There it grew a green manure to be turned under and improve the corn crop. The effects of the annual applications of this coarsely ground material were additive or accumulative and increased with time during the eight years. Also where the use of two tons of these different finenesses was under comparison, the effects from the coarser stone were increasing with time while those from the pulverized stone suggested a decrease with it.

5. a new principle is involved in limestone application

These results suggest that the nutrition of the plant with its roots taking calcium from limestone applications to the soil is not a matter of providing the equivalent of a solution of calcium in the same sense as we see it in the laboratory. Rather, it is a matter of having some calcite particles, or limestone fragments, scattered throughout the soil toward which the plant roots will grow and from the surrounding soil of which they can take the necessary calcium by ionic exchange. These calcium-providing areas persist in the soil during the two years needed by the sweet clover only when the limestone particles are larger than the 100-mesh. These finer sizes and the more soluble salts of calcium pass this nutrient element about quickly. But yet the crop results suggest that their contribution was either leached out or absorbed so completely on the clay that the crop was already without sufficient calcium before the second of its growing seasons had come along. Better delivery of calcium as plant nourishment is seemingly not a matter of providing only pulverized limestone of high solubility, but of scattering through the soil some of this finer material and some coarser particles as well, as was done in using the 10-mesh, mill-run stone.

This principle is quite in contrast to the chemical one considering only nutrients of high solubilities as fitting root environment. Such a new principle takes to the concept that the nutrient delivery comes from a heterogenous mixture of rocky fragments through which by search, resulting from certain unknown responses, the root can provide itself. Soil is serving more effectively in feeding plants according as it represents the rock as having made only a smaller part of the journey toward solution and the sea. This is a far different concept from the belief that every plant nutrient must be blended with every other one into a homogeneous solution. When the coarser, slowly-acting limestone fragments serve the crop more effectively than the highly soluble powder forms, such evidence supports the newer concept of the mechanisms involved when plants are being fed by the soil. It suggests that other mineral or rock fragments besides the contributors of calcium may well be tried for their fertilizer values in terms of these mechanisms.

6. providing agricultural limestone becomes a special business

If these are the facts, as we are encouraged to believe they are, those who crush limestone will readily gear their thinking toward bringing about the economics in grinding that these principles so readily invite. Producers of limestone may well plan on agricultural limestone as a standardized product in a regular business rather than one of dumping on the land and on its owner a by-product of making ballast or other forms of coarsely crushed stone. When for a state like Missouri, it is estimated that 3,000,000 tons are needed annually, merely to hold the lime at the level of its removal, and when the maximum annual use to date in this one state has been one and three quarter million tons, it requires no great imagination as a quarry man to see big opportunities ahead by directing the thought and study toward those in agriculture. To date, Missouri has been one of the largest users of agricultural limestone. So with its own opportunities so large, these will necessarily be relatively larger for agriculture as a whole.

We are just beginning to think and act on the need to return fertility to our soils. We are becoming concerned about maintaining them in this respect. Calcium is only one of the many nutrients involved, but is removed from the soil at the most rapid rate. It is therefore at the head of the list of those constituting our soil needs. It is also recording its deficiencies there in the failing qualities of feeds and foods that register rapidly as irregularities in animal and human health.

In the light of some of these facts, the fineness of grinding for agricultural limestone has implications much more far-reaching than at this moment may be evident.

13.

Dr. William A. Albrecht's radically different interpretation of the function of acid in soils became the subject of countless talks before trade assemblies. Yet today farmers still lime to combat acidity. Schoolmen and testing stations still counsel, "When soil is acid, lime." In many cases the advisors do the right thing for the wrong reason, and often they do the wrong thing for the wrong reason. No less than four elements affect the pH of soil: calcium, magnesium, potassium, and sodium. Magnesium can raise pH up to 1.4 times higher than calcium on a pound for pound basis. A soil high in magnesium and low in calcium can test 6.5 pH, but be inadequate for growth of alfalfa, proliferation of legume bacteria, and maintenance of an environment suitable for decay of organic matter. This short paper is a recap of a speech before a Growers School at the University of Missouri, January 20-21, 1948. It was published in Southern Florist and Nurseryman *January 30, 1948.*

Soil Acidity Is Beneficial, Not Harmful!

Only recently have we come to appreciate the services by soil acidity in mobilizing—making available—many of the nutrients in the rocks and minerals of the soil. When we learned that soils are less productive in giving us legumes and other protein-rich forage, according as they become more acid either naturally or under our cultivation, we came to the conclusion that soil acidity is the cause of this trouble. We now know that a plant puts acid into the soil in exchange for the nutrients it gets. It is the same acid held on the clay that weathers the rock fragments and serves to pass their nutrients on to the clay, and from there on to the plant.

The coming into the soil of excessive acidity is merely the reciprocal of the going out of the fertility. Nature's process of feeding the plants, and thereby the animals and us, is one of putting acidity into the soil from the plant roots in order to break out of the rocks what nutrients they contain for nourishment of all the different life forms. Soil acidity always has been the agency in the soil to keep the assembly lines providing the available fertility for crop nutrition, which in turn is nutrition for all higher life.

1. limestone becomes a calcium fertilizer on acid soils

When we put lime rock on the soil as a fertilizer supplying calcium to our legume crops, we know full well that this rock reacts with the acid-clay of the soil. The acid goes from the clay to the lime rock which, being calcium carbonate, breaks down to give carbonic acid while the calcium is absorbed or taken over by the clay. While the calcium goes on to the clay to be available there for the plants, the carbonic acid decomposes into water and carbon dioxide as gas. Since this gas escapes from the soil, this escape takes away the acid, or, as we say, "it makes the soil neutral." The benefit to the legume crops from the application of this lime rock to the soil does not rest in the removal of this soil acidity. Rather, it rests in the exchanging of calcium as a nutrient to the clay which was holding the acidity or hydrogen, a chemical element that is not of direct nutritional service.

2. various other rocks as well as lime rock are fertilizers on acid soils

Soil acidity has been breaking potash rocks down chemically, too. During all these past years the potash feldspars have been undergoing weathering attacks by soil acidity. On this rock the acid clay carries out its weathering effects in the same way as it does for lime rock, except that it trades acid to the feldspar and takes potassium unto itself in exchange. Magnesium rock, as we have it in dolomitic limestone, is also broken down by the acid clay. By this same process the clay becomes stocked with magnesium. This is then more readily exchangeable and available to the plant from the clay than it would be if the plant root were in direct contact with the rock fragment itself. By exactly the same mechanism we can expect phosphate rock to be made available for the plant's use.

It is in these processes by which the acidity of the soil is beneficial. If the soil contains the two colloids, clay and humus, which can hold acidity, and then if that soil has scattered through it fragments of lime rock, of magnesium rock, of potassium rock, of phosphate rock or in fact of any rock with nutrients, it is the soil acidity that mobilizes to the clay the calcium, the magnesium, the potassium, the phosphorus, or all the other nutrients respectively, for rapid use by the plants. This is nature's process of providing plant nutrients on the clay of the soil in available form. By it nature has stocked our moderately acid soils with fertility. It was that condition of our virgin soils that spelled our prosperity. We need to go back and study the soil in order to learn how we can gear the lime rock into our acid soils so it will stock our soils with calcium and still not remove all the acidity needed to make the other nutrients available in other rocks, of which some are giving phosphorus, some giving potassium and some giving one or more of all the other nutrients in whatever rock form nature has them.

3. acidity is a slow mobilizer of nutrients contained in the minerals

Of course, nature's processes do not demonstrate high speed. It takes six months or more for limestone to stock the clay with enough calcium to feed our legumes. It takes longer for others. Consequently, if phosphate rock is slow in demonstrating its effects on the crop, we need not be alarmed. It certainly will not be very active in soils that are not acid. In our western soils that have considerable free lime rock mixed throughout them, the plants suffer from phosphate deficiencies. Rock phosphate may not be so efficient there. Dr. Hutton of South Dakota pointed out that phosphorus delivery to the crop was the major problem on these highly calcareous, or these distinctly neutral soils. It is on soils of that reaction that cattle have been bone-chewers. Apparently they are going to that extreme behavior in order to get their phosphorus, which the forages as feeds on the soils there fail to provide. So when the soils are not acid we may expect shortages of some fertility elements like phosphorus, for example, in the plants and even in the animals. It is on the acid soils where most of our population has always fed itself, suggesting that it is on those acid soils where there is much more speed in making the fertility active and available from out of the reserve minerals.

4. phosphorus deficiencies in the soil suggest possible health troubles

That human health may be related to that deficiency in the soil has come to be more than mere speculation when animal troubles are localizing themselves more and more according to soil fertility deficiencies. If one looks at the map of poliomyelitis in 1946, for example, it is significant that this health trouble was more severe in those states where the less acid soils give deficiencies in phosphorus for the plants and animals. It is all the more significant when we remind ourselves of the fact that the brain and nerve tissues—most seriously affected by "polio"—represent high concentrations of phosphorus. Whether there is any causal connection between this human affliction and these non-acid soils because phosphorus availability there is low is a question that remains for research to answer.

There are also other nutrients in addition to phosphorus that are not mobilized effectively unless the soils are acid. These include iron, manganese and boron. Perhaps these or some others not yet appreciated are causally connected with this baffling disease. Other diseases may be in a similar category. When we see the inroads by brucellosis on our herds of cattle and hogs, and when undulant fever passed seemingly from these is becoming more common, there is decided encouragement to theorize and do some research on the belief that we need acid soils and need to gear more mineral fertility through the acid assembly lines of our soils.

5. understanding soil acidity means appreciating the deficiency of soil fertility

We have used lime rock, phosphate rock, and green sand as potassium rock, in both the experimental work in the laboratory and in the field trials, to convince ourselves that the plant nutrients within these are made available to plants by the acidity of the soils. Possibly as we come to appreciate the benefits of soil acidity we shall no longer fight it with carbonates, but shall guard it and use it to treat our lime rock as calcium fertilizer, our phosphate rock as phosphorus fertilizer, and possibly many other rocks whose fertilizer values we still do not appreciate. In the future more of these mineral or rock types of soil builders will very probably go into the soil to build up a reserve of minerals there. This will occur when it becomes more common knowledge that soil acidity is beneficial and not detrimental.

14.

That you can't farm without nitrogen has become a catchphrase of no small dimension among modern agronomists, the implication being that nitrogen must be purchased and applied routinely. "Hold on a minute," Dr. Albrecht would say. And, often as not, he would fish from his files a manuscript or two and push it into the hands of his "student." When Acres U.S.A. *first asked Albrecht some rather pointed questions about nitrogen, he produced the manuscript that is printed below. It had been printed before in* Philfarmer. *A few editorial changes, a few notations, and it was ready for recycling. This version appeared in* Acres U.S.A. *as a three-part series, December 1971, January and February 1972.*

Fertilizing Soils With Nitrogen

Among the garden crops, the peas, beans, lentils and other legumes are mainstays because, as nitrogen-fixers, they have long been known to be higher in the proteins and all else which such plants must take from the soil as essential elements and compounds to be legumes. The clovers, alfalfa, soybean and others are some high protein field crops. These are the crops which can take gaseous nitrogen from the air and combine it into highly valuable food constituent—the proteins. In feeding animals, legumes are the feeds for "growing" them. They supplement corn, for example, as feed for fattening animals—usually castrated males carrying no load of reproduction or of growing fetal young.

1. the cow makes her suggestions

Because proteins contain the element nitrogen in a particular chemical structure which we call "amino" nitrogen (and some other forms of it) to the extent of about 16% of these compounds, we have been measuring the nitrogen by "ashing" our foods and feeds in sulfuric acid. Then we multiply the measured value of this element by 6.25 and call the result the "crude" protein of the food or feed. The cereal chemist multiplies it by 5.73 for the proteins in the grains. Thus, nitrogen has become synonymous for proteins, but not without danger of serious confusion.

That confusion is becoming more dangerous now that we are applying nitrogen to the soil generously as fertilizer, and such in highly soluble and active chemical forms. The plant may take up these; it may fail to convert them into the proteins; and yet our chemical analysis will credit the plant with high protein when much of that may even be poisonous. There is also the increasing inference that non-legume crops, fertilized generously with nitrogen and consequently brought to contain a high percentage of nitrogen in their dry matter, must therefore be rich in proteins. If that figure for percentage of nitrogen is apt to be drawn that our use of plenty of fertilizer nitrogen on grasses and other non-legume crops will grow proteins in quantity and quality in those as well as when we use the regular legumes.

That such is not necessarily the case follows from the fact that soils must be well stocked with calcium, magnesium, potassium, boron, manganese, copper, zinc, and many other essentials in the list of the major and the minor or "trace" amounts before legumes will grow. Only after legumes have a *balanced* soil fertility in terms of all the essential elements except nitrogen, and those in good supply first, will those plants add the nitrogen of the atmosphere to that stock and carry out this proces of nitrogen fixation by which they grow, protect themselves, and reproduce. The nitrogen is not taken up in that natural process by legumes except as it moves via the protein forms of it.

That the corn plant, as one of the grasses, can deceive us and take up large quantities of nitrogen from the soil without converting this essential element into protein was demonstrated by the drought when the so-called "silo-fillers disease" gave us not only a newly coined term but also some new concepts about the dangerous biochemistry of silage making. The chopped fodder gave off nitrous oxide fumes overnight to kill one man on entering the partially closed silo the following morning.

Chemical analysis of corn stalks suggested that because the heat of the drought had destroyed the enzymes which were normally converting the nitrates coming into the plant from the soil into other plant compounds, the nitrate nitrogen (or even this in the nitrite combination) had accumulated excessively in the corn stalks. Nitrates and nitrites are deadly compounds for animals and man even in very small dosages. Some analyses of corn stalks, reporting 1% of total nitrogen, showed as much as 0.65% of this element in the nitrate form. Thus, we might have believed that the dry matter of the corn fodder had 6.25% protein. But instead it had 4.06% so-called protein—equivalent of the 0.64% nitrogen—which was poison, as the reported deaths of the cows sadly illustrated.

The cow has perhaps been the oldest producer and distributor of nitrogenous fertilizers, putting out some of it in the organic and some in the soluble salt form, *viz.* urea. She demonstrates her activities in that industry every time she voids her droppings or urine on the grassy swards. There we see the markedly green growth of grass encircling the feces and the much taller and probably greener growth of it encouraged by urine.

But if we observe closely and consider farther we will be forced to conclude that such is an excess of nitrogen fertilization in the humid soil areas and is not producing a nutritious forage feed, regardless of how demonstrative in yield and how attractive to the eye. Testimony to that is given by the cow herself when she refuses to eat that green growth but lets it grow taller while she eats the surrounding grass shorter and shorter. Even she tells us that one must be cautious to balance the nitrogen salts with all other fertility in using them as fertilizers. The cow is not classifying forage crops by variety name, nor by tonnage yield per acre nor by luscious green growth. Instead, she is classifying forage according to its nutritional value in terms of complete protein and all else coming with it, according as the fertility of the soil determines. As a biochemist or an assayer, she is not satisfied with the value of the nitrogen in the ash multiplied by 6.25 and labeled "crude protein."

When we fertilize with nitrogen salts, we dare not always say we are thereby necessarily making crops more concentrated in protein because they look greener and more luscious. The growing of plants is a problem in plant nutrition no less complicated than any other problems of trying to feed some life form properly. So when using nitrogen-carrying fertilizers, one dares not operate under the belief that "if a little is good, more will be better," even though nature seems to follow that philosophy when she fertilizes with nitrogen in the form of organic matter which she may pile up abundantly.

2. *the rabbits testify by experiments*

When we observe crops, either in the garden or in the field, they are anatomically very similar. They all have roots, stems, branches and leaves. They are of a luscious green color. They look alike, and—so far as we can see—we might expect one to do what the other does when we fertilize the soil under them with extra chemical nitrogen. We are usually delighted when both the garden and the field crops improve in their production of more vegetation and that of a greener color. We measure their contents of nitrogen and find that the concentration of this in the dry matter was increased decidedly by the fertilization. We believe we have thereby made them richer in protein as food and feed.

We do not appreciate the fact that, in terms of their physiology, the plants may be as widely different as the talents of folks are when *some can* and *some cannot.* Two plants looking alike are not necessarily of equal feed value. The grazing cow confirms that when she may choose one in her grazing and disregard the other. In the form of hays separated in the feed-rack, she may consume one entirely before even touching the other. Under more delicate testing by using weanling rabbits in feeding experiments, the fertilizing of a grass like timothy with nitrogen increased the "crude" protein in the hay crop decidedly. But even then, this was not necessarily good nutrition. This suggests that the physiology of the non-legume, timothy, in its use of fertilizer nitrogen does not necessarily give us the nutritional values for rabbits in the crude protein or even in the amino acids which are equal to those in a legume like the red clover.

The timothy hay fertilized with both nitrogen and trace elements was fed to weanling rabbits in some experiments during the summer heat wave of 1954, only to have the heat wave kill some of the rabbits with a very severe rise in temperature until 70% of the rabbits were dead. Then since the experiment had gone its planned time and the rabbits added from the stock supply at fortnightly intervals had maintained the number, some dried skim milk powder was added as a protein supplement to improve the quality of the ration in that respect. From that date forward no more fatalities occurred even though the heat waves continued.

Then the original experiment was repeated. There resulted again the repetition of the deaths by the heat waves which were accepted as part of the test until 30% of the experimental rabbits were killed. Then the timothy hay was replaced by some red clover hay grown on Sanborn Field where the clover has been in rotation systems since 1888. From the date of this substitution, and with the heat waves holding on, no more rabbits died while feeding on the red clover hay.

It is interesting to note that the timothy and the red clover hay under chemical analyses for total nitrogen chemical analyses for total nitrogen and under microbial bio-assay for the essential amino acids would not tell us all the reasons why the clover hay saved the rabbits from death by the heat but the timothy hay did not. It is interesting, nevertheless, to note the differences between the two hays so far as the chemical analyses for these essentials could give suggestions. They are given in table 1.

From such data one cannot see differences wide enough to make a guess as to the particular amino acid deficiency which killed or the particular sufficiency which saved. It is significant to note how similar the two hays were in totals of the ten and the nine amino acids respectively, but yet were more widely different in the total nitrogen. In this latter respect the red clover was almost a third higher than the timothy.

As an additional experiment, the red clover hays grown on eleven plots with that many different soil treatments and history were each made up into diets for test rabbits so that each of the series contained 1.31% total nitrogen, or 8.18% "crude" proteins supplied by the clover from the particular plot. All else, as dietary factors, were also brought to a constant as nearly as possible. Then this set of supposedly uniform diets was again fed to carefully selected weanling rabbits and their gains in weights taken as the index of the diets' efficiencies even when all were the same in "crude" protein.

It was startling to find that the gains per rabbit by lots per four weeks varied from a low of 34 grams to a high of 241 grams. The gains as grams per milligram of the nine essential amino acids consumed, varied from a low of 10.6 grams to a high of 77.9 grams. Then, finally, in examining the livers of the rabbits, these varied widely in weights and ranged in color from almost black to the usually expected normal liver color.

Here then was evidence to tell us that timothy hay fertilized with ample chemical nitrogen and given also trace elements, failed to keep the rabbits alive under heat waves while red clover with a variety of soil treatments, ranging from none upward, saved them from death under similar heat conditions. We are not able as yet to catalog all of what the soil gives to the crops by which one kills, and the other keeps alive, the animals feeding on them. Nor can we fertilize the soil with nitrogen and say much for the quality of the protein the plant makes even if we are pleased when it give more "crude" protein. Fertilizing for more feed values is not necessarily also fertilizing for better nutritional value when we use the highly active chemical fertilizer salts.

3. we may use too much salt

Animal manures were the first soil treatments to give improved growth to crops. The accumulated manures of sea birds collected and imported from the arid Pacific Coast of northern South America were some of the first commercial

Total Nitrogen and Essential Amino Acids In Timothy and Red Clover Hays (Mgm per gm dry matter)

	Hays Timothy	Red Clover
Total Nitrogen	13.10	17.40
Methionine	0.61	1.31
Tryptophane	3.11	*
Lysine	1.66	5.26
Threonine	2.97	4.51
Valine	3.49	5.70
Leucine	14.00	6.07
Isoleucine	7.12	4.43
Histidine	0.81	1.37
Arginine	3.64	9.59
Phenylalanine	2.79	4.86

*Not measured.

Total Amino Acids (mgms. per gm. dry matter)	40.20	43.10

fertilizers. They were sold to our cotton farmers under the trade name of "guano." From near that same source and representing probably the evaporated leachings from the guano, there came later the nitrate of soda or the Chile saltpeter.

Thus, history gave us a gradual shift from the fertilizer nitrogen in the organic manures to that in the form of the inorganic chemical salts. But this shift was not accepted without serious protests by the southern farmers. They contended that guano gave better results on their crops than the saltpeter. Perhaps, even they recognized some fertilizing helps from the organic compounds in the guano which were not given by the simpler, purer, more concentrated chemical salts. Many are still contending that organic manures are better fertilizers than pure chemical salts.

Not too long after sodium nitrate became the common nitrogen fertilizer, the ammonium form of nitrogen combined as the sulfate came into the fertilizer trade. This was collected, along with coal-tar wastes, as a by-product of the destructive distillation of coal to give coke for the reduction of iron ores. This was a form of nitrogen different than nitrate. Chemically, it was a positively charged ion. Consequently it is held in the soil against rapid loss to the rain-water going through it. It undergoes a slow, mi crobial conversion there into the nitrate form which has the negative charge, and when united with a positive ion, like calcium, for example, is not so firmly held by the soil. It is then a highly active salt and may be readily washed out. It is taking other ions of nutrient value and of opposite charge, as well as calcium, out of the soil. While this all happens, the nitrogen shifts itself from the positively charged ammonium form, held by the soil, into the negatively charged nitrate form not held. If it is applied as a sulfate, this latter—as a negative ion—also carries calcium or other positively charged ions out. The ammonium ion as a positive one tastes hot. The nitrate, as a negative one, tastes cold. They are opposites in more ways than one.

There comes, then, the suggestion that nitrogen ought to be used along with crop residues, and with applied organic matter if the depleted soils are to be restored and the nitrogen held to the best advantage in the soil.

Composting either within the soil, or in the special compost pile above it, seems to be the prevention against fertilizing with too much salt. This seems to be no small danger when the concentrations of the fertilizers are mounting and the larger applications commonly applied are both crowding under increased economic demands for higher yields per acre. It is these conditions which portend a compulsory reverse of history with its shift from the nitrogen salts back to more of this element in the form of organic compounds.

When these various salts of nitrogen were combined with superphosphate as mixed fertilizers, it was soon discovered that while one could make heavy applications of phosphates with the seeding of grains, the nitrogen fertilizers could not be so used without injury to germinating seed and serious reduction in the stand of the crop. Then, when a potassium salt, like the muriate, was also included in the fertilizer salt mixture to give the so-called "complete" fertilizer, the damage was still worse.

4. on sanborn field

On Sanborn Field at the University of Missouri one plot was started as early as 1888 with heavy applications of commercial fertilizer salts of nitrogen, phosphorus and potassium—enough to represent those three elements taken off by the grain and straw in a 40-bushel crop of wheat. Beginning with the virgin soil, that much of chemical salts drilled with the wheat seeding—where this crop was grown annually—was not damaging to either the germination or the final stand until after nearly 20 years. But during that period, with all the straw and grain removed and with no organic matter returned, the soil lost much of its humus and organic matter, which represents its capacity to "buffer" the shock of so much chemical salts. Ever after that period and even today, the total fertilizers can be applied safely only as a divided application, one part with the autumn seeding while the other parts must be applied in the spring or later.

In general, most fertilizers are now applied on soils unable to handle these more concentrated forms and heavier salt treatments, unless by special placement of them at some distance from the seeds. We are faced with the mechanical problem of proper placement of those chemical salts to escape their early salt damage. That was no problem with nitrogen in guano or in its organic combination.

Such facts are turning our thinking back to organic matter to consider the problem of handling nitrogen and potassium, or other salts, so highly active and thereby so readily damaging to seeding. We are taking to the suggestion that, perhaps, a kind of natural composting of the nitrogen fertilizer by the help of the microbes in our virgin soils of high organic matter contents was nature's method that saved us when we first used it without damage from the excessive salts.

15.

If calcium is the prince of nutrients, then what about NPK? Students of Dr. William A. Albrecht often found that questions rated answers only after they had been straightened out. Management of nutrients in response to cation exchange capacity came first in Albrecht's thinking, and this reality belonged in the forepart of the question. Proper levels of the several cation and anion nutrients, after all, accounted for the natural nitrogen cycle. In his many papers, Albrecht wrote about "maintaining nitrogen" more than about "supplying nitrogen." This paper was first published in 1946 by the Victory Farm Forum.

The Soil Nitrogen Supply Is Still A Big Problem

Among the many soil problems that have come on the stage to hold public attention, perhaps that of maintaining its nitrogren has been the one factor most persistently before us. Even though soil erosion has held such a prominent place in our general public as well as our agricultural thought during the last decade, that fact does not mean that the decreasing nitrogen in the soil has not also been playing for attention right along with it. Our mental gaze needs to be fixed on the deficient nitrogen while this may be prompting the erosion, since both are playing villainous roles simultaneously on the same agricultural stage. We need to see not only the erosion, but also the declining soil nitrogen. It is this that has reduced the soil's capacity to grow its own cover quickly as protection against erosion. At the same time, there is need for some extra soil nitrogen to render services in sustaining more than just any kind of plants merely for soil covers. Vegetative covers of our soils must now be plants producing more proteinaceous foods.

1. protein deficiencies

In our food shortages, it is the proteins of which the deficiencies are most acute. Proteins are some of the many possible chemical combinations of amino acids.

These building stones out of which proteins can be constructed are characterized by their contents of the element nitrogen linked with carbon, hydrogen, oxygen, and, in some cases, with phosphorus and sulfur. The amino acids are forms of nitrogen-containing, chemical provokers of growth. They are synthesized from the different chemical elements into these life-carrying combinations only by plants, and not by the animal body. This synthesis of nitrogen into amino acids is one of the signal services for animals and humans by the plants since neither man nor beast can synthesize them so far as we now know. It is from these highly elaborated compounds that man and animal, as higher life forms in the biotic pyramid, must construct the particular proteins necessary for their body growth and repair.

The efficient utilization and elaboration of nitrogen by plants, both legumes and non-legumes, calls for a liberal supply of calcium, phosphorus, and other mineral fertility coming from the soil, as well as for the necessary supply of nitrogen. The physiological demand for calcium by the plants has been almost blanketed out and put into lesser significance in our thinking by our believing that lime was beneficial for crops only through its reduction of the degrees of soil acidity. Proteins, especially such as meats, eggs and milk, have long been considered a major problem in the struggle for food. We have not, however, led our thinking about meats to connect them as proteins with the nitrogen, with the calcium, with the phosphorus and with other items of soil fertility. We have been content to think of meat as coming from the animals. We have not thought of the animal's source of protein going back to the soil's source of nitrogen and other fertility supplies. Food shortages putting meat into less, and cereals and vegetables into more prominence in our diet will remind us forcefully how essential the nitrogen in certain particular protein combinations is, and how serious the shortages of it in some of them may be.

2. knowledge still lacking

Unfortunately, the word "protein" is still not a completely known term. We still have no chemical measures of each of the many proteins. In measuring the amounts of protein we determine the element nitrogen, multiply it by 6.25 (the cereal chemist multiplies it by 5.7) and call that mathematical result the amount of the proteins. We know that proteins are chemical combinations of different amino acids, but which of them and how much of each amino acid you and I must ingest daily is still not so definitely known, even though there have been cataloged from the 23 known proteins, 8 for man and 10 for animals, as specifically essential.

We know nitrogen quantitatively only in terms of this simple approach of analytical chemistry. We still know little in terms of its synthesis within the animal body and less of its synthesis within the plants. Consideration of protein merely as nitrogen by analyses for this element through combustion of the substances in sulfuric acid, and the multiplication of the amount of this ash form by a mathematical factor is evidence of the pronounced ignorance of its chemical behaviors in plant physiology, to say nothing of them in animal and human nutrition.

3. *different compositions*

As combinations of several different amino acids, proteins suggest different possible compositions varying widely according as the plants are of different species and according as they are nourished differently by the soil. When, for example, by using different ratios of the potassium to the calcium—both in exchangeable forms on the colloidal clay—Dr. Graham reduced the crop yield of soybean forage by 25%, but yet synthesized in it more protein as measured by the total nitrogen, and yet increased the concentration of phosphorus by 100 and of calcium by 200%. Through the modification of only the ratio of these two fertility elements, can we feel that the physiology of the soybean plants was a constant behavior and that it was building the same kind of protein in the larger as it was in the smaller soybean crop? The need to differentiate between the kinds of proteins synthesized by the plants under varying ratios of the elements of fertility and under the variable supplies of different items of fertility in the soil poses one of the nitrogen problems that is high on the research addenda. This is especially true if we are to shift the protein-providing responsibility in our diet away from meats and milk and more toward the foods of distinctly vegetable origin, and thereby a step closer to the soil fertility as the foundation of the entire biotic pyramid.

Plants are not so directly synthesizers of the amino acids and their combinations but rather indirectly via their carbohydrates. The biosynthesis, or synthesis by the plants probably through the use of stored chemical energy, of the proteins is very probably a sequel only to the photosynthesis of the carbohydrates by sunshine energy. Then, too, plants are providers of calcium, phosphorus, and other nutrient elements. These elements are the soil fertility that controls not only the photosynthesis and the biosynthesis, but also, apparently, the construction of the catalyzing agents, like the vitamins, hormones, etc., that are provided by the plant for us as essentials along with other compounds classified more distinctly as foods. Dr. S. H. Wittwer [Wittwer, S. H., Vegetable Crops in Relation to Soil Fertility IV, Nutritional Values of New Zealand Spinach, *Journal of Nutrition* 31 (1946), 56-65] has recently shown that the soil fertility, particularly the nitrogen, is closely connected with the elaboration of some of the vitamins in spinach, for example. If these catalyzers which we call vitamins are connected with the nitrogen as soil fertility, this is another problem for soil research that will connect the soil with nutrition of the animals and humans through a very fine chemical thread.

4. *more study needed*

Nitrogen viewed through the simple chemical evaluation and multiplication by an arithmetical factor is no longer complete satisfaction in understanding the nitrogen in soil fertility. It calls for biochemical measures of its functions in the plant's processes that are building our foods. Nitrogen as a variable is demonstrating too many differing effects and features for us to be contented with an understanding that is no more elaborate and detailed than at present.

Even insects are discriminating to a degree suggesting that they are measuring the effects by nitrogen to a more refined degree. Recently some spinach under study by Dr. Wittwer was supplied with nitrogen at four levels of 5, 10, 20, and 40 ME per plant under ten replications. The crop was initially attacked by the insect known as the "thrips" only in the case of the plants given nitrogen at the two lower levels. Not a single plant at the higher amounts of nitrogen supplied was attacked by these insects. As more calcium was combined with the lesser applications of nitrogen these fertility combinations were more effective in reducing the leaf destruction by the thrips. If the insects are so carefully drawing the line on the different physiological behaviors in spinach of variable nitrogen in association with variable calcium offered in the soil, then there is the suggestion that we need to think about nitrogen within the plants in ways other than mere percentages of the dry weights of it in their tissue.

5. hogs more discriminating

Much more discriminating than the choices by the thrips were the choices by hogs of the corn grain in relation to the nitrogen that was a variable according to the green manure providing it. Plots with increasing mineral soil treatments were grown to sweet clover in one plot series and similarly to red clover in the adjoining one. These two legumes served as green manures to be turned under ahead of the corn planting. The corn grains from the separate plots were put into separate compartments of the self-feeder for the hogs. The protein supplement for them was provided separately.

Quite strange as it may seem, the hogs showed decreasing preference for the grain as more soil treatments were used to grow more sweet clover turned under to produce more corn. But as these same extra soil treatments grew more red clover to be turned under, these animals showed increasing preference for the corn resulting from it.

While we speak of the nitrogen of all the different legumes as green manure for corn in a single category, the hogs, and the thrips, if they could use our language, certainly would speak of it under several categories. Do not such animals and insect demonstrations pose the need for research to help us be more discriminating in the nitrogen of our own food as we are going away from the highly elaborated meat proteins and relying more on vegetable proteins? If the hog is so discriminating in what we are prone to say is due to just a difference in the amount of nitrogen turned under, there is an odious reflection on our contentment with our own ignorance of the relations between the fertility of the soil, the food grown on it and our own nutrition by means of it. The press of food that is now taking on an international scope, and the declining soil fertility so long neglected in our passing the responsibility for it to the farm owner, are making us fully conscious of the fact that the nitrogen is still a big problem in agricultural production on which national thinking needs to be focused more carefully than ever.

16.

Nature's variety, "infinite kinds of chemical linkages," and an animal's ability to select nutritive values, all served as recurring themes in the papers of Dr. William A. Albrecht. Was it not fatuous of man to burn off carbon, hydrogen, oxygen and other "contributions of weather origin," and conclude that chemical analysis of the remainder told the story? The paper presented here was first published in The National Livestock Producer *in 1947.*

Too Much Nitrogen Puts Plants on a "Jag"

That plants may go on a "nitrogen jag" has long been pointed out by the grazing cow when she lets the rich green spots of grass grow taller while she grazes the short grass all around them even shorter. These spots mark the liberal doses of nitrogen in her droppings that results in a luscious, massive growth. But the cow says, "No, thanks. I don't care for it."

She can balance the protein and carbohydrate in her diet as she grazes selectively the different kinds of plants in the mixed herbage of her pasture, or the same plants over different soil areas. That is why she leaves more white clover but less bluegrass. That is why, by June, the pasture is mainly white clover and the horses get the "slobbers."

Those deep-green grass spots, fertilized by much nitrogen applied through her droppings, do not appeal to her as possible of providing a balanced diet perhaps, and so she by-passes them. At any rate her body physiology directs her appetite to balance her ration. The excess of nitrogen represents an unbalance to her, and she says so by refusal.

1. balanced plant nutrients essential

"Too much is too much" only in relation to something else. If too much nitrogen is used in relation to the supply of phosphate, potash, calcium or other growth factors, the unbalanced situation causes trouble. It is the lack of balance in plant nutrition that is disturbing. This is merely saying that there is too much of some, or not enough of some other nutrients. That is the way it must be said when the absolute amount of each nutrient needed for each particular function is not yet known. We do know, however, that when an adequate amount of nitrogen is available, and other necessary factors are adequate, then growth and yield can be truly spectacular.

That animals do instinctively select food which provides a balanced ration was suggested by some work by Dr. George E. Smith. In this test, rabbits were fed grasses grown on soil that had been treated only with nitrogen. This work was a part of the studies leading to the bio-assay of soil fertility, by using the animal to measure the value of soil treatments rather than a mere increase of yield in bulk.

Nitrogen fertilizer on the grass, it was true, made a large and luscious green growth that would seem a great tempter—if only the human eye judged it. It appeared equally significant as a means of growing more grass per acre—if just the growing of more grass were the most valuable part of our efforts to improve pasture production.

2. animals select healthy soils

But the rabbits, when fed the grasses from areas with different soil treatments, had their own criteria for judging the resulting food values. The seemingly beautiful green, nicely cured grass hay from the plots where nitrogen alone was used was not taken eagerly. It was consumed only as a partial defense against starvation and it did not keep the rabbits from getting dangerously close to that before their death was prevented by shifting the ration. Other rabbits given grass hay from plots that had no soil treatment maintained themselves, consuming the ration more completely.

In dealing with a ration of fertility elements for plants, we should consider the plant ration as merely the sum of the separate items. We think of calcium, plus nitrogen, plus phosphorus, plus each of all the others necessary. These are taken into the plant and eventually delivered through it to the manger and thereby to the animal for its use of them as elements. Through chemical analysis of plants—after we have burned off the carbon, hydrogen, oxygen and other contributions of weather origin—we believe that soil fertility is a collection of some 10 or more elements taken from the soil for the use of animal and human bodies.

This suggests that, if that is all we need to do, we might just as well use a shovel and truck to haul calcium as limestone from the crusher to the mineral box. As a curative help to an animal already in disaster this may have some value. It illustrates the widespread failure to appreciate that plant nutrition is not as simple as limestone, plus phosphate, plus potash plus any other thing in any amount merely dumped on the soil to produce crops to haul to the feeding rack.

3. plants can't forage for food

An important factor in plant nutrition is that plants must eat where they are. Unlike the cow, they can't pass up the place where there is too much nitrogen or, rather in the converse, where there is enough of nutrient elements of distinctly mineral origin to better balance the nitrogen in their ration. Consequently, they run their manufacturing business of synthesizing the fertility of the soil into organic combinations by means of air and water the best they can.

If there is much nitrogen, they weave this into chemical combinations with carbon, hydrogen, etc., that builds a lot of green vegetable bulk that may not keep the plant from lodging and may not result in seed to keep the species multiplying. Plants must weave as the woof and warp permit. But the plant is doing as the conditions demand, whether we call it too much of some or not enough of others.

Root growth suggests that plants do make some selective searches through the soil. This is indicated when we find more roots around decaying organic matter, a piece of limestone, a granule of phosphate, or see the high concentration of roots in a fertilized portion or band in the soil. It is by such cafeteria-like "browsing" through the soil by the roots that the plant top is the final blending of fertility elements into the compounds that are built in a major way of photosynthesis. It is this soil fertility that makes the plant more than just sugars or starches—a contribution of sunshine power and the weather.

It is this synthetic performance, controlled and determined by soil fertility, that results in plants of nutritive value for growing bodies which does something more than just fatten them with such energy foods as sugar and starch. Plant growth is a synthetic performance using soil and water, and not merely an addition or collection of these fragments into a common mixed lot.

4. balance the important thing

Here then is the idea of a balance, or the proper amount of each element of soil fertility—nitrogen, phosphorus, calcium, etc.—in relation to each of the others. It is a matter of not too much or of not enough of any one in relation to the others that are needed by the plant. We must not forget that if there is too much of one for a certain synthetic performance, the plant may indulge in performing another way.

A plant's functions are not confined to the use of only one recipe. When carbon, hydrogen, oxygen and nitrogen are involved, there are infinite kinds of chemical linkages and it is these—different amino acids for protein or pseudo-proteins and different kinds of carbohydrates—that the cow is trying to help us appreciate as she selects her grazing according to the differences in soil fertility.

Too much nitrogen, or a "nitrogen jag," you say? Yes, as long as we are not as well informed about differences in synthetic outfits by our crops for nutritive values as are, we must finally admit, the cow, the chicken, or even the hog. After all, it is a matter of proper balance or of being moderate in all things.

17.

That the teaching chore could not be accomplished in classrooms alone became a Dr. William A. Albrecht dictum quite early in his career. Accordingly, he spent a great deal of time preparing releases and articles for the popular press, for farm trade papers, and for those with an interest in soils, crops and animals. It would be difficult to say how many papers such as the one reproduced below appeared in the public prints. Often Albrecht did what platform speakers do: he changed the title and the audience. This one, which appeared in The Weekly Kansas City Star, *June*

16, 1943, is still worth reading. It tells something about associate, Dr. A. W. Klemme, and about the practical search for answers on Missouri farms.

Fertility of Soil Measures the Proteins in the Crop

The present food problem has become decidedly a protein problem. This is not a problem of providing meat, but rather one of growing the protein-rich crops from which animals can make meat. Not all plants can make protein from the fresh air and water by means of sunshine as they can make sugars and starches. But it can be made from this source by the legumes if they are well supplied with minerals and other nutrients coming from the soil.

Calcium in limestone for legumes has long been recognized as the essential help for protein production on the farm by means of these crops. Given the minerals in the soil, they can use the fresh air and the plentiful sunshine to manufacture not only sugar, starch and other carbohydrates, but also protein. Non-legumes, too, can make protein. For them, however, it is a matter of getting both mineral supplies and their necessary nitrogen from the soil. Regardless of whether it is a legume or a non-legume plant that is to manufacture the protein, we must provide the minerals, or the fertility in the soil for them.

Louis Miller and his son, Don Miller, of near Golden City in Barton County, Missouri, have demonstrated the fact that soil treatments are necessary for most efficient protein production by soybeans. In their trials using limestone and fertilizer plowed under for soybeans as a seed crop last year, these soil treatments served as means of more than doubling the protein output per acre.

1. more protein in seed

The Millers' soybean seed crop on unlimed land contained 37% protein. That on the limed land contained 39.7%. This is a relative increase of about one-fourteenth. When superphosphate and potash were added with the lime, the concentration of the protein went up to 40.4%, or gave a comparative increase of almost 10% in the protein concentration of the seed.

These increases point themselves out as significant facts that can help in solving the present protein shortage. Soybeans grown on limed and fertilized soils will carry a higher concentration of protein in their seed. Each bushel of this essential food seed can carry about 10% more of protein because the soil supplies the essentials in the plant's manufacturing process whereby it uses nitrogen from the air to make the protein. Such an increase is particularly significant when it occurs in a food component getting particular attention right now on an international scale.

On most soils, whether growing legume crops or non-legume crops, the protein production is increased by the use of limestone. Corn is greener on limed soil because it gets from that soil more nitrogen which is the distinguishing constituent of protein. The grain also carries the increased protein effect. Soybeans like other legumes gather more nitrogen from the air by the help of nodule bacteria only when they get plenty of calcium and other mineral nutrients from the limed soil.

Now that we are going to produce seed, we are faced with the problem of meeting the shortages in the soil, instead of accepting the deceptive forages measured only in quantity and not in quality. We are now realizing that these same shortages have been giving us poor soybean hay.

2. more hay and seed, too

Miller and his son demonstrated that the lime and fertilizer treatments can change the soil over to one suitable for soybean seed production and better also for hay production. Without treatment of the soil, the crop produced 10 bushels of beans per acre. This is a figure not widely different from that yield given as the experience of many who grew the crop for seed in the early days of the soybean in Missouri. When the soil was limed only the Millers' seed yield was 18 bushels an acre. Where phosphate and potash were turned under on limed land, the yield rose to 26.5 bushels. Soil treatments increased the seed yield two and one-half times.

In terms of the pounds of protein these men produced to the acre, the figures corresponding to the above treatments were 222, 429, and 643. By directing their attention to the soil in terms of the common soil treatments, namely lime and fertilizer, the protein yields per acre were almost doubled and trebled. Surely such improvement by way of attention to the soil cannot be passed up lightly when we are facing the problem of protein shortage on a national scale.

These favorable results were obtained where the limestone had been put on a year previously and where the fertilizer was put into the deeper part of the plant root zone by plowing it under ahead of the beans. Where fertilizer was put into the row with the seeding, or placed in dry soil and above possible root contact during the plant's actively growing period, the figures of protein production per acre were only 243 and 384 pounds. Such experiences point out that if the soil treatments are to be effective helps in solving the protein problem, they must be properly used.

3. how fertilizer pays in cash

Efficiency of production—even without serious labor shortages—demands the increased output per acre wherever possible. The Millers have almost trebled their protein delivery per acre for the small extra input of 200 pounds of fertilizer, the labor cost of drilling it on, and the investment cost of less than $4 an acre. This resulted in 420 pounds of extra protein per acre with a value of about $12.50 as cottonseed equivalent. This says nothing of the acre increase in the oil output which in their case amounted to 181 pounds, for the use of lime and fertilizer, more than that from no treatment.

Soil treatments for the soybeans as the Millers have demonstrated by their trials in cooperation with A. W. Klemme of the Missouri Agricultural Extension Service, can do much to solve the protein problem. This added fertility contributes at the same time to the fat problem, which is one not much less acute. Soil treatments can make these contributions to our national emergency and at the same time relieve the labor situation if less acres can even increase the output. We need to think of soil treatments and use them more effectively as helps in relieving the protein shortage.

18.

Shortly after World War II, Edward H. Faulkner's Plowman's Folly *appeared. It contained a quite valid theory on tillage, but also proposed a dangerous soil self-sufficiency theory. Dr. William A. Albrecht was politely vehement in his refutation of the latter score. For real vehemence, however, nothing could surpass the private disagreements between Albrecht and agronomists who developed cover crops to shield thin ground. Granted that cover is an improvement, but "What's in it?" Albrecht would demand. "Nothing much but animated air and water." Albrecht was known to use the word "scurf" for such cover. Often he suggested that geneticists who juggled genes to get plants to grow under such conditions were simply "hijacking" stores of earth minerals already depleted in order to grow crops that were really not worth feeding. This paper might be read with these thoughts in mind. It became an entry in* Proceedings of the Ohio Conferences on Soil and Health, 1945, *and was published in* The Land *in 1946. Albrecht did not deliver this paper since he was out of the country at the time, teaching at an Army University in Germany. The sharpest point in his evaluation of protein and nitrogen centers on wheat.*

Protein Takes More Than Air and Rain A Key to Failing Fertility

Proteins are not naturally as plentiful as the carbohydrates. Carbohydrates are more nearly the products resulting from air and water brought into combination by the photochemical powers of the sunshine. Proteins are combinations of amino acids, body-building, life-containing portions of ourselves, manufactured in the life processes of the plants from carbohydrates. Protein synthesis is possible only by the help of some ten or more essential mineral elements supplied to the plant only from a fertile soil. Through the proteins, in a particular degree then, our nutrition is dependent on the fertility of the soil.

1. bread is the mainstay

Bread is the mainstay to which starving peoples cling. Wheat varies widely in its protein content. This variation underlies our classification of wheat as either "soft" or "hard," a variation commonly ascribed to differences in rainfall. Only recently have we seen fit to believe that the soil fertility underlies the production of protein in plants. Wheat may well be taken as a pattern of the process and result.

A recent decline in the concentration of the protein in wheat has been alarmingly recognized. For the crop of 1944 the War Food Administration announced that the minimum figure was set at 10.25% protein as compared to 11.00%, which was the corresponding figure for the previous years. This reduction in the specifications upset the millers and bakers. It should not be without reverberations in the minds of all of us. It serves notice to all of dwindling supplies of soil fertility and falling nutritional values.

High wheat yields under liberal rainfalls for the last five or six years in our wheat belt have been rapidly exploiting our store of soil fertility. Isn't it about time that we consider managing the fertility of our soils with food quality in mind as well as mining them continually for maximum production of bulk only?

2. weather and soil

Any crop is the result of the weather and the soil. Weather is beyond man's control. Soil yields to the skill of his husbandry. But it takes so much rainfall and so little from the body of the soil to make a crop! Likewise it is so easy to note differences in the weather and so difficult to evaluate differences in the soil as they affect the plant body. Consequently, one can readily understand why the weather is quickly considered the common control of the yield that is measured as bulk and likewise of the quality that registers itself, for example, only as protein concentration in the wheat grain.

When the soil itself is only a temporary rest stop of rocks while the climatic forces of rainfall and temperature are dividing and hustling them off to the sea, we are prone to believe that the weather and not the rocks determine both the quantity and the quality of the crop. We are not so ready to see that, first, the climate makes the soil, and that second, the soil, by its control of how well the crop is fed, determines how much of crop and what quality of crop is produced. Both of these aspects of a crop are more apt to be controlled by how well the plants are fed than by how wet or dry, or how cold or warm they are. Our failure to understand many of the details by which soils feed the plants has kept the soil fertility aspect out of the explanations of plant behaviors. This failure has put weather, especially rainfall, into prominence as the controlling factor in crop yield and in the particular crop quality, like protein concentration in the wheat.

Of course, rainfall is a factor in making the crop, but so is the fertility of the soil that determines through nutritional influence how effectively that rainfall will serve. At the same time that the meteorological forces provoking evaporation are taking moisture from the plant and thereby from the soil, only relatively small amounts of it are being built into the body of the plant. Large amounts are transpired in keeping the leaf tissues moist, just as we exhale much moisture from the lungs.

Shall we continue in confusion by thinking of moisture of transpiration as if it were in control of, or controlled by, plant nutrition? Is the degree of frostiness on our breath a matter of the liquid portion taken for breakfast? The transpiration rate is not a measure of growth rate, and water loss is too often a liability on account of fertility shortage when we believe it an asset in growth. The rainfall is used most efficiently when there is liberal soil fertility. It is the climate that either provides or removes the latter. The rainfall therefore enters into crop control not directly but indirectly by way of soil fertility.

3. *differences confused*

Differences in rainfall, in soil fertility and in protein concentration are apt to be confused in their causal connections. That the bushels per acre are not entirely a matter of the amount of rainfall is clearly shown by the fact that recently with only 20 inches of annual rainfall the Dust Bowl has had areas with yields of 40 bushels of wheat per acre. Missouri, with 40 inches of annual rainfall, would be doing well to get an average yield of half that much. It is a common concept that more rainfall means more crop yield. This is the truth in terms of the ***bulk*** of vegetation produced. It is well illustrated as one comes from western Kansas with its short grass area in the belt of 20 inches of annual rainfall and moves eastward to the former tall grass or big bluestem region of eastern Kansas and the more eastern "prairie" states. While the production in total bulk of vegetation or grains per acre is going up in this traverse eastward, the *quality* in both vegetation and grain as complete and efficient sustenance for animal and human life is going down.

In wheat, as a single crop covering this entire stretch of wide climatic variations from west to east across the midlands of the United States, the percentage of protein in the grain goes down as we move from low annual rainfall to higher annual precipitation. While pointing only to the fact that protein, which is such an essential for body growth of higher animals, is *going down* as percentage of the grain, we have not been emphasizing the fact that the carbohydrate, which serves mainly as fuel and fattening for higher animals, has been *going up* as component of the grain. Nor have we pointed to the fact that within that same grain the various nutrient elements of distinctly mineral origin in the soil have also been *going down* in concentration at a relatively more rapid rate than the protein concentration has been falling.

So oblivious have we been to this fact of the plant's chemical composition being related to the soil, that when the wheat grain increases its ash content to the point of putting more than 0.5% of this into the flour milled therefrom, we are prone to accuse the miller of getting "dirt" into his product during the milling process. We are about to realize that it is the more fertile soil giving this higher mineral content.

In the variations in the yield and the quality of wheat the variations in the rainfall have not been consistent with them to make the rainfall stand out as a causal factor. More rainfall gives more vegetative bulk as one goes eastward, yet with a constant amount of rainfall across that area there is an increasing grain yield while going in the opposite direction. The straw output is not necessarily commensurate with it. Surely the differences in plant nutrition, in terms of what the plants find to eat in the soil rather than of the amount they drink from the heavens is at the control of the plant behaviors in their ecological array. We must call into this picture of composition of the grain of wheat—variable along the lines of latitude—some other causal factors beside the difference in annual rainfalls. We must examine the soil fertility.

Note that particular rainfall belt into which wheat is ecologically fitted. This crop is not so commonly found at high production either as to yield or quality where the soil had undergone excessive development, or chemical breakdown and subsequent serious leaching. Rather, the wheat grain that is so commonly recognized as the main ingredient of the "staff of life" is located the world over in regions of moderate rainfall. It gives its maximum concentration of protein and conversely its minimum concentration of carbohydrate as it is produced in regions of the "more hazardous" or lesser annual rainfalls.

The rainfall as a force either in developing a soil or in depleting it cannot be measured directly by considering the annual precipitation. Increase in weathering forces may be first constructive and then destructive in terms of the soil's reserve minerals and the nature of its clay.

4. ratio of rainfall to evaporation

If we lay the pattern of rainfall over that of the ratio of rainfall to evaporation, we shall have help in understanding Dr. Marbut's soil map of the United States. Then, when we view these climatic forces of rainfall, wind, evaporation, and temperatures factors, it is not so difficult to see our western area as it represents increasing forces in soil construction and the eastern area as increasing forces in soil destruction.

The soil pattern is the result of these climatic forces. Is it possible, then, to see the belt of soil running north and south through the central United States as the balance between the forces, or with the maximum of soil construction and the minimum of destruction as the acme of good soil formation? Is it not possible to see there the production of a significant amount of clay in the soil, but not such a complete breakdown of the original rocks as to leave only clay in the surface soil, as is true in the humid tropics, for example? In these favored central and midwestern areas, that we may call the midlands, the soil contains reserve minerals of other-than-quartz in its silt ready to weather out now and pass on their nutrients for crop production. Soil construction has gone far enough to have removed the injurious alkaline elements, but not far enough to have taken away the alkaline earths, namely, calcium and magnesium, and other elements following in a chemical property series. It has, however, broken these essential nutrients for plant and man out of their rock forms in sufficient amounts to load liberally the clay complex with them, and in forms readily exchangeable to the plant roots.

The generous presence of calcium and other similar mineral elements in the soils of the midlands has encouraged natural legume growths to fix nitrogen and to build up in the soil the generous stock of black, nitrogenous humus so characteristic of the chernozem soils of the northern extremity of this belt. It is this particular fertility situation that gives these soils their great depth and uniformity. They are characterized not only by excellent granular structure but also by liberal stores of nitrogen and other elements to invite root growth far down into well aerated soil by which these plant parts get significant amounts of effective water even if it is present only at low concentrations. This fertility aspect cannot be so nonchalantly divorced from the chemical composition of the plants produced thereby, whether we consider their forages for animal feed or their grains as human food, more especially in terms of concentration of protein and other aids to growth as contrasted to that of carbohydrates of fattening value mainly.

In the eastern states with their higher rainfalls and particularly with their higher temperatures as one goes south and south-eastward, these climatic forces are severe enough to represent soil destruction. The calcium is readily taken off the clay by the leaching waters saturated with carbonic acid coming from decaying woody vegetation both within the surface soil and as deposits above it. In place of having calcium left on the clay, there is the hydrogen or acid. We speak of these soils that are deficient in nutrients by labeling them as "acid soils." We classify them as producers of "soft" wheats. They give us this crop high in carbohydrates mainly rather than in proteins and all else that is associated with protein production.

In the much warmer south, the more reddish clay is chemically and mineralogically different. It does not have the capacity to hold even hydrogen or acid so generously. Nor does it hold the many other positively charged chemical elements of nutritive value. Wheat grown on such soils is readily classified as "soft" wheat. Rainfall is erroneously considered to be directly responsible for this pronounced inclination toward production of carbohydrates, while we are oblivious to the wide differences in the fertility conditions of the soils. These, as the second factor in plant production, join with the weather to give us all the different kinds of vegetation that serve to embellish the entire land surface of the earth.

Surely the soil, as it is developed by the climatic forces to deliver fertility generously or as it is leached to where it does little more than hold water, cannot be disregarded so completely when the vegetation changes from legumes to non-legumes or even to woody forests in going across the United States. Nor can we forget the soil fertility when the west once supported buffaloes feeding on short grass and today is growing beef cattle naturally, while in contrast, the east and southeast can maintain cattle only by help of the utmost in veterinary and nutritional sciences.

19.

This manuscript turned up among the Albrecht papers dated "1964, reading time, 30 minutes." A much abbreviated version appeared in Agricultural Chemicals, *June 1964. In both cases, Dr. William A. Albrecht made the point that salt fertilizers can function with success only in a living soil. Further, he held, this success is purchased at the cost of chemical shock and wasting away of the soil's humus supply. This discussion on langbeinite followed discovery of deposits carrying three inorganic essential plant nutrient elements—potassium, magnesium and sulfur. In* The Future of Langbeinite in the Maintenance of Soils, *Albrecht noted that "While potassium in plants stimulates their synthesis of carbohydrates, the element magnesium serves as the core, or chelated nucleus, of large organic molecules active as enzymes. . ." As for sulfur, "This element of multiple negative charges stands out in opposition to the single positive charge of the potassium and the two positive charges of the magnesium." Sulfur, as an anion, combines with the latter two cations to form langbeinite, originally the residue from evaporated sea water. Albrecht did not discuss the trace elements in the langbeinite papers, but the implication of trace mineral importance was there.*

Langbeinite—Its Use as a Natural, Potash-Mineral Fertilizer

The fact that agricultural limestone is used as a natural mineral fertilizer supplying calcium (and magnesium) on humid soils, and with so much improvement of the high-protein crops, should arouse the question, "Why not consider other crushed, natural minerals as direct applications for better nutrition of crop plants?" Langbeinite, a natural potash-mineral, carrying sulfur and magnesium also, deserves such consideration.

1. more concentration in less soil means more efficient use by plants

Experiments have demonstrated extensively that the calcium (and magnesium) of limestones are seasonally available to the plant root in spite of their rock nature and their low solubility as calcite and dolomite. Though these minerals give no salt injury when put in contact with the planted seed—for the avoidance of which damage the special placement of commercial fertilizers was designed—this latter practice for applying limited amounts has found itself highly efficient for agricultural limestones. Though they are considered insoluble, they are, nevertheless, *available* within the growing season of application and thereafter.

Turning the research light on the age-old liming practice, by testing it under these limited applications, has established the concept that the soil need not be a uniform medium as to the distribution of all the essential nutrients. Rather, "the soil may be a mixture representing a heterogeneous collection of foci of each of those in the mineral or rock forms, weathering slowly while in contact with the acid clay. Plant growth may, then, represent the summation of root contacts with all these centers of fertility while the roots move to—and get from—them all that is needed for maximum crop productivity. According to this concept, both the very soluble and the less soluble nutrient materials, applied as fragments or in granular sizes by specific placement, would maintain the seemingly beneficial heterogeneity of fertility sources for plant growth better than would any practice aiming to blend the soil to the uniformity with which nutrient solutions are endowed." [Plant Nutrition and the Hydrogen Ion. V. Relative Effectiveness of Coarsely and Finely Pulverized Limestones, by William A. Albrecht]. This concept suggests, also, an economy when other more soluble mineral fertilizers, like langbeinite, will serve as does the less soluble without the extra cost of their finer grinding.

This concept of heterogeneity of the soil as plant nutrition holds for the clay's services as a colloid adsorbing the more soluble minerals, like langbeinite, with its nutrient elements as foci held against loss to percolating waters. The concept applies also to the acid clay's activities giving foci dissolving the calcium and magnesium carbonates of the stone to make those elements exchangeable therefrom to the plant roots. Then, too, the higher concentration in such limited soil volumes makes the fertilizer application more efficient. The plant can take a larger percentage of it than when less fertilizer mixed into more soil allows the clay to hold a higher percentage, and that too firmly for the plant roots to remove.

2. fragments versus fine powders, tested by experiments

With reference to fragments rather than finely pulverized minerals applied to the soil, the above concept of heterogeneity has always been holding true for plant growth by nature. It was tested for agricultural practice by using limestone as finely pulverized rock passing the 100-mesh sieve in contrast to using the rock in a mill-run, 10-mesh mixture with fragments of that maximum size, and with less than one-third of the mass of the 100-mesh fineness.

The limestones of these two finenesses were used to establish biennial sweet clover in the oats crop of one year so that the clover could be the green manure plowed under the next spring ahead of the following corn crop. By this means, the effects of the limestones were measured in terms of the resulting organic fertilizing values, especially nitrogen-fixation from the atmosphere, for the corn crop with that element coming through the sweet clover.

It was most surprising to find that as little as 300 pounds of the mill-run, 10-mesh stone, drilled into the soil in the same manner as soluble salt fertilizers commonly are, increased the corn yields by 8.2 bushels (the mean of eight successive crops) above the yields alongside where 300 pounds of the finely pulverized (hence more soluble) 100-mesh stone were similarly applied. It was also regularly observed that while the latter, the finely powdered stone, started the sweet clover in the nurse crop of oats, it did not carry this legume along into its second year.

Apparently, the active acidity and the high exchange capacity of the acid clay dissolved the finely pulverized stone rapidly to adsorb the calcium from its carbonate combination and to hold it too firmly, or ***unavailable,*** there for uptake by the plant roots. In case of the mill-run, 10-mesh stone, however, the fragments of 1/10 inch dimension supplied enough focal points of high-calcium saturation, of which the effects as calcium for nutrition on the plants carried into the second season to make sweet clover an effective green manure for the above increase in the following corn crop. From these facts, it is clear that the granular particle sizes of the more soluble fertilizers can also be expected to represent this extended, or "sustaining fertility," effects just as nature has been using that principle during the past ages.

3. roots find deeper foci of mineralized organic matter

We have been told that roots may be going through the soil but not necessarily growing in that direction. The "going," or thinner, roots suggests their struggle in searching for nutrition; the "growing," or thicker, ones must be finding it enroute. That simple difference in root behaviors was readily illustrated by barnyard manure mixed with phosphates and plowed under in the experiments by Professor A. R. Midgley of Vermont. By making vertical cuts into the soil, there were exposed the very slender roots finding the clumps of phosphated manure. Also, the thin roots became large ones both as individual roots and masses of them.

Similarly, the fine root that finds the crack in the joint of the tiles in the sewer outlet of your home, grows a root mass within it large enough to stop the flow of sewage for serious sanitary disturbances. Those very thin, but searching, roots go deeply into the soil and extend widely through subsoils of low fertility to find more of it in soil volumes as limited as that of spillage out of, and around, the crack in the sewer tile, which spillage leads the root to find entrance into the sewer.

In the light of such struggles and successes by the plant to search and to find, we should have no hesitation in applying either the less—or more—soluble fertilizers deeply enough to put them below the more organic surface soil, i.e. on top of the less fertile sub-soil, and to expect the plant roots to find them for improved nourishment thereby. Plant roots less successful as searchers for foci of higher fertility are *going* but *not growing*, mainly at the expense of low yields of plant tops. Roots, more successful in finding those foci numerous throughout the soil volume, grow themselves, and the yields of the plant tops, both larger as the result.

4. microbial cooperation instead of competion

The placement of fertilizers has made only its first step as improvement in fertilizer practice when it drills fertilizers, irrespective of the less—or the more—soluble forms, at close proximity to—not in contact with—the seed. The practice of "plow-down for potassium," recently adopted by farmers, points to the need for a second step of improvement; namely, that of not mixing it so much with surface soil. That improvement calls for placement *below* the surface soil where there is less competition by microbes because of less organic fertility on which they depend for energy.

By their ample energy supply from the sunshine, the plant roots travel through the surface soil readily. Soil microbes are not similarly endowed, hence abound mainly in the surface, organic-rich layer. The fungi, consuming the highly carbonaceous organic matter and abounding in the uppermost surface soil, extend themselves by their thread-like growths, or mycelia. Their activities of decaying the crop residues are most prominent in and near the upper horizon of the surface, thereby narrowing the carbon-nitrogen ratio of those organic materials for their fuller subsequent use by the bacteria. Both fungi and bacteria consume the soil organic matter more rapidly (growing themselves faster accordingly) when nitrogen and other inorganic nutrients are amply available to represent their narrower ratios to the carbon as better diet for either one of the microbial life forms. Due to inorganic nutrient supplements from the soil, the decay of organic matter put into the soil is more rapid than when it accumulates on the soil surface. [See *Methods of Incorporating Organic Matter With the Soil in Relation to Nitrogen Accumulations*. Missouri Agricultural Experiment Station Research Bulletin 249, December 1936, by William A. Albrecht.]

It is very probably in relation to that aspect of a balanced microbial diet offered in the vicinity of the plant's root surface or in the rhizosphere, that we find a few species of fungi with their mycelia, not only closely entwined with the plant roots, as the so-called "mycorrhizal" association, but also entering them to a shallow depth with improved growth-effects on crop plants. This may be an unappreciated symbiotic interrelations, or one beneficial to both plant and fungal species in the upper horizon of the surface soil. It is another one, emphasizing the organic matter of the soil and a relation apt to be disturbed by mixing salt fertilizers with the surface soil.

But the diet of the bacteria is favored by a narrower ratio of carbon to nitrogen, and to other soil-borne nutrient elements, than fungi require. Consequently, the bacteria operate well in the lower part of the surface soil. They, too, speed up their growth and consumption rate of the carbon of organic matter to bring about its escape as a gas into the atmosphere when nitrogen and other fertilizer elements are increased by their salt applications. While fertilizers give increased crop yields, that economic advantage is too often gained at the high cost of the speedy destruction of the soil's reserve organic matter. The rate of that tells us that our reserves cannot "buffer" the microbial shocks from many salt treatments regularly repeated over the years. There is the cost, also, of a competition between the microbial crop within the soil which takes up the salts ahead of the crop above it. But that occurs only when soil organic matter provides the microbes with the energy from the sunshine and nourished by our deeper placement of fertility below the surface soil, will reduce the competition, or that cost, and favor the cooperation of the "life of the soil," with the crops we grow above it.

5. sanborn field gives evidence

While the reserve organic matter of our soils has served as "shock absorber" in the past century against our mistreatment of "the living soil," a much more carefully considered soil management must be followed right soon for soil fertility maintenance as the data of the second quarter century from Sanborn Field at the Missouri Experiment Station show, by citing changes in their amounts of total nitrogen per acre of soil under different treatments.

Among four plots grown to wheat continuously, of which the soils were analyzed chemically at the close of the first 25 year period and then again after the second one (1938), plot 2 contained 2,140 and 2,000 pounds (respectively at those dates) of nitrogen per acre (2 million pounds) where commercial fertilizers were applied annually equivalent of the fertility removed in a 40-bushel crop.

Plot 5 contained 2,800 and 2,380 pounds of nitrogen for the two successive quarter-century periods, respectively, including first the treatments of six tons and second of only three tons of barnyard manure annually.

For plot 20, the nitrogen inventories reported 2,900 and 1,620 pounds per acre after six tons of manure annually and then after the change to ammonium sulfate for the next 25 years.

For plot 30, the corresponding figures were 3,420 and 1,880 for the shift in the periods of treatments, first, of six tons of manure and then of sodium nitrate, for the second quarter century.

The startling facts shown by those figures are the losses of 44 and 45% of the soil's total nitrogen to represent that much loss of the reserve organic matter which was originally present in the soil before nitrogen as commercial salts was mixed into the surface soils.

In case of two plots in continuous timothy during the half century, one with no treatment and one with the six tons manure, both during all that time, the former, plot 23, had quarter-century inventories of total nitrogen of 2,820 and 2,700. Plot 22 with six tons manure annually had 3,540 and 3,900 pounds nitrogen per acre. The former was a loss of 4.2% of nitrogen as organic matter for no soil treatment under grass sod. The latter plot had a gain of 10.1% for manure annually put on as a surface application.

It is evident that the three cases of applied chemicals, particularly the nitrogen alone, were rapidly destroying the soil organic matter when the microbial action of balancing the salts called for carbon as energy to effect that, and consequently, bring about its oxidation rapidly. That is a soil process much as is that for the urine of livestock, mixed with the straw-bedding and piled, which heats rapidly in destruction of the straw with its wide carbon-nitrogen ratio suddenly given extra nitrogen and other salts for microbial balance of the less active carbonaceous organic matter, preserving itself against decay in absence of nitrogen.

6. deep placement—practical, and biologically advisable

That the principle of deep-placement for higher concentration of fertilizers in limited soil volumes is a feasible and effective practice was demonstrated by the Missouri Agricultural Experiment Station. For this there was used a TNT plow, designed by the John Deere Company for test purposes. It had a small subsoiler following closely behind the regular 14-inch moldboard plow. Dropping of the fertilizer closely behind the subsoiler was accomplished by means of an International Harvester hopper attachment. The fertilizers applied on four plots were a) 8-8-8, b) 8-0-8, c) 8-8-0, and d) 0-8-8, with the spring-plowing for corn. That plan gave fertilizer streaks at depths from 10 to 12 inches and at spacings of 14 inches to be searched out by the growing plant roots.

The significant observations made, and the facts established, were not the varied yields of corn grain by the four plots, but the advent of a late summer undergrowth of weeds, as a sequel to August rains. There was also the early seasonal moldiness and rotting of the tops of the standing stalks. But those appeared where the nitrogen, or the first of the three figures in the combination, was applied. The plot given no nitrogen continued its exhibit of clean cultivation, and the bright amber, mold-free dead corn stalks, lasting well into winter.

The uniformity of the stand of young weed plants over the deep and wide placement of that nitrogen or without green fertilizer streaks so common when those are due to shallow fertilizer applications in the surface soil, was most significant in telling us that the seedling plants will penetrate deeply and rapidly to find the extra fertilizer below the surface soil layer. This practice fits the biological behaviors of the root system and seems advisable, not only for providing nutrition in the deeper soil layers which are more uniformly moist in relation to rainfall, but it is also an advisable practice of conservation of the soil organic matter within the surface where the admixtures of salts, especially nitrogen, are so damaging by burning out the surface soil's carbon and nitrogen by the microbial shock of salt additions there.

7. summary

In the light of the preceding observations and experiments, it is evident that a natural potash-mineral, like langbeinite, may be effectively and efficiently used as a fertilizer when due consideration is given to the inorganic and the organic soil factors in relation to microbial and plant nutrition. The successful management of the interrelated activities of those three lower strata in the biotic pyramid, *viz.* soil, microbes and plants, calls for full appreciation of the following facts:

1. Higher concentrations of either the more-soluble or the less-soluble fertilizers in limited soil volumes will feed the plant for more efficient use of the applied minerals than when the latter are intimately mixed with more soil.

2. The above practice makes the clay efficient in holding the more-soluble materials for ready exchange and the less-soluble for more complete mobilization.

3. Fragments and granules of nutrient minerals in the soil represent focal points of *sustaining fertility,* and more than mere bits of *starter fertilizers.*

4. Deeper placement is not apt to put inorganic fertilizers beyond the prompt reach by plant roots, according to experiments and observations. This holds not

only for greater depths, but also for wider spacing as much as 14 inches according as placements are deeper.

5. Placement as deep as below many of our surface soils, or into the upper part of the subsoil, represents efficient use of the fertilizer by the plants. It represents, also, the elimination of microbial competition for it since they burn out much of the organic matter in their struggle for energy or carbon to balance their use of the more soluble minerals.

6. Nature's maintenance of productive humid soils under her climax crops suggests practices for us, when in their evolution the crops and soils conformed to two major conditions. In the first place, there were reserves of the requisite nutrient elements in weatherable rock minerals throughout the profile and/or they were regularly added to the surface by inwash or wind. In the second place, the organic matter was contributed most completely in place, both as root remains within, and as crop residues on top of the soil.

7. The above facts and factors demand consideration not only for an economic soil management looking toward fertility maintenance, but also for the most complete soil conservation through which alone we can hope for foods and feeds of nutritional quality supporting all strata in the biotic pyramid by which man at its apex is sustained.

When the use of agricultural limestone as a less-soluble, natural mineral fertilizer, supplying calcium and magnesium, has served for so many years in soil restoration and greater agricultural production by following the above, we have an excellent precedent telling us that a natural potash-mineral, like langbeinite, supplying also magnesium and sulfur, can be similarly used to good agricultural and economic advantages.

20.

The relationship of soil fertility to human nutrition received a block-buster treatment in one of Dr. William A. Albrecht's first papers on trace elements. In this case he wrote his own preface, which was also a summary.

• • •

Trace elements are focal points of research in the biotic behaviors and biochemical reactions with which they are connected. That trace element function in the activities of the proteins, and possibly vice-versa, is the suggestion from recent research.

They are tools in the plant's synthesis of carbohydrates and conversion of these into proteins and similar compounds. Hence, the magnitude of the plant's ash content of manganese, boron, zinc, copper, cobalt, chlorine, molybdenum, and even iron, does not represent the magnitude of their separate biochemical services. Elucidation of their creative functions is therefore difficult.

As cations, they may be adsorbed on, and exchanged from, the colloidal organo-clay complex of the soil to the plant root for the hydrogen or other ions offered in trade. As anions they may be active much like any other anion within the exchange atmosphere of the colloidal molecule.

The root-hair cell, another colloid, but composed of living protein, represents the dynamic potential for transfer of nutrient elements from the clay into the life processes of the cell protein and all others of the plant. Trace elements are part of the larger suite of ions required as a balanced ration for plant nutrition according to the laws of chemical behavior of the elements regardless of trace amounts or larger.

Higher concentrations of protein in legume plants, resulting from bacterial inoculation affected more ionic exchange from a standardized soil colloid into the plants. Since copper seems necessary for the plant's synthesis of its own protection against fungus attack, and since balanced suites of major nutrient elements grew scab-free potatoes while the unbalanced suites failed, the theory is ventured that the balance of the major nutrients may mean plant protein in quantity and quality for the root-hair's more effective movement of trace elements from the soil into itself and the plant.

Better balance of nitrogen and phosphorus in the soil growing corn protected the stored grain against the lesser grain borer. Better balance of calcium and nitrogen in the soil growing spinach protected it from the leaf-eating thrips insect. More calcium on the colloidal clay growing soybeans protected the plants from fungus attack. These were demonstrations under controlled experimental conditions.

Selection of crops for large vegetable yield as replacement for higher-protein ones failing on the same exploited soil may have increased the deficiencies of trace elements in animal rations and human diets. Less of proteins in quantity and quality because of deficient major nutrients, or their imbalances, may bring failure to mobilize trace elements from the soil into the crops. This is a sound postulate when the feeding of a non-legume hay grown separately on a soil gives trace element deficiency of the animals, but feeding it when grown in combination with a legume like alfalfa on the same soil does not.

We are slowly recognizing the problem of the proteins. It has become one of having soils fertile in (a) the major nutrient elements, (b) in organic matter, and also (c) in the trace elements required for the synthesis by the microbes and by the plants of all the essential amino acids in truly complete proteins. Trace elements apparently contribute more to quality than to quantity of nutrition in what we grow, especially the proteins.

• • •

Trace Elements and the Production of Proteins *was published in* The Journal of Applied Nutrition, *volume 10, number 3, 1957.*

Trace Elements and the Production of Proteins

Extensive research attention has been given during the last two decades to the trace elements and the proteins, the two parts of the subject accepted for discussion here. Both have been the focal points of intense efforts to understand the natural biotic phenomena with which they are connected. As the result, nature's secrets are being revealed about the connection between them as they may be reciprocally causes and effects in diverse behaviors by the many forms of life.

The trace elements represent the increasing refinement of our knowledge about the inorganic contributions by the soil to the support of life processes, both simple and complex. The proteins offer themselves in the many biochemical behaviors to be comprehended in their normalities and abnormalities, since proteins alone are the compounds which carry life. They are the ultimate detail by which microbes, plants, animals and man are chemically organic, living matter. Hence, while the trace elements are the area of maximum inorganic detail in our efforts to understand "the handful of dust," the proteins represent the corresponding degree of curiosity about what the resulting creation is when the warm, moist breath of the climatic forces is blown into that bit of rock converted into soil. Those of us who are concerned with soil as the means of agricultural production visualize the numerous trace elements no less essential in the production of the many kinds of proteins than we consider the more gross soil fertility, the sunshine, the air, the water and all else contributing to plant nutrition and plant growth.

1. trace elements via proteins are tools more than building materials

Very naturally, if any element enters into the chemical structure of proteins, it will be essential for life. Hence, we readily understand why carbon, hydrogen, oxygen, nitrogen, phosphorus and sulfur are required for the growth of crop plants and of all other life forms. Magnesium in chlorophyl for photosynthesis, iron in hemoglobin for oxygen-carrying service in respiration, and calcium in the bones of our skeleton give us nine of the ten elements considered essential for many years. The essentiality of more than these of the nearly hundred or so known elements was not so readily granted in terms of their recognized functions. This was true of potassium as the tenth on the list. We do not yet know of any organic combinations of potassium in living forms comparable with compounds of phosphorus and sulfur, nor the indispensibility of it in the biochemical processes. It can be partially substituted, but never completely, by other monovalent elements. Only recently has it been accorded more careful quantitive essentiality in matters of human health.

The ten elements so far listed were thought for many years to be all that is required for plant growth. But in recent years, more elements required in amounts equivalent to mere "traces" have been established as essential for plant growth. According to the amounts required, iron may have been the original among the trace elements and was classified long ago as essential because of its known function. Those recently established for only traces needed for plant growth are manganese, boron, zinc, molybdenum and chlorine, or but seven, including iron. Very recently cobalt has been reported essential for the nitrogen-fixing, blue-green algae, which are single-celled bearers of chlorophyl, but not classified so regularly as plants. Since plants absorb about every kind of soluble, inorganic element, the burden of proof is one for us to establish the essentiality for the required functions in which each element plays a role and without which element, therefore, the function fails and the plant cannot survive. The basic problem is one of setting up the conditions for control, and demonstration, of the function in plant growth failure under certainty of complete exclusion of the responsible element aimed to be tested. This involves chemical purity to a degree which we can approach only most laboriously and not always with utmost certainty.

That the list of trace elements will be extended may well be anticipated. This has been the experience even during the last couple of years when chlorine was added for plants, and cobalt was included for blue-green algae, especially for the latter's function of nitrogen-fixation. In order to establish essentiality of only trace amounts of an inorganic element, we must measure and with higher degrees of refinement than ever measured before, some function for which the smallest variation in amount (usually absence and presence) of the element is responsible. In the legume plants, like the subterranean clover, the symbiotic fixation of nitrogen by the microbial Rhizobia species in the nodules on its roots failed, and consequently the plant growth failed, unless the molbydenum deficiency in the soil was corrected by amounts as small as one-sixteenth of an ounce per acre. Here the readily observable symptoms as gross as crop failure, indicated the essentiality of a trace element in a function representing both microbes and plants in their synthesis of nitrogen into the proteins.

In case of the trace element iodine, demonstrated as essential for animals, its service is in the body's synthesis of thyroxine, the hormone produced by the thyroid gland without which metabolism and growth of the body fail. Iodine is associated with nitrogen in this chemical compound which, like the many other hormones, suggests itself as made up of protein molecules traveling through the blood from one organ or tissue to another and serving by catalytic action, or some similar way, as regulators of the many body functions. It is in connection with the functions of proteins that trace elements have demonstrated their essentiality, hence it is in the multiplying research on these life-carrying compounds by which we may hope to elucidate the roles and essentialities of the trace elements coming as inorganics from the soil and contributing themselves as fertility to the synthesis of the organics representing the growth of all life forms.

According as more means for detecting and measuring smaller and smaller amounts of elements are designed and discovered, the list of those essential trace elements found in living tissues and fluids will be a constant challenge to determine their essentiality. In the case of milk, the list of elements known earlier to be in it included calcium, magnesium, potassium, sodium, phosphorus, chlorine, iron, copper, manganese, and iodine, both major and minor elements all recognized as essentials. The natural deficiencies of iron, copper and manganese in milk were a decided help in the research in the trace elements. But when the spectrograph came along, this list of trace elements in milk was extended to include also silicon, boron, titanium, vanadium, rubidium, lithium, selenlium, barium, strontium, chromium, molybdenum, lead and silver. That these are all essential in feeding mammalian offspring remains a challenge to our lack of knowledge as to what inorganic elements, and how much of each, milk must supply to give complete nourishment in its protein transfer from mother to young. For these items we have no complete theories regarding how or where they function. Nevertheless, as we use some of them as soil treatments, as feed or as food—even in excess and with resulting damage—doubtless our observations will provoke theories and thus advance research, accordingly, to extend the list.

2. chemical cations and anions interacting on clay and plant roots include the trace elements

Enough has already been said about this subject to make the essential point, as H. A. A. Schweigert puts it [in *Discussion of the Mechanism of the Reaction of Trace Elements*] that the "trace elements function in biological systems and events, no matter whether the processes are in the nature of energy transformations, a stimulus, a synthesis, or a destruction; whether detoxification of an organ, or part of an organ, takes place; or whether substances must be translocated, transferred, split-up, or reconstructed. The trace elements are mainly concerned with the activation of enzymes." In Schweigert's listing of some 150 enzymes, five of the seven trace elements (those other than boron and iodine) are connected with many of them serving (a) as activators in ways not understood, (b) as integrants in some function, (c) as active in the enzyme, or (d) as the added part of some compound by which it becomes an enzyme. The functions in which the enzymes serve include oxidation, reduction, hydrolysis, transferences of parts of compounds to others, lysis, synthesis, and various other vital performances.

Lists of the same trace elements, and many others, are active in the root cell. Their activities include not only the cell's metabolic functions, but also absorption of the nutrient ions through the cell membrane by contact with the colloidal clay of the soil from which the inorganic ions are exchanged via the carbonic acid surrounding the root as the result of its excretion of respired carbon dioxide. It is the well-considered assumption that the root-hair proteins play their role as the chemical means of moving inorganic fertility elements, both cations and anions, through the cell-wall of the root hair, and that these are moved within the plant similarly through their absorption on the colloidal proteins. Thus, since proteins are amphotyric, *i.e.,* both negative and positive in their ionization of the carboxyl and amino groups respectively, proteins within the plant move the trace elements, cations and anions, from the soil into the plant. This is the postulate by means of which the explanations were built to equate the root hair as one cell against the clay molecules with active cations adsorbed there on a much less active silicate anion. Thus by single cell concepts we visualize how major and trace elements are held on the soil and moved from there into the root as nutrition for it and the entire plant. With our thinking reduced to the single cell scale, we can see the root hair as the equivalent in biochemical and biophysical principles of all the plant cells combined. Trace elements must function in connection with protein production in the cell of the root hair first and in other plant cells second.

Trace elements going from the adsorption atmosphere of the soil's colloidal particle into the plant root are each a part of a larger suite of cations and anions. Their complete effect on the plant is not an addition of their separate effects. Rather, it is an integration of the possible interractions of all the ions on each other, both cations and anions, and both nutrients and non-nutrients, in the colloidal atmosphere of the clay surface-root surface contact. Then, finally, it is the resultant effect of all that on the protein of the root hair cell. As cations (iron, manganese, zinc, copper and cobalt) and as anions (boron, molybdenum and chlorine) they must be viewed also in relation to the corresponding ionized groups of the major elements.

Among the latter, we commonly emphasize the interactions and interrelations between calcium and potassium, magnesium and potassium, calcium and magnesium, nitrogen and phosphorus, phosphorus and iron or aluminum, and between many others. Among the trace elements, the interactions are equally as pronounced, with some elements counteracting, whereby one prohibits another's injurious effects at higher concentrations. In the latter there is the illustration of arsenic offsetting the damage by selenium, and of copper counteracting excesses of molybdenum and zinc, but even not then unless in relation to the inorganic sulfate. In animals, the studies of the metabolism, according to E. J. Underwood [*Trace Elements in Human and Animal Nutrition*] of molybdenum seem worthless unless the copper and the inorganic sulfate in the diet are known. In the rat, the copper requirement is affected by the level of zinc. Copper and iron are interrelated in their activities in the building of red blood corpuscles. The level of copper modifies the effect of iron, and the level of iron, in turn, modifies the effect of copper.

Iron going into the plant and likewise manganese is modified by the level of phosphorus. All of this tells that trace elements, like any other inorganic contribution from the soil to the plant's nutrition, is a part of complex chemistry and biochemistry of which the separate parts cannot be varied without corresponding variation in many or all of the others. One factor cannot be varied while all the others remain constant. Older research tactics must be replaced by approaches more inclusive than the single factor techniques.

3. interactions by pairs are magnified in behaviors by swarms

If the problem of plant nutrition is one of soil fertility elements for plant diets as delicately balanced as the nutrition of humans and animals is, then the matter of deficiencies in trace elements for plants might be considered the result of imbalances often in the major elements. Research on these tells us that both imbalances and deficiencies of the major elements may bring failure in the plant's equipping itself to take trace elements from the soil to the same degree as when those are properly balanced.

If we grant the truth of the plausible theory that copper is needed by plants to grow their antibiotics protecting them from fungus attack, then a demonstration of balanced fertility (calcium in relation to potassium) growing potatoes free from scab, while its imbalance did not, suggests that the balanced fertility may have produced compounds in the root hair moving the trace element, copper, into the potato for its protection against the scab fungus.

It has been demonstrated [in *Nitrogen Fixation, Composition and Growth of Soybeans in Relation to Variable Amounts of Potassium and Calcium,* by William A. Albrecht and Herbert E. Hampton] that inoculation as a means of helping soybeans produce more protein, in contrast to non-inoculation and bacterial absence with less protein in the plants, increased the extent to which the constant stock of exchangeable major cations moved from the refined clay into the plants. Since copper is a cation, we may well postulate that the higher concentration of protein in the potato plants by better balance of fertility was moving more copper into those plants, too. Thereby they might be scab-free under balanced fertility but not under the imbalance of the major elements. According to such thinking, then, the trace elements may actually be present in the soil when imbalanced plant nutrition in respect to major nutrient elements keeps them out of the plant and leads us to believe the soil deficient in trace elements. By applying the latter, their higher concentrations may or may not move enough of them into the plant to exhibit their effects—either direct or indirect—in protecting it.

If the varied interrelation of calcium in potassium in growing potatoes either protects from, or invites, the fungus scab to suggest the plant's taking, or failing to take, possibly copper from the soil for antibiotic service in the potato plant, in like manner the interrelation of nitrogen and calcium also suggested variable plant protection against insects in the growth of spinach. Whether trace elements were involved was not tested. But with the effects by calcium of mobilizing other elements from the soil already granted and with the effects also by proteins for the synthesis of which nitrogen helps, there may have been more of some trace elements moved into this crop to create the specific protective proteins rather than only more crude proteins in general.

In another illustration, the interrelation of the two major elements, nitrogen and phosphorus in better balance in the soil growing corn, were the protection of the grain against damage by the lesser grain borer. Grown on a highly fertile soil of South Dakota treated with applications of nitrogen only, the sample ear of corn grain under observation in storage was badly damaged while the sample ear grown on the same soil fertilized with both nitrogen and phosphorus and stored in contact with the infested ear was not.

Here again we have the demonstration exhibiting no clearly specific cause, save the suggestion that higher protein possibly of a particular quality in the plant was probably a cause per se, or was a help within the plant root for moving more fertility in kind and quantity, including more trace elements, from the soil into the plants and was causally connected only indirectly. Again, then, the soil with balanced fertility may have mobilized more of the inorganics into the plant or may have facilitated syntheses of special protective organics. The services by the soil directly and by the plant directly, or by the two in their interrelations, brought about the protection of the plant via nutrition for which a trace element like copper has long been believed of service only in terms of a poison sprayed on the leaf surfaces.

Magnesium and calcium in their interactions with the protein of the cells of the intestinal wall of our own bodies may be theoretical grounds for interpreting the purging effects by Epsom salts as action other than merely dehydration by a salt. When salts other than the magnesium sulfate taken in equal quantities do not purge, it seems more logical to consider the Epsom salt effects as a case where magnesium is exchanging places with the calcium in the colloid of the intestinal cell wall and contents to make that membrane, and more, a magnesium colloid. As such, it would not necessarily keep water, calcium, and other matters from flowing out through the intestinal wall and from its blood stream into the intestinal tract by which the purge would be brought about. The duration of the purge would represent the time required for the blood stream to remove the magnesium from the intestinal wall and to replace it with calcium by which the former normal, non-purging intestinal wall is restored. That such exchange of magnesium for calcium takes place is suggested by the increase of the magnesium in the blood during the purge. That blood calcium is lowered is suggested by some types of arthritis for which such purges have been considered hazardous.

The plant root hair in contact with a colloidal clay of low saturation by calcium has demonstrated the losses to the clay from the planted seeds' supply of nitrogen, of phosphorus and of potassium. It has been possible, thereby, to grow what looked like a good "hay crop" of soybean plants on so-called "acid" soil which was low in its calcium and other fertility elements while the crop was building up the soils' supply of nitrogen, phosphorus and potassium (none others tested) through losses of these from the supply in the planted seed. What seems like a good fortune in building up the fertility of the soil by planting and growing a legume for hay is a misfortune for the survival of the soybean species. Grown on such soil it could not make seed enough to procreate itself in the next crop. There is still greater misfortune for any poor cow that consuming the crop, almost like a hay-baler, could not get the equivalent of protein and its accompanying essentials she would have obtained had she eaten the seed originally planted. Root hair cells in contact with the clay illustrate the many possible interactions between all the inorganic elements on the clay, and then between these and the organic contents of the root hair cell. All these illustrate the complexity of the biochemical requirements for protein production in growing healthy plants, animals and man.

4. plant products, microbes and trace elements even in the animal rumen

That the deficiency of trace elements may be due to the kind of crop which we select for its bounteous growth rather than one of higher protein content, not so easily grown without soil fertility additions, was emphatically exhibited in a report by Dr. H. J. Lee, of Australia [in *The Toxicity of Phalaris tuberosa to Sheep and Cattle and the Preventive Role of Cobalt*]. Now that the popular grass plant, *Phalaris tuberosa,* after its introduction there some 30 years ago, has extended itself in pure stands and in combination with other species to over 200,000 unirrigated acres (annual rainfall 17-20 inches), the sheep and cattle become afflicated and die with a staggers syndrome ascribed to the toxicity of this popular grass species. Diluted sufficiently with lucerne, subterranean clover or other more proteinaceous fodder species there is no toxicity. The animal damage and deaths result from some toxic principle within the plant—not stored in the tuber—during its active growth and made harmless as the plant dries off.

The trace element cobalt, administered as a soluble salt to the grazing sheep frequently, Dr. Lee reports, "will completely protect the animals against phalaris staggers even when they are grazed continuously on pure stands of phalaris. During the course of our various experiments, 77 sheep have been dosed at least once each week, with a total of 28 mg. of Co (as $CoCl_2.6H_2O$) per week. Not one of these animals developed staggers, whereas 52 of 78 comparable untreated sheep developed the malady under the same grazing conditions." Also, "when the frequency of the administration is increased, minute amounts of cobalt have proved entirely effective. Thus as little as .05 mg. of Co, when drenched each morning and evening, completely prevented the onset of staggers when all the untreated controls succumbed to the malady."

That the cobalt functioned as prevention via the microbial processes within the rumen of the digestive tract is suggested when "in another experiment it has been shown that frequent intravenous injections of cobalt chloride solution had no effect whatsoever on the incidence of staggers. Even though the amounts injected were sufficient to load the tissues of the sheep about ten times the normal concentration of this element."

Further, "it is known that cobalt injected intravenously is excreted in part into the small intestine, being conveyed there mainly via the bile, and that it is to be found in appreciable concentrations in the large intestines. From these considerations it is apparent that cobalt does not act systemically nor yet within the intestines, but rather that its preventive effect depends upon the maintenance of a certain minimal concentration of cobalt within the rumen."

Still further Dr. Lee reports that "This is a specific activity of cobalt. We have shown experimentally that sheep that were dosed twice each week with a composite mixture of the approximate soluble salts of copper, zinc, manganese, iron, molybdenum, boron, magnesium, nickel and titanium, at the rate of 35 mg. of each of the elements developed severe symptoms of staggers while confined to a phalaris pasture, whereas comparable animals in the same flock which were drenched with cobalt chloride at the same rate remained unaffected."

That the health of animals depends on delicately adjusted interactions between fertility of the soil, microbes, and plants is well illustrated in the reports from research in Australia and other reports on trace elements. There is the further significant suggestion that the deficiency of trace elements coming from the soil through the plant may not necessarily be remedied by using the hypodermic needle to put the deficient trace element into the system when such an introduction of it bypasses the alimentary tract, the portal vein, and the liver or the combination for normal physiological censorship before introduction into the bloodstream. Bringing trace elements along nature's assembly line from soil, to plant, and to animal or man is a procedure not yet understood fully enough to believe we are as completely in control of it as we are in the production of some non-living industrial output.

5. trace elements for humans and as soil treatments for animals give protective proteins

Elements deficient or out of balance, regardless of whether they are trace or major elements, will bring on failing functions with the resulting poor growth, insufficient protection against "disease," and poor reproduction. Such conditions mean deficiencies and imbalances in the essential organic compounds which are synthesized by the help of the elements. They mean less organic compounds in the soil from decay of the preceding crops grown there which may be taken directly by the root hair cells as organic "starter" compounds from which the plant's biochemical construction takes off in building all else that is the plant tissue, growing in mass and protecting itself against digestion by some other life forms like bacteria, fungi and even insects, if not also viruses.

When in the ecological climax of any crop in nature, we observe that the plants are a pure stand without crop rotations, without crop removal, and are without weeds, fungus diseases, or insect attacks, where no poisonous chemical sprays have ever been used, are we not compelled to believe that when conditions other than those prevail, the plant nutrition must be falling much lower than that of the same kind of plants in the climax? Must we not grant the failure of the plants to synthesize the proteins and all else by which they protect themselves in the climax? Should we not believe, also, that in failing to protect themselves they are correspondingly lower in the nutrition they represent in our use of them as our food? Fighting diseases and pests with poisons is a case of calling in pathology to explain where the failing physiology because of its failing support by complete nutrition is unknown and not considered.

Brucellosis in animals (and humans) suggests itself as an illustration of our undue emphasis on symptoms and pathology when the cow's failing conception, her abortion of the embryo calf, and all else of symptoms associated with this so-called "contagious disease" are ascribed to the Brucella microorganisms harbored initially in the vaginal tract. Much is made of the protein by-products from this microorganism taken into the blood stream and their provocation there of the body's production of counteracting proteins which agglutinate with laboratory cultured products from the isolated microorganisms. It is not commonly questioned whether the contemporaneous association of the vaginal infection and failing calf crop are truly causally connected; whether the disease is really contagious; or whether both might have a common cause in the nutritional deficiencies, including deficiencies of the trace elements. Yet, the fertilizing of the soil with all elements considered helpful, including the generous application of the trace elements, shifted a badly infected dairy herd from its failing reproduction to complete conception under minimum male services and with calf crops nearly 100%.

Seventeen heifers and their calves, all of which were produced during four years of contact with the marked herd after proper feeding by soil treatment for it, remained free of the microorganisms for which the other animals were about to be condemned to death. Here is a case where a "contagious disease" apparently was prevented by the best of soil treatments including the trace elements. At the same time, patients with undulant fever given trace element therapy under guarded diets and controlled activities soon shifted from ailing bodies to those of buoyant health and ebullient enthusiasm and testimony for the virtues of trace elements as food supplements.

Since protection of the body depends on specific proteins (antibodies) the synthesis of which is brought about by the introduction of the correspondingly foreign proteins (inoculation for a desired immunization); and since any proteins are built up by combinations of amino acids as proteins taken into the body and digested out of food; we are dealing, in the main, with the problem of always having ready, or providing quickly, the required amino acids. Since the human body does not synthesize these from the elements, it cannot build antibodies (compounds also of amino acids) hastily unless ample supplies of the required compounds are in the body's stores. If the body cannot build antibodies or immunity, must that not be due to shortages of the amino acids required for this construction?

When an allergic reaction occurs, must we not interpret that as due to the shortage of the amino acids (proteins) required to help the body handle the foreign, or causative, protein and thus preventing that severe reaction? Such shortages may result, seemingly, from the body's failure to have equipped itself earlier in life for such occasions. When a baby cannot take foods of promiscuous kinds, but must take each new or foreign one in the order of development of physiological processes or the body mechanisms to handle it, does it not seem probable that when the body does not meet a certain protein early in life, it is passing up the occasion for establishing the necessary mechanisms for counteracting such foreigners, and that those mechanisms may never be established? It seems clear, then, that the animals and humans must depend much on the amino acids delivered to them, while the microbes and the plants may synthesize them from the elements and simpler compounds. Accordingly, then, the protein problem is one of the major elements, the trace elements and the organic matter in the soil and above it by which the amino acids are synthesized by the growing plants and the microbes.

6. management of soil and crops for growing required amino acids offers hope

If crops are to be managed for their output of amino acids for us, then attention must go first to the amino acids about which we need be concerned, namely, the three most commonly deficient, tryptophane, methionine, and lysine; second, to the choice of those crops given to higher protein delivery and such in a wide array of amino acids; and third, to the balance of soil fertility in both major and trace elements, through which those three commonly deficient amino acids may be increased in the plant's output of protein for life processes of higher order primarily for its own survival and secondarily for ours.

Research work directed along these lines of modifying the plant's output of amino acids by using the colloidal clay technique for accurate control of the offerings in exchangeable soil fertility, and in nutrient solutions, has demonstrated that more methionine, the sulfur-containing amino acid, could be grown into legumes like soybeans and alfalfa by increasing the sulfur in the soil when all other elements were ample. Alfalfa pushed its concentration of this amino acid higher also on taking more sulfur from the soil.

Tryptophane was related to the magnesium, the increase of which in the soil served to increase the concentration of this amino acid in the legume crop. But it has not yet been demonstrated clearly that tryptophane's concentration could be increased unless most of the amino acids (those measured readily) were increased to keep them in a more nearly constant ratio. The concentration of this amino acid was not related necessarily to that of the nitrogen in the plants. Tryptophane in red top hay, a non-legume, increased decidedly as the calcium offered was increased. That concentration was higher as the variation in calcium was demonstrated at higher levels of phosphorus in the soil. Both alfalfa and soybeans carried higher concentrations of tryptophane as the soil offered more boron. Other trace elements were also effective in the increase of tryptophane synthesis.

Lysine has not yet been fitted into a relation to any factor suggesting the latter's significance in modifying the concentration of this one of the diamino constituents of the proteins. The trace elements were prominent in their effects toward increasing the amino acids in these legumes. To date these correlations have not become specific enough to connect one trace element with one amino acid necessarily more significantly than with any others in each case.

In alfalfa, sulfur deficiency in the soil has long been known to be disturbing to its growth. This deficiency, studied in relation to 18 amino acids in alfalfa, demonstrated the reduction in their concentrations in the dry matter for 16 of them. Sulfur deficiency in the soil increased the concentrations of but two of them, a doubtful amount of arginine, a six carbon, diamino acid, but a decided amount of aspartic acid, a four-carbon acid. The dry weight of the alfalfa represented by these two acids which were increased was the equivalent of the combined reduction of about ten of the other amino acids. This would suggest the role of aspartic acid as possible source of the others if the enzymes for its conversion into them were activated by the higher sulfur in the soil. Sulfur as a deficiency may be responsible directly for the failure of this conversion, or indirectly through some other role of sulfur.

The agricultural production of all that is required for nutrition of ourselves and our livestock, via plants and microbes by wise management of our soil already mined for many years, emphasizes the production of the essential amino acids as the major struggle. For this, not only the major fertility elements and gross yields of any kinds of crops will be the true measure or guarantee. Instead, soil fertility refined to the degree of trace elements, both those known and possibly many unknown, must be considered in relation to production of the complete array of amino acids essential for the nutrition of the particular crop, and in turn for the nutrition of animals and man. Trace elements can help as tools. Proteins serve similarly in many cases as enzymes, hormones, antibodies, etc. Trace element behaviors need to be fitted into the chemical order characterizing major elements, with reference to one element's activities influencing those of any others. If all these are studied by means of bioassays using microbes, plants and animals, then the chemical situations will be elucidated to degrees of refined clarity far beyond possibilities by common laboratory measurements.

Integration of the inorganic and the organic, of the chemical and the biochemical, and of other reactions must be extended. Such integration is nature's behavior as illustrated even by trace elements influencing the effects of organic compounds through microbes in the animal rumen. Also much is not recognized in the human body's reactions to "traces" of elements considered essential, to say nothing about the many others present in the body regularly but not catalogued for their essentiality.

Should we study ecological climaxes more carefully, possibly the essentiality of organic compounds of the soil would be emphasized more. Those might be connected with trace elements, both essential in soils growing foods for better human health through possibly more complete proteins for microbial life in the soil. We would see organic microbial by-products contributing nutrition to plants for better human health as well as for their own.

With increasing allergies, caused supposedly by our failing mechanisms for handling foreign proteins, we need to study the amino acids grown in plants other than crops. Now that we are using only a few trace elements as fertilizers on the soil, while we are dumping the major essentials, calcium, nitrogen, phosphorus, and potassium on generously, we should see the need to connect trace elements with complete proteins. Such knowledge would emphasize the growth of more complete proteins as our food, and for their protective helps and other incidental services in which they and the trace elements connected with them biochemically seem to perform widely.

All of this tells us that "ash" element deficiencies are only a part of nutrition of man and animals. That is true also for nutrition of plants and microbes, even

Calcium, in either the chloride or the acetate form, was more active in mobilizing the growth activities of these young soybeans into a healthy start than were potassium and magnesium.

Grown on soil fertilized with nitrogen only, the representative hybrid ear of corn (center) was taken by the Lesser Grain Borer, but grown on the same soil fertilized with both nitrogen and phosphorus the representative hybrid ear (left) was attacked only where the two ears were in contact for two years. The open pollinated ear in contact for six months had only one borer hole as damage.

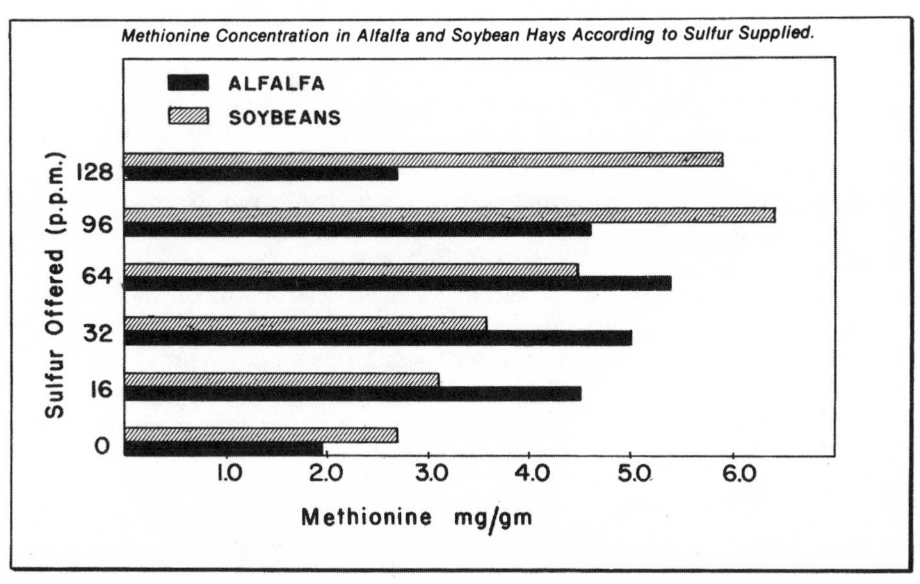

Sulfur as a neglected soil fertility element demonstrated its effect on the concentration of the commonly deficient amino acid methionine in the two legume crops, alfalfa and soybeans.

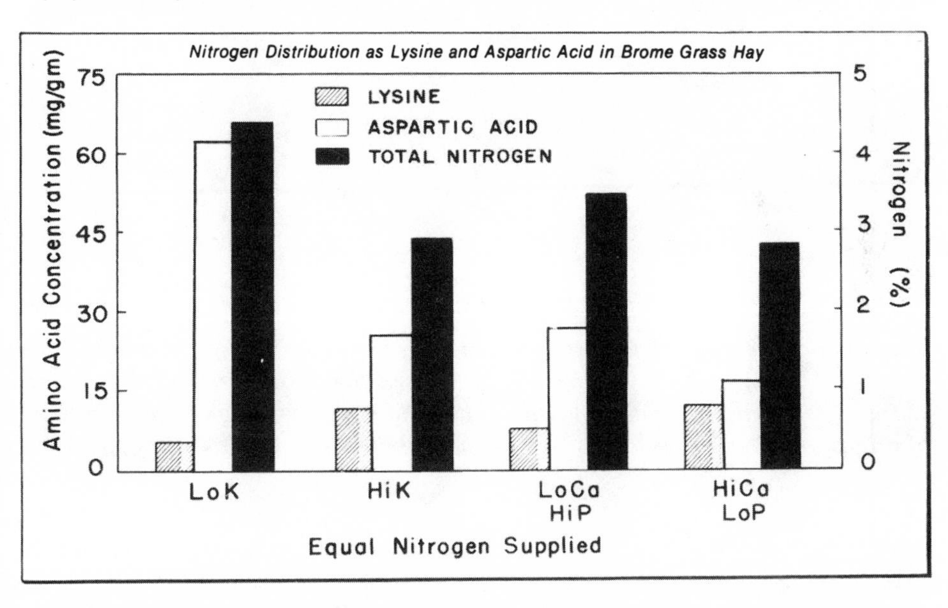

That the nitrogen in a non-legume like brome grass hay may vary without correlation to some of the different amino acids, especially the commonly deficient lysine, has been demonstrated. Total nitrogen as the index of "crude" protein does not give any suggestion as to the quality of protein in terms of the respective amino acid.

though they synthesize their necessary proteins from the elements. All lower forms of life can use organic compounds and grow more rapidly as those compose the larger share of the feeds. We are not yet well enough informed of all the items our bodies use when we consider their nutrition, hence even knowledge of trace elements and protein production—ever so challenging in their complexity of detail—leaves much that is still unknown if we were to attempt to design nutrition by assembling the separate chemical parts. Regardless of a profound respect for nature's contributions to our foods via the soils that grow them; for all the various crops that will serve as food; and for our judicious selection of substances either organic or inorganic with which to supplement these, there is still much to be learned if we are to build the best of health possible by means of proper nutrition.

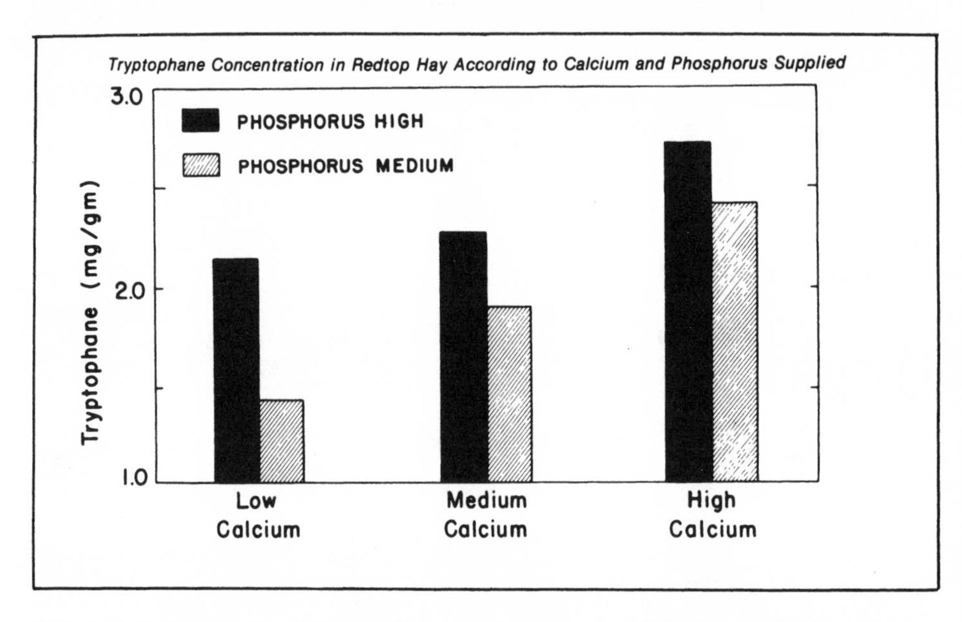

The major elements in their relations to each other modify the synthesis and concentration of tryptophane in non-legumes. When the phosphorus offered by the soil was low the increased offerings of calcium did not give as high a concentration of this essential amino acid as when the phosphorus was also high.

BACKGROUND

It's the Soil That Feeds Us

"You have got to have a vision," Dr. Albrecht used to say. "Unless you do, nature will never reveal herself." Quite early in his career, nature gave up rare secrets to Bill Albrecht simply because he was a curious farm boy who liked logic and adventure, in that order and in that order of importance. "As a boy—before I left country school—I told my mother I'd learned something. There are no hoop snakes. And I said, Mother, I'm going to study snakes. I got myself an Osage orange cane with a little fork at the bottom. And a cane is longer than most snakes. That snake has to keep at least half of its body down to get the leverage for the other half to strike. It has to have an anchor. When I finished my graduate school work, I had over 200 specimens of various things of that nature preserved and put away on the stockboard nailed on the joists in the basement, all cured. Alcohol only cost 50 cents a quart. And I knew the saloon keeper. The thing that disgusts me is that your scientists go to technology instead of teaching. They patent everything and make it secret. I don't like that. So I decided that I was going to study and learn. If you analyze what I've done here that they've paid me for, it's nothing but learning what nature did which had never before been recorded." In finding out what "had never before been recorded," Albrecht soon developed the outline for his career. In a simple form, this next entry states that outline. Beyond that, all else is elaboration. Dr. Jonathan Forman, the Ohio allergist and editor of State Medical Journal, once captured a career outline for Albrecht by building on a statement the Missouri scientist often made himself: "Food is fabricated fertility." Forman went on to point out that "within the plant, carbohydrates are built of air, water and energy from the sun operating through the chlorophyl within the leaf. This fabrication is the great agricultural chemical industry that is agriculture itself." Here, then, is the rest of Forman's statement. . .

• • •

Protein building, on the other hand, Albrecht says, is a constructive performance, more deep-seated in the plant. Proteins are, roughly, some carbohydrates into which nitrogen, phosphorus, and sulphur have been dramatically compounded. Just as potassium coming from the soil seems to act as a kind of carbohydrate catalyst, so calcium plays a catalytic role in the formation of protein. All life—plant and animal—analyzes about 5% mineral. In other words, all living things are 95% air, water, and sunshine and 5% soil. Since the part from the soil furnishes us with our skeletons and the spark plugs (catalysts) for the most of the vital processes in our bodies, soil fertility, or the ability of the soil to deliver minerals is of the greatest importance in all considerations of our national nutrition.

The mobilizing of soil fertility for plant nutrition, and thus for man indirectly, is graphically portrayed by Dr. Albrecht. The sunshine, the rain, and the water come to the growing plant, but the plant must send its roots down after the minerals. The hungry roots of the plant come in contact with the colloidal clay or the soil humus upon whith are held the nutrient ions like calcium, magnesium, potassium, etc., in forms not removable by water, but exchangeable by similarly charged ions. The clay colloids act as "food brokers" who can, from their stocks, offer quickly this adsorbed supply of calcium, magnesium, and other, for the hydrogen from the root surface offered in exchange. That is the reason why "exhausted" soil happens to be acid. It follows that the exhaustion is of significance, not the acidity as such, Albrecht insists.

Soils, Dr. Albrecht points out, are made by climatic forces. This accounts for an east and a west in the United States, as well as for the fact that our eastern states, long ago divided into a north and a south. Their natural divisions must not be ascribed to personal or political difference; but to those in the soil produced by climatic activity upon the native rock formations.

Plant health is dependent upon soil fertility, i.e., upon the availability of these essential minerals. Further nutrition for both plants and animals needs to look to clay soils and their essential mineral resources. Civilizations on sandy soil have always been short-lived. We in the United States need to take inventory of our soils to determine their potentialities for future nutritional service. We need to know how much we have in the way of mineral resources. We are about to enter upon international responsibility, a situation which makes this knowledge all the more imperative.

Fertility differs according to soil, depending upon whether it is under constructive or destructive climatic settings. This is the reason, Albrecht offers, for the fact that alfalfa grows without extra soil treatment in Kansas, Nebraska and Colorado where there is less leaching and less destruction with more construction than in many places. Cotton grows in the red, more highly weathered soils of the south. It is these different levels of plant nutrients that bring about correspondingly different chemical composition of the crops and move them from the category of animal forage feed in the case of alfalfa to those of no feed value in the case of cotton plants.

So in our crop juggling to overcome the loss of fertility and to maintain our yield, too much emphasis has been put on volume rather than upon value. This tendency has been decidedly aggravated by the fact that the advent of the machine has made food a market commodity and farming a business.

These new crops, Albrecht warns, may help to give us a supply of energy, or "go" foods, but they are not as helpful as "grow" foods which carry the essential soil elements elaborated into complex organic compounds of high nutritive value. We can see what soil erosion has done to our land, but we fail to appreciate what soil depletion has done. We do not seem to realize that our stewardship of the land calls for the return of that which we have borrowed.

Unconsciously, therefore, we have gone from a proteinaceous condition and a high mineral content in plants growing in soils under construction to a carbonaceous condition with a mineral deficiency increasing in soils under destruction. The nutrition of our animals and of ourselves at the same time tends to go from a level of bone-building, sound construction of good teeth of superior quality, and good muscle-building to obesity, weakened bones, and flabby muscles to say nothing of decayed teeth, alveolar bone disintegration and other troubles. So far as the people of the United States are concerned, the most common sign of malnutrition is overweight and obesity. We need to see the connection between soil and nutrition.

• • •

This entry was published as a series of 12 articles by Natural Food Associates in 1960, all headlined It's the Soil That Feeds Us. *Subheads used here designate the original titles for individual entries in the 12-part series.*

It's the Soil That Feeds Us

1. climates make soils to feed or to fail us

To the pioneer the climates were considered "good" or "bad" in relation to health. He often considered a "change in climate" as a remedy for bad health. Warm climates were the "cure" for tuberculosis at one time. But such temperatures with less rainfall and the foods on those less-weathered, mineral-rich soils were much better for health than warm climates with more rainfall and the resulting mineral-poor, protein-deficient foods grown there.

The food, in terms of the more or less fertile soil which is the result of the climate acting on the rocks, is what makes or breaks the body rather than the comforts of the climate in terms of how wet or how dry, and how cold or how warm. The differences in the climate make differences in the fertility of the soil. The soil differences make the differences in the foods which either feed us or fail us. Those differences are in line with the proteins which the soils produce, or fail to produce.

Our own United States are a good case for study of differences in climate. If one excludes the narrow Pacific coast area and starts from the east side of the coast range with its desert to go eastward, there is a gradual increase in rainfall. There is a gradual increase in natural vegetation and in crop production. As a result of that observation, one is apt to emphasize the increase in water as responsible, rather than point out that the rocks are weathered more to make more soil and more nutrition from that source. Starting with the rock, this increase in rainfall represents increase in weathering of it to make more soil with more clay in it. As there is more rainfall so is there more *soil construction* in terms of its having enough clay to hold the essential nutrients, and to have enough of all of these to stock the clay for plant growth of the high protein nature, like legumes. This is our west. It is dry too often to support trees widely. It can grow grass since this crop can stop growing in a dry spell and then take off again. So our west is the plains country with its grass because the soils make it and not vice versa.

More significant in our west is the *soil construction* in terms of its being *well stocked with fertility*. It has mineral-rich, lime-rich soils. Our west, beyond about the 97th meridian of longitude, has the soils with much lime. This nutrient is in the surface soil, often in the subsoil as a layer of "caliche." Other nutrients too are still there if the lime remains. Rainfalls are high enough to give *soil construction, but not soil destruction* to wash out the lime and make the soil acid. Rainfall isn't enough to make massive grass growth. That very fact has left fertile soils which cause every little bit of grass to be protein-rich, mineral-rich, and highly nutritious for protein production within the grass, in the wheat grain and in the bone and brawn of every little buffalo that soon became a big one and was able then in turn to make many more little ones.

In our east, that is, east of the 97th meridian which divides our lime-rich, protein-producing soils from our acid, carbohydrate-producing soils, the increasing rainfall represents *soil destruction*, even though the clay content is high. Unfortunately, the soil is weathered so highly that the carbonic acid from decaying vegetation had already taken most of the nutrients out and put acid, or the hydrogen ion, in its place. Plants putting their roots into that soil and offering to exchange their root acid, or hydrogen, for calcium, magnesium, potassium, ammonia, and other nutrients get little but their own acid back in the exchange. As a consequence, they can make carbohydrates—woody, starchy growth—but they do little to convert those into proteins. Protein-producing crops, like the legumes, starve unless we fertilize the soil.

The Pilgrim Fathers found our east covered with forest trees. This woody crop was all the Creator Himself was making of those washed-out, so-called "acid" soils. More washed out as they were in New England, the forests were conifers, not even hardwoods, giving a little protein crop in their seeds. And still more washed out by the higher temperatures on going south, again they made only a coniferous forest crop possible even when nature was dropping all the leaves and their fertility reserves back annually to decay there and to rotate through succeeding crops.

Much rain to grow much vegetation may mean soils so washed out, or so low in fertility, that more bulk means less protein. It means less of the foods that really build the body by making muscles and other proteins. Big yields of vegetation may therefore be deceptive in terms of building the body and guaranteeing the reproduction. Higher temperatures added to much rain aggravate the situation all the more even though we may point with pride to big yields of tons and bushels per acre.

The pattern of woody growth of forests in the east, and of nutritious grass in the west, is an expression of the soil as fertility for protein production rather than an expression of more rainfall for bigger yields in the east and less in the west. That pattern puts our preferred high-protein meat animals for beef in the west. It puts our short-lived, fat producer, the pig, in the east. "Grow" foods in the west, and only "go" foods in the east is an expression of differences in climate, yes, but more because these differences in rainfall make differences in soils representing either construction or destruction in terms of their growing crops that feed us (protein) or fail us (fat).

Soil for creation of new life more than for the fattening of an old one is a challenge to our knowledge of managing the soil fertility. Nature suggests that if grass is to be nutritious, this crop cannot be moved about and serve to truly feed us merely because it grows. It will fail us if we merely move it from the west to the east. If we move the grass that once made the buffalo, and makes the beef today, we must duplicate the soil fertility to feed the grass as well as this forage plant was nourished by the soils in the west. Plants, like seals, will not perform well unless we feed them well. According as the soils feed or fail our crops, so will those crops feed or fail us in good nutrition.

2. more fertility means move cover, stable soil structure and less erosion

When soils erode, our first reaction prompts us to take up the fight against running water. Much like when some disease comes over our body, we think first about "fighting" the microbes. When we break a bone, we put the limb in splints. Similarly when a field is broken down by gulleys, we line it up with terraces.

Whether it is our soil or our body that is in trouble, we fail to realize the preceding but gradual weakening of our body or bones and of the soil body, too. The weakening occurs long before the noticeable disaster of the fracture or the gulley befalls us. Broken bones too often are the result of malnutrition for a long time ahead to make them weak. Coffee and toast don't maintain bone strength. Unsteadiness in muscle may have come along with the weakening skeleton to bring on the fall as well as the weak and broken bones. In like manner, the exhaustion of the strength of the soil, its fertility, weakens the soil body to make erosion the consequence.

That such are the facts for the soil body is suggested by the experimental plots on Sanborn Field at the Missouri College of Agriculture. That field, after 62 years [in 1960] of its recorded behaviors, is a sage in telling us what the experiences of the soil body mean in bringing on what can be "old age" of it.

Two plots have been planted to corn each year since 1888. Professor J. W. Sanborn outlined the use of six tons of barnyard manure annually on one of these, while the other was expected to go forward in corn production with no soil treatment. Fortunately these two plots are along side each other. There is a good sod border on three sides, or in the direction water might run on these seemingly level areas. All of the crop, namely grain and fodder, is removed. Outside of the return of the fertility in six tons of manure on the one plot, the management and history of these two classic soils has been exactly the same.

That the removal of the fertility without return of any on the "no treatment" plot has weakened the soil body to make it erosive is now clearly evident. Had the sod border not protected this plot, its soils—like so much from the rest of Missouri—would now be resting in the Gulf of Mexico near New Orleans. After that soil body is turned by the plow, a single rain is enough to hammer it flat, to seal over the soil's surface, to prevent infiltration of the rainwater, and to bring on erosion of that fraction of the surface so readily and so highly dispersed into slush by the raindrops.

Where manure had been going back regularly each year, naturally there was a different soil body. It stood up under the rain and maintained its "plow-turned" condition in spite of the rain. It was the same rain that was so "damaging" to the other plot. One could not blame the rain for any damage here on this manure plot. Instead, the rain brought benefit. Its water went into the soil. It soaked a deeper layer and built up the stored water supply for the summer. This surface soil is cooler by 10 degrees in the summer than the companion plot. Here is a different soil body that behaves different under the same rainfall. It doesn't erode. The rills of running water begin at the line that divides the two plots. Narrow as these plots are, there are rills on the "no treatment," but none on the "treated" one. The former might seem to be a call to "fight" the running water. The latter is not.

Fortunately the "strength" of the soil body against erosion in this case is also the "strength" of the soil for crop production. It is also the "strength" for soil granulation or good soil structure. The corn yield is still twice as large on the plot with manure as that on the plot without it. Weeds grow on the former after the corn roots are deep enough to be beyond their use of the nitrates which accumulate on the surface to invite the weeds. These weeds are a nice "winter cover." They are one that comes there without cost. The granulation of the soil of the manured plot is so much better under laboratory test than that of the unmanured one, that water goes into the soil three times as rapidly. Also, it moves about four times as much volume of water down through and does not plug itself up quickly to stop water movement into the soil.

Here is "strength" of granulation. It is the "strength" of the soil body under the hammering effects of the falling rain. It is the "hidden" strength, and the very same strength that gives the bigger yields of crops. That "strength" is the fertility. This fertility is distributed within the inorganic as well as the organic fraction of the soil. Here is quiet testimony that we ought to see that the weak soil body, and the erosion of it, are brought on because we have removed the fertility, or the creative power, by which any soil naturally keeps itself in place and grows nutritious crops at the same time. Our weakening soil body is suggesting that gradually weakening human bodies are resulting from it.

3. limestone mobilizes other fertility too

Putting lime on the soils of the humid region has been practiced under the belief that removal of the acidity of the soil was the benefit from such a treatment. We now know that liming an acid soil is helpful because of the nutritional value of the calcium and magnesium supplied to the crops by it, and because it helps to mobilize other nutrient elements into the early plant's growth.

Experiments with a crop like soybeans demonstrated the need by the young seedlings for calcium early in their life if they were to survive. Any forms of calcium salts showed their benefits. These benefits were the same regardless of whether these salts reduced the soil acidity or whether they increased it. If the soybean seedlings were planted in a lime-bearing sand for no longer time than 10 days and were taken up, washed and transplanted into a soil, the plants were taller, grew better and gathered more nitrogen from both the soil and the air ever after, than when the first ten days of their growth were in a lime-free sand. Additional trials with other seeds have demonstrated the earlier emergence and better stands of the crop when the seeds were coated with lime or when this plant nutrient was dusted into the soil along with the planting of the seeds. All of these demonstrations indicate that the calcium of the lime is beneficial by the entrance of the calcium early into the seedling stage of plant growth.

More refined experiments were required to demonstrate the fact that lime as calcium, not as carbonate, serves to mobilize or move other nutrients into the crop. Korean lespedeza, originally imported and claimed to be an "acid-soil crop," showed very clearly its higher concentrations of nutrients other than calcium, when the soil was given this element in the soil treatment of liming. By growing test plants in a colloidal clay-sand mixture, it was shown that calcium was required to a relatively high degree of saturation on the clay if the plants were to grow. As this degree of saturation was increased, or as the amount of clay with any degree of calcium saturation put into the sand was larger—to give the plants more calcium—there was more potassium, more nitrogen, and more phosphorus taken into the plants. Lime was the leader, apparently, of the nutrients and was bringing them into the plants.

Quite unexpectedly, it was discovered that when the calcium supply going from the colloidal clay into the plants was very meager, then the nitrogen, the phosphorus or the potassium might even be going in the reverse direction. This was taking place when plants like the soybeans seemed to be growing fairly well. In no case were any plants grown unless they were increasing their calcium content by its migration from the soil into the plants. Growth was impossible except as calcium was mobilizing itself into the crop early. Soybeans plants that would look like a possible hay crop—but could not become a seed crop—had less nitrogen, or less phosphorus, or less potassium than the seed that was planted because the soil did not offer enough calcium to mobilize these essential elements from the soil into the crop.

Here was ample reason for one to become "lawyer for the defense" of the unsuspecting cow that would be asked to consume a soybean "hay crop" grown on a lime-deficient soil. This would be the case on soils for which the early propagandists for this imported legume said, "This is a hay crop if not able to be a seed crop." Fed on hay from this crop grown on such soils, the cow would gain less nitrogen, and less phosphorus, for example, on eating the hay crop than if she had eaten the seed that was originally planted. That a plant may be growing and making vegetative bulk while it is losing nitrogen or potassium from the planted seed back to the soil my still be doubted. But when some of our animals demonstrate their health disasters on much that is called "feed" because it is "plant growth," we ought to suspect that something like nutrients going in the reverse direction might be taking place.

Lime as a helpful soil treatment is quickly indicated by the animal's selection of the vegetation growing on it. All of this may be telling more than just more calcium recognized by the dumb beasts. It may be the indication of the better nutritional values created, or synthesized, within the crop because calcium has mobilized other fertility elements as well as itself into the crop more effectively. Plants create nutritional values by means of the calcium's nutritional service, and not by its removal of soil acidity.

4. thin roots are searching for, thick ones are finding, soil fertility

Root behavior in the soil still challenges our understanding of it. Folks commonly speak of "deep-rooting" or "shallow-rooting" crops as if the behavior of roots spreading out in the immediate surface soil, or of roots going seemingly straight down deeply into the subsoil was a predetermined matter controlled by the plant species. Quite the opposite is the case. It is not the crop that is deep-rooted. It is the soil and the distribution of its fertility that invite root behavior in that particular aspect. Roots don't go joyriding. They are in search of plant nutrients.

One needs only to dig into any soil under cultivation with occasional manuring for a long time to see the manifestations of the roots in relation to the manure and general fertility in the soil. Roots are extremely thin when they are only searching. In the soils under an old vineyard in France, where they cultivate several times annually and plow under manure at least once each year, it was an interesting experience to dig a trench and study the roots of some more than hundred-year-old vines. Cutting through the clumps of buried manure revealed a thickly packed mass of big roots. However, the attempt to find the root leading from the base of the vine down through the soil to the manure clump was quite another matter. Only by digging carefully with a penknife or washing the manure-clumped roots out with water could one discover the extremely fine, hair-like roots connecting the mass of big roots with the grapevine to which they were serving the nourishment. The fine root was the scout out searching. That was the root *going* through the infertile soil to become the root *growing* in the more fertile, manured portion of the soil.

An observation of the sewer plugged by the growth of a tree root was further support of the fact that the roots searching through the soil are thin but become thick ones once they are inside the sewer. When the sewer was dug out, there was but the one single crack in the tile connections of over 60 feet of it.

How did the root come to find that, especially when it was 30 feet from the nearest tree and about 7 feet down in the soil? This is the perplexing question. Many folks answer quickly, "It was searching for water." But the root had water all along the distance of about 30 feet of its travel from the base of the tree to the break in the sewer tile. It was drinking all the time while it was going that distance in search of nutrition. It was led to the crack in the tile by the sewage-fertilized soil around it. The more highly concentrated fertility as it approached that opening led the root right to the one opportunity in the entire sewer to get inside for some liquid fertilizer. Once inside, the root was soon growing.

The tree roots inside were a mass about the size of a horse tail. It was sufficient growth to cause the troubles in the plumbing. Was this root mass taking merely water? Surely not, since the water intake by that root surface would certainly be at a rate far too great to be carried through the crack by the small root, and back to the tree. It was getting nutrition. This was so dilute in the sewage that even the horse tail mass of roots would absorb this no faster than the small, thin roots could carry it up to the tree. Outside of the sewer, the root was only *going*. Inside of the sewer, the root was really *growing*.

Fertilizer placement in the soil may also demonstrate the roots' small size or the slender, thin roots when they are "searching" for nourishment. By putting the fertilizers down below and to the right or left sides of the seeds of beans, for example, the roots would be few leading out in all directions. But once they entered the soil zones into which the fertilizers were placed, there the roots were literally in clumps.

In the case of sweet clover seeded on a shallow surface soil underlain by a "tight" acid clay subsoil, the roots scarcely entered the unfertilized and unlimed subsoil. If they did, they were too thin to be pulled out. It was necessary to dig them out most carefully to even discover them. But if the subsoil was deeply limed, then the roots of the sweet clover were a single major one of unusual thickness growing straight downward.

Root behaviors, then, are not determined by the species of the plant. They are the reaction by the plant and this soil part of its anatomy to the fertility of the soil. It is not the crop that is deep-rooting, but the soil that is such if the roots are big and numerous to mark themselves and their size out so prominently. By the nature of the roots, then, one can judge the fertility of the soil. Thin roots tell us that there is so little of it that the roots are "searching" and finding all too little. Big, thick roots tell us that the soil is growing such sizes of them because they found plenty of fertility there.

5. quality as feed, not only quality as crop, demonstrates effects of soil fertility

Crop yields have long been measured by the amount of the plant mass. If crops are grown for the simple purpose of selling them, and if we measure the returns by the greater monetary income from more pounds or bushels sold, then bigger plants may be the reason for making soils more fertile. In a broader sense, it is the fertile soils that make the plants grow bigger. Though there may be big plants of some kind that grow big on soils that are not very fertile for other plants, one cannot conclude, therefore, that any big vegetative mass proves the soil to be fertile. One can reason the converse of this quite safely, however, and say "small, spindly, sick plants are indicators of an infertile soil."

Certain "big" plants are more apt to be indicators of fertile soils than others. But for that, one must know something about what the plant is creating or making while it is growing. Crops, like the legumes, which are said to "hard to grow," are usually indicators of fertile soils when they are making big plants and especially a big output of seed. It is this reproductive aspect, the activity of making new cells, of creating proteins—through which alone life keeps flowing—and of multiplying its parts and its species that really reports the fertile soils. Plants in nature are big and numerous because they have been multiplying themselves via production of more protein. The production of protein by plants is the real index of the fertility of the soil under the plants.

Perhaps you have never thought that what we consider plant growth is not necessarily multiplication of cells, for which more protein must be created by the biosynthetic, or life processes of the plant. Instead, it may be only a case of blowing up to larger volume the cells laid down in embryonic age. It may be making bigger those cells by putting in more water or more sugar and other carbohydrates of photosynthetic origin. In the watermelon and other cucurbits, this is the case. So the so-called "growth" reflects mainly the air, water, and sunshine going into the resulting products of sugar equivalent and not the fertility coming up from the soil to convert those carbohydrates into proteins, as we expect it to be done by legumes.

But when the plant's embryo is the place where the cells are all laid down, even in very miniature, then the true growth process there calls for much soil fertility and that in balanced proportions of the different nutrient elements. Even a watermelon requires a fertile soil in a certain sense, but we are apt to be misled in believing that the increasing size of the plant or the increasing amounts of plant product are necessarily proof of fertile soils. That fact has been demonstrated for soybean plants readily by shifting the fertility ratio, or the balance between calcium and potassium. We associate the function of the latter with the plant's production of vegetative mass; the former is connected with the output of protein by the biosynthetic conversion of the photosynthetic products or carbohydrates into protein. Thus by unbalanced soil fertility we may be misled to believe that big plants mean big soil. But with balanced fertility, we are correct and fertile soils really mean big plants in terms of much protein.

There is no fallacious reasoning in saying that less fertile soils give us the small, sick plants, when on other soils along side the plants are large. In the same field one can demonstrate this fact easily. Applications of the elements deficient in the soil soon show their effects in terms of bigger plants. Nitrogen on grasses in the pasture as a result of urine droppings is a well-known example in the spots of tall-growing plants. Lime on legumes like the sweet clover and alfalfa, magnesium on soybeans, zinc on fruit trees, copper on clovers in Australia and numerous other soil treatments draw their lines of differences clearly out in the field.

The sickly conditions of the plants give various signs and symptoms. Celery with its dark areas in the stalk when boron is insufficient; white-colored soybean leaves except for green veins when magnesium is deficient in the soil; reddening leaves of cotton under potassium deficiency; clustered small leaves in rosette-like forms of fruit trees needing more zinc from the soil or from spray applications, are all telling us that infertile soils make not only small but also sick plants.

We are gradually coming to realize that plant growth is a creative activity by a life form demanding proper nourishment if that growth is to go forward effectively. Bigger plants generally testify to a fertile soil, especially if bigger yields of seed and more extensive cell reproduction rather than just cell enlargement are the reasons for bigness. Conversely, small sickly plants are true indicators of poor nutrition of them by the soil. Closer observation of our plants and more knowledge of the symptoms of plant hunger are bringing us around to feed our plants via the soil rather than turn them out to rustle for themselves. On fertile soils alone can our crops be truly big in the food-creating services we expect of them.

6. weeds, as the cows classify them

Weeds are commonly defined as an undesirable and worthless crop. But when we think a bit deeper than the crops, and when we study the soils under them, we must use another definition. When weeds are left to grow in the pasture, that is, grow bigger because they are disregarded by the cattle, then we must define weeds as a crop so poor in nutritional values, because of the poor soil under it, that a cow has sense enough not to eat it. Have you ever thought that weeds left in the pasture are pointing to the need to treat that soil with fertility additions if it is to keep good feed crops growing? The cow has never learned the names of plant species nor memorized the "manual of weeds," but she knows the nutritional quality of the vegetation according to the fertility of the soil growing it. She demonstrates that very accurately whenever she has a chance to choose.

More than a hundred head of beef cattle gave such a demonstration on the Poirot Farms, near Golden City, Missouri. What those cows called "weeds," namely, a worthless crop as they judged it, was in decided contradiction to our customary classification of certain plant species as weeds. They refused and disregarded bluegrass, white clover, and even some soybeans in virgin prairie that had never had any soil treatment. For them the bluegrass and white clover were weeds, namely, plants that they had sense enough not to eat, because the herd marched right across this large field and through the gate on the opposite side to eat what had grown up in the previous year's cornfield left unused because of labor shortage.

On that abandoned field of choice forages as demonstrated by these cattle, there were only plant species which the bulletins and books call "weeds." There were cockleburrs, nettles, plantain, cheat, wild carrots, butterprint, wild lettuce, berry vines, and a host of others. Strange as it seems, all were eaten by the cattle and kept down to a short growth during the season while the adjoining virgin grass area, which they traversed daily for water, grew taller and taller.

"That temporarily abandoned but well fertilized corn field in weeds," you would be compelled to say, "was good pasture in the cows' choice despite plant species that we call worthless, but the cows select because of the higher fertility of the soil growing them." For the cows, the adjoining heavy sward of bluegrass and white clover was a worthless crop and thereby they classified it as weeds according to this demonstration of the herd that was expected to eat them. The cows defined a weed, then, as any plant, regardless of species or pedigree, growing where the fertility of the soil is too low, or too unbalanced, to let the plant create what is nourishment for the beast.

Not so long ago, an able botanist surveyed and listed the plant species in the flora on the western plains where a large herd of cattle was grazing. He reported 65 different kinds of plants in the herbage and, strange as it may seem, not a one of them was refused or left untouched by the cattle. On those more fertile, less leached soils under lower rainfall, the cows literally made a clean cut of the forage, irrespective of plant species. Differences in the so-called "palatability" of the plants according to their names are seemingly unknown on those mineral-rich soils.

So when we characterize certain plant species by calling them "weeds" or by saying they either are or are not "palatable," we are giving more or less academic definitions. These definitions apparently need drastic revision if they are to agree with what the cow "calls" them in her classifications.

When we talk extensively about going from corn farming, with so much plowing and its erosion, to a grass agriculture for less soil loss, the cow's judgment of that shift may well be consulted. We need to make that decision with the cow sitting at the conference table as the foremost member of the planning committee. She will recommend that we forget the idea of merely shifting to another combination of crops that we expect her to eat. Instead, she will point out that erosion has increased because of declining soil fertility, and that such lowered fertility will make any combination of crops merely weeds as she defines them. She will tell us that it is not any particular system of agriculture but the higher fertility of the soil under it that makes any kind of crops worthwhile feed for her. Going to a grass-agriculture that will *feed* and not *fool* the cow calls for improvement of the *soil fertility first* and then the selection of *the crop combination* second.

The cows on the Poirot Farms voted against a grass agriculture of bluegrass and white clover on soils of neglected fertility. On the contrary, they preferred well fertilized land under corn farming and the temporary neglect provoked by economic pressure, that could grow only the crop combinations which quickly seeded themselves and we would call weeeds. Cows know their crops, even weeds, better than we do. They judge crops not by pedigrees, species names, nor even tonnage yields, but by the fertility of the soil growing them and thereby the nutritional values as cow feed. The cows recommended more attention to fertilizing the soil and less to fighting the weeds. As we make the soils more fertile in our pastures, we shall have more nutritious forage and fewer weeds as the cows classify them.

7. *more fertile soils prohibit insects*

It was an advertisement for tobacco a few years ago which kept reminding us that "nature in the raw is seldom mild." That was a way of saying that not only the strong individuals within a species survive at the expense of the weak, but also that the weaker among one species may fall victims to other species. That is true not only when we know that weak, poorly nourished plants fall victims to bacterial and fungus attacks but is true also for weak, poorly nourished plants falling victims to the attack of insects. Most of us have given but little thought to the possibility that our increasing troubles with insects may be the result of the declining fertility of our soils. Such are the facts demonstrated by some recent experimental work.

Spinach was grown at the planting rate of two plants per pot on soils of which the clay had been given uniformly specific amounts of all the nutrient elements except calcium and nitrogen. These two elements were arranged in a series of 5, 10, 20 and 40 milligram equivalents (ME) of nitrogen per plant, and each quantity of this was put in combination with the corresponding series of amounts of calcium per plant. During the mid-mature stage of growth of the plants, there was a scourge by the thrips insects. These "pests" eat the green portion of the leaves and allow clear areas of the skeletonized tissue to remain. The severe attacks on some plants and the complete abstinence by the insects on some others, marked itself out in a pattern according to the variable soil fertility applied as different amounts of nitrogen and calcium. This arrangement of the insect attack on the plants according to the soil fertility was all the more striking since the treatments were replicated ten times, and the arrangements of the insect behaviors were thereby repeated ten times.

Insect attacks were numerous on the spinach plants given only 5 and 10 ME of nitrogen. There were no recognizable attacks by the insects on any of the plants given 20 and 40 ME of this nutrient valued for its services in protein synthesis and appreciated for its ever-presence in the chemical makeup of protein. When we remember that calcium, or lime, is a necessary treatment on humid soils to grow the protein-producing legumes, it was especially significant to note that even when the insects attacked the plants given the smaller amounts of nitorgen, those attacks were less severe according as those small amounts of nitrogen were combined with increasing amounts of calcium as additional soil treatments.

Here was the suggestion that the nutrition of the plant, that is, the better balance in the fertility of the soil for production of more protein in the crop, represents better protection by the crop of itself against insect attacks. "Strength" of the plant in terms of guarding itself against insects is also "strength" for reproduction when we remind ourselves that it is only protein that can reproduce itself. These plants using more nitrogen and more calcium to make themselves more proteinaceous were producing within themselves, or building their own defense against the thrips insects. Perhaps you might say "they had built up a resistance" to this trouble. This "resistance" could scarcely be viewed as resulting from a tougher or more fibrous plant tissue, since fertilizing with nitrogen would be expected to do the very opposite, and make the spinach more tender, or luscious.

Here was a case where protein was protection, much perhaps as we use special proteins to be protection for our own bodies or for ourselves. This is much like the situation when we inject serums of various orders, carrying so-called antibodies, as special proteins for protection obtained from animals made immune to the specific disease in question by first having the disease. Immunities in our bodies to attacks by microbes and viruses are not obtained by introducing sugars or fats into our blood stream. The protection is always obtained by the use of proteins.

Perhaps we shall begin to realize that insects take to plants because the plants are poorly nourished, that is, growing on poorly treated soils. If we realize that insects disregard plants on soil well treated for production of proteins, perhaps we shall fertilize the soil more carefully. In so doing we shall feed ourselves better, because of more proteins in the crop. In obtaining protection against insects by such methods we shall also obtain better nutrition of ourselves on more fertile soils, and probably better protection against irregularities in our own health.

8. livestock inspects feed for quality, not for quantity

In the production of feed for livestock, we have thought little about catering to the cow's taste, as an example. The provision of each of the different feeds in quantities for balance as a good ration, and then all in total quantities sufficient to carry out our purpose, has been about the major thought given to animal nutrition. We have dealt in hay, grain, concentrates and mineral mixtures with little regard for some of the more refined nutritional contents of each of these, that is, as these contents may be contributed or denied by the soil growing the feeds.

There is danger in thinking of the cow merely as if she were a mowing machine or a haybaler. She isn't just a labor-saving method of harvesting some vegetative mass we call grass. Instead she is a capable inspector and judge of the nutritional values of what she eats. She cannot handle unlimited bulk to get what she needs as nourishment when that bulk is too low in concentration of nutritional elements and compounds. Moving her from her harvesting of green rye in part of the spring to bluegrass for another period, then to lespedeza for the summer, and back to grass again in September is merely a mowing or harvesting procedure with no thought as to the truly balanced diet in each one of those. Should we not wonder about the "why" of it, when the cow risks her neck in going through a barbed wire fence to get to the highway and its grass growing on soils that have not been exhausted of their fertility by continuous cropping and removal? The cow with a yoke on her neck to keep her within the fenced area is merely mute testimony from her that she knows more about nutritional values of grasses grown on different soils than her yoke-providing master.

Missouri farmers trying to work with their cows so these would work better for them have been reporting numerous cases telling us that cattle select feeds according to nutritional values. The kinds of plants don't seem to make much difference to the cattle as long as the plants are growing on fertile soil. The term "weeds" as particular plant species is still unknown to cows.

One Missouri farmer with considerable virgin prairie as pasture found his beef herd going through the extensive bluegrass and white clover areas of this vegetation to go into the weed patch consisting of the corn field abandoned when the labor shortage of war demands took all his help away. The unfertilized bluegrass pasture was not catering to the fastidiousness of those cows. This soil had never been given any lime or other fertilizers. They went straight through it from the water to the corn field with its cockleburrs, nettles, foxtail, etc., to eat all of what we call "weeds" and they called "feeds." It was the fertility of the limed and phosphated cornfield and not the species of plant that entered into the discriminating behavior of that beef cattle herd.

In another case it was not the green plants but their dried condition as hay that served to let the cattle demonstrate their discriminatory senses for feed quality. In 1936 four acres of meadow area on virgin soil were given fertilizer top dressing of nitrogen, calcium and phosphorus to the extent of no more than 600 pounds total fertilizer per acre. This small area was part of one hundred acres of permanent meadow serving as winter feed in four haystacks each containing the hay from 25 acres. In the summer season the hay made from the four acres was swept in with that from the rest of the 25-acre area into the stack as one of the four in the field.

The demonstration of the cow's selection according to nutritional values occurred the following winter when they were allowed to go into the meadow to eat the four stacks of hay. They soon were all collected around the haystack containing the hay from the four acres of fertilized soil. This was at the opposite end of the field from the supply of salt and water. So the cows went back and forth daily from this preferred stack to the water and disregarded the other three stacks in passing them.

Even though no later soil treatments were applied on this meadow, and hay was made annually with the cows consuming it in the field, eight years later the cattle were still consuming first the one haystack containing in its 25 acres of hay that from the four acres given a surface application of fertilizers years before.

When farmers report that the cattle grazed out first the strip of green barley where the fertilizer drill turned over the previously fertilized drillings to double the fertilizer applied; when hogs will select the corn in the field where the soil was given treatments like lime; when they will select corn in one compartment of the self feeder in preference to another corn in the feeder regardless of position in the feeder; when chickens discriminate between different lots of butter offered them; and when numerous other demonstrations like these by the "dumb" beasts come to our attention; shall we not believe that the animals have uncanny means of knowing more about the quality of their feed in terms of bodybuilding than in terms of fat and energy?

Gradually we are realizing that it is not the carbohydrate and fat values that bring the animal choice into prominence. Instead it is the proteins and similar complex compounds entering into reproduction that are marked out by animal choices. These depend on the more fertile soil. Better animals to make better foods for us will be more common if we are guided by their choice of feeds under proper soil fertility conditions. They judge for quality, not for quantity.

9. wildlife also struggles for its proper nutrition

Because our wildlife cares for itself, we do not appreciate how it, too, struggles with its problems of nourishing itself completely. Only as wild animals accomplish this do they live and multiply. Otherwise, they become extinct. They live, then, in the areas which are nourishing them properly.

In our domestic animal pattern we might at first imagine that animals are merely scattered according to our accidental distribution of them. Quite to the

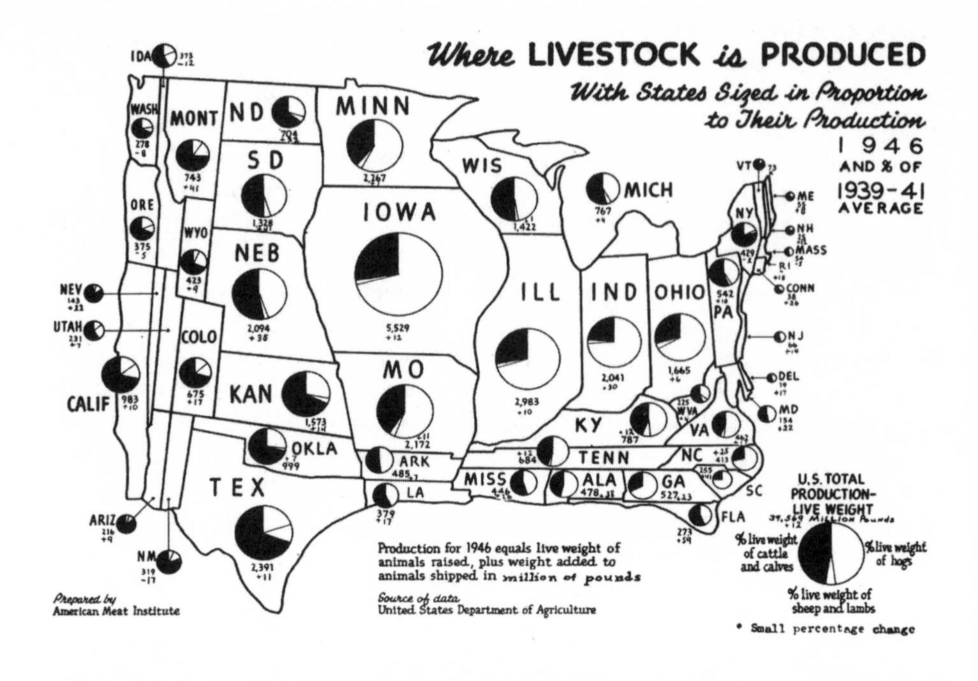

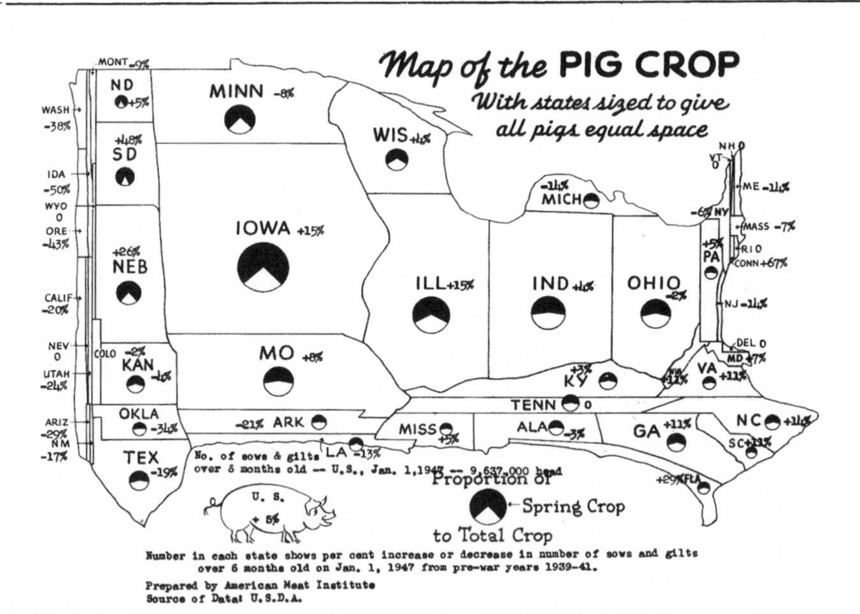

The history of livestock production before World War II is amply illustrated in this series of charts. Before synthetic economics took over, livestock grew and fattened according as soil fertility provided either protein-rich grasses under moderate rainfall for body building and reproduction, or carbohydrate-rich (protein-poor)

BEEF COWS* - *Where they are... How they have increased since pre-war*

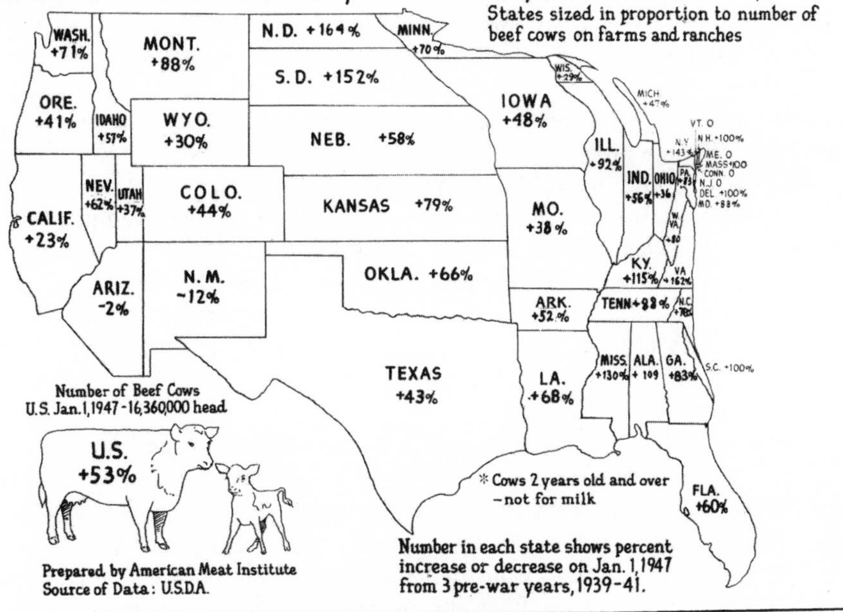

STOCK SHEEP

Where they are ... How numbers have changed since pre-war

States sized in proportion to number of stock sheep on farms and ranches

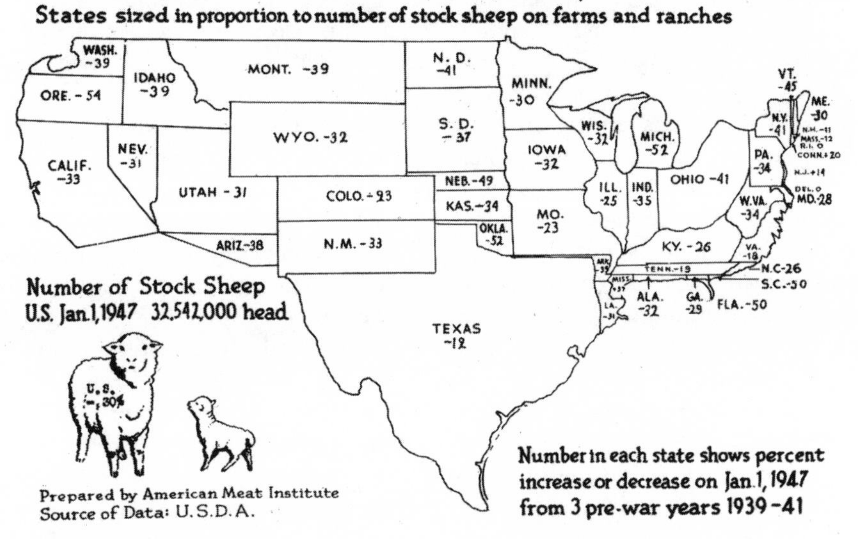

crops under higher rainfall for fattening. Note how swine production located itself in areas quite different from cattle production. Dr. Albrecht often used these illustrations with articles prepared for farmers.

contrary, we have pork as a fat-producing animal in eastern United States where carbohydrate crops, like starchy corn, are its major feed. This is on soil that was originally forested. In its virgin condition that soil was producing mainly wood, or fuel, as its crop. Today it is producing mainly fuel foods and fuel crops. These are high in caloric values. They are low in protein. Even as feed for hogs, corn must be accompanied by some protein supplements.

On those soils, the original wildlife had its problems of nutrition. Wildlife in the forests tells us of its struggles for calcium and phosphorus. The porcupine in the woods of Minnesota was observed by some artists, and so sketched, as it was eating the antlers of deer dropped there. Antlers disappear quickly in the forests on lime-poor soils because the other rodents, like mice and squirrels as well as the porcupine, supplement their calcium-deficient diets by calcium concentrates in the form of bones of preceding life forms. This is their struggle in order to guarantee complete nutrition.

In the same area today, cleared of most of its forests, the gray and red squirrels in our city residential areas are commonly noticed to be eating bones. Seen on the window ledge, they have been carefully observed eating old dry bones, so aged and so dry that obviously it was the bone itself that appealed to the squirrel. This suggests that the feeds produced in the neighborhood for these animals are an incomplete diet in terms of the lime and phosphate which we expect the soil to supply.

It is suggestive that it is during the season of pregnancy that most of these reported observations have been made. Perhaps during part of the season the struggle is not so marked, hence the animal needs only supplement by such perverted behaviors of eating the remnants of its predecessors' bodies during a limited period of the year. Crops too low in lime and phosphate and with only fattening power as feed for hogs call for bones as supplements in case of their feeding wildlife, just as legume crops on those same soils call for lime and phosphates if they are to produce the protein-rich feed values and successful growth.

In the western United States where high-protein wheat has been a common crop and where grasses were once the virgin vegetation, antlers of deer accumulate. It was in that area where the buffalo roamed. With the summer he moved northward, and with the winter he moved southward. He did not move much to the east of his chosen area, not even into the eastern limit of the big bluestem except possibly for Kentucky. In that state he was in limited numbers and limited area originally where the race horse have made themselves famous and established the annual derby. He was also in some of the more productive valleys of Pennsylvania to suggest that he too was struggling to find his complete nutrition. This was according to the pattern of the soil.

It has been reported that deer quickly take to burned-over forest areas on the first signs of growth recovery there. Can they recognize the ash as soluble fertilizers making the grasses and other growth more rich in the inorganic elements and thereby more proteinaceous? Can this burning get rid of the excessive woodiness in the decaying organic matter that keeps the inorganic essentials tied up in microbial life in competition with plant life? Does not the deer get out into the burned-over clearing and risk its life in the struggle for complete nutrition? Deer as marauders in our fertilized field and well-kept gardens are risking their life in their struggle to find what they need to survive, all their fear of the human enemy notwithstanding.

The wildlife in its struggle for complete nutrition has set up its population pattern as the result of—and thereby in agreement with—the soil fertility pattern. All this tells us that our wildlife is just another crop of the soil and only as we manage the fertility of the soil to feed the wildlife will we have this animal crop in abundance.

10. wool differences reflect different soil treatments

The wool of the sheep, much like muscle meat, is a protein of animal growth. While fat is associated with wool fiber in what is commonly called the "yolk," this fat is secondary rather than the primary part in the growing of the fleece of wool. In the production of wool, we have for too long a time been concerned with the quantity of it the sheep produces. Only recently has the quality of the wool fiber come under closer attention. Much more recently have we begun to consider the quality of the wool in relation to the fertility of the soil growing the forages on which the sheep are grazing. We have now begun to realize that the fertility of the soil may come in to control the wool to degrees not generally appreciated.

In some trials with hays grown on soils with different treatments, the wool of the sheep could be scoured and carded out in good order when the hay was grown on a soil treated with both lime and phosphate. It could not be carded and the fiber was broken in the attempt at carding when the hays fed the sheep were grown on the same soil type given only superphosphate. In addition, there was little yolk in the wool on the sheep in the latter but a generous amount of yolk in the former. Here the quality of the wool fiber of protein makeup, and the amount of fat associated with that fiber, both varied widely simply because the calcium-magnesium deficiency in the soil had or had not been corrected by an application of limestone.

More recently, the regularity in the crimp, or the wave, in the wool fiber of the Merino sheep has been correlated with the presence or absence of some of the trace elements, especially copper and cobalt. In South Australia this was very noticeable in the wool from the sheep grazing on the grass on some of the calcareous sands. The drenching of the sheep with from seven to ten milligrams of copper associated with one milligram of cobalt daily restored to its uniformly wavy appearance the so-called "steely" or wire-like wool fiber.

The color of the black wool of black sheep is also related to the presence of these trace elements, especially with the copper. Australian black sheep grazing on pastures on these coastal soils lose the black color to become gray. When given copper and cobalt in the feed, or the drench, the black color is restored.

In many parts of the United States, it is commonly said "Black sheep never stay black. They always turn gray." It might well be questioned whether this is not merely saying that our soils are copper deficient, and possibly even cobalt deficient. We have only recently begun to think of these so-called "trace" elements and their importance in the body's production not of fats, but of its proteins, even of the different amino acids constituting the proteins.

Some studies on the essential amino acids making up the proteins in alfalfa, for example, suggest that the trace elements put into the soil are effective in pushing up the concentrations of these protein constituents created by the plants. Sulfur put on the soil suggests that this element is too low in supply in the soil to let the plants make all the sulfur-containing amino acids they would otherwise make. Magnesium as a soil deficiency also comes into prominence now in connection with possible failure of our forage crops to build the proteins complete, so far as the required amino acids are concerned.

Now that we are measuring proteins more commonly in terms of their nutritional completeness with reference to all the amino acids, rather than by calling all nitrogenous products proteins, the soil deficiencies for plants creating these amino acids are coming to the foreground. We need to recall the fact that animals don't synthesize proteins from the elements, including those in the fertility of the soil. They only collect them from the amino acids created by the plants. Sheep and their protein output in the form of wool are telling us that the fertility of the soil, and our treatments of the soil, reflect themselves quickly in this product. All this tells us that the creation of what carries life itself, namely, the proteins, goes back to the soil for its start.

11. fertile soils make the earthworms, not vice versa

Earthworms, night crawlers, fishworms, or whatever you may call them, are commonly associated with fertile soils. To say that this may not be a case in which the earthworms made the good soil is disturbing to many folks who defend the efficacies of the lowly worm as a soil improver. Because two happenings are closely associated or occur simultaneously, that fact does not necessarily prove that one is the cause of the other. Such performances may result from a common cause. In the good soil and the worms going together, it may be the fertile soil that grows good crops that feed earthworms well with organic matter so they multiply profusely. This is a case in which the fertile soil grows what it takes to make earthworms. The presence of the worms is merely an indicator that the soil is fertile enough to feed them well.

That chemical fertilizer treatment helps in making soils fertile to invite the earthworms was well demonstrated recently in the lawn around a new property built in 1940. This lawn was one so characteristic of most lawns made hurriedly by building contractors, rather than by one with a full knowledge of the requirements for establishing good grass stands. When the final amount of unweathered clay had been thrown out while excavating the basement, it was this which made the leveled soil on the top of which a shallow covering of black soil was put and seeded as the lawn. The owner used limestone and fertilizer with rye as a nurse crop for the first grass seeding. Since the condition of the grass indicated the shortage of nitrogen, the property owner undertook an annual fertilizer program of heavy top dressings with considerable nitrogen, along with phosphorus, potassium, and trace elements on the lawn. But since this did not carry the grass well through every summer it was decided after seven years to spade up the entire lawn and incorporate the fertilizer down deeply in the soil. This was the decision after countless moles in their runs had literally churned up the entire lawn in the spring to kill the grass in streaks.

It was a decided surprise to find that the moles had discovered (and were feeding on) what the property owner found spading up the lawn, namely that there were myriads of earthworms in this lawn that was scarcely more than a clay pile but a few years before. The upturned soil was profusely perforated to reveal the worm holes even in a photograph that caught some of the exposed worms.

Here was a case, quite the opposite of what is the common conception, namely that by using chemical fertilizers on a soil initially free of earthworms, this form of life had been encouraged to come in and multiply profusely. The worms were working down in the clayey portions originally excavated from the bottom of the basement. The soil treatments with the chemicals were the only initial factor inviting the worms to take over so profusely. It was the factor that made the moles disturb the lawn severely when they neglected the adjoining lawn entirely. The adjoining lawn had not been treated, nor had it been growing the green grass, as was common on the heavily fertilized one when rainfalls were not so erratic as to hinder the grass growth.

Worms may go along with fertile soils just as you may be going along down street with a policeman. The worms may not be the cause of the high fertility of the soil, just as we hope the policeman is not the cause of your going in the same direction with him. It is the fertile soils that make the earthworms and not vice versa.

12. soil treatments change concentrations of soil-given nutrients of plants

When we think of vegetables we recall them by name, but with little thought as to their chemical composition. When we eat the spinach we talk about "getting our minerals." We imagine that vegetable greens, like spinach, kale, turnip tops, Swiss chard and others are serving to bring to our digestive tract some iron, some lime, or some copper, as if these inorganic and metallic substances delivered in those metallic forms were rendering nutritional service. Because we measure them chemically in the plant ash, or only after the plant has been burned, we have failed to realize that chewing copper wire, nails and lime rock, or eating them in powdered and encapsulated form would not be the equivalent of the services rendered when these elements are brought to us in their organic combinations by the plants. Nevertheless, even if we measure these inorganic elements only as ash forms, their amounts in the same plant species grown on different soils may vary widely enough to be scarcely believable.

In the case of a simple vegetable green, like kale, for example, one might not realize that the soil treatments would give us plants looking so much alike and yet so different in content of an inorganic element, like calcium. In some experiments with that greens, which belongs to the cabbage family of vegetables, only the clay portion of the soil was modified by letting that clay filter out of solution, that is, absorb onto itself different amounts of calcium and different amounts of nitrogen while it was absorbing equal amounts of the other applied nutrients. When the kale was grown to good size in some of the pots, the plants were still very small in those pots given less amounts of either calcium or nitrogen. But when the plants were put to chemical analysis for the amounts of calcium which each might deliver for bone-building services, let us say, it was surprising to see the wide differences in calcium contents for plants that were not only good looking but even when they were alike in those appearances.

The growth of kale, like that of spinach or other leafy vegetables, responded markedly to the increasing amounts of nitrogen. Better appearances as bigger growth resulted from fertilizing with this element. This better growth and extra green color are the common criteria of our success as growers of vegetables and other crops. That success, however, would be supported mainly by appearances and vegetative bulk produced for sale as pounds of greens. Such might not be success in terms of nutritional value of the kale greens. That success might not stand up when faced with the question, "How much calcium as a mineral nutrient does the kale provide us on eating it?" or "Into what organic combination is that calcium put by the plant for service in our body through ready and complete digestion?"

As for the first question, the experimental trials pointed out the variation in percent of calcium in the dried kale, namely a variation going from 1.98, to 2.26, to 2.46, and to 3.1 merely because the soils in that series were given 5, 10, 20, and 40 milligram equivalents of calcium per plant, respectively, in conjunction with liberal amounts of nitrogen (40 ME). This was supplied in a form that would be readily exchangeable to the plant roots coming along to exchange their hydrogen ion, or acidity, for it. Here one might not be able to select one plant as different from any of the others in the group of four, yet one of these might give us three percent of calcium and the other only two. One plant is 50% higher in bringing to our digestive system the inorganic supplies so far as calcium goes. Looks were nicely improved by the nitrogen fertilizer. The plant's content in calcium, however, was dependent on the amount of lime or calcium put on the soil. Looks may be deceiving in vegetables, as in many other things.

As for the second question, namely the organic combination of the inorganic elements brought to us by plant growth, this is related to the plant species. It is also related to the fertility of the soil, especially to the balanced or unbalanced diet for the plants offered by the soil. The spinach is a kind of plant that puts its calcium, for example, into the chemical combination with oxalic acid, giving us calcium oxalate. This is highly insoluble even in acids and is thereby indigestible. This oxalate forms into needle crystals and irritates the mouth tissues, as one well knows who has been treated to a small slice of the so-called "Indian turnip;" or has eaten spinach that "puts an edge on the teeth." Spinach grown on one combination of soil fertility for its nourishment may be correspondingly distasteful because of its excessive oxalate content. Other spinach grown on other soil fertility combinations have a pleasant taste. It is this variable fertility under the spinach and not the plant species that has engendered love in some cases and hate for itself in others, "Popeye" notwithstanding.

Kale, however, does not put its calcium into combination with oxalic acid and does not make it so indigestible. Here a small percentage in the greens delivers much digestible calcium. This is quite the opposite of spinach. Organic compounds of the calcium into which processes of the plant synthesize or convert this essential element and not the amount in the plant ash, become the criterion of nutrient values for minerals as well as the amounts of these in the plant. Most significant, however, is still the big fact that while plants differ as species in delivery of nutritional services through the inorganic elements they contain, it is the fertility of the soil that comes in to hinder or help in these values more than any of us are wont to realize and to appreciate. The creation of the plants for us still depends on the handful of dust and the creative services in it.

Studying Nature

We are prone to destroy the beast when it aborts, when it gives midgets or when it contracts a disease common also to ourselves. Destroying the evidence is apparently a more common practice than diagnosing it to find the cause of the abnormalities.

• • •

Our agricultural crops illustrate the fact that an evolution of species for speculative economic values only through man's management has increased pests, diseases and extinction rather than their healthy fecund survival.

• • •

Man's failure to maintain such a flow from, and return to, the soil of both inorganic and organic fertility under his crop removal from the soil rather than complete return, has been the quiet force pulling down to a lower and lower level the protein potentials of soils with each crop succession.

• • •

When man's production of crops depends more on blind faith in survival because of a certain pedigree of the seed than on undergirding the potential crop with nutritional security through the fertile soil, the evolution so managed invites pests, diseases and crop extinction.

• • •

Depletion of soil fertility cannot mean successive crops of the same protein and nutritional potentials equal to those grown on the soil when first broken out of the virgin sod under a natural plant climax. Depletion of the soil has reversed the natural evolution which built the climax.

• • •

For the plants, the declining soil fertility functions like a kind of fattening and growing that transcend even those for the pig. It is declining soil fertility, then, as it is giving plant values of only fattening potential for animals, that is undermining the warm-blooded segments as well as the plant segments of the biotic pyramid, including animals and man.

• • •

When the farmer says, "I must get some new seed. My oat crop is running out," he is merely reporting that the regular use of some of his own grain as seed for the next crop, while depleting the neglected soil fertility, has demonstrated the extinction of that species.

• • •

The sudden ravages of crops by insects suggest a sudden shift in the chemical composition of their new crop victim, representing those resulting in a particularly suitable insect diet when formerly the victim's chemical composition was unsuitable for survival of the particular insect. Those shifts in the plant's chemical and biochemical composition result from unappreciated changes in the available fertility of the soil.

LESSON No. 2

Soil System

21.

During and after World War II, Dr. William A. Albrecht fell hard to the task of telling it the way it is so that laymen could understand. His articles drew nourishment from the scientific literature he and his associates had accounted for. Sometimes a single line or paragraph of these papers wrapped into one whiplash lesson was a finding equally as important as the discovery of the wheel. One such series came styled, How Soils Nourish Plants. *This series appeared in several publications, sometimes with changed titles, sometimes as a column,* School of the Soil. *Some parts were recycled in* Acres U.S.A. *with Albrecht's last minute changes. Today, as then, these articles make good reading and good learning. They are linked together here under a single title both to conserve space and to retain continuity.*

How Soils Nourish Plants

We have come to realize that the speedy delivery of nourishment to plant roots is brought about by the finer fraction or colloidal part of the soil. This includes the clay as the mineral part and the humus as the organic part. With the clay content in a silt loam approaching 20% and with the humus as low as only 1 or 2%, the clay naturally stands out as the main dynamic force in storing nutrients and exchanging them to the plant roots. It is true that humus or colloidal organic matter has more exchange capacity per unit than does clay, but since we have been "burning out" the organic matter of soil, dependence is going to the clay as the "jobber" that can hold what is "available" or exchangeable nourishment for the plants.

Clay may be readily frowned upon when the soil has nothing but clay, as in exposed highly weathered subsoil. Clay becomes very desirable, however, when in proper amounts (with some humus) it is balanced with sand and silt minerals as complete soil. Thus balanced, it gives good tilth, and particularly so when these coarser separates are minerals other-than-quartz. It is when these minerals are nutrient-bearing that they become the reserves from which the infertile clay can restock itself regularly after crop production has pulled down its available supply.

1. size of stock of exchangeable nutrients on clay

For many years it was customary to ask the question, "How much of the nutrient supply in the soil is available during the growing season?" It was common to speak of the "available" and the "potential" supplies. More recently, however, we have come to speak of the "total exchangeable amounts" and their "suite of nutrients" per unit of soil. We now use soil tests to measure the amount of calcium, for example, that is readily exchanged from 100 grams of soil by extracting the soil with some solution of other positive ions by which the calcium is exchanged. We now speak of the soil's "exchange capacity" in terms of milligrams or thousandths of a gram of hydrogen (acidity) per 100 grams of soil. This amount of hydrogen can readily be converted into the equivalents of other positively charged ions like calcium, potassium, magnesium, and others, *by using their corresponding weights by which they replace hydrogen.*

We are therefore getting a measure, for example, of the "available"—but more truly exchangeable—nutrients possible in a silt loam when it is tested and we find that it has an exchange capacity of 18 milligram equivalents per 100 grams of soil. This would be 18 pounds of hydrogen in 100 thousand of soil or 360 pounds per plowed acre weighing approximately 2 million pounds. If it were all expressed as calcium, it would be 7,200 pounds. But usually calcium is only about half of it, or 3,600 pounds, while the balance of the clay's adsorbing and exchanging capacity in the acre of silt loam is taken by the potassium, magnesium, iron and all the other adsorbed ions including all of both nutrients and non-nutrients possible in the soil.

2. jobber seldom sells stock out completely

When the plant root grows through the soil and trades its hydrogen or acidity to the clay and takes nutrients from the clay in exchange, the clay does not give up the different elements with equal readiness. Some are held by the clay more firmly than others. Calcium, for example, is not moved from the clay into the plant as readily as potassium or nitrogen in the ammonium form.Hydrogen or acidity is held most firmly of all the ions. This explains why it accumulates on the clay to make acid soils when they have given up their fertility. It tells us that declining soil fertility is what gives low production and soil troubles commonly attributed to the increasing soil acidity. Of the common nutrient elements, calcium is held most firmly by the clay. Even when a soil has become significantly acid, or taken on much hydrogen, it is still holding considerable exchangeable calcium, too. Hydrogen and calcium are closely associated in the exchange activities or in the "jobber" services by the clay. From that we can get the suggestion that soil acidity and liming are closely connected.

In these reactions of the clay and chemical elements we can also see reasons why we don't get all the necessary information by merely testing soils for their degree of acidity. *To say that a soil is highly acid is not telling how much calcium it will have in stock to be exchanged to the growing crop. It says nothing about all the other necessary supplies.* We need to measure the exchangeable calcium in the soil in order to learn how much of this nutrient the clay has in stock. Even then it is well to know whether the given amount of calcium is on a large amount or small amount of clay and thereby learn how highly saturated the clay is with calcium. As it is more highly saturated, the calcium moves into the crop more readily. Thus in testing the soil one may well measure not only the total amount of the nutrient like calcium that is readily exchangeable, but also get some idea of the degree to which it loads the clay's capacity. The clay is a kind of "jobber" that trades readily when its stock is large but less readily as its stock becomes smaller. In fact, it seldom sells out completely in some items. Calcium is foremost among these.

3. *clay must restock itself—this is possible from decay of organic matter*

When once we measure the supply of exchangeable nutrients on the clay, we are soon set to inquire, how long will that amount serve to keep up our crops? That is an especially challenging question even for calcium in the humid region where rain water going through the soil may take out of the surface layer alone at least 100 pounds per acre annually. At the same time a crop of clover may be hauling off 125 pounds of calcium. Even 7,200 pounds, if the entire exchange capacity were taken by calcium, as suggested above, would provide for such annual removals during a period of only 32 years. If the clay's exchange capacity has been loaded to only 75% by calcium, this nutrient supply would last only 24 years.

On cornbelt soils, which were originally plowed out of prairie sod and have been producing crops over several 24-year periods, it is evident that the calcium removal by the crops represents several times one single or total stock on the clay. Logically, then, if the clay cannot be completely exhausted through crop growth, and if the amounts delivered by the clay are several times its total exchange capacity, we may be sure the clay has been restocking itself while it has been producing the crops. What has been the source from which the clay has replenished its stock?

The decaying organic matter in the soil is of decided importance in keeping the clay supplied with exchangeable plant nutrients. Virgin soils in particular are active in keeping up crop production by this means. While plants are combining the chemical elements of soil origin into complex, insoluble combinations with carbon and depositing them as organic matter on and within the surface soil, the microbes of decay are behaving like fires to burn that carbon out of combination and to send the ash as soluble nutrients to be adsorbed and to restock the clay.

4. *clay must restock itself—this is possible from decomposition of mineral reserves*

In the productive soils that are not so well supplied with organic matter, we can recognize that the nutrient stock on the clay must be replenished by breakdown of nutrient-bearing mineral particles in the silt and sand. This may not be such a speedy performance in case the soil has been under humid conditions and cropped for many years. It is very slow in contrast to the restocking byorganic matter in decay. When the silts and sands are made up mainly of quartz, they have little or nothing that can be broken down by the clay acid. In such soils the hydrogen becomes more highly concentrated on the clay. We say these soils have a higher degree of acidity and are therefore not productive. Such soils are not unproductive because of the acidity, but rather because the clay as the "jobber" can not trade its acidity to the mineral reserves for some of their nutrients. While acidity is seemingly detrimental under such conditions, it can become beneficial as a chemical agent to decompose minerals in the silt and sand separates of the soil if these carry nutrients as a reserve.

5. reserve minerals of soil may be brought from our west by wind and water

Perhaps you have not thought that in terms of restocking the clay of our soils "it is an ill wind that blows good to no one," especially if it blows dust from the western United States. Then, too, when the Missouri River and others flowing out of the arid west carry minerals of silt size as unweathered rock fragments to leave these as exposed, dry bars in their river beds to be carried by the winds, and to be deposited over the cornbelt, isn't it possible that this fresh, or less weathered mineral addition to the acid clay of our more humid soils may be equivalent to restocking the depleted clay with plant nutrients? Windblown minerals may be more than aggravating dust. They may be new supplies of soil fertility. They may be means of restocking the clay and a help in maintaining the productivity of some of the more humid soils.

6. inventory of soil capabilities must consider the mineral reserves as well as "available" supplies on the clay

We may well consider the possibility, then, that prairie soils or semi-humid soils of the United States are productive, not only because they carry considerable amounts of readily exchangeable nutrients on the clay, but also because their silts and sands are minerals other-than-quartz, and thereby minerals that contribute plant nutrients to the clay by their decomposition. They are weathered out of their rock forms by plant growth that supplies acidity from the roots to the clay so that it brings this weathering about.

If these are the facts, then, as we attempt to take an inventory of our soils, attention must be given to the plant nutrient supply in the mineral reserves as well as to that in exchangeable or available form on the clay. In measuring the soil capabilities we need to make petrographic inventories of the particular minerals in the silt and sand, as well as to make quick soil tests of the available or exchangeable nutrients held in the stock by the clay. As the organic matter is more completely burned out of our soils and as productivity comes to depend more on the mineral fractions, the mineral reserves will be more highly appreciated for their help in restocking the clay with the chemical elements needed for crop production. We shall understand better and appreciate more fully that in the last analysis it is the mineral reserves in the soil that supply the nutrients to keep up the stock on the clay and thereby keep up crop production.

7. soil fertility is needed, so is soil acidity

Even though the soil may not contribute much more than 5% toward the bulk of the plant in contrast to the 95% coming from the air and water, it is that 5%

coming from the soil that controls the growth of the crop. While four elements are coming from the air and water there are ten, or possibly a dozen, essential elements in the contribution by the soil. A shortage or absence of any one will change the manufacturing business carried out by the crop. It may influence the plant's production of its body by modifying size or bulk that we recognize as tonnage of vegetation, but a more highly significant influence is exerted on what the plant manufactures or synthesizes. We recognize that partly in the decreased yield of seed which is a product synthesized by the plant more specifically than by just sunshine. We are coming to recognize it more generally in the feeding or nutritional quality of that seed, and of the vegetation when these are consumed by animals and man.

8. evidences of declining fertility are numerous

Our soils have not been cropped so many years and some of us may be old enough to recall—or some of us were told—how well they produced certain crops when they were first cleared out of forest or plowed out of prairie sod. Heavy yields of high protein wheat or "hard" wheat, red clover with good seed yields from the second cutting, big litters of pigs, rapid growth of young animals, and generally healthy livestock were common not so long ago on soils where today they seem scarcely possible. We are just coming to realize that we have been taking from the soil much of the fertility, the declining and neglected supply of which may be the main disturbance and may be provoking such troubles.

Soil erosion is evidence number one, though at first thought one might not see declining fertility as the cause of this alarming and seemingly sudden departure of the soil body. In Missouri, for example, the equivalent of one-half of the fertile soil of one-half of the state, has been eroded. This is a tremendous loss of the body of the soil that has come as a sequel to the loss of the fertility through cropping and cultivation. This loss in fertility has weakened the body to where it cannot quickly grow cover for itself; where it does not rebuild its humus supply; where it is not in the stable granular structure that stands up under the beating rain; and where the falling water cannot enter readily but dashes the soil flat instead and then runs off to take the dispersed soil with it. Erosion is evidence of the dwindling supply of fertility, and of the increasing nakedness of nature because of a soil body too weak in good granular structure to hold up against the falling raindrops.

Another evidence, which has so often been seen, but not observed with its interpretation going to soil fertility as the cause, is the increasing failures of legume crops provoking the search for other kinds of them. Instead of ascribing the failure of red clover to its needs for a soil that is fertile in the many mineral elements like calcium, potassium, phosphorus, or even in a good supply of nitrogen, and then instead of going about curing the soil situation by adding these to help grow this crop, we have ascribed the trouble to soil acidity and have applied a carbonate to neutralize it. We have searched for substitute crops. Those that have been found to make vegetative bulk on "acid" soils do not produce the equivalent in feed of those they replaced. They do not grow with the nurse crop any more as the red clover did, but rather they grow after it. Our soils are said to be "too acid," when in reality they are too low in the power of fertility to push the nurse crop into good seed yields and to start the legume crop at the same time. Our soils can no longer grow two crops at one time. These crops must follow each other with a time interval between them for the soil to recover a stock of fertility in some measure by mineral breakdown and organic matter decay.

Other evidences include declining grain yields in spite of the search for, and introduction of, new varieties; increasing weather hazards under such limited plant nutrition; grain crops displaced by forage crops and renewed hope placed in pasture and grass systems of farming; and deficiency diseases of plants and animals creeping in more widely to make animals take to the mineral boxes in desperation when their essential mineral elements are not amply synthesized by the crops into the complex organic combinations that put nutrition of high order into the herbages and grain feeds. While our soils are taking on these symptoms that we include in saying "they are becoming acid" they are giving up their fertility that was accumulated in the organic matter and humus by virgin vegetation; held in available or exchangeable form by the colloidal clay; and retained as reserve in the mineral crystals and rocks that had not been completely exhausted when the virgin vegetation kept the fertility removal at such a slow rate.

9. soil acidity and soil productivity go together

Such soils were not too acid when the pioneer took them over. He was not discouraged by the presence of soil acidity. He grew his crops by encouraging unwittingly the development of a higher degree of acidity in them. He drained the surplus water to advantage, not only of better health in escaping malaria (once known as "ague"), but of plowing to put drafts of air into the microbial fires to destroy some of the preserved organic matter, and to release carbonic acid profusely. Acid production in the soil preceded crop production. It was the process of making available at higher rate for nutritious feeds the fertility that was so slowly available as to make only wood.

More and heavier yielding crops on those soils have hauled off nourishment for animals and man. But in order to do so the growing crop traded acidity to the clay rapidly for the store of available or exchangeable nutrients it had absorbed on itself so gradually during the many years of slow mineral breakdown. That acidity so taken on by the clay has been passed to the reserve minerals and rocks of silt- and sand-size and has served as the chemical reagent to break them down so they would nourish the crops.

Thus, natural conditions coupled with our use of the soil, where rainfalls are high enough to grow crops abundantly, have caused the acidity to accumulate on the clay of the soil. This accumulation resulting from cropping is more acute where there are less mineral and rock reserves left in the soil to remove the acidity accumulating on the clay. This accumulation is then a reciprocal or counterpart of decreasing fertility. It is natural, therefore, that cropping will make an acid soil of any one that has a supply of fertility and gives it up to make crops. It was the supply of nutrient ions in place of the non-nutrient hydrogen or acidity that made it less acid and also fertile originally. It is natural then that while the soil is becoming more acid it is simultaneously becoming less fertile and less able to produce more crops. Soil acidity, therefore, is natural, but while acidity or hydrogen comes from the crop roots or from the percolating water, there must be fertility on the clay to be passed to the crop or to the leaching waters in exchange. There must also be some minerals in the soil to remove the acidity from the clay and to restock that with the plant nourishment if the soil is not to reach the high degree of acidity or the fertility exhaustion readily considered as dangerous.

10. acid soils suggest their need for fertility

During recent years we have been attacking the accumulation of acidity on the colloidal humus and colloidal clay complex in the soil by applying the carbonate of lime. While the carbonate was removing the acidity and getting our enthusiastic applause for what we thought was the benefit to the crop, the calcium, quite unappreciated, was nourishing the crop as the real benefit. Legumes were said to be acid-haters or *acidiphobes,* when in reality they are calcium-lovers, or *calcophyles.* Because liming helps to improve the legume crops we are now liming the soil to give them calcium as nourishment. We are liming to fertilize the crop more than to fight acidity. This soil treatment is benefiting both legume and non-legume crops because it feeds them the calcium, which is the first deficiency element in their diet when the clay has taken on much acidity and sold out most of its fertility.

Acid soils are deficient not only in calcium but may be short in many other nutrients which, like calcium, are positively charged, as for example, potassium, magnesium, and others. Merely providing the calcium heavily and neglecting to supply these others may be only a temporary solution of the present problem and may be the aggravation of a larger and more perplexing one later. We may be learning about soil fertility by meeting each trouble singly and thereby recognizing and remedying only one nutrient deficiency at a time.

11. soil acidity needs to be supplemented by soil fertility

When acidity in the soil develops naturally and through the neglect of soils under cultivation, and when this soil condition is truly a decline or exhaustion of soil fertility, we must anticipate legume crop failures as the result of using our virgin soils as we did. But this does not mean that the legume cannot do well under acidity, where the necessary fertility has been supplied or accompanies it. Recent studies with soybeans at the Missouri Experiment Station found them taking as much as 25% of their total nitrogen from the air, when they are planted in a very acid soil with an initial pH of 5.6. Here was a legume crop using the fertility of the soil just as all crops do, namely by trading the acidity for it and making the soil more acid by its growth. Yet we have been prone to believe that such crops cannot grow if there is any significant degree of acidity.

Experiments showed, too, that in the presence of some acidity the crops make better use of the calcium, manganese, phosphorus, and other nutrient elements on the fertility list. Acidity is a kind of mobilizer, as it were. It is nature's method of giving more relative activity seemingly as the fertility supply becomes smaller in the soil. Acidity makes the movement of it into the crop and also the crop's manufacturing business more active so far as different nutrient elements are concerned. We are just coming to appreciate the beneficial effects to the crop from some degree of acidity in the soil.

Soil acidity is no longer the bugaboo it once was. It is no longer an enemy. Now that we have gotten acquainted with it and understand it, we are no longer fighting it. Instead we are fertilizing for it. We have learned that while our main need for better food production is for more soil fertility we have also learned that we need soil acidity too.

11. balanced diets are required by plants

We have all heard much about balanced diets, especially those containing the seven essential groups of foods, and while farmers have used balanced diets for farm livestock, we have not given much credence to the belief that plants, too, require balanced diets from the soil. For the young or growing animals we recognize readily less need for carbohydrates in balance with the proteins and minerals. For the older or fattening animals the carbohydrates may be more in relation to the protein. The diet is balanced according to the physiological objective. We are just coming to see that the plant, like the animal, will emphasize some one physiological performance over others according as the nutrient supplies are high or low in some elements in contrast to some others. We are coming to see that the food synthesis for us by the plants must be managed by the fertility management in the soil, or the diet we feed the plant.

12. the plant is fixed—its nutrient choice is limited

Since the place where the seed is dropped limits the growing plant to feeding in the soil at that spot, the fertility of the soil volume penetrable by the roots constitutes the diet of the plant.

This is the concept of the situation when plants are grown in solutions, or in the practice of hydroponics. Likewise it is the situation for the plant when the nourishing medium for the crop is a well blended colloidal clay with all the nutrients adsorbed on it. It is a case in which every clay particle is like every other one in terms of what it can give to the plant root for the acidity the root has to exchange. It is much like a human diet of hash. All is so well blended that the dietary constituents lose their identity and their separate effects on our taste. In the case of the plant, every soil area has the nutrients in the same ratio to all others and in no place does any one of them exercise any different relation to the others.

With such a uniform blending of the soil, and with the plant in a fixed position, the synthetic performances by the plant must of necessity give a chemical output that is determined closely by its chemical intake from the soil. In our thinking about putting fertilizers into the soil to feed the plants, much has been said about using a particular ratio of nitrogen to phosphorus or to potassium in that fertilizer.

That these ratios may or may not be proper for a particular crop on a particular soil has long been recognized. This problem was the occasion for the myriads of ratios of fertilizers once available. They were much of a trouble as long as fertilizers were applied directly with the seed, a practice which encircled the young fixed plant by a fixed diet which it could not escape.

But now that fertilizers are being plowed under ahead of the crop and are being located at a distance from the seed to allow the roots some chance for escape to clay areas of different dietary offerings, the numbers of fertilizer ratios have been reduced.

13. variation in fertility is quickly reflected in plant growth

Water cultures as used in hydroponics must always be very dilute, hence these cannot provide a plant diet with very wide ratios between the different nutrient elements in the solution. In the soil, however, where larger amounts of nutrients are stored as adsorbed forms on the soil, there are possible very wide ratios of the amounts of the elements. Then there are also variations in their relative activities. This soil situation gives occasion for what may well be termed "imbalance." It represents difficulty in finding the proper balance and all the more so when a dozen or more items as nutrients rather than only seven as suggested for the human diet, are involved.

This problem of putting soluble substances uniformly on the clay was studied by using tomatoes as the plant performing in its early growth as the diet assayer. A wide variety of ratios was used, but for purpose of illustration tomato plants are shown as they respond to differences in the ratios between only nitrogen and phosphorus—the usual first two items in the common fertilizer formula.

That the plant growth varies widely as these two elements vary in ratio to each other is clearly evident. These variations occur when all the other nutrient items on the clay were provided in constant but liberal amounts. Had any of the others been varied to the point of shortages, the variations in the crop growth might still be greater.

These results as crop growths suggest that, as we increase the amount of phosphorus applied, the plant responses to this nutrient increase will depend much on the amount of nitrogen present. When nitrogen is ordinarily required in so much larger amounts than phosphorus, when nitrogen is relatively much lower than it is in most soils; and when the mass production of crops is so responsive to the amount of nitrogen given, there is the suggestion that our fertilizer applications have not been too well balanced as plant diets, except as some unknown fertility of the soil has covered these irregularities for us. They suggest also that we shall have much to be learned about providing diets for our crops when the natural nitrogen in the humus of our soils goes still lower and we are pushed to feeding our crops on what we put into, rather than what nature left in the soils we farm.

14. plants probably feed cafeteria style

Perhaps it is no fantastic concept to visualize the roots as selecting and balancing the nutrients in the plant diet much as our domestic animals select and balance their diets in pastures of mixed herbages, or as wild animals roam in search of their dietary essentials. Perhaps plants do not do so well on perfect soil fertilizer blends just as we do not do well (in disposition at least) on hash or, as Dr. Richter has shown, rats do not do as well in terms of feed consumed per unit of gain when required to subsist on a feed mixture as when permitted to select and build up their diet from the separate items constituting the mixture.

There is the suggestion by such plant behaviors that *the root may be exercising some influence in balancing the plant's diet,* first by growing into different soil areas, and second, by developing more roots within certain restricted soil areas. This latter has been commonly observed when some small, root-laden portion of manure previously plowed under is dug up under the growing crop, or when a single small root enters a crevice in the sewer pipe to make almost a horse tail brush of clustered roots as an extensive absorbing area for gathering nutrients from the solution so dilute. Here is also suggestion that fertilizer plowed down as streaks in the furrow bottom, or distributed through the soil as larger granules of the separates may be better practice than mixing soluble fertilizers uniformly through the soil as a perfectly homogeneous blend.

Plant diets must be delicately balanced for most effective growth as plant bulk, and more so very probably for the particular physiological functions by which their proteins as final seed crops, and other food essentials for us, are synthesized. We have come to realize that our skills in feeding plants leave much to be developed. Up to this moment we are moving more toward the belief that we will do best in balancing plant diets by providing the nutrients in scattered areas of the soil within reach of the plant roots rather than by compounding them as a balanced mixture of ready solubles to be adsorbed by, and given to the plants from a homogeneous blend on the clay. Perhaps we shall learn that plants given the chance of balancing their diet will make better crops than if we balance it for them, much as the hog producers have profited in following the advice of Professor Evvard, inventor of the self-feeder, when he said, "give the pig a chance and he will make a hog of himself in less time than you will." Perhaps the plants like the pigs can teach us how to balance their diets.

15. plants "select" their diets from the soil

Just how plants can nourish themselves from the nutrient supplies hidden away in the soil has been a challenging question since the early part of the 19th century. The accumulated information to date on this question has listed, as essential for the plants, more than ten elements taken from the soil. It has also listed some other elements not considered necessary for plant growth, but delivered by the plants and demanded for the nutrition of animals and man. It has illustrated the sources of the nutrient elements in the rocks and minerals. It has brought us to understand how the plant roots and the soil get together so the plant can obtain the nutrients it needs and in such balanced amounts that permit it to make of itself the particular products for which we grow it.

When we select the various vegetables and meats for our diet; when animals select some particular plants as their feed and discriminate among them according as the plants are growing on particular fertilizer treatments of the soils; is it too much of a stretch of the imagination to believe that as the plants are sending their roots through the soil they are selecting the elements by which they balance their diet? If your imagination is equal to it, will you not follow through with us in thinking about that possibility?

16. soils nourish plants by exchange processes between these two colloidal substances

That the plant nutrients may be taken in the form of water solutions is a concept that was once helpful, but one that has been found insufficient to explain how plants are nourished by the soil. We know, of course, that plants grow afloat on water. They take their nourishments from solutions there. In botanical class demonstrations the plants have been grown with the ten or more nutrient elements dissolved in water. Then when the new practice of hydroponics came along, it proposed the growing of crops extensively in circulating dilute nutrient solutions. When such have been our observations and academic practices in providing nourishments for plants, it is not surprising that we should incorrectly imagine that water solutions of the nutrients must be flowing into plants from the soil as are nourished by the soil fertility.

Some recent studies used only a suspension of the clay fraction of the soil as the nutrient medium for plant roots. These trials demonstrated that plants will grow well on a clay with nutrients adsorbed on it, and from which none can be washed or filtered out even under high pressure. Such facts tell us that we cannot logically hold to the belief that water going into the plants from the soil is sweeping the supply of their nutrients along with it.

It is now more nearly the truth to visualize that the water is moving up through the plants from the soil according as the meteorological forces causing its evaporation from the surfaces of the leaves are stronger than the forces of the soil holding the water against its movement into the roots. This water movement is normally upward when the forces of evaporation exceed those of adhesion of the water to the soil. But its movement can be reversed and water can be made to move from the plant back to the soil if the soil becomes very dry and the active leaves are removed.

Movement of the nutrients, too, can be in both directions—that is, from the soil into the plant and from the plant back into the soil. The directions of such movements by the nutrients are not determined by the direction of movement of the water, seemingly no more than fish in a stream should swim only with the current. Instead, the movement of the nutrients is determined mainly by differences in concentrations of these items of fertility as they are adsorbed on the clay-humus colloidal complexes of the soil.

This photograph illustrates how roots become thicker as they grow down through** the deeper soil horizons, but only when ample plant nourishment is found there. Such plants will help build up the soil if the farmer builds down by shattering the **subsoil to put lime and other fertility down more deeply.

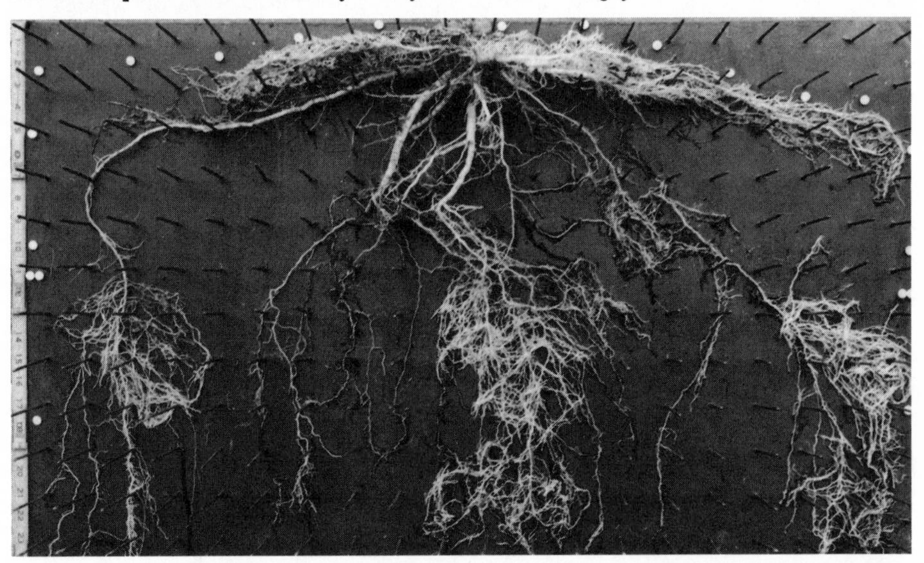

***"Roots may be exercising some influence in balancing the plant's diet,"** according to Dr. Albrecht. Plants probably feed cafeteria style, rather than on balanced hash. This section illustrates how roots reach out, proliferate in the proper nutrient area, and reach out some more.*

The rate and amount of movement of the nutrients are influenced by the composition of the plant roots. It has been shown that the more proteinaceous roots of inoculated legumes took more nutrients into the plant from off the clay than did the roots of plants with no nodules. Nutrient movement into the plants is determined not by water movement into them. Rather it is determined by the root as it is one colloid in contact with another, namely, the soil colloid, and by the concentrations or activities of the exchangeable nutrients adsorbed on each at, or near, the points of contact. The nutrients in the soil move, therefore, under laws controlling their specific behaviors. They move quite independently of the movement of water and of the laws controlling water dynamics.

If nutrients were merely swept in and left within the plant by the water transpiring or evaporating from the leaves, it would be impossible to explain the much higher nutrient mineral contents of the short grasses under the low rainfall of the drier western prairies in contrast to the lesser fertility contents in the tall prairie grasses under the higher rainfalls of the areas much farther east. It is under the higher rainfalls, or where the maximum of water has gone from the soil through the plants, that the plants have the minimum of nutrients per ton of plant mass. Surely we cannot hold to the old idea that the plant is getting its nutrients in simple water solutions passing through and depositing them from the transpiration stream. We must understand the clay of the soil and its chemico-dynamics by which this soil separate helps or hinders the nutrition of our crops.

17. the clay of the soil is the main "jobber" of many plant nutrients

It was some research on the problem of soil acidity that brought the discovery of the clay as a great filter taking the nutrients out of their dilute solutions and concentrating them on its surface. We now know that nutrients are held in the soil by the clay. We know, too, that they can be exchanged to the roots of plants as these are trading acidity for them.

Fortunately, the clays of the cornbelt and other regions of moderate temperature and no greater rainfall have a large exchange capacity. It is this large capacity to carry on exchanges that makes these soils so very acid when they are stocked with hydrogen, a non-nutrient, in place of other positively charged elements which are nutrients. But this large capacity for acid is a correspondingly large capacity for fertility. We are just coming to realize that the natural soil acidity resulted because the clay gave up the fertility to the growing plants in exchange for the acidity they offered.

The acid clay can readily be restocked with nutrients. In doing this, the nutrients must replace the hydrogen or the acid. Nitrogen put into the soil as ammonia is taken on by the clays. In being so taken, it displaces some of the acidity there. Potassium, as a fertilizer, behaves similarly. Calcium as a salt of sulfuric acid, or gypsum, is taken and held by the acid clay only if the calcium pushes some of the weak acid off the clay to make of it a stronger sulfuric acid in the soil. And yet, in spite of this resulting stronger acid, the calcium sulfate, or

Roots multiply profusely to form dense clusters in the centers of a nutrient concentration as illustrated here. Roots selected the clumps of phosphated barnyard manure scattered through the body of the soil. This photo by Dr. A. R. Midgley of Vermont Agricultural Experiment Station was often used by Dr. Albrecht to illustrate articles and talks.

gypsum, as a soil treatment has long been recognized to be beneficial to clover. Calcium as a salt of carbonic acid, or as limestone, has also been more widely used to grow better clover. This calcium as a carbonate also pushes acidity off the clay when it is adsorbed there. This produces carbonic acid in the soil. But this weak acid decomposes leaving only water while the accompanying carbon dioxide escapes to the air Thus we see that both fertilizing and liming have been removing acidity from the soil.

We have, however, emphasized liming for clover while erroneously believing it beneficial because the hydrogen was driven off the clay. Rather, we should be more correctly emphasizing it because the clay was being restocked with calcium as a nutrient for the clover crop. Liming the soil is a case of fertilizing or restocking the clay with calcium. In like manner, putting on potash fertilizer is a case of restocking the clay with potassium. It, too, just like calcium or lime, replaces some of the acidity there. Fertilizing then, like liming, also reduces soil acidity. But we have used so much less fertilizers in comparison to lime that we have not appreciated the importance of acidity in the dynamics of the soil for plant nutrition.

18. soil is a collection of nutrient-bearing mineral centers undergoing mobilization by acid clay

We now know that soil acidity is beneficial. When the less soluble limestone is put on the soil, it is the soil acidity that breaks the limestone down. The acid clay trades acidity for the calcium and then is ready to trade this calcium to the plant root for which it will take acidity in return. Around each particle of limerock put into the sour soil, there is a thin clay layer that has become loaded with calcium by this exchange process. Should we put calcium phosphate rock into the soil, it would similarly give calcium to the clay for the soil acid, and would consequently make acid phosphate. If a potash-bearing rock were distributed as fragments through the soil, the acid clay layer around each of these would become loaded with potassium. Thus it is the clay acid that is the active rock-decomposing agent in the soil. It is the clay that is trading this assembled supply of available nutrients to the crops. It is becoming more acid again as it does so. This clay acid is then the soil reagent to break down, or to decompose the unweathered minerals containing the nutrient reserves. It is the soil acid that changes these from the unavailable rock crystals to the exchangeable forms held on the clay and thereby made available for the plants.

When most of the soils are mainly a mixture of fragments of the original rocks with a small amount of clay or humus colloids, the soil is certainly not a uniform mass. The clay at one point may be in contact with a small limestone particle and be saturated there with calcium. At another point within the soil, the clay may be touching a feldspar or potassium-bearing rock which stocks this clay portion with potassium. At another center there may be a calcium-phosphate rock which would give acid phosphate and have both phosphorus and calcium available on a clay layer for the later-coming plant root. Thus any mineral crystal or rock combination of nutrients at any point within the soil is being slowly weathered out by the soil's acid clay as this is made acid by the growing plant roots.

The clay is the "jobber" then, taking acidity from the plant roots according as it has given them nutrients. It passes the acidity to some original mineral and remaining rock fragments as we plow or reshuffle the clay, the silt, and the sands about in this and other tillage operations. By those happenings in the soil, acidity goes from the plant root to the clay, and from there to the nutrient reserves stored in the silt and sands. These reserves will be broken out and activated by this acidity to move them to the clay and finally on from there to the root. Around each rock particle, then, there is the clay layer loaded with particular plant nutrients which that rock contained. We now can visualize soil as a mixture of mineral centers, each offering to the roots the particular nutrient it has given up to the acid clay surrounding it.

19. roots grow where the supplies of soil fertility invite them

As these roots are going through the soil they may also be growing through it. Doubtless you have noticed that occasionally one root may be very thin while another is very thick. Can't we imagine that the thin root was merely *going* while the thick one was *growing* through the soil by nourishing itself enroute? If it were getting only water in passing we might expect it to be thin. This difference between a root *growing* through rather than merely *going* through the soil will emphasize itself if you will study the thin corn roots where they are in the acid, tight clay subsoil and getting only water in contrast to the much thicker root portions in the more fertile surface soil giving it nourishment. If the root finds much fertility in one of the many mineral centers it will suddenly grow not only thicker and longer, but will also suddenly develop many branching roots to increase the total absorbing surface areas in that center at which so much more fertility is available for it.

Have you ever dug into the soil in a cornfield, for example, to turn up a clump of manure plowed under some time before? If you have, you probably recognized the many roots of the corn plant clustered in the manure clump while unnoticedly you had cut off the fine root leading from there to the plant. You failed to see it because it was so tiny. Seeing no significant root, you wondered about the connection between the plant and these many and thick roots in the clump of plant nutrients. If a tree root ever clogged your sewer, you were amazed at the tiny crack in the tile through which a very small root went. Yet it made the enormous root collection, like a horsetail brush, inside the sewer tile that was plugged by it. These are illustrations of the roots growing enormous absorbing surfaces in order to collect nutrients from the flowing, dilute sewage or from the manure clump. Yet all the nutrients so collected from the sewer or from the manure clump could be passed to the tree or corn plant through but the very small roots coming out of the crack in the sewer tile or through a slender corn root almost unseen.

20. is plow sole and sub-surface fertilization practiced?

When the tight clay subsoil is broken up in the bottom of the furrow only as a mechanical attempt to make a deeper surface soil, this operation does not make the thin roots going into the subsoil suddenly grow into big roots there. Mixing the infertile acid subsoil with the surface soil has the opposite effect in that it makes the main plant roots stay at shallower depths. But if we will put some calcium and other nutrients down into that shattered clay subsoil, then a crop like sweet clover will send its roots growing down as thick ones rather than merely the thin ones of mainly water-delivering value. Soils must have many centers with nutrients if roots are to *grow* rather than merely *go* through the soil, and thereby render nutritional services to the plant.

Plowing down the fertilizers as streaks on the plow sole and the excellent crop improvements resulting from it have demonstrated the wisdom in our belief that plants "select" their diets from the soil as a kind of cafeteria offering a collection of centers of the separate nutrients and not from the soil as a uniformly blended mixture like hash. By using a T.N.T. plow with its subsoilers going at a depth of ten or more inches so as to reach into the subsoil clay, and by putting a nitrogen fertilizer at that depth behind these subsoilers, this particular form of plant nourishment was placed well down into the soil as narrow streaks fourteen inches apart running across the field. That the plants sent their roots down there to find the nutrients was shown by the greener corn crop and its larger yield later. But more significant as demonstration of the plant's selective activity was the advent on this nitrogen fertilized plot of a weed crop after the last cultivation of the corn when the adjoining plots given the other fertilizer nutrients but no nitrogen were almost perfectly free of any weeds. Still more significant was the fact that the weed plants were uniformly distributed over the soil surface and not in rows directly above the nitrogen streaks put down deeply by the subsoiler parts of the plow.

When coarse lime rock and the insoluble rock phosphate mixed into the soil are made available for plant nourishments, and when coarse limestone is more effective over longer time intervals than is finely powdered stone, there is the very strong suggestion that the plants must be searching through the soil with their roots and finding there in the scattered soil centers the essentials for their nutrition and growth. Until we know more about the soils and their services in feeding plants, we may well plan to handle our soils so as to encourage our crops to "select" their own diets from them.

22.

That plants are protected from fungal and insect attack by fertile soil surfaced early in the scientific findings of Dr. William A. Albrecht and associates. Plants with balanced hormone and enzyme systems did not turn out to be fit food for the lower forms of life in the biotic pyramid. Both higher and lower animal forms have anatomical and functional designs of the alimentary canal that are quite different. How the several life forms pass food through their systems for preparation, digestion, absorption, chemical censoring, metabolism, excretion provides a clue to one of nature's great mysteries.

Since the requirements of higher life forms are different from those needed by lower life forms, it stands to reason that unfit food for one becomes a delicacy for the other. The use of powerful poisons to destroy lower life can proceed only at great risk to the life form at the apex of the biotic pyramid—man!

This report was presented in The Journal of Applied Nutrition, *1970.*

Plants—Protected by Fertile Soil

Plant pests and diseases, much like the size of the crop yield, were once considered purely a matter of chance. But we know now that the crop yields demand the management of the fertility of the soil for maximum use of the rainfall, the air and the sunshine.

We are also overcoming the plant diseases and the insects, in some measure, by special efforts in combatting them with poisons. We have not yet, however, given much thought to the possible prevention of their damage as a by-product of managing the soil for bigger yields; for crops of better feeding qualities for our animals and ourselves; and for plants growing their own immunity and self-protection.

1. only properly nourished microbes can defend themselves by making antibiotics

Since we now know that microbes make the antibiotics by which they protect themselves; and which are also nutritional helps to those higher animals fed them; or are even helpful to us when we get them into our bloodstream by other less natural means than the digestive tract; it is dawning on us that the life of the microbes and the compounds they create are a matter of the soil and the level of its fertility for the nourishment of these lowly life forms.

Since the men of the research laboratories have isolated these antibiotics and suggested their chemical formulas, we have discovered that the creation of them within the microbial cell is determined by specific nutrition. Only according as the proper food permits, can the microbe make the chemical compounds by which it keeps other microbes from consuming it. On finding these chemical protectors against microbial attack also serving for improved nutrition of pigs, we see the fertile soil as the food for the microbe, or as the food responsible for making the protection possible for this single-celled plant.

The soil fertility is also the food for protection of the single cell like the root hair. It is thus in control of the creation of the chemical substances by which this underground part of the plant—and the entire plant—grows and simultaneously protects itself from microbes. Thus, with the root hairs and the roots pushing themselves into the fertile soil, their good nutrition, which means good growth (not necessarily measured by size) is also good health and good protection against microbes at the same time. Now that we know that antibiotics are some of the compounds giving growth and protection all in one, we need no longer consider the microbes a case of one fighting the other with this special equipment; rather, we may consider it a case of each microbe feeding itself for proper nutrition and growth. Along with that good nutrition comes protection from attacks by other microbes, and against what is commonly considered microbial "diseases." For the microbe, it is true that "To be well fed is to be healthy" or to be well fed is to take no bacterial "diseases."

2. plants also have their antibiotics for protection against bacteria and fungi even as early as the germinating seeds

We have also discovered that germinating seeds produce antibiotic compounds. The chemists have isolated them and tested their effectiveness. When the mother plant lays down the seed, she gives the potential offspring not only its life via compounds of food value, but also by those same nutritional values guaranteeing life and growth, she is guaranteeing protection against the young plant's troubles with bacterial "diseases." The radicle of the germinating seed protects itself because it is growing. It is growing because proper nutrition was packed into the seed by the mother plant getting her help from a fertile soil. While the root gathers its nutrition, it builds protection against microbial attack for the new plant. Thus, by means of proper nutrition more than weapons for interspecies warfare, nature has given us the surviving species.

3. *animals knowing their own nutrition and medicines make antibiotics which protect us as well as themselves*

Much like the plants, our animals have already been making their own antibiotics, too. Microbes and plants were making theirs long before we discovered these compounds. We could not have discovered them if they had not been there before. Animals must know their nutrition, or what to eat to make theirs. Wild animals certainly know their "medicines" to have survived all these ages in the absence of what we would administer as drugs and veterinary services. The so-called "depraved" appetite of the goat is certainly not bad etiquette in good goat society. The "filthy" chicken and the "dirty" hog running for the fecal droppings of the cattle are animal behaviors shown recently by the scientists to be animal wisdom of high order regarding proper animal nutrition. Those supposedly "filthy" animals know their vitamins, in this instance vitamin B_{12}, and did so long before we knew ours, especially this one among the last to be discovered. It was because they knew theirs that we discovered them. The animals have demonstrated refinements in their chemistry much beyond ours in the laboratory.

These animals are apparently not worried about fighting microbes in order to be healthy. They consume millions of microbes. They scratch for them and root up the soil for them. They waged no struggles against microbes by means of chlorinated and sulfonated carbon ring compounds as the deadly weapons. Instead they went about their business of feeding themselves properly by their own instincts. The prevention of microbial invasions or "diseases" has been a resulting by-product. Good feed has always been the animals' own "medicine." If now new "diseases" are rampant, shall we not begin to consider the possibilities that too many poisons scattered so freely about, too low quality of feed, and too much of its production on the more infertile soils, may have some causal connection?

Our animals have been making antibiotics not only for themselves but also for us. The horse can be given our typhoid bacteria or our typhoid fever and by no other feed than what is customarily good grazing or good horse feed, it can create the protein compounds in its blood to be transferred to our bloodstream as protection against the typhoid bacteria. The cow can take smallpox and generate protection for us, yet both these animals live through these diseases and services of providing antibiotics for us as protection against microbial and virus invasions.

3. *virgin vegetation grew its own immunity to diseases when the fertility was annually put back on the soil rather than hauled off to the market*

If these various life forms below man in the biotic pyramid, namely the microbe, the plant, and the animal were all once able to protect themselves; and if our crop plants were not wiped out by diseases, as their evolutionary survival indicates; why don't they protect themselves more effectively today? Have the "genes" of protection, or of resistance, been lost in the many generations? The answer is in the negative if subjecting the plants to rigid situations should make them grow stronger and develop the necessary characters. The many plants destroyed by fungi and insects tell us that if subjecting our crops to "harsh" or "tough" situations ought to "breed" a resistance to, or a toleration of, disease and insects, then our plants really ought to have come through as "toughies."

But in spite of the belief by some folks that we can select or breed legume plants, for example, to *tolerate* soil acidity, and wheat to *resist* smut, such beliefs rest on fallacious logic. It will not stand the common test of *reducto ad absurdum*. If that reasoning is carried to its final conclusion, then we should be able to breed plants to tolerate starvation. An experiment aiming at that objective would carry the plants no farther than one generation. It would go no farther than one aiming to develop a race of men immune and resistant to women, or a race of bachelors. That immunity, and resistance (or what have you) to diseases and insects are something the plant gets by way of its genetic potential we breed into it may perhaps be a belief to which you hold. But the plant's survival by evolution tells us that it always had that potential and therefore it cannot now be introduced by the plant breeder's art or science. The plants may be failing to demonstrate the presence of those characters because we have not permitted them to be nourished well enough, when growing as they are, on soils much lower in fertility than were the virgin soils in the course of the plant's evolution.

Under virgin conditions each crop dropped itself, its contents coming from the soil, and its own organic creations, resulting from air, water and sunshine, back to the soil in place. Its ash or inorganic elements increased those in the surface soil by the extras brought up annually from the lower soil strata. Its organic residues kept much of nutrient value in the less soluble but microbiologically active condition in that surface soil layer. More inorganic nutrients were being broken out of the mineral reserves by the acidity of the roots and of the respiratory carbonic acid from the soil microbes decaying the abundant organic matter. After the many generations of the same plant species had worked over the minerals and the organic matter and had dropped them back, the soil was offering better nutrition in consequence of the generations of nutritional heritages collected there. This secondary assembly line of rotating organic matter, in addition to the primary one of breaking out the inorganic elements from the rock minerals, made big healthy growth of virgin crops possible.

Here nature was putting back into the soil not only a rotating, balanced diet of the plant's inorganic needs, but she was also building up a bigger supply of the plant's organic needs. Plants of higher order making more complex compounds for their own nutrition—and for better animal nutrition too—use organic compounds from the soil. Virgin crops were putting these organic compounds back into the soil regularly for the next plant generation. Can it be possible that immunity to diseases and insects was greater when the higher content of organic matter of the soil offered more organic compounds as nutrition for the crops to build their "resistance" to these troubles?

5. *experiments demonstrated plants' immunity to fungus "disease" when ample inorganic fertility was put into the soil*

That even a soil more fertile in more inorganic elements alone is better plant nutrition to build immunity to disease was demonstrated in some soils research as far back as 1936. Colloidal clay was freed of its active nutrients by electrodialysis. It was then given calcium to the extent of about 25% of its adsorption capacity. This left the clay extremely acid, with a pH of 4.4, when the figure pH 3.6 represented its complete saturation with hydrogen or acidity, and the figure of pH 7.0 represented the absence of acidity. By merely putting increasing amounts of this acid clay into sterile quartz sand, soybean plants growing in it were protected to an increasing degree against a fungus attack that resembled what is commonly called "damping off."

Here was a legume crop taking its nitrogen from the air. It was carrying enough of the inorganic elements (both major and trace except calcium) in the larger seeds. It was carrying there also enough of the organic reserves for the physiological mechanisms that would build a root system which, if given contact with more calcium clay, would ward off the fungus attack or the "damping off" disease. More calcium in the soil meant more good growth and more immunity, if we view it with "disease" in mind.

6. *increasing plant "diseases" result from lowered soil fertility bringing on poor plant nutrition*

Reasoning in the converse, then, if virgin plants could build antibiotics on higher fertility levels in the soil, shall we not believe that the increase in present plant diseases is premised on declining inorganic and organic fertility, and failure of nourishment of plants to produce antibiotics? Should we not test this hypothesis by making our soils more fertile in every possible way to see if that wuld not help our plants ward off diseases and insects? In test of that hypothesis it would be necessary to fertilize not only with nitrogen, phosphorus and potassium, but there should be included calcium, too, long applied erroneously for its companion carbonate ion in limestone as a neutralizer of soil acidity. We should consider sulfur as necessary. It has been unwittingly applied for many years with phosphate, made more soluble by treating the pulverized phosphate rock with equal quantities of sulfuric acid. We should add magnesium, and the host of so-called "trace" elements, including boron, manganese, copper, zinc, molybdenum and possibly others not yet even listed as needed in "trace" amounts.

In testing the hypothesis whether we cannot ward off diseases and insects from plants by soil fertility treatments, we should apply organic compounds as fertilizer, too. Suggestions are coming that if our bodies provide proteins and other similar complexities equally as valuable in their effects for self-defense as those by the antibiotics of fungal origin, then it may be the proteins and protein-like compounds which the plants synthesize and which serve for the protection against these enemies. Horse serum, as a protein brought up to the higher life level of man for the latter's protection, is a protein. It is not a fat suspension that protects. Nor is it a carbohydrate. It is proteins that we put into our bloodstreams as protection against the invading foreign proteins which we call "diseases."

7. fertile soils building more plant proteins help to ward off insects too

That soils of higher fertility are protection for plants against insects as well as against diseases was also demonstrated by the Missouri Agricultural Experiment Station. Spinach was grown in the glasshouse. Again there was a very delicate control of the variation in the fertility by means of colloidal clay, which was mixed into highly pure quartz sand. Only two nutrient elements were varied in the amounts offered each plant. These were nitrogen, and calcium adsorbed on the clay. Nitrogen is naturally the common symbol of protein. Calcium is readily connected with the synthesis of proteins by legume plants. It is also the element serving closely with the non-legume plants' efficient utilization of nitrogen from the soil. Nitrogen and calcium were each used in amounts ranging through 5, 10, 20 and 40 milligram equivalents per plant in the many possible combinations of the two elements used in these amounts.

While we do not ordinarily spray spinach plants to ward off insects, and do not expect insect attacks on this plant regularly, it happened that the thrips insects invaded this experimental arrangement. But strangely enough, they attacked only those plants given no more than 5 and 10 milligram equivalents of nitrogen. These were attacks on only those plants of low protein contents because of limited nitrogen as fertility for protein synthesis. Still more strangely, however, even in their attacks on the plants on these soils less fertile in nitrogen, the increasing amounts of calcium accompanying either of the lower amounts of nitrogen, served to give decreasing attacks by the insects.

Here was the suggestion that the soil given more nitrogen for more protein production by the plants gave them immunity against the insects eating the leaves. Likewise, giving the soil more calcium for more protein registered a similar effect. These results suggest that we would not be correct in believing that it is the element nitrogen per se, or the element calcium per se, that protects the plant. It seems much more nearly correct to believe that the plant carries on its biosynthetic activities more effectively, through the extra nitrogen and calcium, in making those proteins by which it is protected against microbial invasion or insect attack. It suggests that the nitrogen and calcium are doing more than hitch-hiking. They are help, seemingly, in the plants' creative services by which it makes the proteins

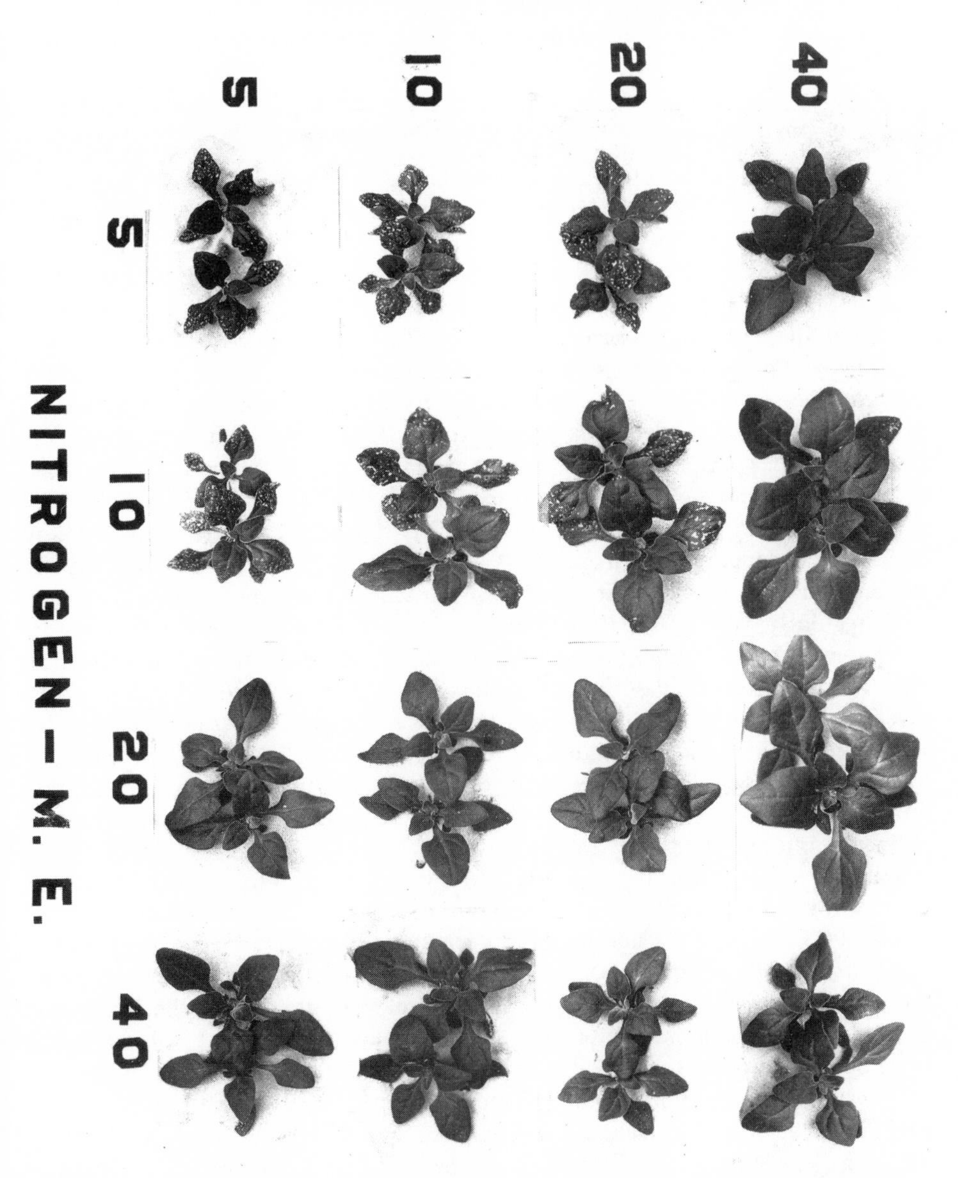

Smaller or larger amounts of nitrogen as nutrition for the spinach plants made them either victims of or victors over the leaf-eating Heliothrips haemorrhoidalis insects. Apparently the nitrogen undergirded the plants in making their own protective antibiotics. Insects took those plants which were struggling for their proteins but disregarded those producing more of this growth substance by the help from extra soil treatment.

supporting its own life and growth with the protection as an incidental. By those same helps it makes the many kinds of proteins that nourish us and our farm animals. It is the fertile soil, then, by means of which we can hope to make our plants immune.

8. *use soil treatments for prevention; insecticides for relief*

While insecticides are a help in the fight on the insects once these destroyers are present, or may be the "cure" for the invasion by them, it is the better plant nutrition via better soil fertility supplies which comes in for prevention. At the same time, the more fertile soil is insurance for a crop of bigger yield and of better quality. Since we grow crops for their earnings determined by yields and by nutritional qualities contributed according to the fertility, then if the soil is improved by treatments to bring about these two customarily desired objectives, we may obtain the protection against plant diseases and insects at the same time as a by-product. By this procedure we need little special knowledge about insects and plant diseases, and little special effort with special expense for a "fight" on these two commonly listed enemies of our food and feed crops.

Our knowledge about soils and their fertility management extends over many more years of experience than our knowledge about the myriads of different insects to plague us and all the unheard-of and unnameable chemical poisons coming forth regularly with their claims as ammunition to fight microbes and bugs. In using the extra soil fertility for prevention, we will also move forward much more effectively in feed and food production than we shall by dependence on poisons to save crops that are already of low yield and low nutritional values even if saved by victory in a war on the bugs and microbes. We are slow to believe that much of what we call "disease" originates at the cellular level because of deficiencies, malnutrition, dysfunction, etc., and that the microbial invasion is a symptom of, and a sequel to, but not necessarily the cause of it.

With the serious hazards to our own health in using the poisons so extensively, to say nothing of hazards to crops in some combinations of them, shall we not try to avoid all these risks if the non-poisonous fertility additions to the soil may be the way to escape all the poisons? Working from the ground up in this case is prevention, of which an ounce is worth a pound of cure. If we are to solve the insect and disease problems most wisely for the plants, animals and ourselves, it is not going to be by a call on only the insecticides and the drugs. Rather it will be by also rebuilding our soils in their fertility. By that means we shall starve the insects but feed our livestock and ourselves just that much better. Plant diseases and insects should be less as the fertile soils become more, and as we demonstrate for plants the age-old truth which says that "To be well fed is to be healthy."

23.

Are the meteorological conditions changing for the worse, or are the biological manifestations of weather, labeled as drought, merely intensified and on the increase as reciprocal to some other factor under serious decline. . .This opening question in Droughts—The Soil as Reasons for Them, *both asks the question and states the probable answer. The inability of soil systems to hold their water, and the greatest Mississippi floods in history after 25 years of the most feverish dam building in history, both point to the formidable perception of Dr. William A. Albrecht. This paper was read before the 11th Annual Meeting of the American Institute of Dental Medicine, Palm Springs, California, 1954.*

Droughts— The Soil As Reasons For Them

When one follows the meteorological reports rather regularly since most of us talk about the weather, at least when the radio reports it for us daily, one might well be asking with serious concern, "How come that we keep on breaking flood records, heat records, past records for drought or for extent of long-time rain free periods and other weather records?" Are the meteorological conditions changing for the worse, or are the biological manifestations of weather, labeled as drought, merely intensified and on the increase as reciprocal to some other factor under serious decline through which the same meteorological disturbances are magnified in their detrimental aspects? We have larger floods and we have more severe droughts as the records truly report. But should we not examine these in relation to the soil for possibly more comprehensive explanations of them and our reduction or prevention of the disasters?

When "droughts, unlike rain free periods, may not be defined from standard meteorological observations, since the intensity and the length of the drought depends on genetic characteristics of crops, soil water, soil fertility conditions, and meteorological parameters," [quoting Wayne L. Decker, University of Missouri Climatological Research Project], it will be evident that we need to recognize the soil as a major factor in the disturbances to crops which we call "droughts." These are in reality dry periods extending themselves to lengths of time that bring about crop disaster. The drought then is more a performance measured by damage to crops than by meteorological indexes.

Drought is then a time period during which there is a serious shortage of water by rainfall for the biological services it usually represents in crop plants. Since the water's services to plants are exercised mainly after the rainwater has entered the soil, then the soil, which is more than merely a water reservoir, may be considered as influencing the effects of the shortage of stored water over an extended rain free period through all of its services to crops beyond that of holding a supply of water. Those many services need to be considered before we use water shortage per se as the alibi for poor crops.

1. the law of continentality vs. the law of averages

The geographic climatic settings for most of the droughts are the area between the humid and the semi-arid soil regions. These represent mineral-rich soils in general, since the low rainfall has not developed them excessively or removed the calcium, and minerals of similar soil behavior, from the profile and replaced them by hydrogen as acidity. These are the soils where agriculture grows protein rich forages; where soils are windblown and where animals grow more readily on what is apt to be called the prairie and the plains soils.

Droughts are also geographically located in the midst of larger land areas where the effects of what is called "continentality" are more pronounced. This represents the degree to which the weather or the daily meteorological condition varies from the climate, which is the mean or the average of the weather for the longer time period of records considered. The larger the body of land, *i.e.,* the more continental the area, the more the weather or the daily condition will vary from the climate or the average. This is "the law of continentality" in brief. Droughts then may be more commonly what we call "continental" manifestations. They are a variation from the mean and the expected since climate is reported as the mean of much meteorological data. It is in the midcontinent of the United States were droughts may be expected more commonly.

When the average, or mean, of weather records is used to describe the meteorological conditions of a region—Columbia, Missouri, for example, is reported to have an annual rainfall of 29.33 inches. For Springfield, Missouri, the same figure is 41.42 inches. This rainfall figure which is a mean total for the year obtained from records of nearly a half century says nothing about how high or how low the amounts for any single year may be. Because of the continentality of Missouri—its location a thousand miles from any seacoast—we formerly considered that from the previous data, Columbia, Missouri, had a continentality effect of 50%. That says that while the rainfall is reported to be roughly 40 inches, the precipitation might vary over a range of 50%, namely, 25%, or 10 inches, below 40 and 25%, or 10 inches above 40. It might range from a low of 30 inches to a high of 50 inches of precipitation for the different years.

But that figure, once established for continentality, is no longer the fact. That record of continentality was broken in 1953 when, because of the drought of that year, the annual rainfall was but 25.12 in place of 39.33 inches. This annual weather in terms of a rainfall of 25.12 is 36.1% below the mean of 39.33 inches, or that much below the climate. Hence, we may expect excess rain also of 36.1%, a high sometime of 53.54 inches, or a continentality effect of twice 36.1, which is 72.2%. For Springfield, Missouri, the drought of 1953 gave but 25.21 inches of rainfall annually, or a deficiency of 39.1% to suggest a continentality effect there of 78.2%.

If one considers the rainfall for only the summer months—May to September, 1953 inclusive—when the effects of the extended rain free period on vegetation are exaggerated by high temperatures, Columbia, Missouri suffered under a continentality effect amounting to 86%. At that same time, Springfield, Missouri suffered one amounting to 135%. This latter was a most severe disaster to an agricultural area given largely to the dairy phase of that business with so much dependence on grass as the crop. Thus the law of averages applied to Missouri may leave one content with averages, but the law of continentality is truly disturbing but highly revealing when such droughts as 1953 are experienced in record breaking dimensions.

The Missouri Drought of 1953 Emphasized Continentality When the Records of the Weather are put in Contrast to Those of Climate

	Columbia, Missouri	Springfield, Missouri
Annual rainfall — mean	39.33 inches	41.42 inches
Annual rainfall — 1953	25.12 inches	25.21 inches
Annual deficiency — 1953	36.1%	39.1%
Continentality effect — 1953	72.2%	78.2%
Summer rainfall — mean*	21.26 inches	21.62 inches
Summer rainfall — 1953	11.94 inches	7.02 inches
Summer deficiency — 1953	43.0%	67.5%
Continentality effect—summer	86.0%	135.0%

*May to September, inclusive

2. droughts are becoming urban—no longer remaining rural

With 85% of our population collected into urban centers, while only 15% are still rural, we would scarcely expect the urban group to appreciate droughts, exhibiting the effects of rain free time largely through the water shortage within the soil bringing crop disasters and livestock troubles. But droughts as water shortage through falling water tables and failing wells are not only rural troubles for thirsty livestock. They are coming to be serious troubles also for urban centers and areas of congested peoples. Where the per capita water consumption per day was formerly a few gallons—a pail full carried from the spring—it is now estimated at 700 gallons per day on a national scale. When this increase per capita is coupled with the increase in population our water consumption since "water pail days" represents an increase of several thousand percent. This supply comes mainly from deep wells. For this the soil is the filtering, clarifying and bacteriacidal agency in most cases which gives us clean, health-supporting water to drink. We have taken water of this kind for granted. We have not seen the soil's services connected therewith. Droughts are making us become water-conscious, not only via disasters to crops as feed and livestock as drink, but also even to the value of water as the major liquid mineral we all drink. When eastern Kansas in 1954, following a rain free period of serious shortage in 1953, had 26 cities critically short of water, we have reason to become conscious of droughts of larger significance than of such to the rural population only.

3. water shortage—a result from excessive erosion and drainage

Our urban centers are coming to see the soils as reasons for droughts in broader meaning of that loosely used term. We realize that we must either limit water consumption per capita, or we must raise the level of the groundwater, *i.e.,* the water table, by getting more water per rainfall to enter the soil. *The shortage in soil-stored water is a sequel to soil erosion.* As the surface soils become shallower they are less of a blanket to hold larger portions of each rainfall for increased amounts of it to filter or to soak down more deeply into the soil to raise the water table there. Every little rill of erosion is a drainage ditch to hustle the rainfall off just that much more rapidly and to leave that much less to enter the soil for storage there. With erosion, too, the structure of the remaining cultivated surface soil has become less granular and less stable. Less infiltration of water per rain is possible for that reason.

Our excessive drainage, increased more recently by the excessive surface runoff bringing about erosion is now magnifying the shortage of water taken from and given out by the soil. In the mind of the pioneer it was the surplus water and not the water shortage against which he waged a constant struggle. He used drainage ditches, tiles, and all possible means of getting rid of what he considered too much water. We seem to have inherited the pioneer's animosity for water and delight in the extra speedy drainage. Instead we now should encourage more standing water for infiltration because we have too much drainage for sufficient of that.

We have apparently lifted our soils too high out of the water when now nearly every acre is drained. Also when all-weather roads are considered a necessity almost every section of land is encircled by such. Each roadway under concrete cover is allowing no rainwater to enter that much soil. Also by its sloping shoulders and parallel drainage ditches each highway is hustling off to the rivers the rainwater falling upon acres and acres in roadways, while draining also more quickly the arable land adjoining them. When we are bringing about all these changes which reduce the rate and total of water infiltration into the soil while rates of water consumption are increased to lower the supply stored, both in the soil profile for our crops and in the water tables for our livestock and the people of our population, is it mysterious that droughts are getting worse and floods more disastrous? Are these new records other than man made? Are they coming other than by way of the soil?

We have then been bringing our droughts as they represent shortage of supplies

How plants withstand drought depends more on soil management than on timely rainfall, according to Dr. William A. Albrecht. In this Sanborn Field test, the center plant received no treatment. The left plant was given manure only. The right plant received the full treatment of nitrogen and organic matter turned under. There was no irrigation. Center corn had two dry bottom leaves (fired). It showed no other visible plant irregularity. Corn at left had five dry bottom leaves, but the upper seven and tassel suggested little damage except rolled green leaves. Corn given full treatment (right) had two bottom leaves dried, but the upper closely bunched dozen leaves were dried from the tip end back to suggest that excessive temperature injured the growing part. This suggests death of the cell's protoplasm much as high temperature in the incubator stops the hatch, but does not visibly change the egg protein.

of water upon ourselves. Droughts are disastrous in terms of deficiency of that liquid mineral in the soil and of the food it grows. The more fertile, high protein-producing soils are exhibiting the more serious drought disasters. Man is thus pushing himself off the soils which are better for nutrition. He is crowding himself to areas of higher rainfall and to soils giving feeds and foods of high-fattening rather than high-feeding values. He has not noticed that he was moving himself out of quality foods by soil exploitation, since hidden hunger is registering itself all too slowly. But now that he is crowding himself out of drink, that will register more quickly since thirst is more speedily lethal than hunger, droughts take on more meaning. They, too, are moving from the country to the towns and the cities. Droughts register as disasters regardless of whether via humans or via vegetation.

Since both routes for troubles of this nature go through the soil, they will finally lead us to the soil as the basis of creation in terms of both drink and food.

4. confusion in considering water shortage in the soil but not recognizing fertility shortage there

In seasons of water shortage for our crops, that shortage in the soil has too commonly been mistaken for the shortage of plant nutrition there. When the farmers said, "The drought is bad since the corn is *fired* for four or five of the lower leaves on the stalk," they were citing a case of the plant's translocating nutrients, especially nitrogen, from the lower, older, nearly spent leaves in order to maintain the upper, younger, and growing leaf parts of the plant. Now that we can apply fertilizer nitrogen along with other nutrient elements, we know that in the confusion and lack of knowledge about plant nutrition we made so much of the drought in many cases where it was not the direct shortage of the water as liquid for the plants, but rather the more common shortage of nitrogen entering into protein and all it represents in crop production.

In this case the soil as shortage of nutrition and not of water was responsible for what was called "drought." With a shallow horizon of surface soil to which the fertility of the entire profile was confined and with an acid, infertile clay horizon beneath it, the drying of that surface layer compelled the roots to go out of the drying horizon originally providing both fertility and water, and into the subsoil where only water but little or no fertility was present. That shallow surface layer was dried, not only as the result of the heat from the sun but also because the roots of the growing crop like corn are estimated to be taking from .15 to .25 of an inch of water by transpiration alone per day. (Some folks consider .10 inch of water transpired daily by a corn crop and define a drought for corn as rainfall of less than one inch every ten days. This gives no consideration to the soil concerned.) To miss recognizing the fertility shortage when emphasizing the water shortage in the surface soil during drought is a mental behavior of long standing. In that error of thought we have been blaming the drought via the soil water for a "fired" crop when it was plant starvation via that route from which also insufficient fertility for plant nutrition was coming.

If these fertility conditions cause the lower leaves of the corn stalk to "fire," in the case mistaken for drought, one needs only to note the growing tip of the corn stalk. If water shortage is responsible, then the growing tip of the plant will be wilted. If it is fertility shortage, the growing tip of the plant will not commonly be wilted since the roots going deeper into the subsoil are delivering water to maintain the active plant tip without its wilting. It is the wilting of the growing tip of a plant which tells us when water is needed, a question to which most any housewife knows the answer who cares for her house plants. Droughts may often be a case of infertility of the soil, or one of imbalanced plant nutrition apt to be mistaken for shortages of rainfall and for bad weather.

5. plants spend most soil water to keep leaf tissue moist for gaseous interchange with the atmosphere—this loss represents cooling effects

Should we clarify some other confusions connected with the properties of water and its biochemical services to plants, animals, and man, we may simultaneously clarify more effectively our understanding of the soil's significance under what we call "droughts." In connection with our own body comfort during times of higher temperatures and longer rain free periods, we appreciate the help by speedy evaporation of water from our own skin as a means of keeping us cool. It is a fortunate property of water that a tremendous amount of heat is taken up when water changes from its liquid form to a gas, or when it vaporizes. We can use melting ice to cool ourselves since about 85 calories of heat are taken up in melting one gram of solid water as ice into the liquid form at the same temperature. But nature has been more efficient in using vaporization of water from our skin as a means of offsetting high temperatures or heat, since about 585 calories of heat are taken up when one gram of water is vaporized from the skin, or in the breath as discharged in the form of water vapor from the lungs.

This property of water, namely, its high heat of vaporization, holds down, and to a considered regularity, the temperatures of small bodies of land surrounded by water. It offsets the effect of continentality as illustrated when Great Britain has a continentality of but 10%, or the Hawaiian Islands have almost none. Vaporization from the surrounding water mass spends the sun's heat which would otherwise raise the atmospheric temperature over the adjoining land were the air from there not exchanged by air from over the water. Soil water vaporizing from the soil's surface is then a cooling agent of the soil and the air above it. So is the water vaporizing from the plant's leaf surfaces. Trees bringing up water stored much deeper in the soil to be vaporized from the tree's leaf surface are a means of spending the heat from the sun and thereby of cooling the atmosphere. Clearing areas of forests has done much in bringing about wider fluctuations in temperatures, first because trees are helpful in getting rainfall into the soil for increased storage and less sudden fluctuation in soil temperature and moisture, and second, in lessened fluctuations in atmospheric temperatures within considerable heights from the soil because of their transpiration or vaporization of water from within their leaves. As crops grow taller they ameliorate for themselves the effects of variations in heat from the sun by means of the water evaporated through them from the soil. (A medium sized hardwood tree may lose 50 gallons of water through its foliage in a day, reports William B. Love, Michigan State College Specialist in municipal forestry.)

The water of transpiration from the plant's leaves demonstrates another of its vital biochemical properties, namely, its services as a solvent of gases as well as of salts for their ionization. Water is lost from the leaf of a plant because the inner, moist tissue of the leaf is exposed to the atmosphere for the exchange of the gases. Those gases are mainly carbon dioxide and oxygen. That inner leaf surface must be kept moist since gases will not exchange through a dry one for help to the plants in taking in carbon dioxide for photosynthesis or oxygen for respiration. Plants lose water by transpiration according to the meteorological conditions vaporizing that water from the leaf surface much as water from any moist surface. The stomates of the leaf, through which gases exchange, may be partially but not completely closed for a living plant. The plant leaves may roll themselves for reduction in transpiration before they wilt. But moist leaf tissue exposed for exchange of the gases must be losing water to the atmosphere, or plants must be transpiring, if respiration and photosynthesis are to continue to keep the plant alive under most common conditions of the water. Only an atmosphere of humidity at 100% or one completely saturated, eliminates transpiration. In nature, this condition does not occur often.

6. plant's transpiration ratio is not an index of efficiency of use of water: soil as plant nutrition determines that

It was the classic work of L. J. Briggs and H. L. Shantz [Relative Water Requirements of Plants, *Journal of Agricultural Research* 3:1, 1914] that measured the water of transpiration of crops in relation to the amount of dry weight in plant tissue resulting as growth. The same soil was used for the many different crops under experiment. At that time, and by many folk today, it was believed that the kind of crop determines this relation and little significance was given the soil as control of it. Their work gave us many "transpiration ratios" apt to be called "water requirements" of different crops. These values are the pounds of water transpired to the air by the plant taking it from the soil to produce a pound of the crop's vegetative dry weight.

Unfortunately, crops have been classified by means of these values into different "efficiencies with which they use water from the soil to give us yields of crop," *i.e.*, only vegetative bulk. "What difference is there in the quality of crop yield per pound of dry matter produced?" was not the question raised even when the transpiration ratios were widely different and the final figures were an average of them over wide ranges. Photosynthesis by the sorghum and sugar cane piling up rapidly their photosynthetic products, namely, sugars and starches as energy food for the plant, was emphasized. Biosynthesis, the production of the compounds like proteins which takes place without the direct service of light and uses some of the sugars and starches for starting compounds and for energy sources or fuel for the synthetic processes, was not considered. The ratio of the pounds of water transpired to the pounds of complete protein produced would have put this thinking on a truly nutritional basis. It would let us see water of transpiration used highly efficiently by alfalfa making a pound of very good protein per 8,000 pounds of water transpired. This is high efficiency in contrast to sorghum, making a pound of incomplete, or very crude, protein per 10,000 pounds of water of transpiration. Alfalfa, a quality feed producer, is more efficient in using water for this purpose than is sorghum.

But the crop specialists interested only in vegetative mass as a service by transpired water, remind us that sorghum uses water at the rate of 275 pounds per pound of dry matter grown, while alfalfa transpires 850. In his mind, which has not yet envisioned nutritional services by crops grown but clings to the criterion of vegetative mass produced per acre as the criterion of crop yield, the sorghum surpasses the alfalfa as the crop for droughty areas or those of lower rainfall. According to these folks using such simple transpiration ratios as their judgment of the crop's efficiencies in using water, low rainfall areas would call for growing bulky crops that starve our animals and ourselves rather than call for making the soils fertile in those low rainfall areas to use that water more efficiently for the creation of real nutritional values. Speculation in agricultural crops on the level of simple arithmetical thinking is more universal than is the creation of real food value demanding our thinking in terms of the science of physiology and all the other forms of organized knowledge undergirding growth, protection and reproduction by the life forms that live to feed us.

Any crop uses water inefficiently for the possible biosynthetic services when the fertility supply in the soil represents an imbalance for the support of the physiological processes required for the maximum of nutrition of that crop. In that nutrition of the crop, any one element in low supply in the soil may cause inefficient synthesis, while the stream of water loss as transpiration runs on just the same. Elements like calcium, magnesium, potassium, phosphorus and others held in place are *soil* fertilizers. Nitrogen, so mobile and not so held, is the *crop* fertilizer. Thus the confusion in this regard occasions inefficient use of the transpiration stream under nature's control, because we fail to keep the supply of nutrients in the soil up to the high level, and in the proper ratio, for the biosynthetic processes of the crop functioning at high efficiency in giving us nutritional values in itself as our food.

The transpiration stream flows according to the meteorological conditions favoring evaporation of water balanced against the soil's conditions representing forces holding the water as a thinner film around the soil particles. The plant and its open, internally exposed wet cells in the leaves are atmospherically exposed water surfaced connecting themselves through the plant and its roots contact with the water film around the soil particles. According as that soil has less water and the film is thinner, the water is held there more firmly against liquid and gaseous transfers of it to the atmosphere via the plant which is the equilibrator of the atmosphere's taking water by evaporation from the leaves and the soil's holding it by surface adsorption. The poor plant is merely the innocent equal sign between the two opposing forces. Even though the plant's leaves may roll, and stomates may nearly close, they must still permit carbon dioxide to enter and escape, and oxygen to do likewise for the continued respiration if the plant remains alive. Its wet, living tissue exposed cannot prevent the water loss any more than you can live and prevent the moisture loss in your breath by stopping your breathing. Plants lose water under variable weather according to the soil and meteorological conditions and not according to the plant species or plant pedigree.

7. transpiration stream of water from soil to plant vs. nutrient movement along that route

The transpiration stream of water moving from the soil through the plant to the air obeys the meteorological conditions of the atmosphere controlling it. The nutrient elements move from the surface of the colloidal clay holding them to the colloidal surface of the root according to the energy changes required to bring that transfer about. This chemo-dynamic performance of nutrient activity follows its set of laws and conditions, including the presence of water but not the movement of the water. The nutrient, inorganic elements within the soil, like the fish in the stream, are not victims of the current. They move with or against it according to forces controlling them.

Experiments using colloidal clay to measure more accurately the soil's stock and changes in the nutrient cations have demonstrated that nutrient ions could go from the plant back into the soil while the plant was increasing its mass by growth and was having a normal transpiration stream of water flowing from the soil to the atmosphere. As a second case, using the seed planted into moist sand, for example, growth occurred with the transpiration stream moving water out of the sterile sand but no fertility elements from there. It was an empty transpiration stream then so far as nutrients hauled in by it are concerned, but it was nevertheless a flow of water. It was a moistener of the leaf tissue only for exchange of gases there which is the normal function of transpiration.

In the desert where the soil is so dry, to cite a third case, the moisture condensing on the plants at night is enough to moisten the soil around the plant's roots by reversing the stream of transpiration. But this does not necessarily reverse the movement of fertility, which continues to go from the soil into the plants. Desert plants take fertility regularly even if the transpiration stream should be a diurnal reversal of its current. As another good case, one can demonstrate plant growth and nutrient movement into the root from the soil when the transpiration stream is not flowing. One can demonstrate growth by putting a potted plant under a glass bell jar into an atmosphere laden with moisture and carbon dioxide with a humidity of 100% and no transpiration. Given plenty of carbon dioxide and sunlight, we can have both plant growth and nutrient movement from the soil even when the transpiration stream is at a standstill. These four cases are the evidence that the transpiration stream is one activity, while the movement of the nutrients is another quite independent of it.

Our failure to study the plant nutrition within the soil, and our contentment with complaints about droughts, have left us growing bulk of plants rather than nutritional values in our agricultural crops. Water has been the great alibi. We have believed the plant concerned only about its drink. We have simply not seen the soil and the plant's concern about something that is truly plant nourishment for biosynthesis by it of proteins and higher food values. We have simply not diagnosed each specific case. We have been content with propagandized practices by the majority of prescribers. We have been running within the pack of humans in place of smelling out the trails of the things of nature.

8. drought, excess of temperature as well as the deficit of soil water

When the absence of water for its services in vaporization from the soil and the vegetation as a cooling effect allows the temperature of the air to rise high, shall we not expect the plant's processes of life, centered in the proteins, to be disturbed by the increased heat? Those processes are doubled in their rate of activities for every 10 C. increase in temperature according to the Vant Hoff Law, until the protein itself may be destroyed by it. We may well expect many life processes to be interrupted long before the protein is coagulated or changes are visible. Eggs incubated near 100 F. give a hatched chick, but if they are held at a few degrees higher than that for even a short period of time, the physiological processes are so disturbed that the normal hatch of the healthy chicks does not result. The protein of the egg need not be coagulated or even coddled to upset the process. Life processes in the plant come under the same category as those within the egg. They deal with proteins within the plant cell. They are concerned also with enzymes which encourage the processes. These delicate catalytic combinations resembling the proteins in many instances, fit into the same pattern of temperature requirements for regular, normal life processes. Low rainfall and accompanying irregular temperatures, then resulting in a drought may be effects of the heat wave as well as of the shortage of the water.

In the ecological pattern of plants distributed over the world starch production and its storage in the seed occur under limited temperature ranges at certain physiological stages in the plant's growth period. Corn grows for example in the temperate zone for high starch output in the crop and at certain months within the year. Other seeds of high starch delivery are seasonally located similarly. For starch producing crops in the tropical zone those seem to be given to storage of this compound in the roots or underground at lower temperatures. Seeds there seem to store their reserve energy supplies as oil. Shall we not visualize the plant injury, during a drought with the dry surface soil going to the higher temperatures, as an effect of the excessive heat changing the physiology of the plant rather than the effect of only a storage of water or this liquid nutrient?

Among the other plant manifestations suggesting effects of drought by high temperatures rather than by water shortage, there is the common change in a bluegrass lawn to one of crab grass or other species, for example, when the lawn owners persist in keeping their lawns watered during the hot summer months. Where the lawn is dried and the bluegrass has disappeared in going dormant, this same grass species comes back with the break in the drought, namely, with the rain again and the lowered temperatures. Such is not the case of the watered lawn, shifted by that watering treatment during the heat to a crab grass flora. That flora persists and excludes the bluegrass during the rest of the season. A Bermuda grass lawn is undisturbed by the drought which displaces the bluegrass. Bermuda grass stays green during both the temperature and water shortage.

Observations on Sanborn Field, Columbia, Missouri, under experimental soil studies since 1888, suggest that corn plants at a low level of physiological activity because of low soil fertility were not visibly injured by either the water shortage or the heat wave of the drought. But as more fertility, including nitrogen, raised the levels and diversities of the plant's activities, the drought damage became more severe. But this suggests itself as the result of the high temperature damaging the plant parts commonly rich in nitrogen and most active in tissue growth. The injury occurred in plant leaf parts where damages from nitrogen deficiencies are commonly observed, but the appearance of the plant parts injured was decidedly different than that exhibited under starvation for nitrogen. This suggests the simple fact that vegetation doing little but the elaboration of cellulosic mass is not subject to drought injury, but plants elaborating compounds of much nutritional value for animals are injured by the heat wave of the drought as well as by the soil's shortage of water.

As another biological illustration of the effects of drought, let us recall that the races of pheasants introduced into the United States came from a range of conditions quite unlike for example those in Missouri in which state the introductions of this game bird have not been so successful. These birds lay their eggs and incubate them too late in the season, or when the high temperatures we experience in the early summer have an adverse effect on the hatch. With the clutch of eggs on the ground, the soil temperatures rise too high and injure the incubating processes guaranteeing a good hatch. For this biological process, the "drought" damage results from the heat wave and not from the deficit of water as drink.

As still another biological demonstration of the heat wave aspect of the drought of 1954, a hatchery reported the death of many chicks, and of more mature chickens and turkeys on its poultry farm during the high temperatures accompanying it. Likewise in some of our experiments using rabbits for biological assay of the differences in grains and forages resulting from soil treatments with different trace elements, the first heat wave in late June and early July took over 70% of the rabbits in one set of the feed arrangements while it took none of another set. All the animals of these two sets were in the same room and temperatures. This mounting of the fatalities of the one set was gradual and persistent as the drought continued and the temperatures mounted in killing even the replenishment of the dying stock from the adjoining surviving stock moved to the fatal feed. When the high percentage of fatality on this dried feed had been reached with eight deaths in one day of record heat, the assay was terminated with a shift in ration emphasizing dried milk proteins. This shift prohibited any further fatalities and stopped the disastrous effects by the heat wave on these animals when considerable publicity of animal death by drought was common.

A repeat trial on the effects by the high temperatures on the rabbits, according to the original ration increased to a grain mixture, duplicated the previous results. This trial was carried on for only three weeks or until only 31% of fatalities resulted during the succeeding heat wave. Here the deaths suggest themselves as due to the high temperatures, but only when the poor nutrition suggests itself as the route through which the high temperatures worked their damage. It also casts reflection on the quality of the feed offered the public by some hatcheries along with their baby chicks.

9. superficial postmortems of crop failures blame the weather; accurate diagnoses point to the soil as help to avert them

It is only slowly that the factors in the agricultural production of our feeds and foods are being tabulated and evaluated. For too long a time has weather, especially rainfall as the supply of water for plant growth, been the alibi for irregularities in crop yields. "Drought" as a term including rain free periods of extended time has now broken all past records and become a national disaster. As such it deserves analysis of the problems it presents. Such analysis establishes the soil as a major factor in determining the severity of the disturbances to the plant's growth and reproductive processes by the water shortage and the high temperatures through which the plant is injured under the composite of conditions included in that term.

More soil knowledge through research progress has now pointed to a better understanding of the facts about soil water and the aspects through which some of the injuries by drought can be mitigated. The fertility of the soil as plant nutrition is decidedly significant in that respect. Now that we are separating the nutrition of the plant by the soil from the storage of water in it for the plant, the drought as water shortage is no longer so much of an alibi. Rather drought is more a damage by deficient plant nutrition. In soil management, which may include irrigation, the economy and sound service to plant production demand that the supplying of the soil fertility should be the first concern and the addition of water the second.

Analyses of the problems of drought establish the fact that excessively high temperatures per se as disturbers of the physiological functions of the plants, and even of animals, are factors perhaps more lethal than the water shortage. Even when the high temperature is segregated as a factor of damage, it is significant that this is increased by imbalanced nutrition, or conversely, improved by proper nutrition.

Thus the problem of drought damage moves itself into the lap of agriculture as a problem either to be solved—at least in part—or tolerated with reduced disaster, via the management of the soil for better nutrition of the plants and the animals fed by means of it. In the case of what we call "droughts" we need to view them for possible prevention or reduction of damage via the wiser management of the soils under them.

24.

When Weston A. Price, D.D.S., wrote Nutrition and Physical Degeneration, *he structured a volume so devastating in its proof that it seemed certain no one would ever again doubt the link between poor soil and physical ills. As Chapter 23 of that book, Dr. Price included an abstract in depth of a talk given by Dr. William A. Albrecht before the Regional A.A.A. Conference, Durham, New Hampshire, June*

1944. In referring to that address, Dr. Henry Bailey Stevens, Director of the Extension Service, University of New Hampshire, made these comments: "Sometimes a powerful thought, like a flash of lightning, throws a great area from darkness into sudden light. Such an illumination was experienced by those of us who heard the two addressed by Dr. William A. Albrecht. I confess that I am not sure yet how far or along what strange paths this message will take us. Apparently we can no longer think of foods as having a fixed value; for such value varies according to the soil content. . .It would seem that nutrition is a much more profound science than has been generally recognized and that in studies heretofore its surface has perhaps only been scratched."

Nutrition and Physical Degeneration *is being kept in print by The Price-Pottenger Foundation. All informed nutritionists must read it. In the meantime, here is the chapter Dr. Albrecht contributed to that work.*

Food Is Fabricated Soil Fertility

Food is fabricated soil fertility. It is food that must win the war and write the peace. Consequently, the questions as to who will win the war and how indelibly the peace will be written will be answered by the reserves of soil fertility and the efficiency with which they can be mobilized for both the present and the post-conflict eras.

National consciousness has recently taken consideration of the great losses by erosion from the body of the surface soil. We have also come to give more than passive attention to malnutrition on a national scale. Not yet, however, have we recognized soil fertility as the food-producing forces within the soil that reveal national and international patterns of weakness or strength. Soil fertility, in the last analysis, must not only be mobilized to win the war, but must also be preserved as the standing army opposing starvation for the maintenance of peace.

What is soil fertility? In simplest words it is some dozen chemical elements in mineral and rock combinations in the earth's crust that are being slowly broken out of these and hustled off to the sea. Enjoying a temporary rest stop enroute, they are a part of the soil and serve their essential roles in nourishing all the different life forms. They are the soil's contribution—from a large mass of nonessentials—-to the germinating seeds that empowers the growing plants to use sunshine energy in the synthesis of atmospheric elements and rainfall into the many crops for our support. The atmospheric and rainfall elements are carbon, hydrogen, oxygen and nitrogen, so common everywhere.

It is soil fertility that constitutes the 5% that is plant ash. It is the handful of dust that makes up the corresponding percentage in the human body. Yet it is the controlling force that determines whether nature in her fabricating activities shall construct merely the woody framework with leaf surfaces catching sunshine and with root surfaces absorbing little more than water or whether inside of that woody shell there shall be synthesized the innumerable life-sustaining compounds.

Soil fertility determines whether plants are foods of only fuel and fattening values, or of body service in growth and reproduction. Because the soil comes in for only a small percentage of our bodies, we are not generally aware of the fact that this 5% can pre-determine the fabrication of the other 95% into something more than mere fuel.

1. history records changing politics rather than declining soil fertility

Realization is now dawning that a global war is premised on a global struggle for soil fertility as food. Historic events in connection with the war have been too readily interpreted in terms of armies and politics and not premised on mobilized soil fertility. Gafsa, merely a city in North Africa, was rejuvenated for phosphorus-starved German soils. Nauru, a little island speck in the Pacific, is a similar nutritional savior to the Japanese. Hitler's move eastward was a hope looking to the Russian fertility reserves. The hoverings of his battleship, Graf Spee, around Montevideo, and his persistance in Argentina were designs on that last of the world's rich store of less exploited soil fertility to be had in the form of corn, wheat and beef much more than they were maneuverings for political or naval advantage. Some of these historic material events serve to remind us that "an empty stomach knows no laws" and that man is in no unreal sense an animal that becomes a social and political being only after he has consumed some of the products of the soil.

Geographic divisions to give us an east and a west, and a north and a south for the eastern half of the country, are commonly interpreted as separations according to differences in modes of livelihood, social customs, or political affiliations. Differences in rainfall and temperature are readily acknowledged. But that these weather the basic rock to make soils so different that they control differences in vegetation, animals and humans, by control of their nutrition is not so readily granted. That "we are as we eat" and that we eat according to the soil fertility, are truths that will not so generally and readily be accepted. Acceptances are seemingly to come not by deduction but rather through disaster.

2. patterns of nourishment are premised on the pattern of soil fertility

We have been speaking about vegetation by names of crop species and by tonnage yields per acre. We have not considered plants for their chemical composition and nutritive value according to the fertility in the soil producing them. This failure has left us in confusion about crops and has put plant varieties into competition with—rather than in support of—one another. Now that the subject of nutrition is on most every tongue, we are about ready for the report that vegetation as a deliverer of essential food products of its own synthesis is limited by the soil fertility.

Proteinaceousness and high mineral content, as distinct nutritive values, are more common in crops from soil formed in regions of lower rainfall and of less leaching as for example the "midlands" or the mid-western part of the United States. "Hard" wheat, so-called because of its high protein content needed for milling the "patent" flour for "light" bread, is commonly ascribed to regions of lower annual rainfalls. "Soft" wheat is similarly ascribed to the regions of higher rainfalls. The high calcium content, the other liberal mineral reserves, and the pronounced activities of nitrogen within the less leached soil, however, are the causes when experimental trials supplying the soil with these fertility items in high rainfall regions can make hard wheat where soft wheat is common. The proteinaceous vegetation and the synthesis by it of many unknowns which, like proteins, help to remove hidden hungers and encourage fecundity of both man and animals, are common in the prairie regions marked by the moderate rainfalls. It is the soil fertility, rather than the low precipitation, that gives the midwest or those areas bordering along approximately the 97th meridian these distinctions: its selection by the bison in thundering herds on the "buffalo grass"; the wheat which, taken as a whole rather than as a refined flour, is truly the "staff of life"; animals on range nourishing themselves so well that they reproduce regularly; and the more able-bodied selections for the military service of whom seven out of ten are chosen in contrast to seven rejected out of ten in one of the southern states where the soils are more exhausted of their fertility.

Protein production, whether by plant, animal or man, makes demands on the soil-given elements. Body growth among forms of higher life is a matter of soil fertility and not only one of photosynthesis. It calls for more than rainfall, fresh air and sunshine.

The heavier rainfall and forest vegetation of the eastern United States mark off the soils that have been leached of much fertility. Higher temperatures in the southern areas have made more severe the fertility-reducing effects of the rainfall. Consequently, vegetation there is not such an effective synthesizer of proteins. Neither is it a significant provider of calcium, phosphorus, magnesium or the other soil-given, fetus-building nutrients. Annual production as tonnage per acre is large, particularly in contrast to the sparsity of that on the western prairies. The east's production is highly carbonaceous, however, as the forests, the cotton and the sugar cane can testify. The carbonaceous nature is contributed by air, water and sunlight more than by the soil. Fuel and fattening values are more prominent than are aids to growth and production.

Here is a basic principle that cannot be disregarded. It has signal value as we face nutritional problems on a national scale. It is, of course, true that soil under higher rainfalls and temperatures still supply some fertility for the plant production. Potassium, however, dominates that limited supply to give prominence to photosynthesis of carbonaceous products. The insufficient provision of calcium and of all requisite elements usually associated with calcium does not permit the synthesis, by internal performances of plants, of the proteins and many other compounds of equal nutritive value. The national problem is largely one of mobilizing the calcium and other fertility elements for growing protein and not wholly of redistributing proteins under federal controls. The soil fertility pattern on the map

The cow "clings to her instincts of selecting particular grasses in mixed pasture herbages." She "strikes up a partnership with the microbes in her paunch" where they synthesize essential vitamins for her.

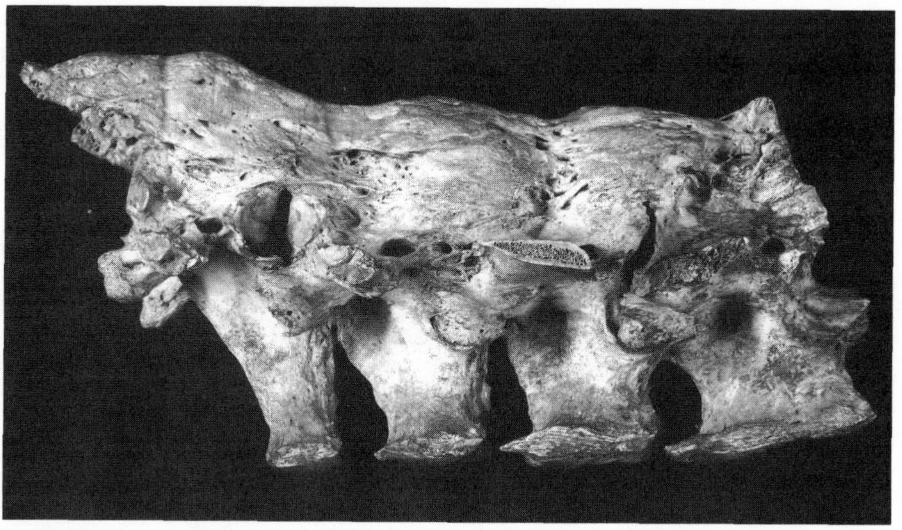

This backbone mid-section of an eight-year-old beef cow of noble pedigree exhibits how she was compelled to rob it of calcium and phosphorus while growing her calves in fetus. Then she restored it as a solid, non-jointed and inflexible one, later to be a poor breeder for but a short life span on a fertility deficient farm.

delineates the various areas of particular success or particular trouble in nutrition. It marks out the areas where, by particular soil treatments, the starving plants can be given relief.

3. the fertility pattern of europe is a mirror pattern of our own

The more concentrated populations in the United States are in the east and on the soils of lower fertility. For these people, Horace Greeley spoke good advice when he said, "Go west young man." It was well that they trekked to the semi-humid midwest where the hard wheat grows on the chernozem soils, and where the bread basket and the meat basket are well-laden and carried by the same provider, *viz.*, the soil. It was that move that spelled our recent era of prosperity.

In Europe the situation is similar, but the direction of travel was reversed and the time period has been longer. It is western Europe that represents the concentrated populations on soils of lower fertility under heavier rainfall. Peoples there reached over into the pioneer United States for soil fertility by trading it for the marked "made in Germany." More recently, the hard wheat belt on the Russian chernozem soils has been the fertility goal under the Hitlerite move eastward. Soil fertility is thus the cause of no small import in the world wars.

4. calcium and phosphorus are prominent in the soil fertility pattern as it determines the pattern of nutrition of plants and animals

Life behaviors are more closely linked with soils as the basis of nutrition than is commonly recognized. The depletion of soil calcium through leaching and cropping and the almost universal deficiency of soil phosphorus, connect readily with animals when bones are the chief body depositories for these two elements. In the forest, the annual drop of leaves and their decay to pass their nutrient elements through the cycle of growth, and decay again, are almost a requisite for tree maintenance. Is it any wonder then that dropped antlers and other skeletal forms are eaten by the animals to prohibit their accumulation while their calcium and phosphorus stay in the animal cycle? Deer in their browse will select trees given fertilizers in preference to those untreated. Pine tree seedlings along the highway as transplantings from fertilized nursery soils are taken by the deer when the same tree species in the adjoining forests go untouched. Wild animals truly "know their medicines" when they take plants on particular levels of soil fertility.

The distribution of wild animals, the present pattern or distribution of domestic animals, and the concentrations of animal diseases, can be visualized as superimpositions on the soil fertility pattern as it furnishes nutrition. We have been prone to believe these patterns of animal behaviors wholly according to climate. We have forgotten that the eastern forest areas gave the Pilgrims limited game among which a few turkeys were sufficient to establish a national tradition of Thanksgiving. It was on the fertile prairies of the midwest, however, that the bison were so numerous that only their pelts were commonly taken.

Distribution of domestic animals today reveals a similar pattern, but more freedom from "disease"—more properly, freedom from malnutrition—and by greater regularity and fecundity in reproduction. It is on the lime-rich, unleached, semi-humid soils that animals reproduce well. It is there that the concentrations of disease are lower and some diseases are rare. There beef cattle are multiplied and grown to be shipped to the humid soils where they are fattened. Similar cattle shipments from one fertility level to another are common in Argentina.

In going from midwestern United States eastward to the less fertile soil, we find that animal troubles increase and become a serious handicap to meat and milk production. The condition is no less serious as one goes south or southeastward. The distribution patterns of milk fever, of acetonemia, and of other reproductive troubles that so greatly damage the domestic animal industry, suggest themselves as closely connected with the soil fertility pattern that locates the proteinaceous, mineral-rich forages of higher feeding value in the prairie areas but leaves the more carbonaceous and more deficient foods for the east and southeast with their forest areas. Troubles in the milk sheds of eastern and southern cities are more of a challenge for the agronomists and soil scientists than for veterinarians.

Experiments using soil treatments have demonstrated the important roles that calcium and phosphorus can play in the animal physiology and reproduction by way of the forages and grains from treated soils. Applied on adjoining plots of the same area, their effects were registered in sheep as differences in animal growth per unit of feed consumed, and as differences in the quality of the wool. Rabbits also grew more rapidly and more efficiently on hay grown where limestone and superphosphate had been used together than where phosphate alone had been supplied.

The influence of added fertilizers registers itself pronouncedly in the entire physiology of the animal. This fact was indicated not only by differences in the weight and quality of the wool, but in the bones and more pronouncedly in the semen production and reproduction in general. Rabbit bones varied widely in breaking strength, density, thickness, hardness and other qualities besides mass and volume. Male rabbits used for artificial insemination became sterile after a few weeks on lespedeza hay grown without soil treatment, while those eating hay from limed soil remained fertile. That the physiology of the animal, seemingly so far removed from the slight change in chemical condition in the soil, registered the soil treatment, is shown by the resulting interchange of the sterility and fertility of the lots with the interchange of the hays during the second feeding period. This factor of animal fertility alone is an economic liability on less fertile soil, but is a great economic asset on the soils that are more fertile either naturally or made so by soil treatments.

5. *animal instincts are helpful in meeting their nutritional needs*

Instincts for wise choice of food are still retained by the animals in spite of our attempts to convert the cow into a chemical engineering establishment wherein her ration is as simple as urea and phosphoric acid mixed with carbohydrates and proteins, however crude. Milk, which is the universal food with high efficiency because of its role in reproduction, cannot as yet be reduced to the simplicity of chemical engineering when calves become affected with rickets in spite of ample sunshine and plenty of milk, on certain soil types of distinctly low fertility. Rickets as a malnutrition disease according to the soil type need not be a new concept so far as this trouble affects calves.

Even if we try to push the cow into the lower levels in the biotic pyramid, or even down to that of plants and microbes that alone can live on chemical ions, not requisite as compounds, she still clings to her instincts of selecting particular grasses in mixed pasture herbages. Fortunately, in her physiology she strikes up partnership with the microbes in her paunch where they synthesize some seven essential vitamins for her. We are about to forget, however, that these paunch-dwellers cannot be refused in their demands for soil fertility by which they can meet this expectation. England's allegiance in war time to cows as ruminants that carry on these symbiotic vitamin synthesis, and her reduction of the population of pigs and poultry that cannot do so, bring the matter of soils more directly into efficient service for national nutrition than we have been prone to believe.

The instincts of animals are compelling us to recognize soil differences. Not only do dumb beasts select herbages according as they are more carbonaceous or proteinaceous, but they select from the same kind of grain the offerings according to the different fertilizers with which the soil was treated. Animal troubles engendered by the use of feeds in mixtures only stand out in decided contrast. Hogs select different corn grains from separate feeder compartments with disregard of different hybrids but with particular and consistent choice of soil treatments. Rats have indicated discrimination by cutting into the bags of corn that were chosen by the hogs and left uncut those bags not taken by the hogs. Surely the animal appetite, that calls the soil fertility so correctly, can be of service in guiding animal production more wisely by means of soil treatments.

Dr. Curt Richter of the Johns Hopkins Hospital has pointed to a physiological basis for such fine distinction by rats, as an example. Deprived of insulin delivery within their system, they ceased to take sugar. But dosed with insulin they increased consumption of sugar in proportion to the insulin given. Fat was refused in the diet similarly in accordance with the incapacity of the body to digest it. Animal instincts are inviting our attention back to the soil just as differences in animal physiology are giving a national pattern of differences in crop production, animal production and nutritional troubles too easily labeled as "disease" and thus accepted as inevitable when they ought to have remedy by attention to the soil. The soils determine how well we fill the bread basket and the meat basket.

6. patterns of population distribution are related to the soil

The soil takes on national significance when it prompts the mayor of the eastern metropolis to visit the "Gateway to the West" to meet the farmers dealing with their production problems. More experience in rationing should make the simple and homely subject of soils and their productive capacity household words amongst urban as well as rural peoples. Patterns of the distribution of human beings and their diseases, that can be evaluated nationally on a statistical basis as readily as crops of wheat or livestock, are not yet seen in terms of the soil fertility that determines one about as much as the other. Man's nomadic nature has made him too cosmopolitan for his physique, health, facial features, and mental attitudes to label him as of the particular soil that nourished him. His collection of foods from far-flung sources also handicaps our ready correlation of his level of nutrition with the fertility of the soil. We have finally come to the belief that food processing and refinement are denying us some essentials. We have not yet, however, come to appreciate the role that soil fertility plays in determining the nutritive quality of foods, and thereby our bodies and minds. Quantity rather than quality is still the measure.

Now that we are thinking about putting blanket plans as an order over states, countries and possibly the world as a whole, there is need to consider whether such can blot out the economics, customs and institutions that have established themselves in relation to the particular soil's fertility. Since any civilization rests or is premised on its resources rather than on its institutions, changes in the institution cannot be made in disregard of so basic a resource as the soil.

7. national optimism arises through attention to soil fertility

Researches in soil science, plant physiology, ecology, human nutrition and other sciences have given but a few years of their efforts to human welfare. These contributions have looked to hastened consumption of surpluses from unhindered production for limited territorial use. Researches are now to be applied to production, and a production that calls for use of nature's synthesizing forces for food production more than to simple non-food conversions. When our expanded chemical industry is permitted to turn from war-time to peace-time pursuits, it is to be hoped that a national consciousness of declining soil can enlist our sciences and industry into rebuilding and conserving our soils as the surest guarantee of the future health and strength of the nation.

25.

Among the Albrecht papers examined in preparation of this volume, several stand out head and shoulders above the rest. The entry, Soil Fertility as a Pattern of Possible Deficiencies, *was published in the* Journal of the American Academy of Applied Nutrition, *Spring 1947. Here Dr. William A. Albrecht told a group of informed professionals many of the details uncovered through years of research at the University of Missouri. Here he illustrated how soils dovetailed with the requirements of human health, and he illustrated his thesis via chemical analysis, biological assay, and observations on the national health profile. That federal officials would one day declare as "law" the proposition that there was no relation-*

ship between quality of food and the soil in which it was grown would not have occurred to Dr. Albrecht. When it came near the end of his life, he called it "a retreat from reason." His refutation to Federal Register *offerings in the early 1970s was written some 35 years earlier.*

Soil Fertility As A Pattern Of Possible Deficiencies

It is the purpose of this paper to lead your thinking to the earthy subject of soils, particularly as a pattern of possible diet deficiencies in animals, including man.

It is my good fortune to address you just after you have made a wise application of your knowledge of nutrition. This is my first experience with a group so intent on nutritive contents that the very menus carry not only the listing of foods but their caloric, vitamin and quantitative analysis. I was glad that the arithmetic proved my dinner was as good as it tasted to me.

It is my purpose to lead your thinking, as medical men, as dentists, and as nutritionists, a little further back toward the origin of foods than the butcher's block, the corner grocery or the dairy. There, foods are dispensed, but foods are built upon the farm and it is toward the farm we must look to understand and evaluate them.

In the store, foods are known by the labels upon them. A head of lettuce is a head of lettuce although one head may have been grown in the soil of a western state, its neighbor in the tired soil of an eastern farm. Butter is butter regardless of the fodder upon which the source cows were fed. Ground beef is hamburger though one batch of edemic protoplasm cooks down to nothing as the water boils out, another seems all meat.

On the farm where vegetables are grown and livestock is fed upon them, the relative quality of foods is built into them by the purely local soil chemistry and weather. To paraphrase, by the soil upon which they grew, ye shall know them.

Agriculture, which was originally the industry of growing foodstuffs, and became the business of exploiting virgin soils, now is being forced to learn how to build foods by conserving and reconstructing worn-out soils. The richness of our soils is one of our greatest national resources. For years we have exploited it without thought (or consideration) of the eventual depletion of that wealth.

You, who are interested in applied nutrition, are particularly concerned over the qualitative feeding of our people. It should not be on the basis of stuffing the stomach while starving the body, but rather on an optimum level that will support our bodies, endow our offspring with inherent health and bring the vigor and nervous stability to our mental processes that will enable us to fulfill the responsibilities of world leadership now thrust upon us.

1. clay is the dynamic part of the soil

How is it that the soil, which is supposedly inert, insoluble, and just "dirt" to many people, can provide the essentials that go to make food? The answer lies in understanding the relations by which the soil, as a mass of decaying organic matter and decomposing rock, can contribute what it takes to make vegetable or animal or human tissues.

Many people have the concept that the nutrients in the soil are soluble and that the plant puts its roots into the soil and sucks out nutrients along with the water. We've been in error to believe that might be the way it works. We find that the nutrients in the soil are held by adsorption forces on the clay. While the rock is breaking down, the dilute solutions of it are filtering through the soil to have the essential, positively charged (and many of the negatively charged) nutrients caught up by the clay. When we speak of the soil as sand, silt and clay, we need to visualize that the sand portions, if they are not too tremendously insoluble, are breaking down; that the silt portions are breaking down; and that the clay is the residue which of itself has little or nothing to give by its own breakdown. However, it has the adsorptive force which collects and holds the nutrients in high concentrations so that the root can come in and quickly get its supply. But for each supply that is seasonally delivered by the clay to the plants, there must be time allowed to restock that clay. We might think of soil exhaustion, then, as a case in which the clay has quickly given up what it got and then has not had time to restock itself. We must develop an understanding of the mechanism by which this soil functions as a reservoir of nutrients for the crop, and thereby an understanding of why we might expect some deficiencies in nutrition.

2. Composition of human body, of plants, and of soil suggest deficiencies

Let us look at the chemical composition of the human body, in contrast to the composition of plants. Particular attention is drawn to the fact that, whether plant or animal or human body is considered, air and water constitute about 95% or 96% of each. In other words, with the carbon and the oxygen coming from the carbon dioxide of the air, the oxygen and the hydrogen coming from the water, plus a small amount of nitrogen that in the ultimate comes from the air also (though most plants take it from the soil)—those constituents of atmospheric origin make up the bulk of the plant body and the animal body. They represent the loss by combustion. They are distilled off readily at high temperature. The remaining content of the body, namely, 5% comes from the soil as the ash.

However, that small amount from the soil is significant. The atmospheric contribution may well be taken as the warm, moist, breath of creation, which is blown into the handful of dust. The 5%, though, which is that handful of dust, determines how successful that blowing operation will be in giving something that is more than hot air in the final result. We may well give thought to the problem of getting our nutrients from the atmosphere. The plant contains about 6% hydrogen, the figure for that in the human body is 10%. In the plant we have about as much carbon as we have oxygen, but in the human body there is less carbon and much oxygen which represents considerable oxidation. But there is also considerable reduction when the 6% of hydrogen is increased to 10%. This is illustrated by the conversion of the carbohydrate into fat, since fat making represents the removal of the oxygen from the carbohydrate compound and the substitution of hydrogen in it. This makes more nearly a straight carbon-hydrogen chain of high heat value on combustion. Conversion of plants into animals increases the hydrogen from 6% to 10%. An increase of about two-thirds. In case of the nitrogen, the absolute increase is from 1.6 to 2.4%, is a relative increase of 50%. You and I in building our bodies by eating vegetable matter must struggle to build up those higher concentrations of hydrogen and nitrogen.

Suppose we look at the problem in terms of the nutrients coming from the soil in contrast to those coming from the atmosphere. Calcium is at the head of the list of soil-borne nutrients needed for the human body. At the head of the list for the plant is potassium. These are significant facts. For all the plant operations that produce protein, we must have a high calcium supply in the soil. For those plants, that are carbohydrate makers, we must have a generous supply of potassium. In the plant composition, then, potassium stands highest in mineral elements needed because the plant's main physiological activity is carbohydrate production. In the human, the main physiological activity is one of protein elaboration while carbohydrates are burning or undergoing decomposition to synthesize it. We need to drive that fact home. One can take a plant and by shifting the relations between calcium and potassium amounts given it, shift it from a plant that is mainly a carbohydrate producer and storer to one which is highly proteinaceous in its synthetic and storage activites.

The human and the animal bodies contain as much as 1.6% of calcium. In our nutrition we go back to the plant that has only .6%, and plants, in turn, go back to the soil that has only .3% of readily exchangeable calcium. Do you see the problem of trying to start with a source of only .3% of calcium; then to increase it in the plant to .6%; and in bringing it from the plant to the human body concentrate it to 1.6%? That represents a problem of increasing the concentration of these nutrients from the soil, not by 50% or 60%, as is the case of other elements coming from air and water. It is a problem of increasing by 200% from soil to plants, of roughly 300% from the plant to the human. Finally, in going from the soil to the human, it is a problem of an almost 500% increase. Can you see that the deficiencies in the soil in terms of a simple element like calcium are going to register when they are so highly magnified in this creative process? Nutrition is concerned not merely with the problem of the calcium as a nutrient coming from the plants to the human, but a problem also of its coming from the soil to the plant and thereby from the soil to the human.

When we think of phosphorus we are reminded that it is the companion of calcium for bone construction. Bone is not just a portion of the reinforcement in a flabby body. It is a portion of the essential, physiologically active parts, particularly when we recall that the activity of the bone marrow is one of blood corpuscle regeneration. There is a distinct problem in the provision of phosphorus when there is only seven-thousandths of a percent of it readily soluble in the soil. The plant must bring the phosphorus concentration from this dilute source up to about .56% within itself, and then you and I must literally double that to make our bodies. From such facts you begin to see that in agriculture we have been putting phosphate and lime on the soil with nutritional benefits to the plant, to the animals and to the humans. But up to this moment we have been putting them on largely in order to make more tonnage, and not because we've thought them essential to the animal or human nutrition.

In the art of agriculture lime and phosphate were put on the soil in order to make more food. More recently, seemingly under the science of agriculture, we put them on in order to make more money. Prompted by concern about better nutrition we are beginning to go back to the art and put them on to make more and better food, because those two essential mineral nutrient elements, calcium and phosphorus, play a tremendous role in the elaboration of proteins. They play a tremendous role in the plant's reproduction by seed and likewise in the human and in the animal reproductive processes.

With reference to potassium, the struggle for the plant is a difficult one. It is not, however, such a difficult one for the human. The plant finds .03% potassium in the soil, which small concentration must be elaborated to one as high as 1.68%. And then you and I, as humans, like the other higher animals, excrete the excess potassium given us by vegetation. So as long as we had farm animals, particularly horses, which were consumers of large amounts of vegetation as roughages, the animal was putting much potassium back into the soil. This kept it in rotation. But when the tractor came along to replace the horse it made no contribution to that fertility cycle.

Calcium, phosphorus and potassium are at the head of the list of about a dozen of the nutrient elements that come from the soil. If the soil must provide a dozen; if most of our individual rocks contain at the maximum only three or four; and if the particular soil is developed from one single type of rock, it is logical to expect some deficiencies. If a dozen elements are needed, while that rock contains only four, and that rock is converted into soil, the conversion has not added anything beyond those four. And yet we are prone to assume that every soil contains all of the dozen. Thus in terms of its origin, we ought to expect deficiencies in the soil, particularly when our use of the soil is a mining instead of a managing operation. We must expect deficiencies unless we feed the soil to give us output just the same as we feed an animal for a particular output.

3. soil mechanisms in plant nutrition

In considering the nutrition of plants, it has long been the general idea that plants use the nutrients in solution. That concept ought to seem logical to those who put ammonia into irrigation water as a nutrient. However, if we use calcium as a nutrient in solution and increase the amounts in that form applied per plant to be grown, there is an increase in the crop grown up to a certain limit, and then from stronger solutions the crop growth decreases. These higher concentrations literally "salt out" the plant. However, if one gives the plant its calcium in the form of the calcium mineral, anorthite, in this same test series of increasing amounts of calcium by increasing the amount of mineral, one doesn't improve the crop growth. This is so because the mineral isn't breaking down rapidly enough for the plants to be nourished. The crop fails to live even though the mineral contains more calcium by several times than was used in solution. But if one will take the solution of calcium, trickle it through some clay, which may be even an artificial one like that used in water softeners, namely permutite, that permutite will take the calcium out of solution and adsorb it. Regardless of how concentrated, or how much calcium we provide on that permutite, there is no injury to the crop. The calcium is held on the permutite in an insoluble but exchangeable form. When the root comes along it trades or exchanges its acidity for this calcium.

Thus you can see the principle of plant nutrition by the soil in whichthe clay is an adsorber and an exchanger. Thus, if in regions of low rainfall nature has not weathered the rock down to anything finer than sand, there is need to put something in with the sand to serve as the permutite does. Fortunately, decaying organic matter becomes colloidal in form like the permutite and may serve in that respect. But it is important to note that unless a significant concentration of the nutrient like calcium is put on the clay, the plant gets none of it. This is true because the clay holds it with a force in equilibrium against the plant's forces. Such colloidal activity is always significant in soils where the rock has not been weathered down far enough to make much clay as is true in low rainfall regions. And so when one thinks about this whole matter of plant nutrition it must be approached in terms of these fundamental understandings of how, or by what mechanisms, it is that the root and the soil, or rather its clay portion, can get together and the plant to be nourished thereby.

As an illustration of the clay-plant root interactions, increasing amounts of an electrodialized acid clay were put into sand. The sand had nothing of nutrients on it. It had no exchange power. In such large particle size it had no great surface area. But when that clay was added in the increasing amounts the crop grew better accordingly. Putting it another way, they were diseased plants when they were starving, they were healthy plants when they were well fed. They were well fed by nothing other than an acid clay, but by having more of it for more root contact. As there is more clay developed in the soil through greater rainfall thus making that soil heavier, there is the possibility of greater crop production. Thus it is that the farmer in the Missouri river bottom, with its heavy soil, knows that if he can get his corn crop planted after the June flood, he still gets a crop because the soil has a large clay content to give a large amount of nutrients to the crop in a very short time.

4. plants trade acidity soil for their nourishment

If you will consider the plant root as having itself surrounded with carbonic acid (carbon dioxide in water makes carbonic acid), that acid is the ionized or active hydrogen. Hydrogen is the most active of all of our soil elements. It is traded to the adsorption atmosphere of the clay and there it replaces some of the nutrients, which may be well illustrated by calcium. By such trading or exchanging the clay is becoming acid while it is helping the plant to be nourished or the plant to grow. It is by this bartering that growing plants make the clay become sour or acid. And yet we worry about the soil having become acid. Instead of worrying about the acid

Chemical Analysis of the Human Body in Comparison With That of Plants and of Soils

Origin or Source	Essential Elements	Human Body %	Vegetation % Dry Matter	Soil % Dry Matter	
Air and Water	Oxygen	66.0	42.9	47.3	
	Carbon	17.5	44.3	.19	
	Hydrogen	10.2*	6.1*	.22*	
	Nitrogen	2.4*	1.62*	—	
		96.1%	94.92%		
Soil	Calcium	1.6*	.62*	0.3‡	3.47*
	Phosphorus	.9*	.56*	0.0075	.12*
	Potassium	.4†	1.68†	0.03	2.46†
	Sodium	.3	.43	—	
	Chlorine	.3	.22	.06	
	Sulfur	.2	.37	.12	
	Magnesium	.05	.38	2.24	
	Iron	.004	.04	4.50	
	Iodine		Trace		
	Fluorene		Trace	.10	
	Silicon	Trace	0-3.00	27.74	
	Manganese		Trace	.08	
Body Compounds	Water	65	—		
	Protein	15	10		
	Carbohydrates	—	82		
	Fats	14	3		
	Salts	5	5		
	Other	1			

**These are involved in the plant and animal struggles to find enough to meet the high concentrations needed.*

‡Amounts common as the more available forms in the soil in contrast to the total, most of which is but slowly available.

†This represents struggles by the animals to eliminate it.

having come on to that clay, we should worry about the fertility having gone off the clay. And when we talk about soil acidity being dangerous per se, we ought to remind ourselves that this soil trouble might not be due so much to the advent of the hydrogen, but rather to the exit of the nutrients.

If the soil were made up of only clay, then if a crop were grown on it, it would remain sour or acid as we have demonstrated experimentally many times. But if one will mix some original crushed rock, limestone, for instance, granite, or any of the other rocks and minerals that can be broken down, then the acidity given by the plant to the clay will be transferred to those rocks and minerals. They will be weathered like they are by any weathering agents in the outdoors. The contents of those rocks as nutrients will be moving in the opposite direction to that of the acidity and will serve to feed the plants.It is thus that crops are making the soil acid while they are being nourished.

We have thus come to see that the plant is putting acidity (active hydrogen) which is not a nutrient, into the soil. Acidity is weathering the minerals, and their breakdown is nourishing the plants and by that means nourishing all of us. Now if the soils don't have enough clay, we can substitute humus, or soil organic matter, because the dead carbonaceous material of previous plant generations is decaying into the colloidal form, and it serves as an exchange agency in exactly the same way as the clay does. But humus also decomposes faster than clay. Its decomposition is producing ash in that same manner as the decomposition of rock is passing its nutrients to the colloidal exchanger and then to the plant roots. When you think therefore of the more sandy, open-textured soils of Los Angeles County, that once had the highest agricultural output of any county in the United States, followed by Lancaster County, Pennsylvania, and then by McLean County, Illinois, is its lowered output today the result of exhaustion of the mineral reserves in the sand? Have not these changes in productivity occurred more probably because tillage operations have been burning this humus out of the soil, and nutrient deliveries have been getting down to a limited mineral level, the weathering being too slow under limited rainfall to keep these assembly lines of food production going at the rate at which we like to have them go? It was this breakdown of the humus of the virgin soil that gave the speed to earlier production.

It has been the breakdown of the large supply of organic matter in the virgin soils of the United States that gave us our American prosperity—and we might well consider saying "past" American prosperity, because we have gone to the limit of our westward movement to better soils. During our increasing prosperity we were moving to soils that had more organic matter. We were exploiting a great natural resource. We have not reached the limit in gaining by moving. The problem of organic matter, to which we have been pointing with such emphasis, is illustrated not only in your own county, but it is a universal problem. The importance is not lessened because it isn't quite so acute in some of the other regions where the output of agricultural production is not so nearly complete as food for direct human consumption. Unless we understand these basic principles, by which the nutrients we need from the soil are brought to us through plant help, we are not going to be

More clay in the sand (left to right), even if that clay is very acid, gives better and healthier plant growth. Soil acidity represents deficiencies in plant nutrition.

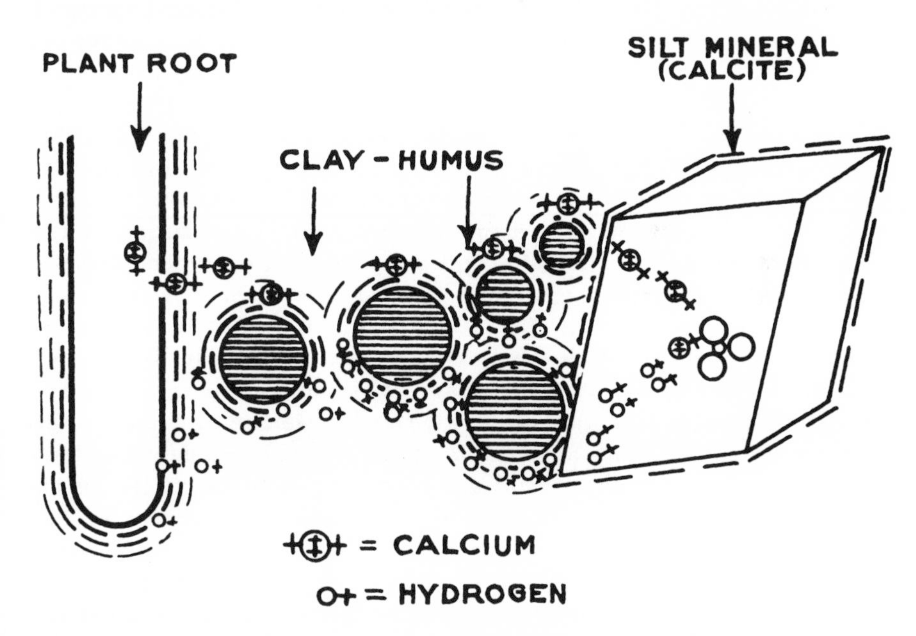

Plant nutrients, like calcium, on the colloidal clay or humus are exchanged to the plant root for the hydrogen or acidity it offers. As the colloid takes on more acidity this goes to break down the minerals and restock the colloid. Acidity goes from the roots to the minerals, nutrients go from mineral to the roots, all through the colloid.

able to manage this business of food production without suffering deficiencies and the so-called "diseases" provoked by them.

5. pattern of climate suggests pattern of deficiencies

Perhaps you have not even thought that it is the forces of climate that make the soil, and that the soil in turn feeds us. California has been telling us about its excellent climate for years. But the climatic pattern is an outline of influences, not in terms of how wet we are, or how dry we are, or how warm we are, or how cold we are, but rather the climatic picture exercises its importance because it determines how well fed we are. Perhaps at first thought you will not agree with that statement. Nevertheless, it is a great truth. This pattern gives order to the whole business of nutrition. It exercises its significance not only in the United States, but has its international implications as well. If we take a look at the rainfall of the United States we see in the rainfall the main part of the pattern. Suppose we disregard the western coast, which has so many variations that one cannot delineate it on the map and make it clear and readily recognized, and start with western United States with its rainfall of less than 10 inches annually. As we go eastward there is the next area with 10 to 20 inches, and then a long area running north and south with rainfall of 20 to 30 inches. Kansas, as an illustration, is covered by three or four of these rainfall strips running north and south. Its annual rainfall varies from about 17 in the west to 37 inches in the east. As we approach the center of the United States, the rainfall pattern shifts from longitudinal strips to more nearly latitudinal ones. In eastern United States we must add temperature as a factor in the climatic pattern since temperature may alter the rainfall's effect.

The rainfall weaves itself into the more complete climatic pattern if we superimpose on it the forces of evaporation. If, therefore, you will take the rainfall in inches annually, divide it by the evaporation from a free water surface also in inches per year and multiply it by 100, you will have the percentage relation of rainfall to evaporation. Areas of constant figures may be marked off. For example, along one line we may have only 20% as much rainfall as there is evaporation. Therefore, as the rainfall strikes the rocks, it soon evaporates. It doesn't go in deeply to break up those rocks or to carry down and through any of the products of that decomposition. The western portion of the United States is said to be "arid," because the rainfall is so much less than the evaporation. There is no washing out or leaching of that soil of great significance. However, in Missouri, Illinois and Iowa, or in the cornbelt, with its rain coming in the summer when there is a high evaporation, the rainfall comes in ample amounts to break down the rocks. But instead of it going down through to deplete the soil, it evaporates and leaves the residues to saturate the clay. These climatic conditions give us a productive soil. Out in western United States, these forces are not in such a fortunate combination. The ratios are not high enough. And then in the eastern part of the United States, the rainfall is far in excess of evaporation and the figures are above 100. They have so much rain there that the soil is broken down and the products removed by leaching them down through the soil.

6. climatic soil pattern gives us our west and east and our north and south

It is the pattern of the climatic forces that puts the cornbelt into the midwestern region and into the prairie group of soils. If we view the map of soils of the United States, which was made before the Russians gave us the understanding of the climatic forces developing the soil, we see that the soil pattern is the same pattern as we get by superimposing the pattern of the evaporation-ratio over that of the rainfall. The soils divide the United States into an east and a west. We haven't been talking about the people of the east and west without some real provocation for it. People are different because they are on different soils. It is the soil which has divided the eastern portion of the United States into the north and south. That division line wasn't drawn by color lines of the different peoples. It was the color lines in the soil because those in the south are red and those in the north are gray. The red soils of the south and the tropics have a clay that does not hold nutrients. It doesn't hold any soil acidity either. So the south has said "we don't have any acid soils and therefore don't need lime to fight it." But they surely have been needing lime badly to provide them with calcium, when so many southern mothers tell you that each childbirth costs two teeth. As we look at the climatic soil pattern, we begin to understand some of our deficiencies. And then if we give close scrutiny to the soils of the central portion in the United States, and likewise to the drier western portion that has allowed the winds to pick up the unweathered mineral materials and waft them eastward as a scattering of blessings regularly on the cornbelt, some of the food situations and deficiencies in feeds calling for supplements will explain themselves readily.

7. soils in the west are under construction, in the east under destruction

In a traverse of the United States from zero rainfall in the west to the east there is an increase in the force of weathering as we go eastward. Weathering starts with the rock and makes soil, which is nothing more than a temporary rest stop of that rock on its way to the sea. As that rock breaks up, more and more under higher weathering forces, it makes more clay and also makes a better soil in terms of nutrition. And as weathering increases there is still more clay, but when the rainfall reaches the amount of 35 inches in the temperate zone it represents the maximum for soil construction. This holds true whether it is in the north with low temperatures, or the south with higher temperatures.

One can go anywhere in the world and use this pattern to guide his understanding of the soils and their value in nutritional qualities of the foods and feeds. When there are about 35 inches of rainfall in the temperate zone there is the maximum of breaking down of the rock and the minimum relatively of washing the products away. This permits the maximum loading of the clay with the exchangeable nutrients so that the plant can come in with its roots and feed itself abundantly. And then as we go east in the United States with still more rainfall, weathering is washing the nutrients off of that clay. The reserve materials and rocks in the soil are so thoroughly weathered that they do not restock the clay and the humus colloids. There is left the sand and silt, but they consist of the insolubles, the permanent quartz. It is in the central United States and east, thereof, that soil acidity comes pronouncedly into the soil pattern. As one goes to the southern states, the soil acidity becomes less because the soils contain a different clay.

That, in brief, is the pattern. With mounting soil construction, there goes increasingly better nutrition. More intense climatic forces represent soil destruction in terms of nutrition even though the body of the soil still remains. In our westward movement initiated by the pioneers, we have been exploiting the accumulated organic matter. It was that exploitation that permitted us to push westward. And so we have left in our wake mainly the inorganic soil residues with dwindling power to produce. It isn't so surprising then that now that we can't go westward much farther, we are come face to face with problems of nutrition in the United States.

8. soil construction favors proteinaceous, soil destruction carbonaceous, quality of vegetation

If one recalls, for example, the areas of the original forests or virgin woods we are reminded forcefully that woody crops were all those soils would grow naturally. Nature was growing wood in the colder acid soils of the northeast. She was growing wood in the leached soils of the south, and in the higher altitudes of the west where the rock has not yet been formed into the soil. And so when there is not much soil as yet constructed, or if there isn't much soil left in terms of destruction, the best that nature can do is to make carbonaceous products. But in those regions where the soils are only partially weathered, and consequently fairly well saturated with plant nutrients, it is there that the grasses abound.

The original productivity levels of our soils were indicated by the experiences of the pioneers. When they landed on the well-wooded eastern coast their search for food was rewarded by the find of a few turkeys. When they found those, they were so thankful that we have had to be thankful for them every year since. But when the pioneers came westward to the grassy plains, they found the buffaloes of massive bone and brawn roaming that region in thundering herds. Low rainfall and soil under construction were growing protein abundantly, high rainfall and soil under destruction growing mainly carbonaceous products.

If one catalogs the virgin vegetation of Kansas, as Dr. Schantz pictured it, by starting from 17 inches of rainfall in western Kansas and going eastward to 37 inches, it is clearly evident that more rain represents more crop. That is a simple fact that every farmer argues, namely that he would grow a larger crop if only he had more rain. Ask yourself, though, what is the nutritional quality of the crop that is growing as the tonnage is increased by more rainfall. It is well recognized that short grass is the crop in western Kansas and tall grass in eastern Kansas which we are prone to attribute to more rainfall. The soil also varies in going from western to eastern Kansas. The lime, which is found at one foot depth and is one foot thick in western Kansas, is no longer visible in the soil profile in eastern Kansas. In eastern Kansas the roots of plants are going down deeply into a wet soil, but they are finding a soil that has been highly leached and the plant nutrients, including lime, have been highly washed out.

It was in the western portion of Kansas that the short grasses grew which were eaten by the buffalo so regularly as to be called "buffalo grass." Those were the soil regions over which the buffalo roamed north and south extensively, but not east and west very far. There were no obstructions against his going east in Kansas when the rivers there run in that direction, and when Kansas was a great prairie. But he wasn't interested in tonnage increase per acre as we are when sales are made by quantity and not by quality. He wasn't interested in the products grown on highly weathered soils. Instead he was interested in high concentration of minerals and protein that are built into the short grass of which every mouthful counts.

Unfortunately, we have scarcely shown buffalo sense in evaluating our agricultural output of food products. We have been concerned with the tonnages instead of the nutritional value. What was the buffalo area is the same area where our cattle today multiply themselves. It is from that area that cattle are shipped eastward to be fattened by crops from those soils that have fattening power rather than much growth-producing power.

The human food product of Kansas, namely the wheat, fits into the same category of proteinaceous or carbonaceous according to the degree of soil development. The protein of the wheat grown in Kansas builds up in concentration as one goes westward across that state. This phenomenon has always been explained in terms of the decrease in rainfall. This might seem a reasonable explanation when according to the survey by the United States Department of Agriculture in

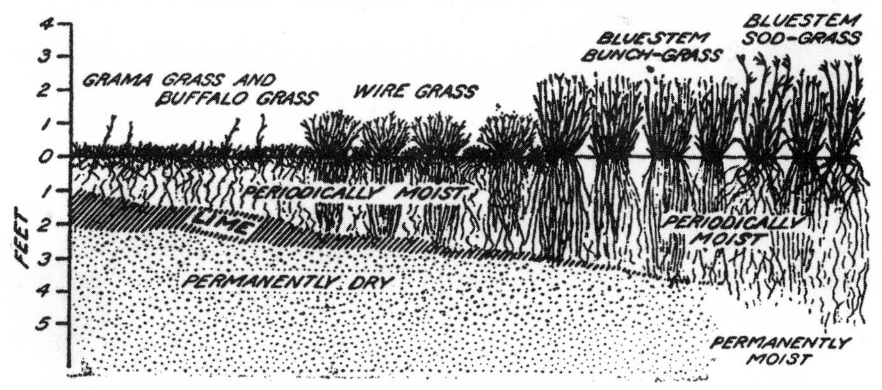

A traverse across Kansas from 17-inch rainfall in the western part to 37 inches in the eastern part, showed its virgin vegetation increasing in bulk with increasing rainfall. The soils were also more highly developed giving different root pattern and different feeding value to the forage when the grass chosen by the buffalo was not that of maximum tonnage yield per acre. The buffalo recognized good qualities and deficiences in his diet. (*Drawn by H. L. Shantz*)

1940, there was an increase in the protein from east to west. The lower or southern tier of counties across Kansas, starting in the extreme southeastern corner showed a steady increase in protein from 10 up to 18% in crossing the state westward.Such concentration of protein is not a matter of the wheat growing in a dry season. Instead, it must find its explanation in the soil conditions that elaborate much starch and convert little of it into protein in eastern Kansas yet give big yields as bushels per acre, while in western Kansas the late delivery of nitrogen in the season converts starch into protein and thereby less bushels per acre.

We have not given much thought to the fact that in going westward it is the soil that causes plant processes to shift from those given mainly to making and storing starch, to those making protein and consequently burning much of their starch in running that process. When plants make only starch they can readily make big bulk as yield per acre. Failure to recognize these facts has been the occasion for a controversy on wheat quality between the producers and the millers of wheat and the bakers in Kansas. The recent past five or six years have given high rainfall for Kansas. They have also given tremendous yields of wheat. But at the same time the fertility of the soil was being exhausted so seriously by those large yields that the wheat was making mainly starch. The farmer was getting bigger yields while the miller was complaining of the declining quality in the low protein concentration. The baker, too, was unable to get the large loaf from little flour, since it is impossible to lighten the loaf of bread by means of yeast gas and at the same time hold the water to give it weight unless the flour is rich in protein. Consequently, the bakers complained that the farmer wasn't growing the proper variety of wheat in order to keep up the baker's volume of business. The farmer reported his volume as bushels per acre on the increase and satisfactory. Here was a controversy between two groups provoked by a problem common to both of them, namely the decline in the fertility of the soil that should call for the interest of both groups in its conservation. Such situations call for attention to our soils in order to grow proteins that satisfy instead of starches that merely stuff the body but give us hidden hunger.

9. pattern of plant composition suggests deficiencies

The pattern becomes a bit more specific when we consider the chemical composition of the plants. It was possible to study 38 different plants and note their chemical analyses. There were 31 cases of different plants that are native to the soils farther east where they are only moderately developed, and 21 cases of plants that are native to the soils of the east, including the soils of the south, that are highly developed. According to the analyses of these plants, their contents of potassium, calcium, and phosphorus added together as an average amount of 5% for the plants on the less weathered and less developed soils. As we go eastward and southward, the contents of these three elements go down to 4% for plants on moderately developed soils, and then drop to the low figure of 2% on highly developed soils. Now we might point only to the mineral situation and say that these plants are hauling more minerals as they grow on less highly weathered and

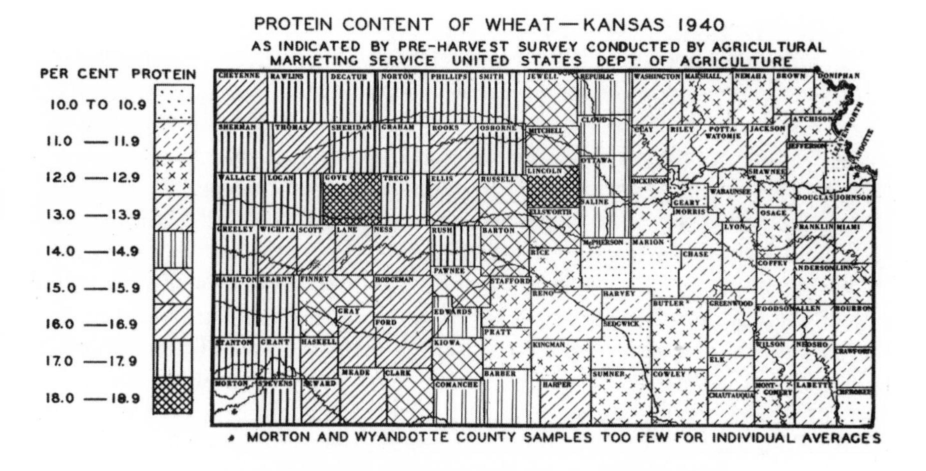

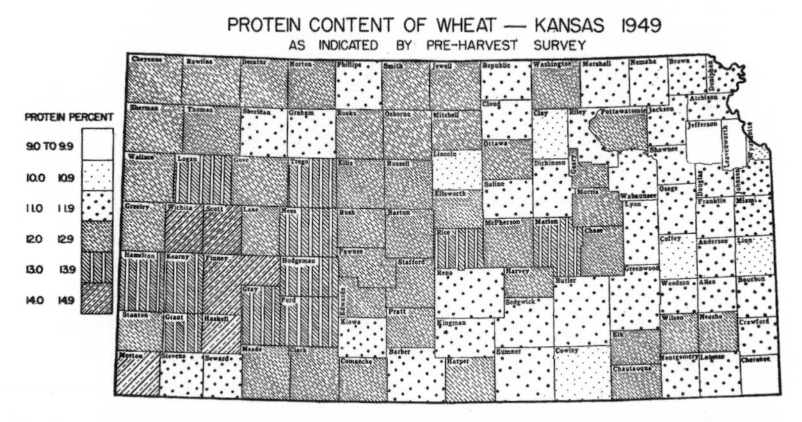

A traverse across Kansas in 1940 (above) from its low rainfall in the western part to the higher rainfall in the east revealed declining protein concentration in the wheat. Below, a survey hardly a decade later revealed the steady decline in protein levels.

more fertile soils. But they are also manufacturing many more of the complex synthetics that you and I, located as we are at the top of the biotic pyramid, need to build our complex bodies. Consequently, we find increased woodiness and increased starchiness of our food crops as the soils are highly weathered under the higher rainfall and higher temperature. If we travel in the other direction, from east to west, from highly weathered to less weathered soils, we find proteinaceousness and mineral richness representing higher food values. As the plants range from soils slightly developed to those highly so, their potassium and phosphorus contents showed a drop from 2 down to 1, but the calcium took a corresponding drop from 7 down to 1.

Here in the changed plant composition in relation to soil composition is reason why in humid soil regions we need to put calcium on soils first in order to have them provide mineral-rich proteinaceous crops. Later we come with the phosphorus and the potash as soil treatments. These are the three mineral constituents that stand topmost in our fertilizer program. We got them, strangely enough, through the art of agriculture long ago rather than through the science of modern agriculture. Calcium is and has been at the top of the list of these necessities. But, unfortunately, we were entangled in the false belief that it was the acidity in the soil rather than deficiency of fertility there that was dangerous. In order to attack that false reasoning, some calcium as a chloride was applied as a streak across a field of soybeans many years ago. Calcium chloride will not neutralize in the acid soil, instead it will add hydrochloric acid to the soil. But even though we put hydrochloric acid along with calcium into the soil, the crop was improved by adding calcium in this form. This did not remove the acidity. In fact, it made it worse. Yet it improved the crop growth. For years we have been led to believe that soil acidity in itself is terrible. It is terrible mainly because so much of the fertility has gone out before so much acidity can come into its place. Where the soil treatment was calcium nitrate, this salt of nitric acid also improved the crop. This and the chloride gave crops as good as where calcium hydroxide was used to neutralize the acidity at the same time that it was providing calcium.

And so, as agronomists, we must confess that we were reasoning wrongly. Because we had some simple little gadget, like the hydrogen electrode, which we could push into the soil and measure the concentration of hydrogen, we were arguing that the soil acidity was the cause of the crop failure because acidity of the soil was going along with it. This is a characteristically common type of fallacious reasoning, namely, ascribing cause to one of two contemporaneously associated phenomena and the wrong one of the two. We were putting on the carbonate to neutralize acid, but at the same time, unwittingly, we were putting on calcium to feed the plant. We emphasized the neutralization of acidity instead of the nutrition of the plant as the beneficial effect. We were therefore delayed many years in our clearer thinking about calcium as a nutrient for plants, and through them for animals and man. Calcium as a deficiency is late in getting recognition.

10. crop bulk as criterion of soil productivity invites deficiencies

The calcium-potassium ratio, to which reference has been made, has given us a pattern of the protein possibilities in the crop. If nature, under less rainfall, has left much calcium in the soil, we have a proteinaceous crop. If the soil is under higher rainfall to give a small amount of calcium in relation to the potash, then we have a carbonaceous crop. The validity of this belief, namely, that a liberal supply of calcium in the soil in relation to potassium represents production of crops rich in proteins and minerals, while the reverse relation gives crops high in carbohydrates—thereby low in proteins and minerals—was tested. Soybeans were grown with increasing amounts of potassium available in the soil and associated with

By keeping the calcium supply constant but increasing the potassium, a 25% increase in tonnage was obtained. However, the protein content of the plants dropped by more than 25%. The small crop had twice as much phosphorus, three times as much calcium as the bulkier crop. Soil fertility sets the pattern of composition.

constant amounts of calcium. Three ratios of calcium to potassium were used while all other nutrients were liberally supplied. Increasing the potassium increased the forage yield to a maximum of 25%. This fact would draw ready applause for an experimenting agronomist. Such work can win funds in support of it as research. But the buffalo of the western plains didn't evaluate herbage in terms of bulk. Our livestock does not use that criterion either. Hence, while increase of bulk may appear laudable, fixing of our attention on bulk in relation to soil fertility has been leading us to grow more crops with serious deficiencies as feeds.

Chemical analyses were made of the forage. The nitrogen content of the smallest of the three crops was 2.8%; of the intermediate crop 2.5%; and of the largest crop 2.19%. While we increased the bulk 25%, we reduced the concentration of nitrogen, and therefore the protein, by more than that figure. So that the greater amount of total protein was not in the largest but in the smallest crop.

If a cow were to get the same amount of protein in the largest crop in contrast to the smallest she would be compelled to take about five mouthfuls instead of four. Most of us are familiar enough with cows to know they cannot increase their intake by 25%. Consequently the cow is really going to suffer some deficiencies. She cannot handle more bulk in order to get the necessary protein.

The phosphorus concentration, by analysis, was .25% in the small crop; .18% in the intermediate; and .14% in the largest crop. Assuming that the cow could digest it completely, she would be compelled to eat approximately twice as much of the larger crop to get the same amount of phosphorus. In the case of the calcium, this was approximately .75% of the dry weight in the smallest crop and only .27% or about one-third as much as in the largest crop. Can any cow increase her consuming capacity by three times? We can't expect her to become a hay baler.

We need to be concerned not only with the bulk of the crop, but also with the synthetic operations of the plant in using the fertility elements from the soil to convert the carbonaceousness over into the proteinaceousness. Those processes make foods of value in terms of growth instead of only fuel. Those are the features that make "grow" foods instead of merely "go" foods. They must be more generally appreciated if we are not to invite nutritional deficiencies more commonly.

Some research studies were made to illustrate this conversion of carbohydrate into protein by the plant. A legume crop was thrice planted on a series of soils with potash increasing in relation to calcium, knowing that more potash in relation to calcium makes the plant a producer of carbohydrates rather than of protein and also increases the yield. The first of the three successive crops on the soil was not given bacteria. Consequently this legume, limited as it was to only the nitrogen in the seed, could not convert air nitrogen into protein. That first crop had no more nitrogen in it than was originally in the seed. The crop built a great bulk, yet the smallest crop in the series grown by less potassium had the same amount of nitrogen as the larger crop. The second of the crop succession was inoculated with nodule-producing bacteria so that it became a protein synthesizer instead of being only a starch producer as was true of the first crop. This growth exhausted the soil fertility seriously as protein producing crops must; consequently the third crop was limited by the soil fertility and, like the first crop, was a starch producer.

The first crop gave the highest tonnage yields, the third one was next in order, and the second crop gave the lowest yields of all. In fact, the third crop was more than 30% higher in weight than the second crop, and yet it was the third successive crop in the course of exhausting the soil. Crops are commonly judged in terms of tonnage yields per acre. Little thought is given to the fact that crops making only carbohydrates build bulk quickly but those converting the carbohydrates into proteins do not. The first crop had a low sugar content, but a high starch content. Those plants were converting the sugar promptly over into starch because in the absence of soluble nitrogen and legume bacteria there was no nitrogen hence no way to convert the sugar into protein. The second crop, which was inoculated, had a higher sugar content, but it was not being converted into starch of which the concentration was very low. These facts suggest that the sugar was being converted into protein. Analyses showed that a goodly amount of nitrogen was taken from the atmosphere. The third crop yielded a large amount of bulk again as did the first crop. This had a high content of sugar. It was inoculated and was making and piling up sugar that seemingly should have been converted into protein. But the two preceding crops had exhausted the fertility of the soil so that the best this final crop could do was store starch. Consequently the third crop was of a high starch content and high yield of bulk.

Crop bulk as the criterion for a crop leads us to choose those crops which are making carbohydrates rather than making proteins and other nutrient complexes with more than fuel-food values. High yields as bulk may therefore give us deficiencies. Crop quality in terms of nutrition rather than mere tonnages should be the criterion for selecting crops.

11. vegetables also invite deficiencies according to soil growing them

Spinach was made the subject for research into its chemical composition as influenced by the fertility of the soil. Spinach as a vegetable green is probably one of the most debated elements in our diet. Some argue that they like it, others that they do not like it. Chemically considered, there is good reason why some people do not like it. Spinach may well be classified as one of the hypocrites of the garden plants. It can put up a fine green appearance and have less calcium, for example, than most any other garden plant used for greens. Therein may be the reason why some people love it and some people hate it. The people who love it probably get the spinach that is high in calcium and those who hate it probably get the spinach that has little calcium but is high in oxalate.

Spinach was grown experimentally in order to get at the question of soil acidity, which in terms of calcium for spinach is not so dangerous. In fact, soil acidity is beneficial. Two soil series were arranged to provide increasing amounts of exchangeable calcium by units of 3 milliequivalents to a maximum of 12. All the other nutrients were offered in constant but ample amounts. In putting the calcium on the soil the first series of soils was given it in the form of salts that left the soil acid. The second series had exactly the same preparation, but the calcium was put on as oxides and hydroxides. These forms made the soil neutral. Thus there were increasing supplies of calcium and constant amounts of all the other nutrients in one series of soils that were all acid in reaction, and another series of soils as a duplicate except that the soils were all neutral.

As the soil was left acid that grew the spinach, the increasing amount of calcium applied on the soil gave a corresponding increase of concentration and total of calcium in the crop. It also gave an increase in magnesium, though the magnesium was applied in constant amounts to all the soils. It may be well to point out here that calcium is the one element which moves many of the rest of the elements into the crop. It moves the magnesium; and it moves the phosphorus because it is the "keeper of the gates," as it were, into the roots of the crop. As a matter of fact, its deficiency lets some nutrients go from the plant back to the soil. Consequently we may be growing crops that have less nitrogen, or less phosphorus, or less potassium in them than was in the planted seed.

Spinach manufactures oxalic acid. This compound contains no elements of soil fertility. Oxalic acid is not desired by all folks because it "puts your teeth on edge." Rhubarb gives the same sensation but we eat rhubarb for its acid. But few people care to eat spinach for that same effect. The oxalate unites with the magnesium and the calcium to make them insoluble and indigestible. In the case of the spinach grown on the acid soil there was so much calcium and so much magnesium in the crop that the oxalate content, even though seemingly high, was not high enough to make all of the magnesium and the calcium indigestible. Therefore there were significant amounts of the calcium and magnesium in the spinach that were soluble and digestible. The spinach grown on neutral soil did not take increasing amounts of lime in proportion to the increasing amounts of calcium applied on the soil. Nor did it take increasing amounts of magnesium as were taken in the other case. Spinach grown on the neutral soil had so much oxalate that this was more than enough to make insoluble and indigestible all of the calcium and all of the magnesium taken by the crop. Had one fed that spinach to an unsuspecting baby and given some milk along with it, that mixture would have not only been deficient in terms of digestible calcium and magnesium from the spinach, but it would even have made indigestible some of the calcium in the milk.

In dealing with this matter of nutrition, our thinking must go back to the soil that grows the crop. We must understand some of these fundamentals by which the nutritive value is controlled. We cannot be certain of vegetables as sources of mineral elements merely because of vegetable name and reputation. Vegetables may encourage deficiencies because of the fertility deficiency of the soil growing them.

12. soil exhaustion means going to starchy and woody food products

So much has been said about the soil-conserving virtues of grass agriculture and livestock farming as to leave the impression that such practices keep lands from serious depletion of their fertility and productive capacity. Yet 40 years of grazing alone in a Santa Rita Mountain Valley were sufficient to change the native grass vegetation to mesquite bush of no grazing value. In 1903 that valley was a temptation to the cattlemen because it had a nice, luscious growth of grass. They brought in their cattle. Instead of allowing the accumulated soil fertility in each crop to drop back as nourishment for the next one, and to bring a little more of mineral decomposition into that cycle and thereby build it larger, they brought the cattle and annually hauled off much of that grass by means of them. By 1943 they marvelled at how one lone mesquite bush present in 1903 could have taken that whole valley as a mesquite infestation by 1943.

By rotating the fertility unmolested nature was producing proteinaceous vegetation, but when man began hauling off that crop nature switched over to the production of woody vegetation. It was one which could send its roots deeply enough into the soil to get the necessary minerals and one which is equipped as a legume to take much of its needed nitrogen from the air. And so while civilization has been moving westward across the United States, the mesquite is moving westward across the state of Texas to take over the prairies. Is it any wonder that in the wake of that fertility exhaustion the soil should shift from producing proteinaceous vegetation to the production of woody vegetation? Is it any wonder that the "soft" or "low-protein" wheat is moving westward with the depletion of soil fertility? Thus we are bringing out and into prominence the pattern of deficiencies because starchy wheat will not feed us any better than wood will feed the cattle.

13. animals are connoisseurs of feed and not mere mowing machines

The grazing cow is not merely a mowing machine. If one observes a pasture it is common to find that she hasn't done a clean job of harvesting the grass crops. Instead she has been selecting and balancing the carbonaceous vegetation against the proteinaceous according to the nutritive ratio that she needs in order to deliver a calf or to provide milk for it. When the pasture crops are not completely taken by the cattle we emphasize the necessity to mow the pasture. We say much about the necessity of having successive pastures during the grazing season. Instead of thinking about juggling the crops as pasture successions, we should be thinking about undergirding the cow's selection of her herbage by producing it with those particular qualities she needs as nutritive values. Our animals by their appetite are searching through the different crops in mixed pasture herbage, balancing their diets in proportion as their body needs may represent a fattening performance in one season, a fetus-developing performance in another, or a milk-giving performance in still another season of the year.

We haven't been ready to believe that the cow will select and graze first the barley where 200 pounds of fertilizer were applied and will disregard the same crop where 100 pounds only were applied. But such was the observation of Mr. E. M. Poirot, a farmer of Missouri. The cow is a better chemical assayer of the soil conditions than we are in the laboratory. It would be necessary to call on the spectograph and other equally refined instruments to find the difference in the soil between 100 and 200 pounds as applications per acre. And yet the cattle detected a case of that kind for several successive years. That fact was demonstrated by the cattle in selecting one among the four haystacks in 100 acres of virgin prairie meadow. In the early spring 1936, a four acre portion was treated with different fertilizers carrying calcium, phosphorus, nitrogen and other essentials. The grass species were enumerated during the summer. Sample yields were taken. Chemical analyses were made. There were increases of about 5 to 7% in some of the different ash constituents, and about that much in yield.

In the late summer the grass crop was made into hay. About 25 acres of hay were put into each one of four haystacks. The four acres that had been given soil treatments were promiscously included in one of the haystacks. The cattle were turned into the field in the late fall to consume the haystacks. The owner reported in 1936 (December) that the cattle had passed up the three haystacks and had eaten first the fourth one in which there was the hay from the four acres of treated soil. There were no additional treatments put in the soil after 1936. The grass was made into hay annually as four haystacks. The hay from those four treated acres, along with that from about 21 acres given no soil treatment, went into a stack at one end of the field annually thereafter. This was the regular one for careful observation. The cattle ate this one haystack first in 1937, and each year thereafter through 1944 for eight years.

The eighth crop provided an interesting discrimination by the cattle. In that year when the stack bottom was started on which to build up, it was not made large enough. After the four acres, along with hay from untreated acres had gone into the stack to the limit of height, considerable untreated hay was left. Consequently the balance of the hay of 25 acres was put at the end of the stack to extend it. The cattle were turned in as usual in November. As they had done regularly since 1936, the cattle disregarded three of the haystacks. They crowded about the one containing the hay from four acres given soil treatment as part of the stack. They cut this stack in two, selecting first the main portion containing the treated hay. After they had consumed this part they still worked over the trampings before they took to the other stacks. When this remnant was all that was left they went to the other stacks. Here then after the ninth successive crop the cattle were still recognizing the quality of the hay induced by the small amount of soil treatment that was not over 600 pounds per acre.

Even the hog, supposedly such a dumb beast, has a very discriminating taste. A Missouri farmer had 40 acres of corn into which he put his hogs to take the grain as feed from the field rather than harvest it in the usual manner. They went into the field and seemingly disappeared. He didn't see any indication that they were eating the corn until a neighbor told him the hogs were doing very nicely in the field corner near his farm. The hogs had been going through the 40-acre field daily from the water to this chosen corner. The farmer remembered that several years before he had tried to grow some alfalfa in that corner. It had been limed and fertilized but had been forgotten.

We commonly think the hog has no discriminating power, but in experimental work the hog's appetite can be a helpful tester. One can put the different grains into the several compartments of the self-feeder and the amounts the hogs take are an index of their choice. Hogs have demonstrated consumption of one kind of corn as low as 10% and as high as 100% of another grain alongside according as the soils growing them were given different fertilizer treatments.

Insects, too, are discriminating in their attacks on vegetation fertilized differently. Varying amounts of nitrogen were applied on the soils for a series of spinach plants. These applications were a supplement going with increasing amounts of calcium. All other nutrients were generously applied. The thrips, which is an insect that attacks spinach, cut the leaves wherever the crop was given low amounts of nitrogen, but left untouched the leaves of plants grown on soils high with nitrogen. The experiment was replicated ten times. Two rows of spinach with attack by the insects alternated with two rows free of insects in repetition ten times. It is an old saying, namely, "to be well fed is to be healthy." Parallel with that seemingly one can say "to be well fed is to ward off the insects." At any rate, there is the suggestion that the pattern of declining soil fertility is the pattern of plant diseases and insect troubles with plants.

14. animal deficiencies go back to soil deficiencies

Much has been said about the mule, especially about its fastidious appetite. Its progenitor, the donkey, is similarly cataloged. In our problem of raising donkeys we have forgotten that the donkey requires feed grown on calcareous soils. The misfortune that befell a young donkey illustrates the case. A young jack of fine breeding was moved from the region of Missouri that is famous for jacks, or from the region of the farm that is known as the "Limestone Valley Farms," and taken to western Missouri that is supposedly famous for bluegrass. As a result of that transfer, this donkey of good breeding or of a fine pedigree developed a severe case of rickets. The pedigree could not overcome poor feeding. While breeding can bring about certain qualities and tolerations, *one can't breed either animals or plants to tolerate starvation.* Breeding for such toleration, much like a race of bachelors, goes on for only one generation.

In the beef cattle business (one of the phases of agriculture in which a state like Missouri is highly active) the fertility pattern exercises decided influence. There are many steers purchased in Texas, in Oklahoma, and adjoining states, and moved to the cornbelt for fattening. All too often the farmers in this business say they had "bad luck" because some of the steers "went down" just about the time they should have been ready for market. Now what's the trouble involved? The experiences of one of our Missouri farmers gave a good illustration of the soil fertility conditions involved. He had brought some Texas steers, weighing about 600 pounds each, to his farm in northeast Missouri. It is in the more level region with some exhausted soils in the area where troubles in breeding herds are also all too common. He had 43 head, and at about the time they should have been going to market, one of them "went down" on its hocks with the symptoms suggesting that it had been hamstrung. This animal lingered along until it was much emaciated when the second one "went down," with what suggested paralysis of the hind legs.

Now this farmer had been feeding cattle for some 40-odd years. His father ahead of him had fed cattle, too. They had accumulated much experience in cattle feeding. But while they were accumulating experience, they were also bringing on soil destruction on their own farm. These animals were fed on home-grown feed. This included a peck of corn, a sheaf of oats, some cottonseed supplement, and some legume hay. The hay was soybeans which supposedly grow well on "acid" soils.

About the time his second steer "went down," calls came for some of the Extension men to go out to help in the situation. Before the veterinarian arrived, the third steer was down. Now what was happening? The tendons had pulled out of the soft bone where the animal was hamstrung. The pelvis had separated from the vertibrae and had paralyzed the whole rear part of the second animal. While these animals were laying on fat, they were not building the calcium and the phosphorus into the bones. Their bone structures were so soft they were letting the animals "go down."

The differences in the soil fertility represented in the life of these animals are as contrasting as the animal behaviors. One region, namely the same region where the buffalo roamed, had grown these animals. That area during the life of the young calf couldn't give enough mineral and protein reserves to permit putting on only extra fat and getting that to market. Such experiences in the so-called "cattle business"—which is mainly a buying and selling rather than a production activity—come all too often. They are serious disasters because we do not realize that even in the fattening of the animal one must provide it with carrying power (for the load of fat) by building well the bones and the protein at the same time. Evidence is accumulating, but unfortunately too often coupled with serious disaster to compel the realization that the deficiency troubles in animal production are not so much a matter of better breeding but one of better feeding. Feeding is not a problem of combinations so much as a problem of quality "grown into" the feeds. Feed deficiencies are showing themselves as fertility deficiencies in the soils that grew them.

15. animal improvement can be brought about by remedy of soil deficiencies

That applications of calcium and phosphorus to the soil transmit their beneficial effects through the feed was demonstrated experimentally by feeding hays from treated soils to some sheep. By combining some soybean and some lespedeza hays, both grown on the same soil treatments, there was sufficient hay to grow lambs for 63 days. Seven lambs were fed on hay that had no soil treatment; seven on hay that had been grown on soil that had phosphate; and seven on hay grown where the soil had been given both phosphate and lime. The sheep ate exactly the same amount of hay per head per day and were given exactly the same supplement.

In terms of bulk the feeds were all equal, but in terms of quality, as demonstrated by the growth of the sheep, these feeds were widely different. Where the feed had been grown with no help in the soil, the lambs made a gain of eight pounds per head in 63 days. Where they had been given the help of phosphate on the soil, they made a gain of 14 pounds. Where they were given the still better undergirding by means of lime and phosphate on the soil, they made a gain of 18 pounds. In other words, between 8 and 18 pounds of lamb growth was the improvement brought about by the mere application of a small amount of soil treatment known to remedy a deficiency. Animal growth is not a matter of bulk of feed, but one of feed quality *grown in* by the fertility of the soil.

We may well think of food quality in terms of larger soil areas, as well as of plots and fields. The soils of Missouri, for example, divide themselves very distinctly into five major regions. There is the southwestern portion of the state that is a duplicate of eastern Kansas. The Eldon silt loam is a typical soil there. A portion of northeastern Missouri is level prairie, and its soil is Putnam silt loam. The Ozarks are mainly Clarksville soils. The northwestern part, or the corn section, is mainly Grundy silt loam. The southeastern lowlands, or the cotton section, are covered by a soil series known as the Lintonia. Experiment fields have been located on each one of those soil types. It happened that lespedeza was growing under the same general conditions one year on each of these fields with both the treated and untreated soils. The lespedeza hay was harvested in these different regions of the

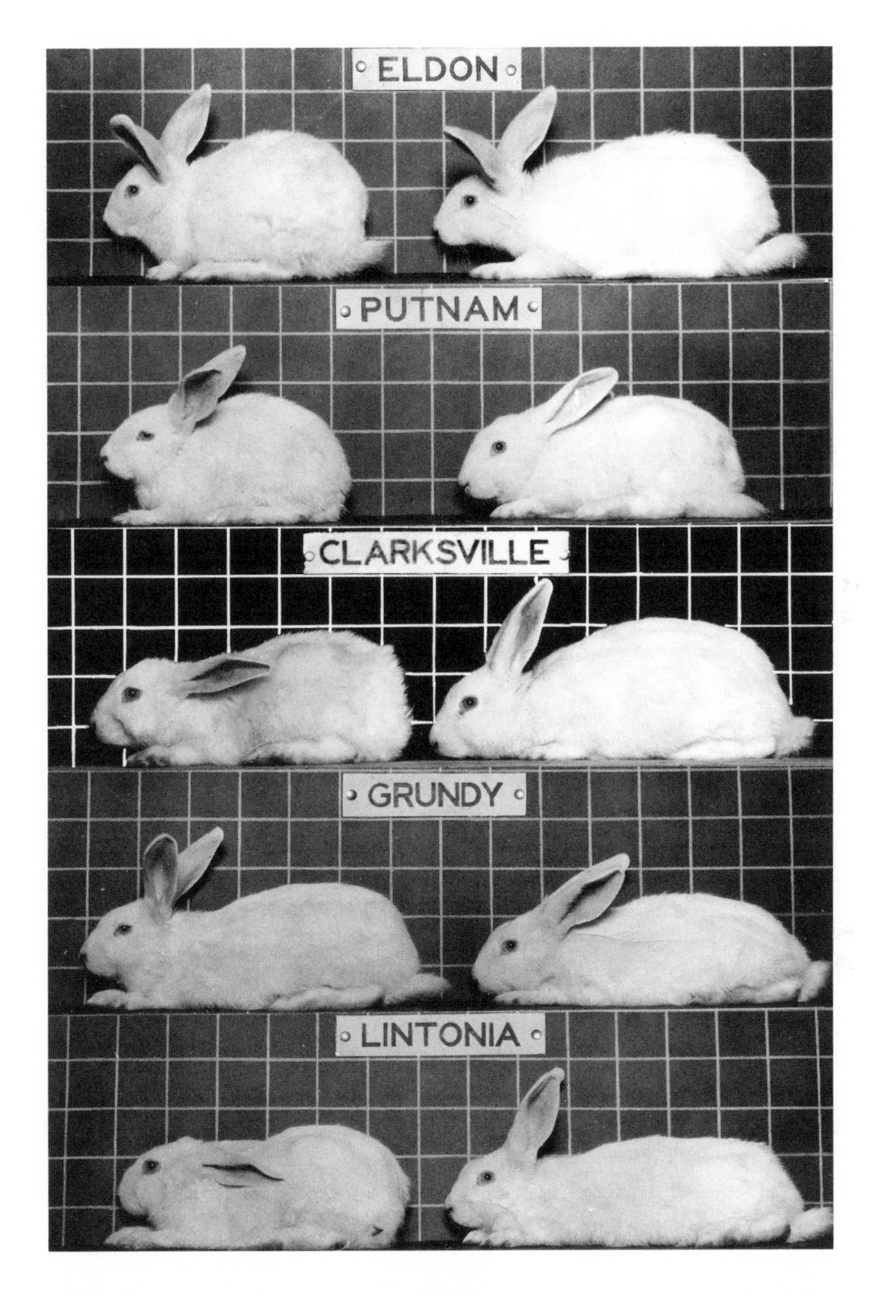

Tests of forages by feeding animals revealed to Dr. William A. Albrecht that soil treatments correct deficiencies, not only in soil fertility but in the diets of animals. The names here indicate the soils. At left, lespedeza without fertilizer treatment. Right, with treatment. Sir Robert McCarrison, working in India, found the same type of results when comparing diets of Hindu races.

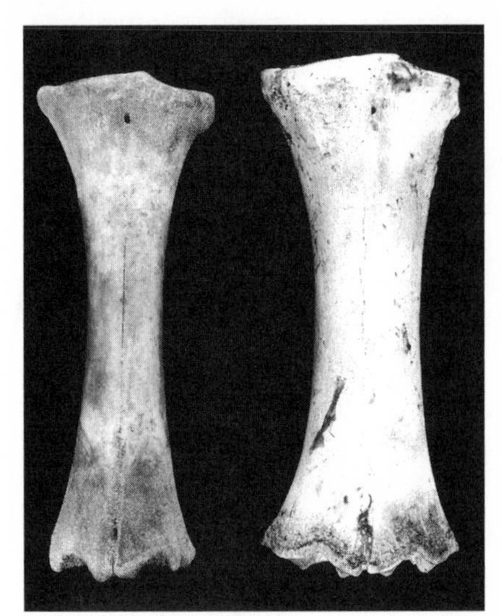
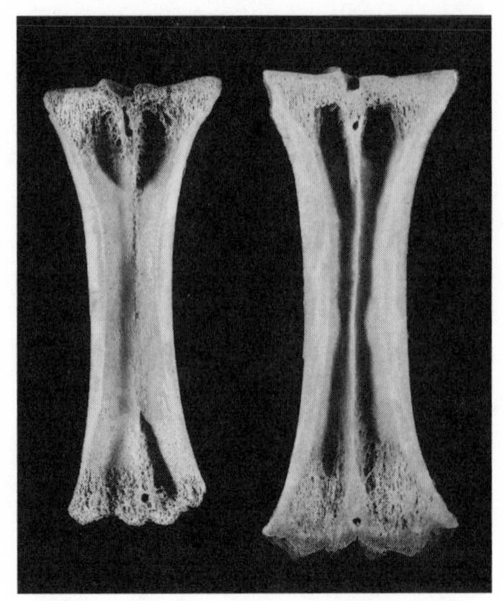

Shinbones of calves reveal differences according to the soils growing them. Deficient soils make small, soft bones (left); treated soils make heavy, strong bones at the same age (right in each picture).

state, brought to the Experiment Station, and fed in 10 different lots of 10 rabbits each. The rabbits were carefully selected and distributed as litter mates and as to sex in order to make the lots as uniform as possible. The 10 rabbits in each pen were like every other 10 as nearly as they could be selected. They were fed the respective hays for a period of about eight weeks.

One might well argue that these animals represented a constant breed at the outset. But the appearances of the rabbits at the close of eight weeks on the hay from different soil regions of a state suggested that they were of different breeds. A wide variation in body size characterized those fed on hay from untreated soils. Such differences suggest that each soil area of the state will bring about certain characters quite different from those of the animals in another area. Hay from the soils given treatments grew animals better in size, more uniform in appearance, and less in difference by soil areas.

In spite of such differences brought on by variation in soils within even a single state, we believe we can maintain a constant set of characters known as characters of the breed. Extensive breed records are being kept in the belief that a constant set of characters is being maintained. All this goes on in disregard of the force of the difference in fertility of the soil which is literally changing the breed in the short time of a single generation.

The body physiology of the animals also was different, according to the soils, if one examines the bones, for example. These showed differences in diameter, density, thickness of the walls, breaking strength, calcium-phosphorus ratios, and many other properties.

If one makes an examination of the wool of sheep fed on feeds grown on different soil, wool differences are evident. The wool is a physiological product of the nutrition of the animal, just as are bones, blood, or other body parts. If the sheep suffers deficiencies in feed, wool fiber is apt to be thin, or will break easily. The wool of sheep was studied according to soil treatments used in growing the feed. The wool of sheep on soils deficient in calcium and phosphorus was almost free of the wool fat. The animals were not excreting freely those fats in the skin that make the wool greasy. Quite in contrast, the sheep which were fed hay from soils treated with lime and phosphate—and were making the better gain of 18 pounds while the others gained only 14—produced a greasy wool, or one full of "yolk" as the sheepmen speak of it. When these different wools were put through the scouring process, the originally more dry, or nearly fat-free wool seemingly gelatinized, cemented itself together on drying and could not be carded without being broken into bits. The wool of the other sheep lent itself to carding and would be normally a regular fluffy wool specimen. These differences in the wool are different physiological manifestations because differences in plant physiology and differences in soil physiology occur farther ahead in the chain of events responsible. The improvement in the animal physiology and animal product resulted from the remedy of the deficiency in the soil.

16. deficiencies in the soil bring troubles in animal reproduction

Reproduction draws heavily on calcium, phosphorus, nitrogen and other soil borne essentials for building bone and muscle. That a cow could sacrifice part of her backbone was demonstrated by one of recorded breeding history. She had become a poor reproducer, and died far too early in her life for the noble line of breeding she represented. The crows had picked her skeleton bare and revealed that the separate vertibrae of the middle of her back were not visible. That section of the backbone was perfectly solid instead of being normally flexible. Can we not see that in developing the fetuses and in giving milk, she sacrificed a portion of her backbone in order to supply the necessary calcium and phosphorus while living on soils deficient in these respects? After each calving to rebuild the backbone it was built as a solid section rather than as the five divided vertibrae. She was a poor producer, not because of poor breeding but because of deficiencies in the soil fertility. These deficiencies made her a "shy breeder," for while rebuilding her backbone she scarcely could be expected to be building ova that would be conceptive and keep her calving schedule on the desired regular pattern.

Deficiencies in soil fertility report themselves in the form of troubles in reproduction not only with reference to the cow as mother, but also the calf in its early life. An interesting case came under the observation of Mr. A. W. Klemme of the University of Missouri, in his agricultural extension work in soils. As a result of his discussion of the subject of soils in relation to animal nutrition at one of the rural meetings, a farmer-owner of a calf reported his troubles with it. It couldn't get up. He diagnosed the ailment as what is commonly called "hollow horn" or "hollow tail." He reported his treatment of it by splitting the end of the calf's tail and putting salt and pepper into it.

Mr. Klemme's examination of the farm and feeding situation revealed that no thought had been given to the soil. The farmer had not yet thought he would follow the "newer" soil practices of his neighbor. The mother cow was of large body but not in good condition. Mr. Klemme's recommendation was that some milk from the neighbor's cow be given the calf and the neighbor was called in for consultation and favorable agreement.

It was ten days later when Mr. Klemme made the second visit only to find the calf up and taking milk from both the neighbor's cow and from its mother. It was quickly recovering from its case of rickets. The calf was still carrying the bandage on the end of the tail as testimony of the farmer's poor diagnosis.

Too often we, like the farmer, let our thinking start from the wrong end. We are diagnosing, as Dr. Price and Dr. Hooton have put it so well, "from the morgue and the grave backward instead of from preconception and conception forward." Too often it is a situation in which we are treating the symptoms instead of removing the causes of the trouble. Animal husbandry is failing to use the maternity capacity to its maximum by failing to treat the deficiencies in the soil and thus fails to obtain more offspring per mother.

17. human deficiencies may also be traced back to soil deficiencies

The work of Dr. Weston A. Price, under the title of *Nutrition and Physical Degeneration,* reports his studies of the teeth among primitive peoples. From it one learns that it is not necessary to go to Hereford, Texas, to find the "town without a toothache." He found native Africans with excellent teeth, yet they had never heard of a tooth brush or paste. He found them suffering with various tooth irregularities when they took over white man's food habits by purchases from his store. Dr. Price points out that teeth, as an exposed part of the skeleton, tell of a corresponding disintegration of the rest of the body as they break down.

Physical examinations of our men for the Army and the Navy revealed that the increasing number of cavities in the teeth fit into the soil fertility pattern of the United States. The smallest number of teeth defects were represented by the men from the mid-continent. The numbers increased on going eastward from there. There is a good indication (with nearly 70,000 cases reported for the Navy) that the pattern of tooth troubles duplicates the pattern of the soil fertility. Deficiencies in teeth, as we may expect of other deficiencies, take their pattern from that of the soil.

Mental deficiency has been related by Dr. Price to defective bone development in the skull. With him we may well raise the question whether defective nutrition may not be provocative of defective minds, since it is as much of a physiological process to think as it is to digest. Insufficient bone growth in the lower skull may give trouble with such vital parts as the pituitary and may occasion troubles in the entire endocrine system. Such irregularities have not been so commonly considered as possible causes of delinquencies and mental irregularities.

Reproduction may be more closely associated with nutrition and the soil fertility pattern than we are prone to believe. Successive children in the family, with increasing body defects, have been pictured by Dr. Price as related to successively lower nutrition of the mother. Have we thought of relating these to soil regions?

18. soil fertility gives pattern to international problems

One needs only to study the soil pattern of the United States to learn how fittingly it may serve to guide our thinking about some of the international problems. The soil map of the world reveals that the hard wheat belt, as a particular combination of climatic forces produces it in the United States, is replicated in similar climatic conditions for fertile soils in Russia, in several of the outposts of the British Empire and in the Argentine. It is also replicated in Northern China but not in Japan or her former island possessions. It is the soil fertility pattern in terms of potency to grow wheat and meat that nominated the great powers of present day international affairs. It is they, with territories in the temperate zone and moderate rainfall, that have soils providing good nutrition. World problems are not delineated so much in politics. They are marked out by the soil fertility in relation to efficient nutrition. Deficient soils mean deficient and delinquent countries in the world family. International problems therefore may well be considered in terms of the map of the soils of the world as the guiding pattern for future thinking and planning for peace.

19. summary

Our knowledge of the soil in terms of its origin, its chemical composition and dynamic aspects can now be coupled more closely with plant physiology to help us understand why deficiencies in our nutrition as well as in the nutrition of the plants and animals may be expected. Some dozen soil borne elements needed by our body are not all commonly present in every soil. Protein is the major problem in more ways than the economic one. Plants struggle for it too, since they require liberal stores of soil elements to build the amino acids from which we construct our proteins. Plants produce carbohydrates generously, but synthesize proteins within the limits of soil fertility. Soil exhaustion, whether by high rainfalls or excesive cropping, invites nutritional deficiencies since plants then synthesize mainly carbohydrates. Animals therefore fit into an ecological pattern according to local soil fertility and humans too must fit unless they move to or obtain foods from distant, fertile soils.

Our soil fertility is giving a pattern via nutrition not only to crops, animals, and human health, but has international implications as well. Food, after all, is the most potent factor of life and it is the soil fertility by which food is provided. Soil deficiencies therefore provoke other deficiencies more far-reaching than we can readily imagine. We shall correct these not by legislative procedures or by political manipulations but rather as we minister to the soil in its wisest use and most efficient conservation.

26.

Fertilization of crops with factory products marched down the insolubility trail starting at approximately the time of von Liebig. By the mid-1940s and well into the 1950s, soluble salts had become the advocated course by almost all dispensing intellectual advice. In this report, Dr. William A. Albrecht explained how the salts worked, and why they sometimes didn't work. Moreover, he made the point that "Since organic matter is the source of energy for the microbes. . .absorption of the chemical shock is bought at the price of some of the organic matter that is rapidly depleted. . ." This entry first appeared in the Fertilizer Yearbook, *1947.*

Fertilize the Soil Then the Crop

As the researches in soil science, plant physiology and other sciences related to plant production make progress, we are coming to a clearer understanding of the behavior of the root of the plant in relation to its supplies of nutrients and their activities in the soil. As the different sciences contribute new facts relevant to the reactions between the soil and the plant, those who direct the manufacture and utilization of fertilizers may well keep alert to all new methods by which these commercial chemical materials applied to the soil can be made to serve for food production more effectively.

It may be in good order, therefore, to attempt the formulation of a clearer mental picture of the relations between the soil as a chemically dynamic body and the plant root as an active biochemical one. We need to understand the chemical and physical forces by which the nutrients move from the soil into the plant—or in the opposite direction—to give us plant production which is the basis of all life. Though we may grow plants commercially in water cultures, most plants still grow by means of the fertility coming from the soil. We therefore fertilize the soil to bolster it as the basis of all life.

1. nutrients serving only as solutions in soil water for plant use as such is no longer a tenable concept

That the nutrients in the soil are taken only as there is a water solution of them going into the plant to carry them along, is an ancient and exotic idea transplanted too suddenly from the laboratory to the field. That a plant nutrient must be readily soluble in water is no longer a requisite for its service in the plant, even though this concept may have held on since the day of the lively discussions by such well known scholars of the soil as Dr. Milton Whitney of the United States Department of Agriculture, and Dr. C. G. Hopkins of the University of Illinois. It is not necessary to believe, as some once did, that ionic nutrient elements must be swept into the plant root by a current of water flowing in that direction. Nevertheless, this older view rendered services until there came the colloidal concept of absorption of nutrients on the clay, and their activities there under their own dynamics, irrespective of the movement of the water.

One of these two pioneers in soil science claimed, in substance, that soil minerals make a saturated solution in water in the soil. Were all of the nutrients of a 2 ton crop of clover, for example, in solution in the soil water at the start of the crop in the spring, they certainly would not be expected to stay there while the heavy rains are going down through the soil to leach them out. We know that we cannot leach much out of the soil, for even the very "hard" water in our wells contains too little to support the growth of household plants in it. Had the soil minerals been very soluble during the long geological ages, would there be even much left as soil at this late stage of time, and in places where we have enough annual rainfall to grow extensive bulk of vegetation? If nutrients are left in the soil in such amounts and forms from which a 2 ton crop of forage can take 150 pounds of them as elements in a 90 day growing season, it seems necessary to discard the belief that they are present there in a simple—and of necessity, very dilute—water solution. Conditions like those which we find in the soil demand the belief that the plant nutrients are possibly first taken out of their dilute solutions and concentrated somewhere in the soil to be held for quick delivery to the roots on their searching advance by growth through the soil.

Under such reasoning, then, soluble fertilizers are not rendering their services merely because they go into solution in the soil water, and certainly not because they are retained in that form. If such were true for phosphates, for example, why would we say that "they stay in the surface soil where they are put?" Would we not expect them to be leached down and out of the soil? If it were as simple as solution in water would there have been occasion for putting fertilizers directly with the seed? Would they not be distributed through the moist soil by their own ready solubility and diffusion, or be moved in the percolating water?

If fertilizer activities were so simple there would have been no occasion to have considered putting them into granular forms. Nor would we find the better crop response when a granular fertilizer is mixed through the soil than when, as an infinitely fine powder, the same amount is completely blended into the soil. Nor would a small application of the mill-run, 10 mesh limestone, for example, carry over its effects in the soil much longer than those from 100 mesh, as we know it does. Small amounts so well mixed through are too low in concentration and too completely held against movement into the root. Granules represent centers of high concentration and ready migration from there.

This emphasis on making fertilizers that are soluble in water has demonstrated, of course, that they are more effective for quick crop returns. But these more rapid returns have not come because fertilizers stayed in the soil in water-soluble form. These returns were not the result of the plant's taking them along with the water, evaporating this from its leaves and thereby having the salt residue from this distilling process left as nutrients in their tissues. This fanciful picture may not have been your idea of how water-soluble, or so-called "available" fertilizers function to more speedy advantage for crop production. It has, however, been the mental concept of some in the past.

2. *high solubility of fertilizer spells rapid absorption by the soil colloid rather than by the plant*

We now have a better understanding of the clays as minerals of known chemical composition, molecular construction, and specific chemical behaviors in relation to many plant nutrients. This points out clearly that water-soluble fertilizers are more effective because they react rapidly with the colloidal clay fraction of the soil. They react rapidly also with the organic matter, or the so-called "humus" which, like the clay, is another colloidal performer. Both of these soil components give speed to any chemical reaction because of their extensive and active surfaces which hold also considerable water. These two colloids have a large capacity to take out of dilute solution and combine or hold on themselves many plant nutrients in exchangeable form. If the colloidal clay and the humus did not have these capacities of absorption and exchange, these nutrients would not be held against loss as leaching; nor would they be concentrated on one or both of these components of the soil in sufficient quantity to supply the plant during the growing season. More clay in the soil means that the soil can hold more exchangeable fertility and thereby have the capacity to produce more crops. The more hasty growth of a late planted crop on the "gumbo" soil after the spring overflow is possible because of this extra fertility on the clay. More clay means speedy and plentiful delivery to the crop roots of the fertility absorbed on it. Absorbed nutrients on the clay and humus cannot be washed off from them by pure water. They can be removed only as some other ions whether nutrients or non-nutrients are exchanged for them.

Water soluble fertilizers are more effective because they go into solution, yes, but more because in that form their contents of ***nutrient ions are absorbed on the clay; are held there against loss by leaching; and are exchangeable to the root as it comes along and offers some other ion with corresponding electrical charge to replace them.*** Hydrogen is, for example, the most common one offered. It is the water solubility of the fertilizers that spells "availability" to the plants above the soil and to the microbes within the soil through absorption and exchange on the colloids rather than through direct use by the plant and microbe as water solution. These services rendered to the plants by soluble fertilizers—possibly also by nutrients not so soluble in water—are dependent then on their reaction with the colloidal part of the soil and not wholly on their degree of solubility in water. In this one respect, then, namely, solubility, we fertilize the soil more than we fertilizer the crop.

3. *soluble salts, as hindrance to seed germination, prohibit heavier applications of fertilizers*

The very fact that fertilizers are such readily soluble salts has been mitigating against more extensive use—and therefore larger sales—of them as long as they

Fertilizer use must consider the soil as well as the crop. By increasing the amount of nitrogen applied (left to right), plants exhibited extra growth when exchangeable calcium was medium. In this series, there was little extra growth when exchangeable calcium was high, but some had less effect than plants pictured here when the exchangeable calcium was low.

were applied in direct contact with the seed. Damage to seed germination by fertilizers is not news to many farmers who became generous in amounts of fertilizers applied in the row. The numerous trials of a specific placement of fertilizers at the side of, or below, the seed in the soil are merely a part of the effort to use more fertilizer but yet escape the danger of injury to the germinating seed, or the growing seedling, in contact with the salt that has been dissolved in the soil water but not yet absorbed out of that solution by the soil colloids. The success of such placements through more complete absorption by the soil colloid of the fertilizer helps us to understand why fertilizers on sandy soils with little colloid may be more dangerous in these respects than on those of heavier texture.

Isn't the placement of fertilizers at some distance from the seed more effective merely because it is a means of keeping the seed out of the salty soil area; of allowing time for the soil to absorb the fertilizer and thereby take it out of solution before the plant roots arrive; and of permitting the roots to act under their own stimulus of growing into the particularly suitable concentration of the absorbed fertilizer? Such a concept of the soil conditions suggests that fertilizer placement is giving chance for the roots to avoid the danger of salt injury and to grow into the soil areas of suitable degrees of its concentration in the absorbed forms on the clay rather than compelling them to be closely surrounded by fertilizer salts. Making fertilizers so highly water-soluble has worked against their own more extensive use so long as they were applied directly with the seed. The concept of fertilizing the crop, rather than the soil, has, as it were, held back the advance of the sound business of wider use of fertilizers for greater crop production. Now that we are thinking more about fertilizing the soil, they are being used more extensively both in new areas formerly not fertilized and in heavier applications for bigger and better crop yields.

Now that we are plowing the fertilizers down, much larger amounts per acre are being safely applied. There is no recognized damage to seed germination under these conditions. There is less damage later in the plant's growing season. Corn can be given more nitrogen for use at the later time in its growth or the time when it commonly "fires." This trouble, so often ascribed to drought, can be eliminated. Fertilizers in this case are put down not only with the immediate crop, but also with the succeeding crops as well in mind. Such fertilizer practice is moving—in thought at least—toward fertilizing the soil more than the crop and toward using fertilizers for maintaining the fertility of the soil rather than only for stimulating the early growth of the crop to which it is directly applied.

4. calcium suggests its protective effects

Fortunately not all fertilizers have been Frankensteins to the same degree as some when used in heavier applications with the seed. Ordinary superphosphate in contact with the seed is less disastrous than salts of nitrogen and potash. Various explanations have been offered for the damage by these salts. We have not been giving consideration to the hypothesis that calcium in the gypsum of the superphosphate may possibly be providing a protective action. But when Professor

Rodney H. True of the University of Pennsylvania [writing in *The Function of Calcium in the Nutrition of Seedlings,* Journal of the American Society of Agronomy, 1921] demonstrated the loss of cations, *i.e.,* the positive ions, by the plant root to a dilute solution of potassium salts into which it was immersed, but an uptake of both potassium and calcium from this same solution of potassium given some calcium as supplement, he was giving possible explanation of these differences in dangers from different fertilizers to the germination of the seed and to the emergence of the plant.

Additional possible support to this hypothesis about the protective effects of calcium may be gotten from the fact that in numerous trials of soybeans growing on colloidal clay, it has always been requisite that calcium should move into the crop in order to get plant growth. But growth has been possible even if too little calcium went from the clay into the root to prevent nitrogen, or phosphorus, or potassium, from going back from the root of the plant to the colloidal clay. Isn't it possible that calcium, and possibly some other nutrients, may be playing a role in the root conditions that move the nutrients into the plant from their absorbed stage on the clay, rather than allowing them to go in the opposite direction as the case of the injury to germination and the injury to even the growing crop may possibly be? Here may be reason, in part at least, why the so-called "neutral" fertilizers, made so by additions of calcium and magnesium, are superior to those not so treated and therefore considered to be "acid" fertilizers. Here may be reason why fertilizers are so much more effective on limed land than on that unlimed.

5. absorbed ions move under their own power

We can now believe that it is not necessary that water flow into the roots to carry nutrients there. We can now believe, too, that the nutrient ions move from regions of higher concentrations of absorbed forms to those relatively lower, much as gases diffuse in the air; or soluble salts diffuse through water; or as water itself moves through the soil under capillary forces. Root surface areas of most any crop are not large enough to give it sufficient nutrients if it were to take only those absorbed on the clay face or that surface in immediate contact with the roots. Since crops under experiments took amounts of nutrients much larger than those on the surface of one molecular clay layer, it became necessary to test whether the exchangeable ions absorbed on the clay could move from soil areas of higher to those of lower concentrations while no water was flowing.

By using a sand mixed with colloidal clay carrying absorbed calcium and by placing this mixture into a pan on one side of a removable metal partition, there could be put some more sand with clay carrying absorbed hydrogen or acidity on the other side. It was possible then to bring these two mixtures into intimate contact by dropping the pan after the partition had been removed. Tests at time intervals of the sand-clay mixtures on the two sides of their line of juncture showed that the hydrogen was moving from the acid-clay into the calcium-clay, while the calcium was going the opposite direction, both at the rates of more than an inch per month. These mixtures were kept covered and lost no water. So without water movement, these insoluble but exchangeable ions held on the clay were moving from higher to lower concentrations. They were adjusting themselves toward their equilibrium as absorbed ions even when no water was moving.

In the light of such facts, fertilizer action in the soil is a case of having the nutrients present on the clay and in the exchangeable forms at concentrations high enough to move significant quantities of their ions toward the lower concentrations within the root while it is growing through the soil. It is certainly not logical to visualize the nutrients being carried into the plant by a flow of water that might in any sense wash them out of the soil and into the roots. It is the concentration of the nutrient ions absorbed on the clay, as given it by whatever kind of fertilizer there is—whether readily soluble or less soluble—and not necessarily the water solubility in laboratory tests that we need to visualize as important in any fertilizer service by way of the soil to the plant. This suggests obtaining a higher effectiveness from fertilizers put into the soil as granules to give a few scattered areas of high saturation, than from a uniform blending of it throughout the entire soil mass.

6. soil acidity as well as calcium makes fertilizers more efficient

The confusion in our understanding of soil acidity has brought confused thinking about "acid" and "neutral" fertilizers. Since we now believe that soil acidity is mainly a deficiency of soil fertility, this acid soil condition which is recognized so readily by simple soil tests, represents in reality a better market for more fertilizer as well as for lime. Then, too, if soil acidity is in reality a benefit rather than a detriment, as we understood it is, acid soil conditions suggest that calcium and magnesium are serving as plant nutrients even when put into the fertilizers for the purpose of making them neutral in reaction. These two nutrient elements are serving for a "synergistic" effect, as Professor True called it, when they are making neutral fertilizers that are better crop producers than are those fertilizers left in an acid reaction.

Nutrient ions coming from fertilizers put on clays carrying some acidity will move more efficiently into the crop than those on clays distinctly neutral. More efficient use will result, therefore, if the fertilizer carries calcium, and if this element is absorbed on the clay fraction which retains some hydrogen, or some degree of acidity, at the same time. It is the fertilizer's content of plant nutrients, including the calcium, and not the reaction or the degree of acidity expressed as pH of its ash content, that comes into consideration in what has been considered a question of fertilizer reaction. It is only the better understanding of the chemical reactions between the fertilizers and the clay of the soil, and between the clay and the plant roots, that will clear up the confusion prevalent regarding the acidity of the soil and that of fertilizers.

7. acidity of soil mobilizes nutrients from mineral reserves

That the acidity of the soil should have been considered such a serious danger to the crop production seems strange when we realize that most of our food production has been on soils that are acid. Then, too, it has been the acid soils that made fertilizer sales and their use in agriculture the big business and the return on the investment that they are. If sulfuric acid in the manufacture of fertilizers makes the phosphorus of bone or rock more available for plant use, can we not likewise picture the acid of the soil, *i.e.,* the hydrogen absorbed on the clay, breaking down the rock to mobilize the nutrients in the mineral reserves of the soil? Through the same manner, we can see that the calcium applied as lime-rock fragments becomes all the more active and the more available for use in nourishing the plant. Has the total fertility delivered by naturally productive soils these many years been no larger than the total exchangeable supply on the clay as we measure it in the exchange capacity at any one moment? Could that supply have been maintaining the productivity during all this time without restocking the clay many times over from some reserve supply? When we use fertilizers in the soluble form to stock the clay with exchangeable nutrients, are we not providing there only the more temporary supplies? What are fertilizer practices contributing toward the maintenance of a more permanent and possibly more wisely balanced soil fertility? These are challenging questions to those who hope that the properly designed soil treatments may not only cure soil troubles temporarily, but may give benefits reaching beyond the few post-application years. They are the more challenging, but also the more hopeful as we think first of fertilizing the soil and then of fertilizing the crops.

When we consider that our soils have been producing crops these many years, surely it cannot have been only the supply of nutrients held in exchangeable form on the clay that was serving. An examination of the output of fertility by a soil like the Putnam silt loam (Sanborn Experiment Field), for example, during its cropping history points out clearly that the nutrient supply in the clay cannot have been the only source of nourishment for the crop. There must have been some activity mobilizing the reserve rock or mineral nutrients to restock the clay and to hold up its supply at some level suggesting equilibrium with the restoring forces. This soil has an exchange capacity of 15 milligram equivalents per 100 grams of soil. Of this capacity, even today scarcely 4.00 equivalents are taken by hydrogen or acidity. This clay still has itself relatively well stocked with nutrient cations that include calcium to the extent of almost 50% of the exchange capacity. If the regular cropping of this soil at the rate of 2 tons of clover hay, 25 bushels of wheat, 40 bushels of oats, and 50 bushels of corn with all crops removed in this four year rotation had been getting only its calcium, potassium and magnesium by trading hydrogen to the clay for them, then the hydrogen accumulated on the clay to date would represent the nourishment of these crops during less than seven rotations or a cropping duration of less than 28 years. Yet these soils have been cropped under record for more than twice that long. If such crop yields continue for such periods of time as we know they have in the cornbelt; and if plants can get their nutrients only by trading hydrogen or acidity into the clay; surely then, if there has been no more acidity accumulated in these soils while we are not adding fertility there must have been a neutralization of it suggesting that this is brought about through the breakdown of the soil minerals representing reserve fertility.

That the acid clay can play a role in this respect has been demonstrated by Dr. E. R. Graham [see *Acid Clay—An Agent in Chemical Weathering,* The Journal of Geology, 1941], when he separated the colloidal clay fraction from the Putnam subsoil, and, after converting it by electrodialysis to an acid clay completely saturated with hydrogen, mixed this with the carefully washed silt fractions of different soils. Observations revealed that some silt minerals were reacting with the clay acid to neutralize it while others were relatively inactive. This effectiveness was correspondingly less as the silt minerals represented soils in an arrangement from lower to higher annual rainfalls across the central portion of the United States west to east [details related in *Soil Development and Plant Nutrition I, Nutrient Delivery by the Sand and Silt Separates,* Soil Science Society of America Proceedings, 1941]. When the mixtures of silt and clay were planted to crop growth, the silts were more productive as they were originally from less weathered soils and as they consisted more of minerals other-than-quartz.

Here is a clear suggestion that we may well think that the minerals of the silt fraction of some soils are being broken down by the acid clay as a weathering agent for them. The rate at which this reaction occurs for the different minerals and the ratio of the clay to the silt fractions of the soil determine how rapidly the clay, once exhausted by one crop growth, can be restocked with nutrient supplies for the next one. Here is a dynamic force in fertility renewal that has not been fully recognized as a variable in connection with fertilizer practices in one section of the country as compared with those needed in another. It is also suggesting that when values of fertilizer materials are determined wholly on the basis of increased crop returns, this criterion may represent undue credit to this soil treatment because fertility is also delivered by the soil's mineral reserve and is not measurable as stock on the clay by soil tests. When natural production has been going forward these many years by means of nutrients taken from the clay in total amounts which no clay could carry in exchangeable form as a single stock, there is reason to believe that the reserve minerals may be significant contributors. The nutrient reserves in the silt minerals of the soil and their mobilization by the dynamics of the acid clay and acid humus as colloids deserves recognition in soil management. This seems wise even before we accept the introduction of water-soluble, concentrated, salts of nutrients into the soil for absorption on the clay as the main concept of the dynamics through which plants are nourished by the soil.

8. high content of organic matter in soils often spells security against bad fertilizer practices

The generous supply of organic matter and probably the reserve minerals, especially the limestone, in our soils have been absorbers of the shock to our soils by bad fertilizer practices. Since organic matter is the source of energy for the microbes, their increased activity may do much to restore the soil from its chemical upset and to lessen the shock on the crop by this soil treatment. *This absorption of the chemical shock is bought at the price of some of the organic matter that is rapidly depleted unless some fertilizers are used to grow crops for its restoration in*

Absorbed nutrients are more effective than those in solution. These pot trials revealed that nutrients absorbed on colloidal clay in dilute suspension and poured over zonolite (right) grew better marigolds than the same nutrients in water cultures handled similarly (left).

the soil. Isn't it probable that the plowing under of fertilizers may be partial prevention against the so-called "burning out of organic matter by fertilizers" if they are put below the humus-rich surface layer and in the deeper, more moist horizon away from some of the microbial competition with the plant roots for them?

Organic matter and mineral reserves in the soil may have been doing more than we realize in preventing troubles in our applying large amounts of a highly soluble single nutrient. Phosphorus has been used in wide variations in the amounts applied. Nitrogen classifies under the same kind of practices. Much remains to be learned before we can apply fertilizers in combinations representing balanced amounts for the particular soil and crop. Much of the confusion about rates of application of fertilizers results from failure to know what the soil has to offer in its decomposed organic matter, and in its weatherable minerals as well as on its clay in exchangeable form. Balanced fertilizer on the soil in the fullest sense of those terms cannot be a ready realization when we are not only juggling the three common elements, nitrogen, phosphorus and potassium to be applied, but when the soil itself is highly variable in its delivery of nutrients from the lime applied, the decaying organic matter, and the mineral reserves. The more nearly correct answer can be given only by several years experience in growing the crops under soil treatments. If the organic matter in the soil could be kept at a high level and if the fertilizer materials applied were less soluble, our juggling act would not be so trying. There is good reason for using a part of the fertilizer to grow organic matter into the soil so that it may be insurance against the possible dangers lurking in our incomplete knowledge at this time about balancing fertilizers more wisely for each soil.

9. concepts to date of reactions within the soil

Fortunately, we are making rapid progress in our understanding of the dynamics of the soil. We are gearing fertilizers into them more effectively. The concepts developed to date of the reactions between soluble fertilizers and the soil, and between the soil and the plant roots visualize these behaviors about as follows:

1. Water soluble nutrients, like the nitrogen, phosphorus, and potassium of fertilizers, are not taken up by the plant as water solutions. They react, rather, with the soil to become adsorbed on the colloidal clay and humus. High solubility of fertilizers means more rapid reaction with these colloids. It is often not rapid enough in drier soils, however, to be without salt injury to germinating seeds if they and the fertilizer are put simultaneously into the same soil area.
2. Nutrients adsorbed on the clay are taken by the plant root mainly as it exchanges hydrogen or acidity for them.
3. Calcium probably plays an important role in the root conditions by which nutrients are taken, or moved, from the clay into the root rather than in the reverse direction.
4. These beneficial effects by calcium are seemingly greater than if some hydrogen or acidity accompanies it.
5. Consequently, soil acidity is beneficial not only in making the calcium more efficient but also in mobilizing other nutrients into the exchangeable forms, including those in the silt mineral reserves.
6. The complexity and numbers of the chemical reactions and interactions within the soil emphasize the insufficiency of considering only the inorganic aspects of them, and suggests that the soil organic matter may represent benefits not yet well understood.

Fertilizing the soil as a help toward its better feeding of the crop is still much of an art. But recent contributions from soil science and plant physiology have been numerous enough to encourage the belief that we are fitting fertilizers more accurately into the conditions dictated by the soil. We are, thereby, moving toward better plant nutrition through an embryo science, at least, of fertilizer use. Much is still ahead as we continue to learn first to fertilize the soil and then to fertilize the crop.

27.

When Edward H. Faulkner's Plowman's Folly *came along in 1943, it received a great deal of publicity. Louis Bromfield write about it and its author in* Reader's Digest, *and he concluded that "Probably no book on an agricultural subject has ever prompted so much discussion in this country." The general idea of eco-farming was making great strides during those days and hours. Bromfield's Malabar Farm was rapidly becoming the Mecca for those interested in preservation of the nation's topsoil. Dr. William A. Albrecht was a frequent guest at Malabar, and the discussions about plowing sometimes became heated.* Is the Plow on Its Way Out?, Go Ahead and Plow, Is Plowing Folly?, Why Plow?, Why Do Farmers Plow? *all became Albrecht entries in the great plow debate. Albrecht did not so much reject the concept of tillage that mixed air, soil, water and trash without the restructure of a plowpan. More specifically, he argued against the concept that there always were and always would be enough trace minerals and enough major nutrients in the soil, and that therefore the farmer could ignore fertilization and let nature do the job. Faulkner modified some of his positions later on, but popularizations that attended publication of* Plowman's Folly *brought Albrecht up fighting from his chair. This entry is a part of one published in* Better Crops with Plant Food, Organic Farming, *et al., circa 1943.*

Why Do Farmers Plow?

This question comes to the fore now because of recent economic disturbances. When natural power in the form of concentrated sunshine collected on the farm and released locally through horses was replaced by machinery using imported liquid power collected in the ages past and stored in the great depths of the earth, the war's disruption of the far-flung distribution of fuels and oils and its deletion of our sources of rubber were not anticipated. These disturbances, both in terms of mechanics and economics, have led some to believe that high costs of plowing would be best relieved if plowing were discarded altogether as a farm practice. This belief is reinforced by successes under reduced plowing in some areas.

In the face of such a rising belief, the practice of plowing deserves a review of both its vices and its virtues. It deserves more searching thought than attention merely to those aspects that are psychological and economic. It deserves more than tabulation of its values, leisurely and short-sightedly considered. Productivity and plowing had many interactions and interrelations for the welfare of humans long before psychology and agricultural economics obtained academic classification of disciplinary mental activities, or a place as controlling forces in national policies. Production and plowing will, in all probability, still be basic when impending international changes bring many of us back to a much closer relation to the soil than we now believe we have.

1. plowing less

We need to plow less on some soils. We need to plow more and deeper on others. We need to learn that the differences in degree of soil development according to climatic differences are factors in determining how important the plow is. The farmers in Ohio haven't invested so much in clod-breaking machinery without provocation. The "one-way" land preparer of Kansas is not so successful purely because of its unusual mechanical design. The soil's physical conditions, premised on chemical aspects controlling them, have some role in these differences between the forest-bearing soils and those of the prairie grass growing areas.

There is need to call out against excessive plowing if it occurs, but it is well to note whether it is the advent of the plow or the exit of the soil fertility that needs correction in improved soil conservation thinking. Certainly soil conservation is more than simple mechanics, simple physics, and simple psychology. It calls for some real friends of the land who will try to understand the soil and crop production therefrom in their fundamental connections, to say nothing of the tillage of the soil in all of its ramifications, even into psychology for all of us so dependent in the final analysis on the productivity of the land.

Fortunately, the plow is merely a tool in this whole matter under discussion. The concern about the practice of plowing is one that brings into question the judgment of him who is using the tool, and the purposes he has for it in relation to the soil as a national as well as an individual asset. One cannot condemn the rifle or the pistol as tools because these are now being used in war, when they can render so many more desirable services. Nor would we condemn the mechanics of the automobile when in its human destruction the fault is not one of the machine but rather one of "the nut that holds the wheel." Our knowledge about plowing and our understanding of soils and not the combinations of simple moldboard, share, and beam, as handiworks of the engineer, are on trial.

Have you ever thought that plowing may be different according as the soils, the vegetation, and even the animals are different? A few wild turkeys and a few squirrels were the population limit in the forest for the Puritans. Those same soils, cleared of the forest and cultivated were soon abandoned as agricultural land by the pioneers who were willing to face the hazardous movement westward. All of these facts have not commonly been related to the low rate, and low total, of nutrient delivery by those soils of the lime, the phosphorus, the nitrogen, and other chemical elements needed to make nourishing vegetation for the building of healthy animal and human bodies.

Soils that had come down to the low fertility delivery represented by the forest level of vegetation before man plowed them are offering so little for animal body-building that the plow must stir them and every posible help is needed to encourage rapid release of the essential mineral nutrients from the meager stock of organic matter within them. Woody vegetation, according to different acclimated tree or shrub species, and a woody composition of any plant species, including farm crops, are characteristic of the "underprivileged vegetation" on such soils unless they are plowed and stirred to increase the rate of decomposition within the soil of residues of plant generations gone before, or are treated by fertility uplifters in chemical fertilizers and other manures.

But on the prairies, where lesser rainfalls have not developed the soil into what is old age, or more maturity, so far as leaching experiences and nutrient losses are concerned, the vegetation is richer in protein. It is also more concentrated in minerals that contribute to bone-building in animal bodies. The soil itself and not the plowing of it determines these conditions. In going from more rainfall to less rainfall or from eastern to western Kansas, for example, the protein concentration in the wheat goes up. We call it "hard wheat" because, as we commonly say, it grows in regions of lower rainfall. More properly it is "hard wheat" because it is grown on those soils that have more nutritional minerals for the micro plants within, and for the macro plants above them. These mineral supplies are producing not only protein-rich forages in legumes like alfalfa, but also protein-rich grain in non-legumes like wheat. Such soils have lime and other minerals nearer their surface where plants can get them to make vegetation rich in calcium, encouraging nitrogen fixation, protein production, and other mineral contributions, all to support animals more effectively than is possible by plants, mainly of fuel value on the highly developed forest soils. We surely cannot subscribe to the belief that all "principles valid for the forest are valid for the fields," when the soils differ as widely as they do under forest and under prairie.

2. *mineral provisions*

Mineral provisioning of the plants by the soils is now more clearly understood. The ideas coming from the soil mineralogists, the colloid chemists, the plant physiologists and other fundamentals of natural laws are helping us to visualize the processes whereby plant nutrition is brought about and what plowing does for it. It was once believed that plant nutrients were coming from the soil minerals in true solution and were caught up as the plant was taking in and passing on this solution as a water stream to maintain transpiration from the leaves. Studies in plant physiology have recently given us the concept that the nutrient ions move according to physio-chemical laws dealing with the kinds and concentrations of the nutrients on the clay; with the different nutrient ions within the roots in terms of concentration, absorption, and the elaboration into the plant compounds; and with a root membrane interposed between the clay colloid of the soil and the complex colloid within the root.

Plowing has been much confused with water movements from the soil through the plant possibly more by imagination than by actual demonstration. Water moving into the root follows its laws of ionic and molecular behaviors. These are quite different from those of capillary movements given by the high water table experiments of Professor King. These laws seem to suggest that there is little travel by water as a liquid and that the plant has little to do in the way of control. The concept of the plant as the channel by which soil forces holding the water are balanced against air conditions dissipating it seems to be logical when we remember that plant stubbles and such dead plant parts transpire water from the soil. Plowing has not been connected with the newer concept that nutrient movement from the soil to the plant may be occurring independently of the Gulliverian wanderings of soil water.

3. plants take nutrients

Plant nutrient ions like calcium, magnesium, potassium and others are held on the finer clay part of the soil in an adsorbed form against loss therefrom by water. They are, however, exchangeable by other ions, particularly hydrogen as an especially active one. That hydrogen is the main item, which the plant exchanges to the clay for what ions the clay offers in trade as plant nourishment, is now fairly well understood as the mechanism of plant feeding. This occurs through a most intimate contact by plant roots with the soil particles. Plant roots extend themselves through the soil to get their nourishment by means of this trading process. Little credence can be placed in the belief common only a decade ago, that the soil gives nutrients to the plant. The performance fits more nearly into the country boy's understanding of how we get milk from cows, when he said, "Our cows don't give milk, we take it from them."

The effects by the root as a nutrient gatherer may extend through a distance from the root of but a few layers of clay particles right next to the root. This is limited probably to distances in millimeters, certainly not to such extensive distances as centimeters. The root systems' effects as nutrition are also commensurate with the total root surfaces. Accordingly, then, the densely matter collection of roots under bluegrass takes more total nutrients from the colloidal part of the soil than does the sparsely rooted crop, like soybeans.

Each root leaves the soil in its immediate zone of activity exhausted to a very low level. The advent of the root has opened channels by which nutrients could go out and energy compounds come in. In fact, it brings about, either directly or indirectly through its own decay and bacterial activities, a reduction of the compounds of the soil about its area of penetration. This reduction may be indicated by a color change from the customary reddish to the drab gray soil, much like we know it to be brought about by water logging. One might expect roots of the next crop to follow successively in these old exhausted root channels, if the soil were not stirred. Plowing serves as a mixing agent to redistribute this reduced clay amonst those clay portions that were not so nearly exhausted of their supplies of nutrients.

That plowing is more essential for this purpose than we commonly believe is indicated by the increasing report of observations of deficiency symptoms suggesting plant diseases of some crops, such as cereals and some of the legumes in such a close sequence as to reduce the amount of plowing. Soils put under fall-pastured barley as nurse crop for summer-pastured lespedeza to be disced and to go to barley again in another annual cycle with only this limited tillage are showing nutrient deficiencies that are not prevalent under plowing. Plowing serves to shuffle the exhausted soil surface into contact with other surfaces not so depleted. It is apparently significant for the crop nutrition that such soils be plowed between even two successive crops. It may be true that the farmer cannot appreciate the colloid chemistry and low levels of nutrients in the soil concerned with the crop disease symptoms, but he does appreciate the improvement in the crops after he plows. He is justified in developing a reverence for the plow much as you and I develop a reverence for the dining room or the kitchen, if reverence of that type is the limit in our thinking.

4. mixing nutrients

Plowing serves for nutritional improvement of the crop by mixing the different clay areas in the soil. Dr. Graham's researches at the University of Missouri have recently pointed out that plowing may be instrumental for better plant nourishment because it shifts the connection between the surfaces of the clay and surfaces of the silt, or the larger mineral particles of the soil not commonly considered so active as exchange performers. He demonstrated that the nutrient ions in the mineral silt moved to the clay in the absence of plants, and that plants picked them from there to their better growth advantage than from the minerals directly. Periodic shuffling of the clay in contact with the surfaces of the silt particle, or after the clay has become saturated during the period of contact for a few months, is the means of keeping more of the clay loaded with nutrients to be passed on to the plant root. Plowing is the means whereby enough clay in the soil picks up enough nutrients from the silt, and other original reserve supplies of fertility, in active forms and in amounts sufficient to give us the quality and the quantity of crops we need to produce.

This then is the picture of plant nutrition as we visualize nutrient elements coming from the soil. It is a chemical performance within the soil to which plowing and other similar mechanical measures contribute speed. The nutrient ions adsorbed on the clay move into the root in exchange for hydrogen ions coming from the plant root to take their place on the clay. The clay on becoming more extensively saturated with hydrogen ions—the active producers of soil acidity—passes them on to the silt and to other mineral soil particles as the means of weathering the nutrients out of these original rock forms. Thus by means of plowing, the clay is rapidly reloaded with a stock of nutrients, or is buffered against what we have been viewing as dangerous, excessive acidity, but which is in reality dangerous soil fertility exhaustion.

As has been demonstrated by Dr. Carl E. Ferguson at the University of Missouri, this exhaustion of the clay's nutrient supply would occur in but a few crops were it not for the silt. It is through these steps, namely, rock to clay, clay to plant, that the nutrients pass. It is in the reverse direction, root to clay, and clay to mineral, that the weathering effects by the plants in the form of hydrogen as acidity travel for soil depletion of its mineral nutrient supply. Plowing increases both of these reciprocal movements of the chemical elements, and thereby facilitates what concerns most of us, namely, food production.

5. plow is not exploiter

Plowing merely hastens many of the same processes that are occurring more slowly when "the land is resting." When land must be allowed to rest in order to boost its productivity back to economic levels again, this is merely proof that the fertility supply on the clay is exhausted so nearly to completion and the mineral reserve of fertility has fallen so low that the interactions between the clay and the minerals are too slow to move enough nutrients on to the clay surface to provide sufficiently for the roots during the growing season. Plowing isn't the cause of the depletion of the fertility supply. Depletion occurs because of the fertility removed within the crop hauled off. The plow is not the exploiter; rather, it is the farmer. The plow is merely the tool that facilitates his exploitation at a faster rate and over more acres than before the plow was given him. The plow has helped him to feed many of us too far removed from the land to appreciate its exploitation.

Some of our plains have been exploited to such an extent that even the plow can't substitute for the time needed to restock the clay from the mineral reserve. These soil processes are too slow in rate, and too limited in amounts of fertility mobilized thereby, to finish, for example, one wheat crop in June and to germinate to a good start another crop by the succeeding October, even with the help of plowing. This is the case of a plot on Sanborn Field in Missouri, where wheat has been seeded annually without fertility restoration since 1888. This plot is now taking an annual rest on its own accord after it produces one crop. It has become a yielder only in alternate years. This is because the soil fertility delivery processes that are moving nutrients from the soil minerals to the clay and from there to the roots in exchange for hydrogen going in the opposite direction are too slow to give ample supplies unless an extra year elapses. Fertility and not water are concerned. Surely, such a biennial performance with regularity over almost 25 years is not a case in which the soil simply takes time out from its business of growing things until the restoration of its normal water supply. Food more than water is involved.

Here is a suggestion that any accusation of the plow as a responsible agent for soil deterioration is a misplaced and unfair condemnation. Such accusation would still seem just, even if by the best of science we should lay bare every principle in only physics that plowing of soil involves. Even if we should dispel the belief that "the exact physical effects that follow the operation of the plow have never been subjected to scientific scrutiny," the plow might still be listed for its exit as an implement. Plant production is more than applied physics and particular mechanics. It is a matter of delivery of the required plant nutrients. The soil processes providing nourishment within the soil are slowing down because of soil fertility exhaustion more than through bad mechanics premised on "what we learned in elementary physics in high school."

Shifts in the kinds of nurse crops and in the kinds of legumes in order that we might accept substitute crops are very striking evidence of soil exhausted of its nutrient reserves. Not only is the slowing down reflected in grain crops by their alternation between a crop and a crop failure, older apple trees become alternate-bearers. Older cows pastured on and fed products from many soils of declining fertility go on similar biennial schedules in reproduction. Surely the plow isn't to be blamed for what happens in the subsoil under the apple trees or for nutritional irregularities on permanent sod pastures that come to light in terms of breeding troubles in cattle.

If we are to bring the plow into this picture of "the debacle into which our American soils have drifted," the case could not be rested on the contention that while the farmer's "reputation for smoothness and neatness of the plowed field was developing, no thought was given to the possible connection between smoothness of the land surface and exclusion of the rainwater from the soil." Antediluvian ideas about water in the soil, about the wet subsoils under freshly incorporated green manure interpreted as interrupted capillary rise rather than "sweat" from the respiring and decaying organic matter—to say nothing of many other ideas almost equally hoary—don't convict the plow except for those unfamiliar with more recent soil science. Plowing and crop production are more than water problems. No one will deny that even these are serious enough. Declining soil supplies of fertility are making the water problem worse as we allow the plants to starve for nutrients while they are wasting their water transpiration and carrying on within themselves little or no construction of the organic, nutritional complexes they are intended to synthesize. Water will be the lesser of the soil troubles when we understand nutrition, and when we feed the plant so that what water we have will be used most effectively for crop production.

Starving plants do more damage than merely that of wasting water. They invite attacks by bacteria and fungi to cause much that is regularly called "plant disease." Starving plants are symptoms of soils that are no longer stable in their desirable structural conditions known as granulation. Their surfaces are hammered flat with the first dash of rain and are moved off in deflocculated condition as erosion in the balance of the rain. To the eye and mind that are observing soil fertility, numerous other plant and soil symptoms are clearly visible. For such an observer the real debacle about agriculture is that we continue to exploit our soil resources without giving the slightest thought to the fact that these unrecognized and unappreciated chemical changes within the soil are basic to erosion, to disturbed agricultural economics, to distorted national economy, and to a disturbed national health, as draftee rejection figures reveal.

6. *the mining performance*

This larger problem is aggravated by the plow, but also by any tool, either mechanical or psychological, that encourages and permits continued exploitation of the fertility of the soil in the same manner as we mine and consume many other resources. The land is the basis for our existence by way of the food it provides for use. The mining performance of it has brought us to where it is difficult to change and to shift into using the soil only as a site for soil fertility "turnover" by putting in about as much of plant nutrients as we take out in crops.

This shift to letting land rest, to putting out the land to grow cover, to encouraging organic matter restoration, to purchasing fertilizers as a definite program of returning almost the ash equivalent of the crop removal demands more than that the farmer quit plowing. This shift to squeezing out the charges assessed against an unearned increment, and to going back to an acre value of the soil as a producer after deducting costs of fertility maintenance, labor, and investment carriage even at the low rate acceptable to the man of the soil whose hope for security is still pinned to the land, is a change that calls for more than invention of a scapegoat in the form of the plow.

The understanding of the processes in the soil as a producer of our foods has become a challenge to an increasing number of people. Friends of the land are multiplying so that with a knowledge about and deep concern for the soil, they will not long leave unanswered the question, *Why Plow?*

28.

One of Dr. William A. Albrecht's great friends was E. R. Kuck of Brookside Dairy Farms, New Knoxville, Ohio. In fact it was a case of animals telling Kuck about vanished nutrition that launched Brookside Laboratories in the first place. After plastering the walls of a new calf barn, Kuck noticed that the animals literally ate the plaster off the walls of their stalls. Calves were scouring at the time. Almost immediately, scouring stopped, and Kuck—consulting with Albrecht—determined that the hungry animals were in fact after the calcium carbonate and the magnesium carbonate in the plaster material. These nutrients had been mined out of the soil of the dairy operation and never replaced.

The sheer desperation of those animals emerged time and again in Albrecht's writings. "Our dietary essential minerals are taken as organo-inorganic compounds. We are not mineral eaters. Neither are the animals. When any of them take to the mineral box, isn't it an act of desperation?" wrote Albrecht.

This paper was prepared specifically for E. R. Kuck and his associates at Brookside Farms Laboratory Association. Kuck later called it a "masterpiece of fundamental logic expressed with extreme clarity." Kuck pointed out that "with each reading I have discovered new thoughts that have virtually electrified my imagination."

Soil Fertility, The Basis for Formulating an Agricultural Policy, *has lost none of its sparkle. Lawmakers who stumble over each other to give depletion allowances to mining interests ought to note how well modern farming fits the definition of a mining operation. Sociologists who are concerned about world hunger ought to read Albrecht's pregnant lines about the wasting away of our capacity to produce. Finally, businessmen who seek to make farming an industrial procedure ought to study this report so that they might discover how distorted accounting principles are really responsible for many profits.*

Soil Fertility—The Basis for Formulating an Agricultural Policy

Much is being said about, and claimed for, our scientific progress. We are about to believe that we are approaching the pinnacle of it, now that the alphabetical progress from the A-bomb to the H-bomb exceeds our expectations. But that *excess* of progress is not in cooperation with the creation of life. Rather it is *progress* in the destruction and death of life. It seems well, then, that we inquire whether we are not allowing ourselves to be deceived by our technological successes. There is a terrific danger in over-confidence because of past success, particularly when that has been obtained mainly in the area or fields of technologies. There is serious danger in believing that such success with transformation of matters that are dead, can be the basis for success necessarily in agricultural production which deals with matters that are living.

1. some biological processes upset or not understood and appreciated

Even though by chemical technologies we have fertilized the soils for agriculture to have produced 300 bushels of corn per acre, let us not believe that we are already controlling nature at the point of *take-off* of her creative activities of life. As farmers, we may well remind ourselves that the study of the soil, from which nature starts her creation, is the study of the starting point of life and the living. Agriculture is biology first. It is not only technology. It is not plows, tractors, seed, soil conditioners and mechanical manipulation, first. It is creative capacity, and creative chemistry first. After that, then, the products which the soil creates become objects for technology. We do not grow crops as a technological procedure. Rather nature and her soil create them by our cooperative and supplemental helps.

We need only to recall the seasons, like 1952, 1953 and 1954 for example, to be reminded of how feeble our efforts are in controlling nature's creation of crops, and how quickly we make the weather the scapegoat when we can't get high crop yields. It is fitting, therefore, to provoke some thinking about the troubles originating in our technological applications to agriculture while disregarding the basic biological nature of the work of growing good crops, good livestock, healthy boys and girls.

Much that is apt to be called agricultural science and applied as technology has seriously upset the biology. Some cases are now bringing us around to reaping the poor harvests, or even no harvest. Our attempts to streamline, to compel, to crowd and to short-cut biology are bringing us to realize that we are mistaken when taking credit for performances which are nature's acts and not ours. We are slow to comprehend clearly the basic principles operating in nature. Our disregard of her laws are not tolerated long before her retaliation becomes costly. The soil under the natural laws operating in and through agriculture is a good case for consideration and illustration.

2. technologies lured man to fringe soils

Man's technologies have put him on lands and their soils where the Creator had no other life forms of similar body demands or corresponding physiological processes. By means of his machines and many technologies man carried himself to places to which he requires shipments of foods from other places to keep himself and his animals living. Can he depend on long lifelines forever to keep him properly nourished when exhaustion of the soil fertility, shifting economics and mounting populations are shortening those lifelines and even cutting them off completely? Have we not upset the biology of human life when we pushed into the regions of higher rainfall and far away from the semi-humid, near-desert soils on which primitive man lived with almost no technological lifelines? Primitive man on those drier, unweathered soils, survived by his own created crops in limited areas where windblown, well-mixed fertility supported him completely. Are our long lifelines not allowing us to push to soils where we are not so well fed, if degenerative diseases now so numerous are giving us any suggestion? Have we not upset the biology of ourselves possibly as much and more than any other biology?

3. nature's monoculture of crops was replaced by crop rotations

The soils which nature uses to grow grasses and animals are no different in fertility from those she uses to grow forests. Yet even with our technologies we try to use those same forest soils where nature grows only wood, *i.e.,* fuels, carbohydrates, to grow a variety of crops in what we call "rotations." We hope to compel legume crops to grow and live where they never grew before. We expect them to create proteins. We speak of those soils as "forest soils," as though the trees made the soil rather than the soil make the trees. We speak of the "prairie soils" as if the grass made the soil and the grass made the buffalo, when it was the soils with higher lime and other higher fertility contents because of the lower rainfall where legumes grew to nitrogen from the air that made both the nutritious grass and the well-boned and well-muscled buffalo or bison. Nature put the animals on those soils. She was in the cattle business there. She didn't put squirrels into the pine and other coniferous forests. If, then, we cut away the coniferous forests where nature herself could do no better than make wood by putting every crop completely back into the soil, can we create nutritious crops with no fertility uplift of the soil? Is not our faith in mere rotations of crops as soil improvement a blind faith? Will we not upset crop biology completely by merely rotating crops and taking all of them off the land?

4. the badly broken soil body goes unrecognized

Now that we are seeing the soil, we realize that depletion of the fertility brought about a soil body so weakened that it cannot stand up under the hammering raindrops without serious erosion. In order to stop that erosion, we are calling for a "grass agriculture" to give grass cover over much of the country where once a forest cover prevailed. Will the growing of grass on those soils without their fertility improvement put nutritional quality into the forage to keep healthy the cows we expect to eat it? Can we make up for our ignorance of quality of grass in contrast to her knowledge of it even when she is confined by a fence? If any rational committee sits down to plan a grass agriculture, it is the cow that should be the chairman and should direct the deliberations and plans. In setting up a grass agriculture to arrest erosion we upset most seriously the cow's biology supplying our food supply in meat and milk when we believe that we are as wise as the cow in judging the feed quality of the grass.

5. biological order has been overrun by chemical treatments

Nature feeds plants by having their roots use acidity, as a by-product of their respiration and growth, to move nutrients from the soil into themselves. Roots grow our crops by making the clay of the soil more sour. Yet we would fight against that root activity by having the carbonate of limestone so plentiful in the soil that it could never be acid. We fail to see that root acidity, and soil acidity resulting therefrom, are helpful in breaking down the lime-rock, and other rocks, to make the calcium and magnesium or lime elements available to the roots as crop nourishment. Excessive liming would upset nature's biological performances for us if it would keep the soils neutral or non-acid. Soil acidity is a benefit if we use it according to the biology of agriculture rather than the technology of the chemical laboratory.

6. the sustaining fertility is not recognized

When the rock minerals of plant nutrient services in the soil decompose to become soluble there are such large quantities that the plant roots will not absorb all of them or the clay's adsorption capacity cannot hold them, then those active fertility elements wash out of the soil and pass into the sea. Yet we make fertilizers of water-soluble salts in the belief that they are taken into the plants more readily from the soil because they are in that soluble condition there. If they remained as soluble salts in solution in water they would soon be gone to the sea, or if the rains were not sufficient to wash them out, the generous use of them would "salt" to death the seedlings from planting with their application.

We use soluble fertilizers with the seedlings in but small amounts to avoid that danger. We place them aside or below the planted seed. They are, therefore, only "starter" fertilizers and cannot supply the needs for the complete crop. The reserve minerals in the soil must supply that, consequently we must follow nature's method of using lime-rock, phosphate-rock, potash-rock, and other minerals as the "sustaining" fertility. Dust storms from the west, or from the Missouri River flood bottoms are the Creator's help on loessial and windblown soils in the form of fresh minerals because of the Missouri River floods. We would upset that geobiological arrangement by putting dams across the river behind which, and under water, there would be the silt deposits now blown from the river on to north Missouri and Iowa to keep those soils productive. Our technologies make mineral fertilizers soluble in the factory. Biology makes them soluble within the soil. Our technologies wear out the soils. Nature's biology builds them up and maintains them in fertility.

7. *nature's criteria for good crops have been replaced by man's*

Biology of nature doesn't seem to strive for big yields as tons or bushels. Rather, growth seems to be a struggle for protection against disease and for fecund reproduction. Technologies under economic controls strive for big yields or mass, rather than of good health and of high fertility in reproduction. Neither does nature make excessive fat. When we manage production we soon suffer from over-production (and overweight of body) because our high yields have brought in the carbohydrate producing crops or shifted them to produce mainly carbohydrates and less of the proteins. Shall we not see declining concentrations and qualities of proteins in our crops as possibly reasons for the increasing insects, diseases, and trouble in reproduction? Managing plant life and animal life as we would make them fit our economic desires, more than their own biological requirements, has upset the biology seriously. Corn hybrids have pushed yields as bushels up now to 300 per acre. But they have pushed concentrations of proteins down. Corn will fatten older animals. It does not serve to grow young ones. We can stay in the cattle fattening business seemingly only as long as there remain in the plains of the west where the cattle can grow the young ones themselves. Gain in weight, as the sole objective, has brought us face to face with a biology upset so badly that midget calves are so numerous as to make the business nearly impossible, in spite of the one implicit—but now shaky—faith in the noble pedigrees of the fathers of the calves.

8. *agronomics have been submerged by economics*

We have not only upset the biology which is the basis of agriculture, but we would even believe we can push out biology completely were we to follow the thinking of the economists who say, "We can now substitute capital for land by the use of fertilizers. We can use capital to make the equivalent of more acres by making each acre produce so much more." While that may be a compliment to those studying soil fertility and plant nutrition, let us not be led astray by the fallacy in that reasoning. Were it sound reasoning, we might use fertilizers on any open spaces (even pavements, possibly) and grow crops. There is so much more in crop production than just fertilizers as we now have them, that we dare not believe we control creation so completely that we can prescribe from the chemist's shop all that a soil is giving when it grows our nutritious foods and feeds. Capital cannot substitute for the soil fertility both inorganic and organic, in the land. All the gold in the bank vaults or in Fort Knox, Kentucky, couldn't give birth to a calf, have a litter of pigs, or even lay a single fertile egg. Money doesn't make crops any more than tractors and farm machinery do. Thinking about money in place of soil has upset biology of agriculture already too seriously during the last quarter of a century. ***Attention to the soil as a biological matter more than an economic matter is the basis from which to start searching for the agricultural policy we seem to be calling for now.***

While we have technological bases for our pride in much that has raised our "reduced-labor" standard of living out on the farm, we have not so much basis for pride in the creative standards, if the problems of degenerative diseases are faced squarely. With 52% of our hospital beds taken by the mentally ill, we certainly are in no position to boast about ourselves. With cold wars the custom, and destruction of life pushed up to H-bomb proportions, isn't it time to confess that only very slowly are we coming to understand, or appreciate, the creation in process that operates as agriculture? We understand some fragments of agriculture as a natural art, which science has lately analyzed. But as yet we have not been able to manage creation which still depends on the soil.

Men of success in agriculture must understand the natural forces of creative production which operate under the power of rainfall and sunshine, but are controlled by the fertility delivered from rock break-down of the soil. Confusion now astir in agriculture is slowly coming to see the soil. Soil conservation is a part of the picture but more of it must be integrated into the farmer's management of his farm so that it will be a by-product of good farming for his living and not as an added expense. Superimposed by those who view agriculture in terms of only technologies of making terraces to stop running water, or in terms of economics of higher yields per acre or per man for less cost per unit of product, the conservation of the fertility of the soil will not be the result. ***Conservation of the soil will result only when the farmer on the land can conserve it with the rewards for that service accruing to him, instead of to all who do not share enough in the work and costs required to arrive at soil conservation.***

9. confusion will prevail until the soil becomes the basis of agricultural policy

Our prosperity in technologies applied to industry moved forward because the application of the same technology to mining the soil fertility gave us ample food under a westward march. But because we disregarded the biology of agriculture and some basic laws of nature, we are now confused about agriculture as we look ahead. Too long have we viewed farming as if it were a case of running a factory. We view it as only another technology. But factories manipulate the dead, not the living, matter. The natural forces of sunshine, rainfall, decay-processes within the soil, the depth to which the fertile surface soil is dried during drought, the duration of a flood, and others, are not unforeseens to upset productions of the finished factory goods. But they are serious upsetters in the plans for the products put out by the farmer. The factory manager can predict very accurately the amount of his output and the sale price required to guarantee his margin of profit. He sells, then, on a seller's market. The public pays his price.

Our urban population is concerned with that kind of economics. That kind of procedure is theirs in making a living. That share of our population now amounts to 85 or more percent [in 1961]. They live by technologies, by figuring costs *plus,* and by the control of their margins as they set them in relation to costs. Their assets are always protected.Even for him whose assets are mainly the brawn he invests, there is the group behavior in strikes, lock-outs, *etc.,* to give him a margin above costs of living. The merchant invests his capital in a stock of merchandise. A dollar sale is not a dollar income subject to tax. Instead, 60 cents, or percent, are exempted to replace depletion of the stock of merchandise; 25 cents, or percent, are exempted from tax as costs of operation. Only 15 cents, or percent, *i.e.,* the margin of profit, are subject to tax. Thus, *the invested capital is always protected, and the business is self-perpetuating.* This is the urban custom whether the investment is brawn, skill, brains or money. The productive capital cannot be lost, but instead, it is guaranteed or protected.

Such is the economic setting within which technologies represent the earning powers. Such is the thinking on which taxation rates, depletion allowances, and the many obligations by property toward the community are assessed. It is from the business groups that our legislators come. Laws are formulated with the urban type of business in non-living matters as the pattern of thinking.

In that type of thinking the significance of soil as a natural resource in food production and in the creation of living things does not present itself. In that type of thinking, food is only a commodity of barter, trade, price, volume, perishability, spoilage, etc. Milk there is a matter of so many quarts with a fixed margin of profit per quart. Milk once in the city is not a matter of failing conceptions by the cows, excessive dry periods of too many in the herd, abortions, mastitis, shortage of protein supplements, failing pastures under drought and all the biological hazards which make milk production as a sequel to the birth of calves an almost unpredictable matter as to the costs. Milk becomes a commodity under technology and control after its creation and delivery by the cow under biology. In like manner, once the animal becomes a carcass, it shifts from a creative result of many biological processes to an object of technology. When 85% of our people see milk, meat, eggs—our major protein foods—as commodities under the same business transactions as washing machines, furniture, etc.; when such a large share of our people have had no experiences in the hazards of the biological creations bringing us milk, meat and eggs on the farm; and when our spokesmen in economics, politics and matters legal come mainly from that majority, can we prevent confusion in matters agricultural? Can we reason out a wise policy coming from those commodities on the shelf? In our research for an agricultural policy should we not start our thinking by considering the soil fertility, which is the beginning point of all the biological creations which are agriculture? Should agriculture not have help in perpetuating the creative, the productive power, in which its capital is invested? *Were we to establish the possibilities and the costs of perpetuating the fertility of the soil* (*or enlarging its possibilities*) *as we protect other capital investments, in our humble opinion, we would be in good position to settle on some semblance of a rational agricultural policy. Should we not start thinking from the ground up?*

10. depletion of fertility capital under guise of faking profit brought on our agricultural economic dilemma

As a result of agricultural research, we have come to understand some of the soil processes by which crops are grown. We have learned that the soils are not self-perpetuating. Soils may be rapidly exhausted of their capacity for quality-crop production. Rotations of crops per se do not build up, or even maintain, the soil fertility. Either the landowner or the operator must do that by returning the organic-inorganic fertility equivalent of the crops removed. He must restock the shelves regularly. In the drier regions or near them, as primitive man illustrated, we may enjoy the good fortune of having fertility blown in as dust. Missouri, Iowa and other states of the mid-continent have enjoyed that good fortune in their loessial or wind-deposited soils. We have learned that depletion of the soil organic matter and of the nutrients exchangeable on the clay, has pushed out some crops. The introduced new ones are making less of proteins and more of carbohydrates. We have weakened the soil body and brought on its erosion. We need grass cover, but the grass, like the hybrid corn, would not grow calves, though it would serve only to fatten older animals unless the soil fertility is improved. We have also learned that *fertilizers must* eventually *be more than starter treatments.* They *must be* the *sustaining fertility* as we learned for limestone and rock phosphate in the case of rebuilding the soil in only three elements, calcium, phosphorus and magnesium. *We are gradually learning that capital assets, namely, the fertility of the soil, in our farms have been under liquidation while no depletion allowance for them has been made.* It is on that fact that the agricultural dilemma on a national scale now rests. We have exploited the soils in a westward march, and can now go west no farther. All that dawns upon us, when we are given *the responsibility of world leadership,* which *is calling also for world feedership.*

11. the challenge comes to the minority

If any agricultural policy is to be formulated, shall not agriculture speak first for such? Shall not the biological processes of the soil managed by the farmer in the rural areas come under the same categories of taxation, self-perpetuation, margins of profit, etc., as the technological processes of the factory managed by the industrialist? When as a farmer you invest $100,000 in a farm, you are purchasing the supply of mineral and organic fertility. That is your productive capital by which alone the rainfall, the sunshine, and the fresh air are fabricated into food for all of us. But with the sale of the products grown on your farm you are liquidating those fertility assets without recovering their original costs in the price of the sale. The rate of depletion of soil fertility per bushel of corn, at present prices of it and fertilizers, is near 20% of the sale price of that grain. In our income tax regulations the depreciations in the farming business consider only buildings and fences. No depletion allowances are made for the fertility of the soil. Yet you liquidate your original investment with every sale, but make no charge therefor. You sell on a buyer's market. You buy on a seller's market. We have been liquidating our national food source and the farmers' capital investments in fertility of the soil, but under the erroneous belief that the farmer was taking a profit. We are now all crying for an agricultural economic policy.

Let us consider, then, the reasonableness of viewing the soil and its mineral contents in the same category as a limestone quarry, as a gravel pit or an oil well. Depletion allowances of 15% are common in the mining of the minerals. If the removal of calcium as mineral from a quarry warrants 15% depletion allowance to protect the investment so as to provide another quarry on the exhaustion of the first one, shall we not view calcium removal from the soil in crops as a deserving depletion allowance correspondingly or higher when it represents potential food? Isn't it fair that the minority protecting the national food source should speak up and protect its investment? Shall the capital assets of the farmer be liquidated for the privilege of protecting the food resource of the country?

The faith of the farmer in his compatriots has not yet fallen so low to let him believe that his voice calling for just consideration will go unheard. The time is now upon us to present the case of the soil, as a national asset for which there is no substitute. We have exploited our forests but have found substitutes for wood for our shelter. We have exploited our wildlife but are content to forget fish and other wildlife. But when the soil fertility is the creative means of food, we dare not forget that for which there is no substitute. As yet, no Congressional debater has been found to take the negative side of the proposition, "Man must eat." On the affirmative side of that proposition, the farmer wins the argument when as a minority he reminds us that our national strength lies, not in the technologies of destruction, but in the biologies of creation of life, all of which takes its start from the fertility of the soil.

29.

Most of the points covered in Neglect of Soil Fertility Reflected in Farm Animals *may have been covered before in this volume. This selection is being included here because it illustrates how tirelessly Dr. William A. Albrecht worked at the task of taking the message of clear science to farmers. This report appeared in the* Kansas City Daily Drovers Telegram *on November 11, 1942.*

Neglect of Soil Fertility Reflected In Farm Animals

Sound bones are the basis of good horses. Such bones can best be made by healthy bodies that get the necessary lime and phosphates—bone ingredients—from the soil by way of nourishing feeds. When limestone and phosphate are soil treatments to improve crops, the question often arises whether horses need dosages of these minerals directly. Our soils have much to do with the delivery of these in forages for efficiency in livestock production, more particularly for those longer-lived animals, the horses.

The fertility of the soil, which includes those plant nutrients that make up mainly the plant ash, is now coming in for wider general appreciation. Nitrogen as the nutrient hidden in the soil organic matter, or humus, has long been appreciated because of its scarcity and of the difficulty in replacing it. As erosion dug deeper ditches that hurried the water off in record-breaking floods and most disastrous droughts, we recently became engaged in one of the largest national action programs outside of war. We have undertaken to keep the body of the soil at home. Even there we are no longer thinking of only dams in the gully to control floods. We have gone out of the ditch and on to the upland to stop much of the rain water where it falls by making more of it go into the soil. This infiltered water is growing more crops as cover, more of them on the contour and otherwise, all to keep more soil at home.

The realization is about to come that it isn't so easy to grow perpetual cover crops successfully. Even before the soil started going, much of the soil fertility—the soil substance that it takes to make plants—had already gone. Not only erosion and difficulties in growing some crops are bearing testimony of this fact, but even the farm animals are reporting their troubles in deficiencies of growth and in reproduction because of our past neglect of soil fertility. These are some of the signals flashing caution and reminding us to look to the maintenance of the soil for the future health and profit of our livestock.

Grass for the grazing horse serves two body functions. One of these consists in supplying the materials by means of which the body is constructed. The other is that of providing energy to run the body machinery and move the animal and its load about. Growth demands protein to build muscle. It demands lime and phosphate to build the bones. Only a proteinaceous substance already fabricated by the plant can meet the protein requirement for horses. They cannot use the simpler elements for making this protein as in the case of plants. The lime and the phosphate also come by way of the plants from the soil. The minerals and the proteins are required in larger amounts by young horses, by mares during gestation, and by stallions for most successful service, because these are the "grow" foods and come from the soil.

Energy foods are supplied in the form of the carbohydrates and fats. It is the carbohydrates that make up the larger part in the bulk of the plants. These compounds are the main product of the plant factory as it takes carbon from the air, water from the rain source, to be combined into carbohydrates by means of the green chlorophyl in the leaves as it catches the sun's energy to do this chemical work.

1. "go" foods and "grow" foods

Starchy grains, saccarine plant compounds, and much of the plant's fibrous structure are the horse's energy or power sources that do not come directly from the soil. They are the "go" foods that plants seemingly make from the weather, or those materials amply present without the soil. Plants won't render this energy-snatching service, however, except through the help of the five or ten percent of their own "grow" foods that are taken by their roots from within the soil and represent the plant's mineral part, or ash.

Horses are power plants to release for our service the sun's energy stored by the plants. Horses haul their own coal when they come from the pasture. They must first be built, though, by means of the body-constructing proteins and minerals that come from the soil. Horses must "grow" first and "go" later. The soil is the foundation of our farm power plants when there are horses.

It is easy to become excessively optimistic as to what can be accomplished by the breeding of horses, now that breeding of plants has given us hybrid vigor, crosses of poultry have served for sex distinction in chicks, and certain hog crosses have given unusual growth capacities. Breeding has its possibilities, but feeding by way of soil fertility treatments needs wider consideration as to what it has done and can do for horses.

One needs only to survey the the different sizes and kinds of horses in different countries and relate these to the soil to appreciate how much the soil may be in control. It is now known that soils in a region limited enough to have almost the same climate are different because of different parent rocks of their origin. As we have learned to recognize these soil differences we are likewise appreciating the influence of the soil on the development of horses.

2. country, soil affect size

Within small ranges of latitude and longitude reaching no great distances out from the British Isles as the center, and all within the influence of the Gulf Stream, one can go from the smallest to the largest of the horses. The Shetland pony, or the midget horse, is at home on the more rocky, less developed soils of the islands at 60 degrees north latitude. The Irish pony and Welsh pony, larger but still in the pony class, are on the granitic and slaty soils respectively, at 55 degrees north. Of about corresponding size are the Russian horses on the gravelly, glacier deposited soils of North Russia, and the Norwegian horses in the rock bound fjords at not much different latitudes.

One needs only to go into Scotland with the greater clay content of its soils to find such active and stylish hulks of horseflesh as the Clydesdales, or east from Wales and its slates into England with its clay soils to go from ponies to the massive Shires and Suffolks. South, a bit farther, there are the heavy, closely-coupled Belgians. Nearby in Normandy of France on soils similar to those of England where heavy clays, heavy plows and heavy horses all go together, we find the original Normans or the Percherons of tremendous body, surplus power, and excellent disposition.

Through this small area—all within the region where woolens are the common wear—the climate is not so widely different. Yet the soils include a great variety because the different rocks in similar climate mean different minerals. Therefore, the soils made from them are different. Different soils make different feeds. Different feeds mean horses differing in size, speed, conformation and disposition.

Horses differ by countries with different soils, not because horses eat the soil directly, but because even the same crop differs in its chemical composition and its service as feed for horses according as the soils are different. What, from general appearances, may look like the same soil may be decidedly different in feeding function by way of its forages. In fact, the same soil in the same place may change with time, or through neglect, enough to make its crops of greatly lowered feed value.

Soil treatments, including manures, limestone and fertilizers, are but small additions to the soil. Yet they may alter most decidedly the nutritive value of grasses thereon. We are just beginning to appreciate different soils as they represent larger or smaller stores of reserve nutrients, and a more or less active factory with an annual output of plant nourishment available according to the crop needs through the course of the season.

Recent soil studies have reported that it is in the clay and humus, or what is spoken of as the colloidal part of the soil, that the plant nutrients are held in the adsorbed condition. This adsorbed condition means in forms available to the plant by exchange. Sandy soils in the humid region, or those with little clay, when broken out of virgin condition, are soon in crop troubles. These bring animal troubles, too. Heavy soils have always been known to hold up longer. The clay is the custodian of the mobile nutrients, while the mineral particles of silt size are usually the reserve of them.

3. *good horses from good soil*

It has also been discovered that the plant gets it nourishing elements from the clay by trading hydrogen, or acidity, for them. The clay in turn, as a kind of jobber, trades the acidity to the silt particles, or the sand, for their minerals of original rock nature—if they have any trading stock left. Thus, calcium, magnesium, potassium and other constituents of the plant ash are removed from the original rock form in the silt via the clay or the humus colloid to the plant roots. In the opposite direction there goes the acidity, or hydrogen, from the plant to be taken first by the clay. It is then passed on to the silt to be neutralized by its rock fragments breaking down slowly in consequence. As soils differ in clay content, in humus content, or in mineral reserves of the silt and sand, necessarily the crop growth will be different. As the reserves are exhausted, naturally the kind and the quality of the vegetation will differ.

The feed quality of vegetation is largely a matter of whether it is mainly woodiness or whether it is rich in minerals, proteins and all the accessories, both known and unknown, that the better forage feeds have. Forages like the legume crops are rich in these latter respects. Can it be mere coincidence that where veterinarians so commonly recommend alfalfa hay as a remedy for vitamin deficiency they are calling on a crop that is at the top of the list for its concentration of minerals, and its heavy demand on the soil? Legumes are universally accepted as effective feed for colts and other growing animals. White clovers in the bluegrass pastures are evidence of lime delivery to the crop by the soil. Can there be any connection between the fact that clovers grow only on soils rich in lime and phosphorus, and the fact that clovers are "growth" foods? This is a question that more horsemen might raise when at last they realize that the Dutch clovers have not been seen in their pastures for so many years.

The lime and phosphate soils of Tennessee and the bluegrass region of Kentucky don't mean fine horses merely because of idiosyncrasies of the people of that area. Can it be a coincidence that the winners of Churchill Downs are seldom imports into that region? Such things suggest that we can connect good soils and good horses with a good likelihood that the former is the cause of the latter. Success in growing colts cannot be divorced from lime, phosphate and generally good fertility in the soil. Quality of forage is more than a trademark stamped on the package. It must be grown into the goods by way of the soil.

4. nature posts warnings

As soils are more weathered in consequence of their location in heavier rainfalls or higher temperatures, or have been more heavily cropped, they are correspondingly more depleted of their phosphorus, lime and other minerals. As this depletion occurs, the crops on them shift to those of less growth value and mainly of a woody nature or of fuel value.

Nature has demonstrated this fact with the forests located on soils low in mineral store. Woods or timber are the last stand by vegetation against the flow of fertility to the sea. Wildlife in the forests is scant, because the minerals in the forage are scant. The pregnant timber squirrel carrying bones is no unusual observation. Well-gnawed or teeth-marked fragments of bones in the squirrel's nests would scarcely be considered as dental artifices for keeping the teeth and jaw muscles up to their maximum. Antlers in the woods disappear because of the struggle by animals to get their necessary lime and phosphorus from past animal life when they can't get these from the plant life.

Soils that are depleted, whether by nature or man, mean crops mainly of fuel value and of less help in animal growth. Not only horses, but other animals reflect these conditions in their bone troubles, teeth troubles, reproductive irregularities and alternate breeding when the more exhausted soils provide them with crops of lowered mineral contents. This principle may well be more widely applied.

5. better soil, better bones

Unfortunately, we are a bit late in realizing that the depletion of our soils is the reason for failure to grow white clover and good bluegrass that once were the delight of horsemen. Mechanical genius may have brought in the tractor, but it is going to take more than the diversion of steel to war needs and rationing of tractors to the extent of their elimination to bring good horses back again. Bone blemishes on horses were all too numerous in the cornbelt even before the tractor. Branded bronchos from the lime-laden soils of the west were excellent examples of soundness in bone.

We didn't associate the declining store of fertility in our own soils with increasing curbs, spavins, splints and sidebones. But now that intensive cultivation by tractor, and diminishing amounts of manure and fertility going back to our soils have depleted them to the point where they won't grow cover fast enough to stop erosion, we can't bring horses back merely by economic necessity. If they are to come back economically (for back they must come under present indications) they must do so by way of better soils and fertility restoration in them.

Pasture research is going forward to give us better pastures. Much effort is being put into the search for substitute grazing crops. To date, as most horsemen will agree, there has been none found to take the place of the combination of bluegrass and Dutch clover. The clover goes out with the mineral depletion of the soil. Departure of the clover means that the bluegrass becomes less nutritious. Might it not be possible that depleted soil fertility is the reason why bluegrass quits so much earlier in the summer and doesn't begin until so much later in the fall? Can't we see the increasing need for so much supplementary summer pasture possibly connected with our neglect of putting some fertility in the form of minerals and organic matter back into the soil?

Substitute crops have come because of neglect of the soil. Now that there are more refined means of controlling the soil as it delivers nutrients to the plant, more careful study of plant composition points out that the soybean, for example, recently introduced, has a remarkable "staying power" on soils where other crops passed out.

One needs to be reminded that if soybeans are making two tons of forage where alfalfa made but a half ton, then the soil-given minerals must be diluted by four times as much woody matter in the soybean forage as in the alfalfa. Soybeans have demonstrated experimentally that they may be growing to good height and yet may contain less protein and less phosphorus in the crop than was in the planted seed. They have also been shown to behave in true legume-like fashion when the soils with ample lime and phosphorus, but behaved like woody vegetation when these two were not so amply provided. Here is the explanation of why one might believe them an "acid-tolerant" crop, when in reality they shifted from a legume crop over into a timber crop.

Substitute crops are bringing on increasing numbers of disappointments for animals. Juggling of crops to maintain tonnage per acre is dropping the animals into nutritional troubles. Wide use of calcium gluconate as a remedy points toward needed attention to lime and phosphate, particularly when pregnant animals can't make it through the winter, as acetonemia and other reproductive troubles indicate. We may be trying to "rough them through the winter" but roughages from the less fertile soils are proving too rough. Substitute crops will continue to emphasize the fact that their production of mere mass is hiding the deficiency in soil fertility, which is the real need in the situation.

When the daily mineral requirements of horses are measured in fractions of an ounce while minerals as soil treatments are measured out in pounds, we readily think of mineral mixtures on the drugstore shelf as feed supplements. Even if such mixtures are helpful, this is no proof that they are a complete substitute for these applied as treatments on the soil and all that they bring with them in traveling to the horse's stomach via green or dry forage.

Recent experiments using sheep demonstrate the fact that putting the lime on the soil makes lespedeza hay, for example, a much more efficient growth producer. Liming increased the yield of the lespedeza crop by about 25%. Each pound of limed hay, however, was about 50% more efficient in terms of lamb growth resulting from consuming it. With the animals eating all the hay they could, those eating the hay given proper soil treatment made 50% more gain. Because of better crop yield, and greater growth-producing efficiency of the hay the limed acre was then about 75% more efficient in terms of increase in sheep weight.

That the lime was effective, not wholly because of the nutrient element calcium, and the phosphate not wholly because of the element phosphorus, delivered by these soil treatments is shown by these hays in digestion trials with rabbits. Contrary to expectation, the hay giving the poorer growth rate was the more completely digested. Therefore, the animal machine was handling the vegetable matter to the best of its ability.

Unfortunately, however, the unlimed hay was deficient in something to help the animal build the calcium and phosphorus into its body. These two bone-building essentials in the animals on the poorer hay were being eliminated by way of the urine just twice as fast as from the animals on the more efficient hay. These minerals were digested, but apparently the plants had not worked them into proper combination, or provided the manufactured supplement for their effective service within the body.

6. store minerals not enough

The mere delivery of calcium and phosphorus to the digestive tract, and a high degree of digestibility of them are apparently not enough. These essential minerals must enter into nutritional service for the plant first if they are to be of nutritional service to the animal. If these are the facts, then drugstore minerals shovelled into the feedbox are not the equal in value to those put on the land as soil treatment and as help in the better output of the many complexes from the plant factory.

As the soils become poorer for certain crops and as substitutes are used, these substitute crops tend to become mere mineral haulers. Unfortunately, the minerals they deliver consist of silica with no feed value, in place of calcium, phosphorus and all else of nutritive value that comes with them. An unbalanced plant diet offered by the soil cannot be offset by minerals added to the vegetable bulk used as feed, any more than wheat straw would be good feed when supplemented by salt-peter, limestone and bonemeal. Synthetic diets at best leave much to be desired before they will be equal to the spring growth of the forage in bluegrass-white clover pastures on fertile soils.

Body processes of horses are not such simple performances. Neither are the processes of plant growth. When calcium in animal ash is 40 times as concentrated as is the mobile calcium in the soil, and phosphorus similarly more than a hundred times, we may expect the animal to be in trouble when compelled to eat herbage getting little of these essentials from the soil. Animals know their forages so well that even a blind horse, according to Doctor Dodds of Ohio State University, will graze to the line of the soil treatments represented by only a few hundred pounds of fertilizer.

We might then expect that the thousands of pounds of fertility hauled off through the years of farming are a decided disturbance in animal behaviors. In place of going to the drugstore for mineral supplements, it would seem better to let the animals make their selection via plants from a liberal variety of them in the form of fertilizers put back on the soil. Animal production is not wholly a matter of short cuts and economics, but a most discerning cooperative effort on our part in the complex performance of nature.

Fortunately, the cornbelt and much of the United States were blessed with good soils, particularly good for horses as pre-tractor days demonstrated. They will be good soils for horses again if we treat our pastures with the proper mineral fertilizers to restore white clover, the bluegrass fertilizing legume. Fertility depletion during the youth period of Americanism toward our soil need not prohibit our handling it from this day forth with the maturer judgment of American adulthood apparently about to arrive. We can hold our soils at the present level, and even build back.

Horses can help us in this program and guide it by their help in their more refined assay of the mineral nutrient levels. Short-lived animals suffering early nutritional irregularities reveal soil troubles quickly. Meat producers like sheep are more responsive than hogs, of which the fat constitutes most of their bulk and hides the trouble more readily. When it comes to the long-lived horses their troubles more slowly accumulate and longer remain hidden as minor defects. For these hidden faults, prevention is better than antidotes or cures. For prevention there is no better means than good feed of animal choice and collection from fertile soils. For horses, as for humans, the way to be healthy is to be well-fed.

Just as good horses are supported by fertile soils, likewise good horses in turn support the fertility of the soil. Horse power is merely the sun's energy released on the farm right where it was collected by means of the soil minerals or soil fertility.

None of this needs to be exported from the farm to pay for horse power as is done for liquid fuel and lubrication. In using horse power, the soil fertility is merely in rotation from the field to the barn and back to the soil. As the amount in this cycle of rotation becomes larger, the yields of crops on the farm go up.

Horses as addition to other livestock can be significant additions to our efforts in soil conservation in a larger way. As more of the minerals mobilized out of the soil into organic matter combination go back to the soil, these are more nearly good growth food for the soil bacteria. This puts "life" into the soil, which with more grass pastures will win its way back toward virgin condition. It was—and will be—on such mineral-rich and humus-rich soils that our lands will do much to conserve themselves. It will be because they can take the rain water and because they have the fertility to grow their own cover to prevent erosion. It was—and will be—on such soils that the production of sound horses need not alarm us much about the necessity of supplying them extra minerals.

30.

Soils, Nutrition and Animal Health *is another restatement of what might well be called the Albrecht Thesis—climate, soils, plants, animals, man. It appeared in the* Journal of the American Society of Farm Managers and Rural Appraisers, *1956.*

Soils, Nutrition and Animal Health

The subject for discussion, namely, *Soils, Nutrition and Animal Health,* as requested for this occasion, raises the question whether the order of habitual reading from left to right might not be reversed to let us consider the order of this title from right to left. It would seem more logical—even though not habitual—for us to think of animal health as it is related to animal nutrition; then of the latter as it is related to plant nutrition; and finally of the relation of the nutrition of both of these forms of life to the soil as their foundation. This reversed order of thinking from right to left is a case of post-mortem. It is suggested because we are more practiced and more experienced in that mental procedure in agriculture than we are in predictions, starting with the soil, building upward from its fertility to establish agriculture of extensive and healthy crops, and thereby producing many and healthy livestock numbers. So in order to be habitual in our thinking, mainly in terms of post-mortems rather than in prophesies starting from the soil, perhaps we may well begin with the post-mortems. Possibly then we shall find our thinking coming around eventually to the prophecies and predictions.

1. past experience, a good teacher

A historical glimpse of agriculture in the old world may be suggestive. The young European farmer, planning the beginning of his farming career starts with the questions, "How many animals can I assemble with the funds I have?" "How many can I keep on that land area I have in prospect in order to know the amount of manure that can be made?" "Therefore, knowing the number of animals, I know then the size of the farm I can manage and maintain with that annual turnover of fertility in the manure." In his concern about the management of the farm, he is reading the items of our title, not as a chain from either left to right or from right to left. Instead, he is considering them as an endless cycle of turnover of soil fertility in connection with healthy animals. If we take that viewpoint, as a challenge to our thinking here, we shall approach this discussion inductively from left to right. But if we start with a post-mortem concept, we approach it deductively by starting with the animals (mainly sick ones), reasoning from them back to the plants or the crops, and from those back to the soil fertility. By that deductive procedure, which has been much of our post-mortem pattern of national thinking, we shall then see the wisdom of future productive thinking which starts with the soil as a determiner of potentialities of our farm management by the individual farmer and thereby of all farmers collectively, or the national result in summation.

Since folks connected with agriculture are emphasizing more and more the troubles in growing healthy animals, these irregularities suggest the dwindling agricultural potential in our livestock. They point to the breakdown in the animals or in the major means of uplift of the monetary returns from farm management. (Some years ago, Professor Mosher, University of Illinois, Department of Farm Management, gave the following factors and their relative values in determining farm income, as the summarized result of the study of many cornbelt farms. Crop yields—1000; Kind of Crop—700; Feeding and Managing of Livestock—1000; Utilization of Machinery—160; Utilization of Labor—160; Size of the Farm—00. Of the total 3270 points, the first three, or nature's management as the biological phase, composed 2700 points or 89%; while man's management composed 320 points, or but 11%. This emphasizes the health and nutrition of plants and animals and their connection with the soil as the major means of agricultural production, or of earnings.) They point also to the threatening shortage in our choice foods, which are, the meat, the milk and the cheese, or the costly proteins. Since the starchy, or carbohydrate, crops and the animal, or vegetable, fats, as the energy and caloric foods, represent surpluses ever since the introduction of economic controls, it may be well to follow the deductive procedure or the post-mortem approach for our discussion. By considering our animals and crops post mortem-wise in the approach to the soil, the consideration of it as the foundation of the entire biotic pyramid might be means of a more predictive or prophetic management of agriculture in the future to a degree of success far beyond that of management by means of the many post mortems of the past with the soil paying the ever-increasing cost of those diagnostic processings of farm management failures.

2. foods and feeds may fill but not necessarily nourish animals

In connection with the production of our farm animals, the economics of the procedure have too often had the first and the only consideration. We have not yet come to view livestock management from its biological base. Feeding trials, reported in the many printed volumes, are given to one objective, namely, making cheap gains or cheap gallons. What price the animal pays in health and reproduction has not yet had thought. When fats, as fuel foods, either on the carcass or in the pail, have been the major objective it is not unexpectable that calories as measures of heat should have long been the major criterion of nutritional effort. Naturally for fuel objectives, the fuel foods, i.e., the carbohydrates, would take major concern. Consequently, the carbohydrates have been at the head of the foods list. In that thought pattern, the building of the animal body was taken for granted and only fuel to stoke it was sought. Proteins have been second on the list. We speak of them as "purchased" protein supplements rather than concern ourselves with protein-rich feeds grown where the animal is. Minerals have had similar attention as the inorganic elements in the ration mixture. This kept them as a part of the ash rather than as part of the organic compounds which contained them or the synthesis of which they prompted. Proteins and the inorganic elements have not commonly been considered together for their close association in the growth of both the animals and the plants. The proteins have been considered only as "crude" proteins. They have been classified as those compounds containing any forms of nitrogen in total to about 16%. It is in this lack of complete understanding of the protein compounds, of their functions in our bodies, and of the services by the inorganic elements connected with them that much of the irregularity in being well fed and healthy arises, whether that be in animals or man. Under no fuller knowledge than this, the feeding becomes mainly a matter of filling with a collection of foods in general, rather than a matter of nutrition with health as the specific function and purpose of what is consumed.

When gains in weight of the animal, consisting mainly of more fat and more water, represent possible sale of the entire animal weight at higher price—and then usually of a castrated male—we are apt to lose sight of the animal health involved. Does a fattened, show-animal suggest buoyant health? Doesn't it suggest the very absence of it? Feeding operations on such a score and purpose ought not to be classified as animal production. Rather, they seem to be a case of mere speculation in a culinary excuse for buying low and selling high. Health would scarcely be an expected associate when the feeding performance suggests its necessity to limit the lifespan of the animal to that of baby beef or of the barrows in the ton-litters, before the animals break down in health disasters under such treatment. One must naturally raise the question whether animal feeding under no more searching criterion than that provided by an ordinary scale is apt to bring good health with an economic margin, or whether it must be bad health and all the hazards and disasters associated with it. Feeding to encourage the building of muscle, to guarantee fecund reproduction, and to protect from the invasion by the microbes or the diseases calls for a more scrutinizing criterion. It calls also for foods and feeds that are more than mere bulk for filling purposes. It demands the appreciation of some physiology, and some comprehension of body functions. It transcends the matters of economics resting on no firmer foundation then simple arithmetic.

3. declining fertility of the soil goes unrecognized when quantity rather than quality rules

The simple fact, that the fertility supplies of the essential nutrient elements in the soil supporting all life have been declining, has not yet been widely comprehended. The dwindling amount of the soil's creative power has encouraged us to search for crop substitutes as soon as a "tried and true" crop indicates its decrease in yield of bulk or of bushels per acre. Rather than rebuild the fertility of the soil to nourish the "tried" crop, we have searched the four corners of the earth for another one to take its place.

By that procedure we have introduced more and more of those crops which are making mainly vegetative bulk but are producing less of real nourishment and nutrition for animals. Those crops have been said to be "hay crops but not seed crops." While juggling the new crops into the particular rotation or the chosen farming scheme, their nutritional quality in protein, and their contents of inorganic essentials, of vitamins and of the other necessary compounds of high value as feed for good health, have been juggled out. By growing more and more of those new crops, the carbohydrates composed mainly of air, water and sunshine are amply produced for fattening services, but proteins are becoming scarcer in the feeds. That is the feed situation bringing about the increasing troubles in the health and reproduction of the animals consuming them.

Crops that create the proteins are considered "hard to grow." The cost of the extra fertility for the soil to nourish them so they can create their own needed health and better chemical compositions for better animal nutrition is side--stepped. A big crop yield, but less protein in it, is thereby produced. While failing to see the declining fertility of the soil responsible for less meat and less milk and poorer reproduction, we are calling for more artificial insemination and other technological procedures looking towards improved breeding more than towards improved feeding as a possible help.

While one generation of us is a sufficient time period to exhaust the fertility of the farm, it is, in most cases, not long enough to convince the owner of a farm of what has happened to the soil. Having never figured the cost of maintaining the fertility of the soil, he is not ready (nor are many of the rest of us) to appreciate the great fact that agricultural products have always been priced under the assumption (or the absence of even any assumption) that what the soil contributes is not a part of the cost of their production.

Depreciation of the soil is not recognized in terms of the income tax question. Only the buildings and the fences are considered as depreciable. Our ignorance of the mineral soil fertility as a nutrient-delivery service still leaves that basic substance as of no value in determining farm earnings, and as of no cost in agricultural production, in the minds of those directing internal revenue procedures. Yet oil wells, coal mines, quarries, and similiar productive mineral resources may be depreciated as much as 15 to 25% of gross income per year. When soil minerals are not minerals for all that unless the political aspect of lobby pressure rather than common sense so classifies them, can there be any other result from exploited soils, abandoned farms, and poor quality in our foods and feeds than the invitation to bad health and poor reproduction in our animals when all the qualities determined by fertile soils are so completely forgotten?

4. animal instincts go unheeded—the plow precedes the cow

One needs only to look at the beef map, or the pork map, of the United States to see that the beef cow has gone west to the soil which the buffalo mapped out for his choice in making bone and brawn, but not necessarily in making fat. Beef cows range, and choose their grazing from soils that once made wheat of higher protein content and also more nutritious grass. The beef cattle grow out west. We fatten them in the east. The buffalo that chose that same western soil area called for no imported protein supplements, no veterinarian, and no mid-wifery helps during parturition. He did not populate our east where the dairy cow in close company with the congested human population is expected to serve as foster mother for that crowd left in the wake of the westward movement on account of soil exploitation.

The buffalo was far ahead of the plow. He was choosing the soil under his feed, rather than worrying over the particular plant species, or the pedigree of some supposedly choice variety of recently imported forage crop. The poor dairy cow has no chance to exercise her unique instinct for selecting the forage of higher nutritional values according to the better soils growing it. It was the plow that took her where she is. The fences confine her and so do the stanchions and the feed mixtures until she is little more than the front end of a milk producing arrangement in which she is a machine for consuming certain alloted daily amounts of feed mixtures according to calculated compulsory delivery of pasteurized gallons of liquid and pounds of fat at the other end of that complicated, highly mechanical rather than biological arrangement. The protein content of the milk, of major value following parturition in the reproductive process that gives occasion for the milk, has been almost completely disregarded and avoided in considering the real values of milk. Seasonal variations in the quantity and the quality of the milk proteins draw little, if any, concern. This body-growing service which the cow intends for her calf has not been guarded for the correspondingly high value of the cow's foster children.

Isn't the protein problem from the animal's viewpoint possibly a large part of the picture when the cow breaks through the fence, or when she searches out certain plants in the pasture and eats them shorter while she lets others grow taller? Are not her instincts given to building her to produce proteins, too, and not just fats? When once we think about more protein for healthy boys and girls as well as for calves, rather than just bottles of milk and pounds of fat for sale, we shall be compelled to think of the complex soil fertility required under the cow in making the former rather than just rainfall and sunshine above her giving us the latter. Apparently only some necessity compelling us to think, some threat of disaster, or some disaster itself will make us appreciate our natural resource, namely, the soil, which we have too long taken for granted.

5. *soil fertility pattern under patterns of animal distribution suggests better health via more and better proteins*

With meat proteins, milk proteins and vegetable proteins now coming into national concern because some few folks are recognizing the fact that our national resources producing them are dwindling, we are compelled by animal troubles to center our thinking on this one food requirement, namely, our needs for proteins, and the problem of growing them. When these requisites in our foods and feeds are not manufactured, but are assembled by our animals only as the plants which they eat have synthesized them from the elements of soil fertility, air, water and sunshine, we may well see that the fertility supplies in the soil mark the possible protein supplies of our country by way of the crops or the plants which the soils will grow.

A look at the soil fertility map of the United States according to the climatic forces that give increasing soil construction on coming out of our arid west to the mid-continent, then increasing soil destruction in terms of protein potential in going from there to the east and the southeast, helps us to realize that our soil resources are already so low as to make the shortage of proteins our major national problem. While there are the cries for, and hopes in, grass agriculture which is being propagandized so glibly for cover of the soil against its loss by erosion, only a few folks are reminding themselves that one does not get a grass agriculture by mail order and spread it over the farm. It must be nutrition for animals in order to pay its own costs. For that nutritional contribution it must be a balanced ration for the livestock. And it can be that only when the soil fertility is properly balanced nutrition for the grass, a creation of complete proteins rather than merely bales of bulk.

The rainfall in total per year, balanced against evaporation in the west and against leaching in the east, gives soils in the west which are under-developed, and soils in the east that are over-developed for protein production. In the west, there is excess of the alkali and alkaline earths, or an excess of soil neutrality in terms of simple chemistry. In reality, that neutrality represents a deficiency both in soil acidity and in the soil fertility in terms of plant and animal physiology. In the east, there is an excess of soil acidity in terms of simple chemistry, but that also is a deficiency of soil fertility in terms of those same physiologies, namely, those of our plants and animals.

The crop pattern superimposed on that of soil fertility tells us that the Creator Himself was making only wood on the eastern half of the United States. Even for no more than the growth of the starchy grain of corn, the American Indian in New England was compelled to fertilize the corn plant with some fish protein, as the Pilgrim Fathers observed but failed to appreciate fully. Grass, and not forest, prevailed under the Creator's agricultural management in the mid-continent and the west where the buffalo roamed. High-protein, or "hard" horny wheat grew recently on those former "grassy" plains. Credit for the high protein grain of the wheat is still erroneously given to the pedigree of the plant or to the particular wheat variety, because we have not looked deeply enough into plant physiology to see the soil fertility responsible for it. Now that most of the Kansas soils have given us bumper wheat crops to exhaust the fertility, especially the nitrogen, to the point of making "soft" or low-protein grain, we are gradually coming to see that the fertility of the soil was in control of the protein that made quality for nutrition more than just quantity for sale. The many flour mills, closing now that the fertility is gone, give occasion for us to really appreciate what we once had.

Unfortunately, it is to just such soils where denser human populations are now expecting to bring the animal agriculture and a diversified agriculture. Where extensive single crop specialization prevails, it is usually the limiting soil fertility that brings such single crop prevalence. Single cropping is evidence that it is the soil, limited in fertility, that holds the agriculture down to crop specialization. From those areas of single crop agriculture there are commonly numerous life lines reaching out to other areas of higher fertility levels, reaching especially to those areas producing proteins. Into the mid-continent, into its soil fertility for protein, and into its livestock markets, numerous life lines come from all directions. From the area of crop specialization known for its cotton farming in the south, one can see the life lines reaching to the mid-continent while the menus of the hotels down south announce "Kansas City steaks," and point back to that present beef slaughtering center of the United States which was once located much farther east.

As we mined our virgin soil fertility, we naturally moved on west. The beef cow with her limited output of milk went ahead of the plow. Instead of her instincts guiding us to the better soils while she assays them for protein production and delivery of high nutrition, or as she would outline for us the soil fertility pattern for production of that high food value, we have put the plow and other machinery ahead to enslave our cows physiologically, while the significance of that soil fertility in terms of the proper soil management for protein production for our animals has not yet been appreciated.

6. *our gadgets measured increasing soil acidity but missed its reciprocal, the declining soil fertility*

Legume plants have long been the growing animal's choice among forages. Most students of animal nutrition and animal health have been ready to believe that the higher concentrations of proteins, of inorganic elements, and of other food compounds in these nitrogen-fixing forages, have been reasons for the choice. These crops have always been the feed desired most by both the cow and her owner. But with the cow on the highly weathered soils, from which virgin forests were cleared, we discovered that the better legume crops failed to grow on those soils except as the soil fertility was given uplift. With the advent of laboratory gadgets measuring the degree of acidity of the soil, it was soon observed that the increasing degree of soil acidity in nature is associated with more trouble in growing these highly desired forages. Consequently, the erroneous conclusion was drawn that acidity of the soil is bad, since it seemingly prohibits many protein-producing crops from growing and hinders livestock production.

Had we studied the physiology of the plant with emphasis on its biochemistry in place of learning no more soil chemistry than that required to send us out to propagandize laboratory gadgets for sale of limestone as a cheap neutralizer of acids, we could have seen that soil acidity is not a detriment but is in reality an asset. It is the soil acidity that is regularly making mineral nutrients within the rocks and the soil available to plants. When acidity accumulates in the soil naturally to a high degree, the resulting injury to our crop is not occurring because the acidity has come into the soil. This injury results because many of the fertility elements replaced by the advent of that much hydrogen, which is a non-nutrient, have gone out to leave this acidic element as infertility to take their place.

Instead of seeing lime on the soil as beneficial because it provided the nutrient calcium (and magnesium), we saw its benefit in the accompanying carbonate as provided to neutralize the acidity or to get rid of the hydrogen. Simple gadgets measuring acidity should have been accompanied by means of measuring the plant's better physiology making more and better proteins rather than just more yields.

Now that we have made so many originally-acid soils nearly neutral by stocking them heavily with calcium while attempting to drive out all of the hydrogen, we find that those soils so highly loaded with calcium are not more productive than those soils loaded to a corresponding degree with hydrogen as acidity. What is needed to grow nutritious forages in acid soils is the balance of all the nutrient elements rather than only to replace the acidity. Getting rid of the acidity by liming with a carbonate, or an alkali, is not the equivalent of providing the plant with a balanced diet within the soil. Feeding the crop via the soil, not fighting soil acidity, is what is demanded for nutrition of the plant. A little science came in to lead astray the art of agriculture that had long been using lime, as far back as the Romans, to grow better feeds but not to wage a war on soil acidity.

7. limited knowledge is apt to propagandize itself too soon

While fighting soil acidity, during the last two decades, unfortunately there was a delay of just that many years in the progress towards better nutrition of better plants of high protein output and thereby towards better animal health. Fertilizing the soil went into vogue on the basis of no more knowledge than that required for one to get bigger crop yields and bigger monetary gains by this practice. The fundamentals establishing the procedure of fertilizing the soils for better nutrition of all life are not yet common knowledge. Nitrogen, phosphorus and potassium became standard fertilizers on their score of making bigger yields. Even calcium going on as lime did not—and does not yet in the minds of some—classify as a fertilizer. Sulfur, applied to the soil unwittingly in superphosphate and ammonium sulfate, has not been credited with its values in plant nutrition and better animal nourishment. Commercial nitrogen was not used until recently, because we did not know that nitrogen fixation by legumes is often only a hope and not necessarily a realization when those crops are grown. Copper, manganese, zinc, boron, molybdenum, and other trace elements are not yet considered by many minds that are closed to the possible services of these as enzymes or protein promoters in plant and animal nutrition.

With increased yields of vegetative mass as a major ambition and the criterion by which we judge the services from soil treatments using fertilizers, much that results therefrom in the physiology of plants and animals is not commonly observed, much less appreciated. We therefore have not seen the decrease in plant disease, the reduced insect attacks, the better seeds for reproduction, the better health of animals, their more fecund reproduction, and many higher nutritional values in plant and animal products used as food, all resulting as we discover the nutrient deficiencies in the soil, and adopt the methods of correcting them for balanced nutrition and thereby for better plant health and better animal health.

8. *inorganic criteria are insufficient for better health on better soils*

To date, it has been impossible to explain the many plant and animal improvements from soil treatments merely by the bigger yields for the animal's consumption of more, by a higher concentration of ash or mineral elements in the feed, or by changes recognizable after the plant has been ashed for chemical study. But even the differences in the ash from plants grown on soils fertilized differently gave differing digestive results in the studies with an artificial rumen. From increasing experimental evidence, there comes the suggestion that the better soils make not only more but also better proteins. Accordingly, then, we may well look to the more nearly complete array of required amino acids composing the proteins as possibly the nutritional improvement in forages and feeds from fertilized soil. Isn't it possible that the instinct of the animals is directing them to recognize these better proteins when those beasts break through the fence from the fertility-exhausted soils of our field and go out to graze the grass on the still-fertile soils of the highways and the railroad right-of-way? Is not the imbalance of too much nitrogen, or too much crude protein, of the grass growing on the urine-soaked spot just as quickly recognized? Cannot the wild animals and the unhampered domestic ones judge the quality of their feeds in terms of health and reproduction more effectively for their survival than we can do it for them? Do they not carry their search for quality of feed as far as they can, namely, down into the fertility of the soil growing it?

9. *proteins for better nutrition and better protection against disease*

It is only when our soils are better in terms of all the essential fertility elements that they can grow the complete proteins. Just when are proteins complete? This is still an unanswered question. They should be complete as regards all the eight or ten amino acids recognized as required for survival. Some recent research, especially with the trace elements, points out that soil treatments may increase the nutritional value of grains and forages by increase in the concentration of some of the amino acids commonly deficient, like tryptophane and methionine. The use of these trace elements on the soil growing alfalfa and corn, points out by microbiological assay that better soils increase the output of these essential amino acids. Rabbits feeding on the corn, balanced with amino acid supplements, suggested that trace elements function, apparently, through the modification of those amino acid values commonly more deficient in the feeds grown on the less fertile soils.

Magnesium and sulfur, not classified as trace elements, come in for similar effects. Magnesium, applied to correct the soil's shortage indicated by soil test improved the tryptophane content of forages. Sulfur, applied even in the elemental form, increased the amount of methionine. When these amino acids are grown to higher concentrations in the feed, may we not expect those better proteins to result in the animal and human bodies by which there is protection against invasions via the microbes? When the common cold and tuberculosis are invasions via the mucous membrane; when both are considered as breakdown of our defense; and when tuberculosis is "cured" by a high protein diet, is it too much of a stretch of the imagination to theorize that mastitis and brucellosis may be microbial invasions through the mucous membranes of the teat and the vaginal tract respectively? Should we not test that postulate by treating the soil with all possible fertilizers to include the trace elements, and by studying the physiology, blood properties and all other animal manifestiations, in order to learn whether animal health is not related to soil fertility? The blood picture of the cow suggests that it should be a warm-blooded body's reflection of the soil fertility connecting with the blood by way of the paunch carrying the microbes and the plants still very near to the fertility of the soil nourishing them directly and through them the cow indirectly.

Studies so far with animals suggest the truth of the old adage which says, "To be well fed is to be healthy." Experimental evidence now lets us say that some cows must be better fed via the soil if they are to reproduce better, to give more milk and to be healthy with respect to some diseases which the cows now have and for which—because of transmissibility to humans—the cows are about to be innocently slaughtered. Surely such a negative approach by which the cow species would become extinct, ought to be replaced by a positive one looking to better proteins via fertile soils, and to better protection against diseases for the animals so as to keep them living and healthy. Proteins complete for this kind of protection may call for a new degree of completeness not yet regularly associated with this organic food substance which we call proteins, much less associated with the fertility of the soil creating it in the grass or the other crops.

The proteins are slowly being appreciated in terms of the struggle required for their syntheses and their assemblages. Plants are literally struggling for their proteins. They make carbohydrates regularly, and quickly as a weight-gaining performance, but fail often in finding in the soil fertility the helps for converting those carbohydrates into complete proteins and much seed to multiply their species.

Animals struggle for their proteins, too. They can easily put on fat, and add weight, but for the proteins needed in their reproduction they go long distances, search over myriads of kinds of vegetation, and are active from dawn till dark on many of our less fertile soils. But when on good feed grown on better soils, they fill quickly and soon lie down in what we call "contentment," but which is maximum in body physiology in action. Man, too, struggles for his proteins. Unfortunately, he fails to see it as a struggle premised on a similar one by his animals; that, in turn, premised on the struggle by the plants; and that, in its final turn, limited because of the insufficient fertility of the soil. We scarcely indulge in even this simple bit of analytical deduction.

When our best proteins, like those in lean meat, milk and eggs must be assembled and brought to us through that long creative line connecting back to the soil, surely, in this post-mortem approach, we shall finally see that the shortage of proteins, which has much to do with our failure to keep well fed and healthy, is not one of economic quarrels between groups of us. There is, rather, a declining soil fertility under all of us and under all of the lower life forms below us as our nutritional support. Soil conservation is not a fad of the 15% or less of our population classified as farmers. It is a necessity, to a far greater degree, for the 85 or more percent of us classified as urban, and too far removed from the place where the proteins can be created by our own individual management.

When we still have two acres per person in the United States and one acre does well to make only 250 pounds of live beef per year, which dresses out to give 125 pounds of meat and fat, we may see no reason yet for concern about dwindling soil fertility as the shaky foundation of that protein creation. But when we drop to the world level, by taking on world feedership under the guise of world leadership, and cut ourselves back to one acre per person, we cannot have other foods as well as our meat and our milk proteins provided for us so generously even at higher prices. Then, we may be content to call for our daily food allotment of a little more than a bowl of rice. Does it require the experience of that situation for us to realize that healthy nations are protein-eaters because they have ample acreage, because they conserve their soils, and because they keep them fertile?

10. health in the positive via nutrition— not in the negative via drugs

In the discussion so far, we have considered our mental procedures as a kind of post-mortem, working from poor animal and human health back to the plants and the soils. Let us now reverse our order of thinking and substitute induction for deduction. Let us consider agriculture a call for judicious soil management; that management premised on critical study and knowledge of the soil; and thereby estimating, if not predicting, what amounts will be produced rather than an approach to agriculture with economic calculations setting goals for the yields required to match them; then holding post-mortems for failure in quantitative yields to say nothing of failing yields as food quality and real nutrition.

Food for the human has long been two or three times daily as a pleasant experience looking towards satiation, rather than as a careful study leading to good nutrition with good health as the resulting by-product. Health has become less and less a positive matter resulting from ample quality of food to keep us buoyantly active in both body and mind. The same is very true of our animals. Health has become more and more a worry about finding the appropriate drugs and using the hypodermic needle practices under professional guidance to relieve us of the misery of pain and to help escape the chagrin of insufficient health even to work enough to care for ourselves. We are gradually coming to realize that bad nutrition and poor health in any form of life, whether the lower or the higher, can result from deficiencies in the quality of food, or from hidden hungers, even when ample bulk of extensively labelled products may be regularly ingested.

Do we manage, do we contribute, or do we merely console when we hold post mortems? Do we not merely analyze and explain deductively from death as a starting point? Is that a form of creative and productively sound reasoning? Can farm management be a success, if like the prevailing veterinarian philosophy, it deals only with sick animals, and their failing health? Are we not merely clinicians under those circumstances, rather than creators? Can we not see agriculture now in its bad health of the microbes, the plants and the animals suggesting serious subclinical conditions due to declining soil fertility when as Professor Loeffel of Nebraska once remarked, "About every two years we report that we have conquered one new hog disease only to have another one break out on us"? During this past year—in fact, during this last half year—two new hog diseases were announced. In the farm magazine, *Successful Farming,* August, 1955, the following was reported: "Pig raisers can add another disease headache to a list that is already too long. The disease called *hemorhagic interotoxemia* was first found in cattle and lambs. Now the disease is reported to infect pigs in Illinois and is suspected at least in one case in Colorado. The disease is characterized by internal bleeding. . .A vaccine. . .appears to be effective in preventing this disease in pigs."

Then another magazine, *Farm and Ranch,* November, 1955, reported as follows:

"Keratosis—a serious skin disease of hogs in some areas—has been prevented and treated successfully at the University of Wisconsin by adding zinc to the hog ration."

By reasoning from the grave and the morgue backward in the case of human disease, we have offered many explanations but few preventions with little attention to nutrition until the human family has more degenerative diseases than we have ever imagined. In connection with the diseases of our livestock and their cure by drugs rather than their prevention by quality feeds, this deductive reasoning backward from clinical cases has developed the veterinary business to the point where a farmer recently reported that he couldn't afford to call the veterinarian because even if the animal was resurrected from the poor health condition, it would not have enough extra value to warrant the investment in the curative effort by the veterinarian representing a fifty-fifty chance only, or one no better than what was represented by the animal itself without a veterinarian.

We have not yet seen the larger pattern of animals and plants in the ecological setting, or each creative life form of agricultural use in the place where the soil development of its fertility according to the climatic forces would suggest what the nutritional potential for a certain life species is. Curative medical struggles have attempted to prop up the misfits and have encouraged our injudicious transplantings rather than help our introduction for agricultural use of life forms ecologically suited according as the soil fertility and the vegetation in terms of biogenesis of protein which supports them. We need to see animal health in the positive according as the soil and its vegetation will grow the animal to protect itself by means of the proteins it makes, and as those protective proteins consume the invading foreign proteins, whether they be microbes or viruses. We need to catch that vision of growing healthy animals in place of merely recognizing the disease and then calling for a hypodermic insertion of technologically prepared protective serums, antibiotics, etc. The horse takes our typhoid fever and by eating grass builds his antitoxin by the wholesale with no more attention on our part than providing plenty of good grain, hay, grass or good feed. Is it not time to see the declining soil fertility, the substitute crops, the poor nutrition responsible for the disreputable animal health?

Farm management may well believe that in studying disease by this deductive reasoning from post mortems we have brought the "clinical bull into the ecological China shop" as one of our doctors of medicine so nicely put it. We are failing to see our agriculture with its success founded in our growing healthy animals and producing crops, all in their suitable ecological setting. We have failed to see those ecological settings as they represent the levels of soil fertility as animal and plant nutrition complete enough for each of the life forms to be not only well filled but to be also properly nourished. We must manage the soil so that any feeding complies with the requisites of complete nutrition and thereby subscribes in the fullest way to the ancient adage which said, "To be well fed is to be healthy."

Is not the day here when farm managers will use inductive reasoning and start it with the fertility of the soil as their first premise in considering their responsibility in guiding agricultural production most efficiently in cooperation with the laws of nature? Are we so self-satisfied with our present production of starchy, vegetative bulk that we shall continue to be deceived by the boast of our expert agricultural management because we have so many surpluses? Our economic criteria and large yields resting on monetary bases are becoming more and more of a disappointment to farm managers. Are we not about ready to see the many sick animals, the poor results from mating, and other troubles, to say nothing of all the plant diseases, as reasons for farm managers to believe that better animal health and more fecund reproduction by them will come to pass only as better soil management is the first and foremost criterion for better farm management on a biological base? Can we expect nature to do the creative work alone while we apply no more than the economic measures of her services?

If veterinarian clinical procedures must be used because we are dealing mainly with sick animals, why not study the chemical and biochemical blood picture of the cow and then by the help of those biochemical data diagnose the failure or the success of our nutrition of her? From that source, namely, the bloodstream itself, we can have pointed suggestions as to what chemical irregularities in her body originate from chemical irregularities in the fertility of the soil. In my humble opinion, a new day will be ahead in agriculture when we view the health of the crop plants and the health of our animals in the positive via nutrition from the soil up, and view our business of farm management as a participative effort in the Creator's processes in place of viewing them in the negative via drugs and concoctions that cover, relieve and cure, but do not prevent. Creation in agriculture, or production in that industry, starts with a handful of dust. Up to this date, in my humble opinion, our agricultural science has not found much that suggests possible substitutions for the fertility of the soil as a basis for agricultural production of healthy plants, animals and people, and thereby a prosperous agriculture and all dependent on it. A new day in agriculture will come if, and when, we get both our hands and our minds a bit deeper into our soils.

31.

Dr. William A. Albrecht's great rabbit experiments, and the general thesis that dumb animals know more about nutrition than man does, both provide backbone thought for this little paper. Soil and Livestock *appeared in* The Land *in 1943.*

Soil and Livestock

Food bulk is registered by satiation and the relief of hunger. Food quality, when defective, remains unregistered by these means, but gives us the hidden hungers that may be lifetime torments.

These hidden hungers originate in the soil and reach us by way of plants that also suffer hidden hungers. So also animals suffer their hidden hungers, and so humans, in their turn, consuming the products of starved plants and animals, suffer. This whole series of torments is caused by nutrient shortages in the soil. It should be exposed and possibly cured by soil treatment.

Proper nutrition is an enemy of "disease," in plants as in people. Fungus attack on plants, the "damping off" disease, has been demonstrated as related to a hidden hunger for lime or calcium. More recently potato scab has suggested its connection with insufficient calcium in relation to potassium in the soil fertility offered the potato plant. The potato plant demands much calcium in its tops which duplicate red clover in content of this nutrient.

Plant health in the humid temperatures is doubtlessly declining as the soil is declining. As we fail to return manures, fertilizers and nutrients in the equivalent to those taken off in the crops, we are invoking hidden hungers in the plant and encouraging plant diseases.

Man's nomadic habits have covered much of what a life in a limited location might eventually reveal. He has moved to new soil. But now that rubber shortage and gasoline rationing are putting fences about us as we have done to our livestock, our own deficiencies or hidden hungers will lead us more quickly to consider the soil.

1. calcium and phosphorus deficiencies

Calcium and phosphorus deficiencies soon show up in livestock. In some localities animals born in early winter develop rickets by late winter or early spring. Their bones break readily and the pelvic-spine joints separate. The animal "goes down" when the farmer believed "it was doing well."

Disturbed reproduction processes are another consequence, the record of hidden hungers for calcium and phosphorus. Shy breeders among the cattle are increasing.

On some of the less fertile soils farmers are wondering why their cows breed only in alternate years. Is it not suggestive that on many of these same soils the calves show malformations enough to make them less true to breed type? The backbone of a cow reared in such a herd reveals she had sacrificed part of her backbone for fetus production and used the succeeding year to replace her backbone rather than to indulge in another reproductive cycle. (She was originally a regular breeder but became a shy breeder and died all too early in her life.)

Males, too, may lose breeding capacity on deficient forage feeds grown on deficient soils. Male rabbits under experiment with forage grown on calcium deficient soils became impotent. Their litter mates, on feeds grown on soil given calcium, retained their capacity to serve as fathers. When the feeds of these two lots of rabbits were interchanged for the second period of the experiment, the situations were reversed. Those originally made impotent by lime-deficient forage recovered their male potency, while the lot put on forage from lime-deficient soil, lost it.

Reproduction, as a delicate physiological mechanism—if its late arrival and early departure in the individual life cycle is any suggestion—may possibly be disturbed by other soil deficiencies not yet considered dangerous to this process of maintaining the species. Hidden hungers by way of decreasing fertility in the soil may be the quiet force by which species of animal life have become extinct. Nature's warnings allow time for us to heed them against our own extinction if we will look to the soil and conserve its fertility against the forces of exhaustion as well as its body against the forces of erosion.

2. one might well say

One might well say that the moods of Mother Earth are not always constant and kindly toward those creatures whom she nurses. The rates of delivery to the plants of exchangeable nutrients by the soil vary with different parts of the year, or the seasons. Because plant growth stops during the winter season, this need not imply that chemical reactions within the soil cease during the same period. In fact the rapid exhaustion of the exchangeable supply by rapid plant growth should occasion restoration of equilibrium with cessation of plant growth. In the absence of plant growth, nutrients may move into the absorption atmosphere of the colloid to accumulate for ready removal by contact with plant roots during the next growing season.

Spring plowing facilitates these adjustments of concentration of available nutrients by rearranging contacts between colloids and mineral crystals. Consequently, the spring plant growth into this larger supply of fertility and through root extension for its rapid absorption, means a liberal supply of soil nutrients in contrast to the lower rate at which the plant is building its carbohydrates. Further, this means luscious plant growth, of which the lusciousness is contributed by the soil more than by air and sunshine. Such plants are mineral-rich and protein-rich, to say nothing of their contents of vitamins or other growth-supporting essentials.

As the plant growth extends into the summer season, the store of exchangeable nutrients in the soil has already been lowered. At the same time, the increase in sunshine and in temperature bring greater rate of carbohydrate production. Thus, spring growth of plants means high mineral and protein concentration of nutritive value as body construction, but later growth in the summer suggests woodiness and fuel value.

When with each mouthful of luscious grass the animal ingests growth-promoting nutrients assembled by extensive root action through the soil, rather than mere woodiness from an extensive plant top collecting sunshine and fresh air, is it any wonder that the shaggy winter coat of hair is shed, and that a sleek animal condition comes on in early May? Should there be any wonder that animal thriftiness is lessened by August, and shall we accept the oft-given explanation that it is wholly because of the heat and the flies? Some of you may have heard cattle feeders remark, of the steers in the drylot in February, "They ought to go to the market soon, because they are licking themselves." Some even say that this licking behavior is an index of feeding efficiency. It would perhaps be better to view it as a danger sign that recommends sale before disaster comes. Have you ever asked yourself why cattle do not continue to lick themselves, or each other, after they have been on spring pasture for a few weeks on fertile soil?

It may not be out of place to give some theoretical consideration to vitamins as a possibility in connection with this particular animal behavior of licking their own body coverings. Might it not be possible that after a long period of winter on dry feeds of low mineral and vitamin contents that the body needs for vitamin D become greater than the ingested supply? Might not the animals' fatty secretions by the skin become activated in the sunshine with sufficient resulting equivalent of vitamin D which the animal by chance has learned to keep going through the cycle of excretion, activation and ingestion? As a suggestion for this theory, it has been demonstrated that the yolk of the sheep fleece for the fat of the wool is more prominent on sheep fed hays grown on land that was limed and phosphated than on those fed hays grown on soil without lime. Sheep growth per unit of feed consumed was also better by from 25 to 50% on the former. The better body growth and more prominent skin secretions may both be connected with the same physiological improvements connected with the soil treatment. Soil fertility is not flowing into the plant at constant rates at all periods of the season. These differences through the year and in variable soils make for different plant compositions but exert still larger influences on the animals consuming the plants. Naturally, the question implied is this, do the humans escape these subtle forces of the soil?

That seasonal differences in soil fertility exert themselves on animals is not so difficult to understand when we recall that nature has put the birth time of animals so commonly in the spring, or the time when the heavy calcium load of lactation can be met more widely by the luscious vegetation for the survival of the species. Unless the soil supports the vegetation, the vegetation cannot support the animals. Human births are not seasonally concentrated, yet the December (1941) issue of *Human Biology,* reports from a survey of ten thousand students of the University of Cincinnati (birth dates 1904-21, inclusive) that those born in the spring are taller, heavier, and smarter than those born during the summer. Here may be further evidence to the more direct relation of soil fertility to the animal and the human species. Here is human evidence that Mother Nature nourishes more efficiently in the spring than in the winter. The fact that life forms carry the reproductive load of fetus development on the dry feeds in the winter, may make many a maternal animal endure hidden hungers, and even sacrifice itself where some help through better forages on improved fertility of the soil might save.

3. leguminous crops

Leguminous crops have been one of the great feeds to bring breeding animals through the winter in good health and with a generous crop of offspring. They have been fine means of growing young animals with good bone and straight back lines. Farmers all want plenty of legume hays but can't grow them, as they say, because their "soils are sour." Legumes need lime on the soil, to remove what is commonly called "soil acidity." It has only recently been demonstrated that the soil acidity is not injurious of itself. The so-called "plant injury by soil acidity" is largely a matter of deficiency of the plant nutrient, calcium. It is, in truth, a deficiency of many items in soil fertility of which calcium is the one so pronouncedly deficient on the list that confusion has long persisted in our explanation of the benefits from putting limestone on our soils. We have believed that liming gives benefits because the carbonate of calcium removes the acidity. In reality the benefits come about because the calcium carbonate puts the calcium into the soil to satisfy the plant's needs for this nutrient in its body building requirements. After we have come to understand the function of liming, we shall be putting calcium back to the soil and drive out some of the hidden hungers in our livestock. When once we are fully accustomed to putting calcium into the soil to feed our plants and animals, we shall be more ready to do the same for phosphorus and all the elements that we will eventually put back to relieve the animals and ourselves of all these hungers. When only a few hundred pounds per acre of limestone, less than that of phosphate and less than tens of pounds of other nutrients are all that must be put back to make normally healthy plants and the resulting healthy animals, we should shake off our pessimistic views of the problems involved, likewise dispel our fears of the hidden hungers and go to work optimistically in restoring the soil fertility that drives them all away.

Soil acidity has so commonly been considered as a disaster that perhaps you may not be ready to believe that the soil acidity, so prominent in the temperate regions, may in reality be a blessing. The highest degrees of soil acidity occur in regions of moderate temperature and moderate to higher rainfalls. Maximum concentration of human population occurs in those same temperate regions. We have been inclined to believe that disturbed body comfort relative to temperature militates against denser populations in the frigid or the torrid zones. But when human movements like those of our armies are so much "on the stomachs," we must look to their increased chances to be fed there as the cause of concentration of people in the temperate zones. Soil acidity represents these greater possibilities. The very property of the clay enabling it to hold much hydrogen to make it acid, is the same property that enables it to hold many kinds of, and large quantities of, the nutrients for plant production. Soils of the frigid zones do not develop much clay content, or a clay that is able to supply much plant nourishment. Soils of the torrid zone, particularly, the humid tropics, have a clay product formed from such complete mineral break-down that the resulting clay compounds have little holding, or little exchange, capacities. So the acidic property of the clay, and of its associated humus, formed in the temperate zone may be the reason why we live in greater numbers in the temperate zone.

Soil fertility, then, controls the concentration and localizations of the human species within the humid-temperate belts of the world, rather than the necessary mass of clothing (or lack of it) required for human comfort. Our soils have the ability to take up and deliver nutrients if we will manage those nutrient supplies by maintaining them through fertility return to the soil rather than mining the soil continually and then moving on.

I say, then, that soil acidity *in the presence of liberal supplies of soil fertility* is beneficial. High concentrations of people and the food to guarantee them have been supported where soil acidity mobilizes the nutrients into the plants more effectively than under soil neutrality. Recent studies, using spinach as the test crop, have demonstrated that this vegetable took more calcium, more magnesium, and others from the soil when these exchangeable nutrients were accompanied by acidity.

We must not console ourselves too quickly, however, with the belief that soil acidity is always beneficial and that we need to do nothing about it. It is beneficial only when accompanied by nutrients. As these nutrients become exhausted, hydrogen or acidity replaces them, and when acidity is all that remains the plants must suffer starvation.

Such was the situation in our virgin soils. They were acid but fertile because of high humus contents in which fertility had been hoarded by virgin vegetation. These stores have now been mined. With the increasing acidity and the declining crop yields, or legume crop failures, the acidity was considered as the cause of them. It was not the presence of acidity but the absence of fertility, that was the cause. We now know that nutrients in fertilizers put into the soil are taken by plants more effectively in the presence than in the absence of soil acidity. The crusaders against soil acidity may now desert that cause and march under the banner of soil conservation by restoring soil fertility.

4. hidden hungers

That some of the hidden hungers in animals and in plants can be driven away by putting the nutrients, calcium and phosphorus, on the soil as fertilizers has been demonstrated experimentally only recently. Such is true if increased growth on the same amount of nourishment or within the same time period may be considered as absence of serious hunger. Sheep, as test animals fed the same amount of hay and of grain per head per day, gained weight differently according as the soil had been given phosphate, or lime and phosphate. The plants had some of their hidden hungers routed, too, as judged by the different yields per acre. The internal physiology of the plants must also have been changed since they were not widely different in chemical analysis, yet served widely differently in the efficiency with which their offerings in bulk served to be converted into meat as mutton.

Viewed more specifically, if the sheep gains are calculated as coming wholly from the two pounds of hay taken a day, the soybean hay from untreated soil made one pound of gain per 9.7 pounds of hay consumed; that soil given phosphate produced a hay of which 5.4 pounds made one pound of sheep; but when the soil was given lime and phosphate as a means of more properly satisfying the soybean plant hunger, then 4.4 pounds of this hay were all that was needed to give one pound of lamb in return. As an economic consideration of the grain needed to supplement the hay, if one puts all the gain on the basis of grain supplement, the grain requirements per pound of animal gain were 4.2, 2.3 and 1.7 pounds according as combined with the hay from the acid soil without treatment, with only phosphate treatment, or with limestone and phosphate as nutrient returns to the soil, respectively. Even if we should not be moved out of sympathy for the animals that are suffering hidden hungers, there are economic bases that bid fair to have consideration for them. In the last analysis, the most economic animal gains cannot be made by animals enduring hunger, either visible or invisible.

Rabbits have been used extensively as forage eaters to test the hay crop for its dangers as an inducer of hidden hungers, or conversely its efficiency as a producer of healthy animals. This miniature form of livestock is serving as excellent biological assay of the soil's store of plant nutrients. Calcium and phosphorus, in their mobilization in the body from the intestine to the bloodstream, to the bones for storage, or in the reverse direction from the skeletal storage into the blood, the body, a fetus, or milk, manifest their deficiencies and unbalanced relations in terms of enlarged parathyroid glands. Here in these small bits of tissue, the soil fertility shortages as calcium and phosphorus are recorded in the same manner as iodine shortages in food are recorded by goiterous developments of the thyroid glands. Post mortem weights of the parathyroid of the rabbit reveal in some measure the magnitude of the hidden hunger of the animal for calcium and phosphorus. They tell of the torment the animal was enduring while confined by a pen or a fence to the quality of forage feed that itself in turn was confined to the low fertility of the soil.

5. *animal hungers*

Animal hungers for energy-supplying compounds are not so common as they apparently are for the growth-promoting substances, such as protein that contains nitrogen and phosphorus, and for calcium and other elements going into the plant from the soil. Plants reflect the same shortages, in their changed composition and lowered seed yields, within which the nutrient concentration must be constant to guarantee the next plant generation. When these soil given nutrients are denied the animal, it may be even possible that the animals are not using their energy foods efficiently. Growth promoting foods do more than serve as building blocks. Energy producing foods do more than deliver heat and power. They have interrelated effects. There are suggestions that lime and phosphate on the soil may play some role in aiding the animal metabolism to use its energy supplies more efficiently. This is the suggestion by acetonemia in the pregnant cow, by forms of acidosis or incomplete combustion of energy foods, by pregnancy diseases of sheep. When some good green alfalfa hay or an injection into the bloodstream of calcium gluconate represent relief for a case of acidosis in conjunction with the animal's body shortage of calcium and phosphorus in fetus building, the question may well be raised why the failure to burn the energy foods can be temporarily relieved by a blood treatment carrying calcium, and cured by feeding a legume that demands so much calcium and phosphorus in making a vitamin-rich forage of itself. Perhaps the efficient body metabolism of burning energy foods is in some way influenced by these growth promoting foods that are so significant in the soil fertility-plant relations. Further research alone can supply the answers to these questions. Possibly the helps from the soil may make a contribution to their solution.

While the *modus operandi* of our hidden hungers may still be partly unknown, or at least not fully understood, it is consoling to know that much can be done to the nutrient quality of forage feeds for animals and of the vegetables for humans by looking to the soil fertility. Wild animals and our domestic ones, where possible, have exercised unique choices in selecting their herbage with fine degrees of discrimination according to soil treatment. Hogs have demonstrated their choices of the grain according to soil treatments, and it took many years of study and is still a confession to be wrung from hog feeders with difficulty as Professor Evvard has put it, that "a pig will make a hog of himself in less time—if he is given a change—than we can."

Applications of limestone and phosphate to the soil have been pointing themselves out as the main nutrients involved in animal troubles. It has often been demonstrated by choices of many different animals that we need no longer hesitate to see in these two soil treatments a significant help in making the forage feeds from many soils more efficient for animals in growing their own bodies and in producing their offspring. By the time these become common dosages on soils to relieve plant and animal hungers in the humid soil areas, there may follow shortages of other nutrient items as potassium, magnesium and others to bring up hungers again. If their addition to the soil can set in motion the microbial service within the soil, and the manufacturing business within the plant above the soil to give complete foods, certainly this method of preventing hunger by natural routes working from the ground up will be far more simple than searching for drugstore dosages as antidotes against hungers, bad health and impending death. We need to see in the restoration of the soil a means of demonstrating the age-old truth that an ounce of knowledge applied as prevention is worth libraries of information serving as the pounds of cure.

Conservation of soil fertility and its regular return to restock the soil to the level where it grows its own cover against erosion before erosion is disastrous, needs to be more than ideas. It must be action out on the soil. Our lands given treatments of lime, phosphate, and, in many instances, complete fertilizers, for one or two tilled crops, must regularly be put up for the rest and recuperation under sod crops with their fertility restoring effects. The metabolism within our soil has been overstimulated. While the soil has been burning itself out, the crop quality has declined to bring on hidden hungers. Judicious attention to the soil for the better understanding of it as a complex biochemical performance, and of the plant as an equally complex phenomenon, can offer hope against what today may seem like starvation by degrees. Soil conservation as a national effort has begun to lead our thinking in that direction. Better nutrition is taking us there also. As we get the fuller knowledge about the soil and the role it plays in making Mother Nature a better nurse of us all, our present prodigality toward the soil will change to a reverent conservation.

32.

Dr. William A. Albrecht often paused to reflect on the scene around him. Although he worked with soils, his range of interest was fantastically wide, touching philosophy, history, all the sciences, education, even religion. When he paused to write Diagnoses or Post Mortems *in 1954, the die had been cast for a changed society. Agriculture was being taken to a world price level in economics, and pushed into diminishing returns technology at the same time. Personally,*

Albrecht's career was coming to an end, the handle for forced retirement being "age," the fact being "he wouldn't go along." Perhaps these things figured in Albrecht's thinking as he posed some rather stark questions.

Can technology save itself? became one of those questions. To support his thinking, Albrecht quoted Friederich Georg Juenger—"Since even the smallest mechanical process consumes more energy than it produces, how could the sum of all these processes create abundance? There can be no talk of riches produced by technology. What really happens is rather a steady, forever growing consumption. It is a ruthless destruction, the like of which the earth has never before seen."

That destruction was starting with the soil as lesser men sought to take a biological procedure, farming, and turn it into an industrial procedure. When Albrecht was head of the Department of Soils at the University of Missouri, he required all students to take a course in formal logic. Logic has a maxim—reducto ad absurdum. *Many of the cherished successes of the industrial society, extended to their logical ultimate, reach the logical nightmare called* reducto ad absurdum. *Perhaps this is what the great scientist was saying when he prepared this paper.*

Diagnoses or Post-Mortems?

"For generations, the conquest of nature has been accepted as man's prerogative. But man is a part of nature, it being his essential environment, and unless he can find his rightful place in it he has poor hope of survival. Man's present behavior often resembles that of an over-successful parasite which, in killing its host, accomplishes its own death.

"Man's environment is the whole natural scene, the earth with its soil, its plants and its animals. In many places these have reached a natural balance which man disturbs at his peril." [So wrote C. L. Boyle in *Mother Earth,* the *Journal of the Soil Association,* England.]

Since July 1, 1958, was the centennial of the presentation before the Linnaean Society in London of the Darwin-Wallace ideas with their emphasis on natural selection and the survival of the fittest through natural adaptation, it is appropriate to note that while the lower life forms fit into their environment to validate the Darwinian postulates, man does not. He disregards the natural laws. He interprets his position at the apex of the biotic pyramid as one of regality with the lower life forms under peonage and in servitude to his dictates. Under his more recently boastful emphasis on economics and technologies for our "high standard of living," he has not only exploited the lower life forms with the threat of their extinction, but he is also mining and wasting the soil to destroy the creative foundation of all life. Like the successful parasite, man is killing not only his organic hosts, but also his inorganic one, namely the soil. He is slowly bringing on threatening death via degeneration (commonly labeled "diseases") of his body while giving the distress call for more monetary support of his research mainly by post mortems rather than diagnoses of the natural errors in his living anywhere on the face of the earth as if all soils could be "Lebensraum."

For those of us concerned with agriculture, the mention of the names like Darwin, Linneus, Walton, Audubon and others as men who truly studied life forms in nature, reminds us that we in agriculture are forgetting to study Mother Nature. Instead, we are devoting our thought to human nature and its behaviors in politics, economics and technologies. Research in agricultural problems fails to refer to the literature reporting scientific works of even a century ago to profit by the diagnostic helps from what was the observations, theories and conclusions of the pioneers. As far back as several millenia ago serious doubt was raised about man's successful fitting of himself into the natural laws and amongst other life forms. Ancient writers raised the question, *Can man save himself?* It is appropriate, then, to raise the larger question whether man can save himself in spite of his ruthless destruction of not only other life forms, originally evolved as living segments supporting him, but also of even the soil as the very foundation of the whole biotic pyramid. Can man survive when the soil is not only being depleted of its creative capacity, the fertility, but is also given poisons—owner protests disregarded—that persist and bring on the death of the soil itself?

1. agriculture forgets biology. . .it emphasizes economics and technology

The present problems in agriculture seem to have arisen because of the disregard of the simple fact that, in the main, agriculture is normally the art of cooperating with the natural behaviors of the many life forms lower than man in the course of evolution. By the natural creation of these simple life forms and their products serving in his support, man and his survival are a consequence and not the creative power, nor the cause. The expanding technology, coupled with economics, in a land of many natural virgin resources, has made the collection of a bounteous living in terms of monetary capital very simple, *e.g.*, by the assembly line of industry dealing with dead materials. Consequently, we have erroneously assumed that agriculture, too, can be wholly an industry. We have assumed that its living forms of plants and animals may be similarly managed with assembly line speed, and economic controls, from nature's raw materials to sales of finished products according to man's economic, industrial and technological planning.

More and more technologies and more and more economic manipulation have crowded into agriculture. The farm today is almost forgotten in its activities as a creative performance starting with a handful of dust into which the warm, moist breath of sunshine, air and water is blown to help that inorganic mass give forth an organic one in the many life forms. The economist points out how, through economic necessity, the acreage of the soil on the farm is becoming less of significance while the buying of necessities and the selling of farm output are becoming the main activities of agriculture. The economist reports also that in the last ten years some 200,000 small farms have been absorbed into an increasing number of larger farms under modern economical management and technological manipulations resulting in less farms needed as he sees it.

The fallacious logic in such thinking about the future of agriculture is readily evident if we will apply one of the age-old tests of sound reasoning known as *reducto ad absurdum.* By carrying such reasoning about prospective extinction of the farmer to the ultimate limit, sound logic points out that through those shifts under economic compulsion, the mere buying and selling would eventually be all there is left to agriculture. The farm as productive and creative soil area would become unnecessary, according to that erroneous contention. The farm would have only site value for business transactions and we might substitute a paved street for it. Then, also, all the farms would be completely absorbed into successively larger and larger ones until there could be only one large farm. That would be managed wholly for buying and selling with the biological services and nature's creative performances of the great out-of-doors deleted entirely. Such reasoning founded on economics that neglects soil and biology reveals its own absurdity for the future of agriculture.

2. can technology save itself?

Technology has demonstrated its success in what is lauded so universally as giving us a "high standard of living." Unwittingly, technology is simultaneously bringing us slowly to realize what the end of technology really is. Again the simile of the successful parasite is called forth.

"Technology, which is an instrument for utilizing the raw material of nature, can move towards its ultimate perfection only by impoverishing nature: the less you have to work with, the more accurate and sharp must be both the machinery used and the technological thinking employed. The zero of nature would be the zero of technology that had reached both its apotheosis and its death. Conversely, technical thinking, as an art of the mind of man, is qualitative and living—but life is foreign to the essence of technology. Hence the full reduction of man to a set of measurable quantities would be the zero of all technics. The zero of human nature would be the zero of technics that had reached both its apotheosis and its death. Thus, the complete technology is a contradiction. It follows, therefore, that as technology approaches its assymptote, it nears its own destruction." [Frederick D. Wilhelmson, *Introduction to English Translations of Friederich Gerog Juenger, The Failure of Technology,* 1949.]

Friederich Georg Juenger has pointed out the significance of technology in relation to man and nature when he wrote the following well chosen words. "The human hand is the tool of all tools, the tool that has created, and now maintains the whole machine tool arsenal." He pointed out further that technology is the antithesis of conservation. "Since even the smallest mechanical process consumes more energy than it produces, how could the sum of all these processes create abundance? There can be no talk of riches produced by technology. What really happens is rather a steady, forever growing consumption. It is a ruthless destruction, the like of which the earth has never before seen. A more and more ruthless destruction of resources is the characteristic of our technology. Only by this destruction can it exist and spread. All theories which overlook this fact are lopsided because they disregard the basic conditions which in the modern world govern production and economics.

"The machine does not create new riches. It consumes existing riches through pillage, that is, in a manner which lacks all rationality even though it employs rational methods of work. . .When economic cries can no longer be overcome by economic means, human hopes turn toward stricter rationalization of technology; the idea of technocracy arises. . .A planned economy goes hand in hand with technology."

It is necessary to point out often that conservation as a philosophy is incompatible with technology and its consequent pressure for economic survival. It is the disregard of that simple fact and of the natural fact that agriculture is dependent on the production of life and of living products first, and on technology and economics, second, which brings on our national inclination toward post mortem thinking. We analyze the failures and deaths rather than take to diagnosis of the natural situations which give the survival of the living through prevention of degeneration and disease rather than through cure. We treat the symptoms, but do not find the cause.

3. natural evolution is displaced by industrial and social revolutions

The term "conservation" calls for emphasis on the continued flow of various life streams under evolutionary guidance by Mother Nature. Extensive conservation of natural resources can scarcely be expected as a sudden accomplishment through revolutionary changes in the past behavior patterns of human nature. Man is now given to much that suggests social revolution. One human segment is concerned about or quarrelling with another one. Natural creative forces are forgotten. Man will scarcely put himself back into the forces of natural evolution to even any minor degree. Nor will he give much hand to the support of it, save as it operates voluntarily through limited numbers of scattered individuals, or extensively only if it becomes part of the social revolution.

The present prominent emphasis on man as social matter of world-size may well be viewed as the consequence of the economics of mass production in industry. Mass production demands mass markets. They, in turn, call for the building of a mass-mindedness for mass sales. As a case in point, human health has long been emphasized as a social and public matter when we use the term ***public health*** with attention to group responsibility more than to the summation of the respective responsibilities of each individual for his own good health, preferably by proper nutrition or even by his own choice of supplementary medication. Bad health because of body degeneration has recently become a pronounced factor in creating mass-mindedness when public water supplies are considered by publicly elected officials as their responsible social area for mass medication. Likewise, the hypodermic needle, operating to bypass the liver as the body's biochemical censor, is treating mass population as a sequel to the mass mental preparation, apt to be called education rather than hysteria. Likewise food sales and supplementary materials in that category also come under mass-mindedness until the degeneration of the human body is exhibiting itself in three phases: namely, failing reproduction with rickets in bastard babies taking on revolutionary social dimensions; failing health in the diminishing ability of bodies to build their own protection, and the resulting allergies, heart failure, defective eyes and teeth, poliomyelitis and others bringing mass movements for economic advantage in businesses like "health insurance"; and failure in proper vigor of body growth, or a perversion of the body's capacity to direct and differentiate cell growth, so well recognized as the phenomenon of cancer.

Such degenerations may well remind us that their beginnings are centered in the disturbed or erratic activities of the body proteins which the human biochemistry assembles from the amino acids offered in the food and processed by the metabolism of them. Economics, technology and mass-mindedness weaken the human section of the biotic pyramid, because they undermine the health and vigor of the animal life, the next and the most significant factor in offering the more complete protein nutrition in support of man. Only slowly are we coming to a post mortem, and then are reasoning in reverse, namely, from the top of the biotic pyramid downward. Only slowly, and after complete diagnosis of the problem, will we see soil exploitation as the basic cause.

The economics of livestock management have been perverting and eliminating the evolutionary chances for the survival of the fittest among our domestic animals. The life streams of these warm blooded bodies have been directed toward the monetary speculation of buying low and selling high. In the economist's concepts, apparently the economic finality of agriculture consists of one that includes only buying and selling. The farm as it feeds the animals is not viewed as animal nutrition for growth, self-protection and reproduction. Instead it is viewed only for fattening, and hydration for the maximum gain in body weight in the shortest possible time. It is not viewed for the health of the beast that would contribute to its extended survival.

The pig may serve as a case for illustrating the weakening health of the animal segment of the biotic pyramid through the shift of the production of this animal from a natural flow of life to an industrial compulsion of it. The hog has been humorously defined as a four-legged, voracious appetite for corn and a high-power for converting carbohydrates into fat. Strangely, there seems to be more voraciousness grown into the hog according as the ration consists more and more of carbohydrate-rich corn and less of the costly—usually purchased—protein supplements. As a consequence of the selection for, and propagation of, more voraciousness, the life streams of the hogs have been perverted into more hog diseases. "There is a new one about every two years," the chief animal husbandryman of one of the midcontinental experiment stations told us in public.

Such economic disasters resulting from that perversion of the animal appetite toward the increase of diseases, are lessened by marketing earlier in the hog's life when ton-litters go to sale in six months. By selling nearer the animal's birthday, one escapes much of the hazard from disease through the long-lasting innate immunities passed on by the mother to the offspring. Such a short lifespan allows the animal, though sickened by the fattening performance that is abetted by castrating and elimination of any of its efforts contributing to species survival, to get to the packer where its premature sacrifice keeps the health irregularities hidden in their sub-clinical stages and prohibits them from reaching the clinical one required for the fullest recognition of bad health. Slaughter is, then, an economic venture in disregard of the animal's poor health under an inspection geared to the clinical cases only while it is unmindful of the sub-clinical and of incipient extinction of this domestic species.

Some recent reports on research in pig nutrition tell us that while a 50-pound pig is becoming a 300-pound one, its fat content is multiplied about 17½ times. But during that same time, the lean portion (the protein) is increased only four times. This simple fact has been the natural biochemical principle undergirding the economic speculation on all fattening activities, but it has also been the reason for the mounting failure of hog production to the point near hog extinction, when protein is the only living tissue and the only compound that can grow, protect, and reproduce. With such perverted animal physiology under the economic speculation of buying, fattening, and selling, we fail to see that while there is the increased enshrouding of every capillary of the blood vessels and every cell with a thickening layer of fat, the cells normally fed by the diffusion of the nutritives from the capillaries to them will become more starved. Their excretionary products will accumulate, since fat hinders the two-way ionic and molecular exchanges between the capillaries and the cells to give hidden hungers and excessive accumulation of metabolic wastes. Such conditions represent a lazy and sick body of our domestic animals. No wild animal chooses to be fattened. Instead each small one aims to become a big one and then exercises its ambitions to make a lot more of little ones. Wild animals choose their own medicine according as the soil grows it, and thereby exemplify better health and survival on their own than our domestic ones do under our management.

The handwriting on the wall reprimanding the perverted evolution of our animals for only economic gains, and predicting extinction of the domestic species, calls attention also to the midgets or dwarfs in the beef cattle, dairy cattle and even horses. These are offspring born with some irregular anatomy, and with a physiology that has lost its power to grow the normal animal. Because of the past emphasis on the male or the bull, for example, as half of the herd naturally the male side of the breeding line was first indicted for elimination and thereby solutions of this damage by the dwarfs. Such thinking about dwarfs forgets that man's management of nature seldom brings into the breed any characters adding to the animal's fitness for survival. Quite to the contrary, man's selection and propagation have been given to bringing about losses of such. We are slowly accepting the postulate that such selection and propagation of species, mainly for economic advantage, may have accentuated successive losses in the generations moving the life stream forward.

Each chromosome and each set of genes of the procreating cells is a protein unit which divides all its parts equally by cell division. Each unit must grow into its corresponding previous size before the cell's next division, if the succession of divisions for growth is to continue. Such growth demands that the protein representing it be supplied in the feed as nutrition of any warm blooded animal. But when we feed mainly carbohydrates to fatten, and supply only the minimum of "crude" protein in the ration, the probabilities are high that deficiencies in nutrition may be hindering the growth of all the genes to their normal size or function. Consequently, the accumulation of deficiencies and imbalances in nutrition may starve out some genes to delete the one or more that might represent the growth potential in which a dwarf is born deficient.

We are slow to see the sins of our omission. We prefer to place the blame somewhere else, even to the point of killing the bull rather than recognizing the necessity for all life forms, including man, to fit into the laws of nature. Those laws, not the animal breeder, guarantee the growth, health and survival of the species. Agriculture is calling for the diagnoses of such disasters as cases of protein deficiency in the lower segments of the biotic pyramid about to topple itself, with man included. We are more prone to destroy the beast when it aborts, when it gives midgets or when it contracts a disease common also to ourselves. Destroying the evidence is apparently a more common practice than diagnosing it to find the cause of the abnormalities.

Wildlife, unhampered by man's selection and propagation according to economic dictates, demonstrates nicely its fitness for the particular soil in its climatic setting. It shows the heavier individuals and the greater fecundity according as the soil pattern demonstrates its higher fertility. Such has been the report of the Conservation Commission of the State of Missouri after they have undergirded their program by the basic concept of all wildlife as just another crop of higher quality in health according as the soil fertility is higher. [See Melvin O. Steen, *Not How Much But How Good, Missouri Conservationist,* January 1955.]

4. low-protein crops invite pests and diseases

Our agricultural crops also illustrate the fact that an evolution of species for speculative economic values only through man's management has increased pests, diseases and extinction, rather than their healthy fecund survival. Careful study of an ecological climax of plants emphasizes the seed's protein supply in the germ as the limited flow in successive crops for their survival and continued creation. Any such ecological plant climax reminds us that its freedom from pests, diseases or weeds, and the dense growth, all occur in the absence of crop rotations, of weed sprays, insect poisons and even of cultivation and fertilizer treatment of the soil. Such a climax is possible only on the soil where the fertility is most accurately balanced to meet the physiological requirements of the particular plant species in question better than of any other. Also, such is possible only because the accumulated fertility annually released by rock decomposition has been taken by the plant, preserved and returned each year in the enlarging annual growth of organic matter returned to the soil to be sustenance through partial decay for the succeeding crops.

Man's failure to maintain such a flow and return to the soil of both inorganic and organic fertility under his crop removal from the soil rather than complete return, has been the quiet force pulling down to a lower and lower level the protein potential of soils with each crop succession. In nature, evolution was a case of construction towards higher complexity through more complicated synthesis of more protein with man as the climax. When man's production of crops depends more on blind faith in survival because of a certain pedigree of the seed than on undergirding the potential crop with nutritional security through the fertile soil, the evolution so managed invites pests, diseases and crop extinction. Depletion of soil fertility cannot mean successive crops of the same protein and nutritional potentials equal to those grown on the soil when first broken out of the virgin sod under a natural plant climax.

Depletion of the soil has reversed the natural evolution which built the climax. It has decreased the chances for survival of the same plants by their high production of proteins, their improved self-protection and fecund reproduction. This fact is established clearly by the decline of protein concentration of corn, more pronouncedly since its general hybridization. In 1911, before hybridization practice was common, the mean concentration of protein in this feed grain was reported as 10.30% for a single grade [W. A. Henry, *Feeds and Feeding,* 1911]. By 1950, the top grade among five, then listed, contained 8.8%, while the lowest had 7.9% [F. B. Morrison, *Feeds and Feeding,* 21st Edition, 1950]. By 1956, among 50 tested corn grains from the outlying experiment fields of the Missouri Experiment Station, one sample of those hybrids reached the low of 5.15% of "crude" protein, or a value just half of what it had been 45 years ago.

Wheat, illustrated by county wide values for protein concentration, has been telling the same story, according to the surveys reported for Kansas by the USDA from 1940 intermittently to 1951. This grain, not under hybridization, has been giving increasing acre yields for that state through which the annual total state yields have approached 300 million bushels. But its protein has dropped from a high of nearly 19% in 1940 to a high of 14% by 1951.

Selection and propagation of corn and wheat, coupled with hybridization of the former, have been proclaimed widely as reasons for their large acre yields, measured, of course, only as bulk and bushels. Little attention has been paid to the declining soil fertility by which the photosynthesis or accumulation of sugars and starches emphasizes itself in that increasing bulk, while the biosynthesis of proteins, which would consume or convert much of those bulky reserves by that process, is decreasing. Hybridization of plants and selection for high yields of vegetative mass represent, for plants, the equal of castration for the animals in that they eliminate the struggle for the survival of the species. Instead, it gives a physiological pile-up of carbohydrates by which the sunshine absorbed by the plant is used with a reported efficiency of 30%, when for protein production that efficiency is reported at only 3%.

For the plants, the declining soil fertility functions like a kind of fattening procedure and with ratios between the equivalent of fattening and growing that transcend even those for the pig. It is the declining soil fertility, then as it is giving plant values of only fattening potential for animals, that is, undermining the warm blooded segments as well as the plant segments of the biotic pyramid, including animals and man. Less healthy plants degenerating, as it were, in their own physiology may well be inviting diseases and pests rather than that these have become such powerful predators.

5. declining soil fertility brings diseases and pests

That we are pushing crops to the fringes of soil fertility for their survival is indicated by the common farmer report when he says, as an example, "I must get some new seed. . .My oat crop is running out." He is merely reporting that the regular use of some of his own grain as seed for the next crop, while depleting the neglected soil fertility, has demonstrated the extinction of that species. It is showing that it can no longer survive in that soil climate setting. If its own seed will not be its reproduction, shall we not see the advent of a failing physiology because of failing soil fertility, that was formerly protection against diseases and pests under natural survival?

Research at the Missouri Agricultural Experimental Station using the colloidal clay technique to vary the ratios of the different nutrient elements for growing plants has demonstrated that the soil, as balanced or imbalanced plant nutrition, may be the difference respectively between plants free from fungus or subject to serious attack while that difference was accompanied by widely different chemical compositions of the plant substance. It has also demonstrated that when the carefully controlled plant nutrition with nitrogen and calcium—connected with fluctuating protein concentration of the vegetation—was varied in respect to these two elements only, the presence and the absence of the thrips insect was demonstrated accordingly. Still further, the attack on stored corn grain by the lesser grain borer was more pronounced and earlier according to the imbalance of the soil fertility and according as the grain was a hybrid rather than open pollinated.

Here are facts telling us that we have pushed crops to the fringes of soil support and even beyond those for crops that may be still making vegetative growth when imported seed grown on more fertile soil somewhere else is used for the planting. Faith resting on the seed is weakening and eliminating this segment in the biotic pyramid to weaken, via poor nutrition, the segment of the animals directly above the plants and of man at the apex. We are, however, gaining respect for the insect portion above and the microbial segment just below the plant. It is giving hope for the soil as the foundation of all life.

Research into the protein requirements has shown that the insect segment in the biotic pyramid is struggling for its complete proteins just as all life forms are. Now that the separate amino acids can be measured more accurately and that they can be supplied in pure form as items in the insect diet, research has shown that insect reproduction varies widely according to the balance of the separate amino acids in the insect's required suite of them. For instance, the mosquito requires a list of a dozen of the amino acids—among which are the eight required by man—if its egg-laying performance is to hold up at the rate of 14,000 eggs per fortnight. This insect can survive on as few as eight of those acids, but by such limitation reduces the egg-laying or even may prohibit it. [See A. O. Lea, J. B. Dimond and D. M. Delong's *Role of Diet in Egg Development by Mosquitoes* (*Aedes algypti*), *Science* 1950.]

The household cricket suggests its struggle for protein also when it is used as an assaying agent for nutritional values of chopped forage of the same plant species grown on separate plots as different treatments of the same initial soil. The same number of nymphs so fed varied in body size, survival number, rate of sexual development, and failure of such, simply because of the variable fertility of the soil growing the forage.

Therefore, insects, too, classify under the category of survival of the fittest. Their sudden ravages suggest a sudden change in the chemical composition of their new crop victim, representing those resulting in a particularly suitable insect diet when formerly the victim's chemical composition was unsuitable for survival of the particular insect.

Since changes in the plant's chemical and biochemical composition result from unappreciated changes in the soil, our neglect of that foundation of all life with its consequent fertility depletion may well be the reason for insect pests on many crops. Should we not approach the insect problem from that line of reasoning and diagnoses before we wage war by means of powerful poisons for insects of which the physiology in terms of protein requirements is not so far different from that of the human body? What can be speedy lethality for the very small insect body may be simply a slower rate of poisoning for the larger human body. Plant destruction may be due to failing plant physiology for which the killing of the insect is only an attack on the symptom and not a removal of the cause.

The differences between no insects and serious ravages by them hangs by, indeed, a slender thread when, for example, the commonly serious attack on corn by the root aphis does not occur on the cornteosinte hybrid growing on the same soil along with the ravaged corn plants. [See W. B. Gernert, *Aphis Immunity of Teosinte-Corn Hybrids, Science,* 1917]. When changes in the plant composition brought about by so little natural plant modification as hybridization in such similar plants will serve as prevention of insect invasions, it seems like fallacious logic to call for cure in the form of wholesale poisoning of the tempted invader with all the dangers to other life forms in the same area, and even to the plant considered for protection.

6. fallacious beliefs in epidemics are correcting themselves. . . microbes are saving us

With respect to the microbial segment just above the soil supporting the entire biotic pyramid, we have long fostered a hatred-and-fear-even-worse-than-such for insects. But more recently, even public health officials have been saying less about "epidemics"; are using the quarantine less; and viewing the situation more as one of weakened, and susceptible patients rather than a case of an all-powerful invading enemy or disease. There is the gradual recognition that epidemics are the result more of hysteria than of proper diagnoses when every extensive disease has its ecological limits; and within the affected areas there are always those who do not take the disease. Epidemiologists are apt to forget that after the presence of the microbe in some carrier has been established, that is not necessarily proof that the microbe is the cause of the epidemic. Contemporaneous association of two phenomena is not necessarily proof of any causal connection. It may be a case of some common cause of both of them.

Serious degeneration of the human body may be caused by nutritional deficiencies due to deficiencies in the soil and in the amount of food; refinements and processing treatments despoiling the food; imbalances in the diet; and additives to the food of drugs and chemicals of poisonous potential not yet measured. We are slow to see the possible invasion by the various agencies, commonly disposing a cadaver—even a prospective one—as due to that victim's behavior bringing on the invasions. Placing the blame on some lower life form and starting a fight on it by powerful poisons suggests our ignorance of the Chinese philosophy which says, "Whenever we strike a blow, such is a confession that we have run out of ideas," especially those we might have had by more careful diagnosis of the basic facts of nature.

The advent of antibiotics reminds us that the life forms as far below us as the soil's cellulose-digesting fungi have natural biochemical protection for us as well as for themselves in those compounds. Also, the antibiotics are produced by those microorganisms only when certain requisites of soil fertility nourish them, as was illustrated for the isolation of *Streptomyces aureofaciens* by its dominance only in the plot under continuous grass some 60 years on Sanborn Field at the Missouri Agricultural Experiment Station. But the tremendous amounts of antibiotics now being used in fighting bacteria point to the desperation with which man is grasping at the self-protecting biotics in the very lowest level of life forms. He has not fostered development of self-protection by his own body through guided nutrition. Instead, he uses hypodermic torment of it with unknowns under *cut-and-try* processes with high economic earnings rather than carefully studied theory based on observation of fundamental principles of nature's performances that guarantee health. Seemingly we are bent on the economics and technologies of treating the body by means which only the life forms as low as the ray fungi can offer. That may be the Frankenstein operating via degeneration. This philosophy of fighting an enemy is contrary to the Darwinian concept of survival of the fittest, namely, by improvement through natural nutrition and not because of poisons scattered over the environment in a fight on diseases and pests removing the unfit.

The lucrative returns via giving relief from bad health have given us many post mortems with clinical conditions of the bodies. But post mortems do not give diagnoses of subclinical ailments of living bodies which are troubles becoming more common in environments containing "smog," 2,4-D, DDT and others of pronounced lethality for lower life forms and of accumulative death potentials for us when present at rates of only a very few pounds per acre. It has taken nearly two or three generations to focus attention on the accumulative disturbances from tobacco, hydrogenated fats, and many other substances of commercial significance only lately geared into health research on a larger scale. While many poisons of the past are being discontinued, the concentration of their use is taking to the most powerful ones, namely, the chlorinated ring-structures or hydrocarbons, especially those of which the beneficial role in nature's great biotic arrangement we do not yet comprehend.

7. we are poisoning the foundation of the biotic pyramid, the soil

One institution in Germany, concerned with the nutritional qualities of the foods we grow and the dangers to health by pesticides has been reporting pesticide residues in vegetables as a consequence of their application to the planted seeds. For black radishes at eight weeks, four of the applied chemicals were found in 0.1 p.p.m. In carrots at 25 weeks, three preparations were present in 0.06 to 0.125 p.p.m., the tolerable limits of 1954. Other radishes, at seven weeks after seeding, had a mean content of four preparations between 0.125 and 0.250 p.p.m., exceeding the tolerable limit. For onions, no residues were found after 25 weeks.

Successive growth weights of rats fed vegetables treated in the customary manner with pesticides and hormones emphasize their damaging effects.

The forerunner of the chlorinated ring structures, or hydrocarbons, commonly listed as DDT, has been accumulating in the soils and retaining its poisonous nature there, when for many organic substances the microbial life in the soil is considered their destruction. But carbon ring structures are even poisons for soil microbes and remain unchanged in the soil to a greater degree than we realize. Not only DDT but its many relatives by other names, including chlordane, aldrin, dialdrin, benzenehexachloride (BHC), endrin, heptachlor, lindane, methoxychlor, TDE, and toxaphene behave similarly in the soil.

Chemical extractions of the soils and bioassays of them by insects where previously treated with some of these pesticides, tell of the seriously poisoned soils with those conditions holding on over many years. Studies in the Department of Entomology of the University of Wisconsin tested turf soils from Cleveland, Ohio in 1955 to which DDT was given as treatment for Japanese beetle in 1945. For applications at the rates of 12.5, 25.0 and 37.5% as pounds per acre in 1945, the amounts recorded in 1955 were 10.9, 14.1 and 17.9% respectively of those applications. For 14 apple and peach orchards "the DDT, recovered from the soil, amounted to 93.5—106 p.p.m. in Indiana, 38.6 p.p.m. in Ohio, 36.6 p.p.m. in

Missouri, and 1.5-38.3 p.p.m. in Michigan. The time periods of the accumulations included one apple orchard for 11 years; two for 9 years; six for 10 years; two peach orchards for 10 years and three from 3-6 years." For crop soils 24 samples listed for DDT in the upper six inch soil layer, 22 contained DDT ranging from 0.38-4.6 p.p.m. "The amounts recovered were 0.53-0.90 p.p.m. in Wisconsin (corn), 0.56-1.0 p.p.m. in Iowa (corn, oats and alfalfa), 0.76-2.22 p.p.m. in the East St. Louis area of Illinois (corn, horseradish, potatoes), 0.6-3.25 p.p.m. in the Rochelle area of Illinois (corn, lima beans, barley and pumpkin), 1.46-4.6 p.p.m. in the DeKalb area of Illinois (corn, peas, barley, lima beans, oats), 0.38-0.55 p.p.m. in North Dakota (potatoes, wheat, flax, barley), and 0.49 p.p.m. in Missouri (potatoes). The average recovery from all 24 crop soils investigated resulted in 15.5% of the total amount applied and 61.2% of the average yearly application. No relationship was established between the amount of DDT found in the soil and the various crops grown on the soils." [See E. P. Lichtenstein, *DDT Accumulation in Mid-Western Orchard and Crop Soils Treated Since 1945, Journal of Economic Enterprises,* 1957].

Some bioassays by means of the common laboratory fruit fly of the Aldrin and Lindane in Soils, also at Wisconsin, were checked against recovery by chemical extraction of these poisons from the soil [C. A. Edwards, S. D. Beck and E. P. Lichtenstein, *Bioassay of Aldrin and Lindane in Soils, Journal of Economic Enterprises,* 1957]· as was done for DDT reported above. Their disappearance period measured was 25 months and was measured for inorganic soils of different textures, namely sand, sandy loam, loam and for the organic ones, namely, muck. The dosage per mortality for both aldrin and lindane was higher as there was more clay in the inorganic soils, and still higher for the organic matter of the muck. Data are given in the table.

Disappearance of lindane from 22 months after treatment

Soil type	Treatment (Pounds per acre)	Chemical Assay p.p.m.	Bioassay p.p.m.	Bioassay Confidence Limits p.p.m.	Bioassay Chem. assay x 100
Muck	10	11.6	2.21	1.87- 2.61	19
	100	118.6	28.4	24.1-33.4	24
Loam	10	1.85	0.5	0.46- 0.56	27
	100	29.8	10.3	9.4-11.7	35
Sandy	10	1.8	1.65	1.47- 1.85	90
Loam	100	29.0	22.2	20.1-24.0	76

It is significant to note that the organic matter of the soil is an absorber of the insecticides to reduce their lethality. With ordinary common soils of low organic

	Treatment (Pounds per acre)	Chemical Assay p.p.m.	Bioassay p.p.m.	Bioassay Confidence Limits p.p.m.	Bioassay Chem. assay x 100
		Disappearance of Aldrin—25 Months			
Muck	20	4.95	2.25	2.08- 2.43	45.5
	200	158.0	57.5	53.2 -62.0	36.2
Loam	20	0.69	1.89	1.75- 2.04	274.0
	200	16.5	28.0	26.0 -30.2	169.5
Sandy	20	0.24	1.25	1.15- 1.36	522.0
Loam	200	22.5	66.0	60.5 -72.0	293.0

matter the absorption is not significant. The percentage of organic matter demonstrating significant protection was much higher than most common soils contain.

Other reports of persistance of these indestructable poisonous compounds in the soil compel us to put them into a different category of most organic substance which "decay in the soil." These chlorinated hydrocarbons do not decay so rapidly but may even be transformed into more poisonous ones.

If then our soils have been depleted in fertility to the point where the crop plants have been pushed to the soil fertility fringes for their survival or are literally "sick plants" at the outset, surely we would not "add insult to injury" by poisoning the soils as additional hindrance to the struggle of the species to survive in spite of us rather than because of us. Can we not see insects as a symptom of the failing crop rather than the cause of it?

8. summary

Consideration of nature's processes in their course of evolution of different species emphasizes them as the positive force which has built up the segments of different life forms, including man, for survival of the healthy and the most fit for each particular climatic setting according as it produced soils of high protein potential for growth, self-protection and fecund reproduction, so well illustrated in ecological climaxes. This was the situation as long as man, also exemplified his behavior according to the laws governing him as simply another biotic form. But after man's shift to his assumption of control of his environment for his advantage, the forces of evolution in the positive, or for survival of the fit, have been thwarted more and more until in each biotic segment we can now read the evidence that many transplanted life forms are misfits in their soil climate setting for the production of the foods by which they can be healthy. This holds true for man, too. In consequence of his faith in his technologies to control the locations and all lives below him, the slow extinction of the biotic pyramid as a whole has gotten underway, including the destruction of the soil itself as the biotic foundation.

The present social revolution is multiplying the hindrances to healthy survival of the various life segments. There is usually hope, however, in the deepest darkness just before dawn. By looking to the ***natural laws and their application*** as helps, ***man may*** rescue some areas to ***re-establish ecological climaxes*** for some of the life forms serving below him and thereby supporting the climax of healthy survival of himself in limited numbers. If such prove to be a vain hope, then the post mortems of these many failing segments in the biotic pyramid will in their summation be the diagnosis telling us of the impending post mortem of the human species itself.

33.

Benzene rings in the food supply, on the crops and on the pastures! Dr. William A. Albrecht both lectured and published on the subject, and he left a legacy agriculture cannot ignore with impunity— his doubts! This short paper on weeds contains great logic—the simple proposition that weeds can survive on less, and that therefore weed proliferation is really related to fertility imbalance in the soil. Poison Weeds or Pamper Crops? *appeared in* New Agriculture, *Spring 1952.*

Poison Weeds or Pamper Crops?

Weeds are commonly considered as a detriment because they are plants that use soil space but do not provide nutrients either to us or to our animals. In pastures weeds are plants that a cow has sense enough to refuse if she can help herself to other more suitable plant growth. The vegetation she eats is not chosen according to its particular species, but rather according to its palatability and the balanced fertility of the spot or area of soil where it grows. Unlike the cow, however, we have not seen the imbalance, or the deficiency, of the soil fertility which still lets some kind of vegetation grow even if failing to make food for us. We have defined a weed as "any plant out of place." It would be more accurate to say: "Weeds are plants making scarcely more than vegetative bulk growth on soils too low in fertility for other kinds of plant growth."

Because of this confusion in definition, we have recently taken it upon ourselves to put those plants, *i.e.*, the weeds, back into place—or out of the wrong place—by calling in either the chemical, or the hormone, sprays. By means of these treatments, the weed (and non-weeds, too) can be killed easily and speedily. Emphasis on the plant species out of place, and our readiness to engage in a fight on some unsuspecting plants, have made us lose sight of the decline in soil fertility that had starved out the food-producing plants but let the non-producer of food, the weeds, still surive. *As the weed plants produce less they can survive on less, at least on less coming from the soil as nutrition for them, for animals and for ourselves.*

1. weeds and low soil fertility

Lowered fertility may bring in weeds; lowered more, it may prohibit even weeds. That such reasoning rests on a logical basis of the soil facts is the testimony of the two classic plots on Sanborn Field at the Missouri Experiment Station. Corn has been grown continuously now [1952] for 62 years on these plots. The entire crop, both grain and stalks, has always been removed. One plot was given six tons of barnyard manure annually. The other was given nothing but corn seed, cultivation, and the opportunity of producing a corn crop, if it could, by the help of the same weather that has been giving almost twice as large a corn yield on the adjoining manured plot.

If these two soils are observed in the spring, ahead of plowing them for another crop in this continuous corn series, one would be apt to conclude prosaically that the use of manure brings in weeds. The "no manure" plot has no weed crop, but it has almost no corn crop either. There is no need to cultivate it to get rid of any weeds. There are seldom any weeds on it now after 62 years of continuous corn. This is true since there is not enough fertility in the soil to keep growing any sprouting weed seed that might have gotten there from being blown in, or brought from somewhere else by other means. Here is a nice clean, weed-free plot. It might win our admiration for the absence of weeds if one were to think no deeper about good soil management and wise farming than just the need to keep the land free of weeds.

On the plot given the six tons of manure each year, there is a heavy growth of weeds to be plowed under every spring. These start coming in soon after the third cultivation of the corn in the summer. They are already a good cover crop against erosion by the time of the autumn rains. They carry over through the winter. They take up the soluble nitrogen from the surface soil in the late summer. This is no harm to the corn after cultivation has killed all the shallow corn roots anyway, and when the corn is feeding by means of roots at much greater depth, and in more regular moisture content, than this upper section of the soil profile represents.

These weeds are nature's way of giving us a self-seeded cover crop to prevent the loss of soluble nitrogen and the damages by sheet erosion. In the face of these extra and beneficial services to the soil given by the weeds, when we neglect to supply the equivalent fertility support to the soil of the unmanured plot, shall we engage in a fight by means of chemical poison sprays on such a service-providing weed crop? Shall we think of weeds as coming into competition with our crops, or shall we think of them as nature's attempt to help improve the soil conditions?

2. new practices should provoke caution

The spraying of chemicals on weeds may well call for some thinking about the nature of the chemical compounds that render this violent death blow to vegetation. It is important to note that the core of the sprays' chemical composition is the benzene, or the ring, structure. It is the compound originating in the coal tar by-products from distilling coal for making coke. Apparently these ring compounds were not destroyed by the microbes active in the anaerobic decay of the woody materials that originally went into the formation of the coal. Such is readily true since bacteria commonly tested in the laboratory do not break down the compounds of benzene or ring structure. In fact, such chemicals usually arrest the microbial processes. They kill the microbes as we know for carbolic acid, or phenol, one of the early ring compounds, was once widely used as an antiseptic.

The wood producing plants are the natural agents creating the compounds with the ring arrangement of the carbon and hydrogen composing them. Lignin of wood, that is, the compound enshrouding the cellulose to make wood of it, is not readily decayed or destroyed in the soil. It accumulates there as humus. Lignin consists almost exclusively of ring structures combined or linked together as many combinations. It is a synthetic output by the vegetation on the soils more highly developed by the climatic forces and on the lowered fertility where wood production is the major crop possibility.

Legume crops, for which a less highly developed and more fertile soil is required, do not produce humus in the soil in such amounts as is the case of woody and non-legume plants. While leguminous crops make cellulose, they do not cover it with much lignin or phenolic compounds to convert it into wood. Lignified cellulose is not so digestible. Not only do bacteria fail to digest the lignin, or ring compounds, but such structures are not broken down during digestion by higher animals either, so far as we know. These synthetic chemical creations by woody plants, growing on the less fertile soils, are in the chemical groups representing little or no food values to microbes or to animals in ways we commonly visualize that foods serve us. Rather they are in the drug group. They are more nearly in the class of poisonous compounds. This can be said of them when they are ingested in significant dosages or amounts equivalent to common foods. Taken in smaller dosages, as we do with many commonly considered poisons, they are stimulants.

It is in this category of stimulants that one of the more commonly used weed killers, namely 2,4-D, is classified. It acts as a growth hormone for plants. It is a substance of which very small amounts serve to increase the speed of the chemical reactions of life, if not to initiate new ones counteracting this drug. It literally makes the plant respire itself to death or burn itself out.

While considering these chemical ring structures in the form of weed killers, acting like terrific stimulants for the plants' body processes, it may not be too far-fetched to remind ourselves that the sex hormones (powerful stimulants) of man and animals are also complex ring compounds. Some of the amino acids, too, have similar chemical arrangements as component parts of their atomic structure. In fact, a small, but required, part of the protein molecule contains this particular arrangement of some of its carbon and hydrogen atoms.

In the light of these facts and of such reasoning, it seems that our own body requires and uses some compounds of the benzene ring structures in its functions. But by what chemical methods it links together the plant-given ring compounds, or transforms them to construct those serving it so specifically in such small amounts, is still unknown. According to present knowledge, the ring structure is not broken while in the body. Compounds of its kind ingested are eliminated as such without chemical destruction of the ring itself. They come in as very small amounts naturally to render their services. In larger amounts they quickly give disastrous effects, as do other poisons.

It is still more significant to recall, in this connection, the fact that experimental skin cancers in mice are produced by repeated applications of coal tar. Cancer was originally named as "chimney sweeps' disease." Cancer of the lips of pipe smokers raises the question of possible causal connections in this case with coal tar distillates from burning tobacco. With ring compounds so disturbing in human and animal physiology and through biochemical ways still unknown but disastrous, it raises some question as to the hazards that might possibly be brought along when they are scattered about so promiscuously as weed-killing chemicals.

3. feed crops rather than fight weeds

If weeds are competitors with our food-producing plants, naturally it makes us think first of getting rid of them. Only after some careful thinking will we take to the idea of tolerating them, within the proper limits. On more careful consideration of the details of soil fertility and plant nutrition, however, it would be far better soil management to consider using not so much and so many chemicals to kill the weeds, as competitors to a weak crop, but more chemicals as fertilizers to increase the soil fertility. By nourishing the desired field crop well, with the proper fertilizers applied deeply and by cultivating it to keep down the weeds until the crop is deeply rooted, then the later growth of weeds would not be serious competition for the fertility. In fact, at that later time, weeds would be conservators of the fertility in the upper soil layer which would not be of use to the corn or other crop plants anyway during that part of the season. The use of more chemicals of fertilizer values, plowed down deeply to nourish our crops well enough and still grow a succeeding weed crop as natural cover, plus soil organic matter, is a sounder practice than the use of more chemicals of the poison class scattered about profusely in a fight on what we call weeds. Instead of trying to cure a failing crop by killing the weeds when deficient fertility is the principal fault, it seems much more logical to practice prevention by strengthening the crop through proper fertilization of the soil. This positive approach tells us that we must pay more attention to the soil and its fertility, for nutrition of the crop, and not depend wholly on drugs for poisoning weeds. Sanborn Field at the Missouri Station is the sage giving this advice after its 62 years of carefully recorded and well-considered experience.

34.

Looking at Nutrition from the Ground Up *required Dr. William A. Albrecht to look at the subject from nature's point of view. As always, the position of each life form in the biotic pyramid figured in the presentation, as did the uncomfortable questions that arose each time man attempted to annihilate life forms that, indeed, supported the apex of that pyramid. This little report was first published in* The Polled Hereford World *magazine, 1964.*

Let's Look at Nutrition. . . From the Ground Up!

When a cow breaks through the fence, you perhaps ask (in one way or another), "What's in her mind?" If you correctly answer that question, there is a mutually beneficial solution to the broken fence problem.

If not, there is conflict between Mother Nature and you. . .conflict in which nature strikes back with a recoil damaging to both you and the cow. For the fence breaking cow more than likely is demonstrating her recognition of the higher quality feed growing "on the other side of the fence." Grass on virgin soil along the highway or railroad right-of-way is better nutrition, in her judgment, than what is growing on depleted soils in her pasture. By such discrimination, animals were surviving naturally and in good health long before domestication.

Careful study of some of the soil and plant facts related to livestock behavior, and the behavior of all other living things, reveals that animals carefully consider their nutrition "from the ground up." And accurate interpretation of such behavior is essential to the progressive stockman.

Organized knowledge of the distribution patterns of different kinds of life on the earth's surface is the science of ecology. The search for reasons why a particular kind of life is in a specific place has yielded a list of possible responsible factors that includes rainfall, temperature, geology, topography, soil acidity and many others. More recently, however, we realized that all factors can be summed up in the word "food" as the major determinant of any ecological pattern. Of the various food components, proteins are the major constituents which life struggles to obtain.

Soil science in its recent remarkable development, with much help from other sciences, has pointed to variations in yield and quality of proteins according not only to the different crops but to any crop according to differences in fertility of the soil growing it. Those differences in soils' capacities to create living proteins in crops and animals result from variations in the climatic forces of rainfall, temperature, topography and other weathering agencies acting on the rocks to produce soil. Consequently, "We are what we are because of where we are" and "We are what we eat" ("Mann ist was er esst"), in the words of a geologist and a German geographer, respectively.

In observing animals, plants and microbes as life forms in that order below man in the biotic pyramid where all rest on the soil as their creative foundation, we need to be reminded that when man came on the scene he was met by each of these forms as healthy "climax crops." By natural evolution, each was probably at its height of (a) growth, (b) self-protection and (c) fecund reproduction. They were discovered and domesticated only because they had achieved natural healthy survival through their own instinctive nutritional struggles. In other words, they had proven fit to survive in a climatic and soil setting responsible for a particular degree of soil development that undergirded their survival with proper nutrition. It was for each a case of ecology at its best.

Unfortunately, we have no "natural climax crops" left as standards of "fitness" and excellence for plants and animals in agriculture. We have transplanted from anywhere to everywhere, and healthy survival has not been among the objectives of such reshuffling. This has brought more life forms into conflict with the natural instincts and struggles that originally kept them healthy. Transplanting, but with neglect of the virgin soil as to its origin and climax, is now gradually resulting in malnutrition and failing health—deficient nutrition from the ground up. Some observations in support of this thesis will be discussed under the headings of interdependence, a natural law, survival depends on selection of foods according to soil fertility and the struggle for proteins and evolution of helpful body organs.

1. interdependence, a natural law. . .

I. Legume Plants and Soil Bacteria. The biotic triumvirate of the soil as foundation, microbes as decomposers and plants as major producers of energy and growth substances emphasizes their own essential interrelations and interdependencies as linkages, cooperations and symbioses by which they are the foundation of all other higher and more complex life strata. But the latter also have interdependencies with microbes and plants, thereby with the soil.

It was as late as 1888 when we first recognized the interdependency, nutritionwise, of legume plants and the nitrogen-fixing soil bacteria in their root nodules as a case of symbiosis or mutual benefit. We discovered years later that legumes are not nitrogen fixers purely because of pedigree. They are such only because of ample soil supplies of their delicately-balanced requisites for many inorganic nutrient elements, of which calcium (not listed among contents of commercial fertilizers) is the foremost.

When the soil failed to supply calcium more abundantly than any of the many other soil borne inorganic nutrients, soybeans when first transplanted to the United States produced much vegetative matter but no seed. Disregarding fertile soils to guarantee reproduction and, thereby, survival of the species, superficial thinkers pronounced this plant immigrant "an excellent hay crop but not a seed crop." They overlooked the undernourished crop's low feed value when the forage plant could not even synthesize the necessary proteins to be mobilized later into seed for reproduction and procreation of the species. It required research by colloidal clay, techniques to demonstrate that such soybean hay crops (roots and tops combined) contained less nitrogen, less phosphorus and less potassium (only ones tested) than were in the planted seeds.

II. Non-Legume Plants and Soil Fungi. The nutritionally advantageous alliance between either non-legume or legume plants and soil microbes (fungi) on soils well stocked with organic matter is not even yet significantly appreciated. Attention has been intensely focused on soil fungi lately because they are the commerically lucrative source of the many antibiotics, death dealers to bacteria. But the bacteria in soil are secondary microbes there. Fungi are the primary ones, more capable of obtaining food for both energy and growth from the woody carbonaceous organic matter of crop residues, the main energy source for all life within the soil. Much like any other higher life form, soil microbes are also struggling for proteins.

Accordingly, since bacteria have narrower carbohydrate-protein ratio requirements than fungi (as we know from manure-making, composting practices and commercial mushroom production), nitrogen will be conserved longer and held in insoluble organic forms rather than in water-soluble forms when there are more highly carbonaceous residues in the soil. While fungi are holding nitrogen in their cellular compounds along with residues of less solubility or wider carbon-nitrogen ratios, bacteria are using organic matter of narrower ratios to make their nitrogen highly soluble. Soil rich in natural nitrogen must also be relatively rich in carbon compared to nitrogen to hold the latter there.

It must also follow axiomatically that the addition of nitrogen salts to the soil works against natural nitrogen conservation in soil organic matter. Soluble nitrogen is used by microbes, both fungi and bacteria, to help build their cellular protein tissues (provided carbon of soil organic matter as food energy is present). But while that process occurs, about twice as much of the combined insoluble carbon in the organic matter is respired or converted into gaseous carbon dioxide that escapes into the air.

III. Microbes Give Nitrogen But First Take It. According to such natural laws, we can understand why 25 years of continuous wheat on Sanborn Field, fertilized by either ammonium sulfate or sodium nitrate only, lowered the totals of both nitrogen and carbon (organic matter) in the soil below amounts in similar soil given no treatment—when in both cases all crops were removed. Simultaneously, the use of 6 tons of manure per acre annually under similar continuous wheat increased both nitrogen and carbon as constituents of organic matter in the soil. We dare not forget that not only plants but all life forms are interdependent in one way or another on soil microbes. All are supported by the soil; but microbes, the lowliest and simplest, always eat at the first sitting.

Just as fungi are more capable than bacteria in decomposing insoluble carbohydrates, allowing bacteria to profit by using secondary fungi products as bacterial nutrition, so we find fungi more capable than bacteria in decomposing rock minerals. Fungi are more powerful in separating the cationic elements like potassium, calcium and magnesium from the anions like silica. Fungi in symbiosis with algae, in the form of the lichen, live on what appears to be clean rock. In this living combination, fungi supply the inorganic essentials and the green algae provide the organic carbohydrates (possibly proteins, too), so this union again supports a distinct flora of bacteria. This is, then, a unique almost cellular or microscopic association of fungi, bacteria and plants (in very close contact) converting rock surfaces into soils to support all three. It is a case of the lowest three strata of the biotic pyramid reduced in concept to symbiosis and interdependence at cellular dimensions. It represents almost twice as much thermo-dynamic potential for chemical mineral decomposition as any plant root contact with rocks.

We apparently have not appreciated the ability of fungi to hydrolyze insoluble cellulose into simpler carbohydrates like sugars, to hydrolyze proteins into available amino acids or to weather rock minerals for plant use, since scientific studies of this order have not seemed potentially as lucrative as many other avenues of organizing natural facts into sciences. Nor have we regarded agriculture as biological performances first and economics second, with the former the cause of the latter.

Instead we have searched for economic opportunities even at the cost of dire conflict with matters biotic. By that view we have created Frankenstein's monsters within biology that are now giving us so-called "health problems." We apparently don't realize that the so-called "dangerous" soil microbes or "germs" are not demons of destruction. . .save as failing health invites their natural acts of premortal decomposition. Instead of recognizing ourselves as responsible for the conflict with nature, we are designing by means of our technologies the most powerful chemical poisons to destroy lower life forms completely. We have most seriously accepted Pasteur, so capable in his public relations during his time, but are not familiar with Bechamp, the scientist on whom the former as publicist depended.

We are trying to destroy the lowest life forms, those next to the soil from which all creation takes off. We are learning that microbes give, but under our seeming ingratitude, we are slow to learn of nature's recoil when microbes also take—and we call it "disease."

IV. Higher Life Forms Dependent on Microbes. Insects, as one of the lower animal strata of the biotic pyramid, are also in symbiosis with microbes when the latter must be harbored internally or be "nursed" by insects. Termites (wood-eating roaches not considered strictly insects though a corresponding low-life form) harbor a number of species of cellulose-digesting protozoa in their highly-developed hind intestine. These roaches depend on the protozoa for their major food supply, while most of the protozoa are dependent on their roach hosts for their restricted habitat and for food of wood particles.

Another group of insects dependent on microbes includes the screw-worm larvae, the sheep maggots of several species and the special ones discovered and used in World War I for cleansing badly-infected wounds, including bones. It has been established that those maggots feed not so much on the infecting bacteria but rather on the protein products resulting from microbial digestion of protein tissue. It is said that the "wounds are cleaned" and the specific fly maggots used do not seriously harm living tissues.

In one experiment, a house fly laid its eggs on wood shavings in the corner of the cage where an experimental cancerous white mouse was regularly urinating and

Carbon and Nitrogen Content of Soils Under Continuous Cropping to Wheat and Timothy
[Sanborn Field, Missouri Agricultural Experiment Station]

Plot Number	Cropping Periods	Crop and Soil Treatment	Carbon %‡	Nitrogen %	Ratio C/N*
		Tilled Soil, Continuous Wheat			
2	1st 25 years	commercial	1.13	0.107	10.5
	2nd 25 years	fertilizer	1.02	0.100	10.3
5	1st 25 years	6 tons manure†	1.52	0.140	10.8
	2nd 25 years	3 tons manure	1.27	0.119	10.6
20	1st 25 years	6 tons manure	1.38	0.145	9.5
	2nd 25 years	ammonium sulfate	1.07	0.081	13.2
30	1st 25 years	6 tons manure	1.61	0.171	9.4
	2nd 25 years	sodium nitrate	1.30	0.094	13.8
		Sod Soil, Continuous Timothy			
23	1st 25 years	no treatment•	1.32	0.141	9.4
	2nd 25 years		1.45	0.135	10.7
22	1st 25 years	6 tons manure	1.69	0.177	9.5
	2nd 25 years		2.04	0.195	10.4

‡Percent of dry soil ‡
**Ratio of carbon to nitrogen in the soil*
†Tons per acre of barnyard manure
•Except for periodic plowing of both plots after plot 23 became foul with weeds.

manure was accumulating so as to result in a composting process. It was observed that the mouse was very cautiously searching out and eating the house fly maggots regularly. Soon thereafter the cancerous tissue of the mouse began to atrophy, with increased activity of the mouse resulting. [This interesting case was reported to Dr. Albrecht in person by Paul O. Sapp, Ashland, Missouri.]

Such natural behavior suggests the stages of dependence and the chain of interrelated struggles for protein building upward to warmblooded bodies by means of the decomposition of organic matter supporting microbial synthesis of their own proteins from the cellulose composted with urine of the mouse, the maggots' required proteins supplied by the microbes and their digested products and those proteins required for the mouse supplied by the house fly maggots via initial urinary body wastes of the mouse in cycle of re-use via that catenation of the several biotic strata. Through them, man at the apex of the pyramid is able to hoist his nutrition from the soil upward.

Like that of the termite, the hind intestine (or the large one) of all higher life forms harbors microbes, particularly bacteria, of many kinds. That holds true for all the strata above plants, including insects, birds, herbivora, carnivora and man. That those microbes render services of nutritional and survival value is a slowly-growing conviction. The "sterilizing" effects of antibiotics taken either orally or hypodermically, often with serious health disturbances, support this fact.

Products of both microbial decomposition and synthesis in the large intestine are currently lending themselves to fuller elucidation and recognition. Those include synthesis of vitamins, enzymes, recycled compounds and other reactions not yet completely cataloged.

V. Microbial Biochemistry Favored in Special Body Structures. The herbivora are a unique case of close connection with soil microbes at both the anterior and posterior ends of the digestive canal. At the anterior end, the true symbiosis of the ruminant with microbes occurs in three alimentary organs especially designed for advanced microbial services before digestion by the true stomach takes place. This arrangement is highly essential for ruminants, which are nourished so extensively by bulky high-cellulose vegetation. The rumen is particularly equipped for microbial digestion and synthetic services by which the microbial substance itself is a very important nutritional factor, digested enroute through the animal and supplying many essential nutrients not initially ingested. Thus the significance of soil fertility is connected with the fore as well as the rear anatomy of the animal in biochemical ways still unknown.

These facts suggest that the ruminant is clearly a warmblooded summation of its creation through a series of all the biosynthetic services from the soil via microbes, plants and the animal itself. Should we not, in consequence, accept the chemical picture of the blood of such an animal as the best index of the fertility of the soil? Should we not expect deficiencies in the latter to be reflected as irregularities in the nutrition and health of the former?

2. *survival depends on selection of foods according to soil fertility*

I. Animals Discriminate Among Compounds Not Elements. For survival, any animal must respect its relation to all other biotic strata on which it either is dependent or with which it is competitive. Evolution tells us that no species can or dares extinguish another completely. Accordingly, an animal must capably assay what it chooses to consume. Since variable soil grows foods of variable quality, the animal's successful discrimination must rest on its highly refined ability as a "connoisseur of soil fertility."

In agricultural practice, then, we can premise better husbandry on our confidence in the cow's choice of feed components and their qualities according to fertility of the soils growing them. Soils must be managed by us as caterers to livestock and not as dictators compelling animals to accommodate themselves to our technologies and economics of simple business transactions.

II. Calcium. Livestock often reach across the fence or graze one certain area in the pasture, telling us that the element calcium is perhaps the first soil fertility requisite for quality forage on soils in humid areas. The importance of calcium in the synthesis of proteins of pulses, clovers and other more nutritious forages (according to the animal's choice of them as supplements to non-legume forages for balancing its diet) was emphasized in the Old World since the early days of the Romans and in the New World since Benjamin Franklin. Unwittingly and for many years, calcium has served to build proteins, though lately it is emphasized mainly for its carbonate serving to reduce the degree (pH) of soil acidity. In both respects, magnesium plays a similar confusing role. . .but less prominently because of the lesser quantities involved.

Animals naturally are not expected to be capable of assaying the ash content of forages. But rather we expect livestock to recognize the organic compounds created from inorganic elements in the soil. Animals in desperation will consume even crushed limestone and other inorganic elements to make up mineral and salt deficiencies in the soil on which they graze. Since calcium plays a major role in the synthesis of required amino acids composing proteins, we are apt to emphasize the individual nutrient element (calcium) of the soil rather than all the synthetic food substances, especially proteins, for which it is responsible.

It has been recently reported [in Melvin Calvin's *From Molecules to Mars,* Bulletin, American Institute of Biological Sciences, 1962] that, by assembly of proper laboratory chemicals under specific conditions, the energy of an electric discharge will synthesize a whole series of requisite amino acids. But it will occur only if calcium, even limestone, is present as the catalyst. The establishment of calcium's role in amino acid synthesis helps us to see the animal's choice as a most important instinct, antedating by eons the knowledge we have of it.

It should then be of no surprise that livestock graze first on the limed portion of a field and neglect that which has not been limed. Nor should it be a mystery why cattle in the southern piney woods will travel for miles to graze along the edge of a cement highway.

III. Magnesium. That animals should discriminate between magnesium and calcium as soil factors responsible for their choice might not readily suggest itself. But such was demonstrated some 15 or more years ago as reported by E. R. Kuck in *How Guernsey Calves Helped Solve a Feed and Crop Fertilization Problem, Better Crops,* December 1946, when a dairyman noticed that his Guernsey calves ate the second coat of plaster, containing magnesium, in a particularly sanitary barn but did not mar the first coat containing the calcium or the lime plaster. Suffering seriously from white scours, the calves (which had to be housed in the autumn before the new barn was completed) discriminated in what might well be considered an act of desperation.

Their observing owner—mentally prepared for this accidental discovery—noticed the damage to the second coat of plaster on the finished stall but none to the first coat in the unfinished one. His analytical mind caught the implication of magnesium missing in the calves' fodder because of magnesium deficiency in the soil growing it. Dosing the calves with magnesium salts and later treating the soil with dolomitic limestone eliminated white scours and calf fatalities.

Magnesium is a divalent alkaline element and a close companion to calcium. It is also the ash element core of some presently known two dozen enzymes (protein-like chemical structures). Accordingly, animals also struggle for life promoting magnesium. This is emphasized even in our thinking of magnesium as an ash element only and as if the animal were discriminating between only the inorganic elements of plant delivery from the soil. We need to see that it is rather a manifestation of needed major synthesis by plants, struggling for organic compounds like the proteins, in which any one of the deficient essential inorganics will lower nutritional quality of the nitrogen carrying compounds apt to be considered as proteins.

IV. Phosphorus. During the early period of increasing use of agricultural limestone and other fertilizing materials, the practice of drilling superphosphate with the wheat-clover seeding brought many reports of farmers observing the livestock taking clover in the part of the field given the phosphate on limed soils but neglecting the same legume where no phosphate was used. Similar observations were reported where rock phosphate was the fertility uplift on unlimed soils.

Acid phosphate supplied not only the required phosphate in such cases but supplied also gypsum, a compound of both calcium and sulfur. Hence, one might believe the animals were choosing in favor of one or both of these last two elements also connected with (or retained in) protein synthesis. But since limestone has already provided calcium and since rock phosphate supplies no sulfur, it was established that phosphorus was the element limiting the nutritional qualities the animals were assaying. Since phosphorus so commonly limits growth, is also connected with the energy-transferring feeds and is becoming more limiting as soils are more depleted of their virgin organic matter by which insoluble phosphorus is made available, animals are more and more emphasizing phosphorus by their discriminations.

V. Sulfur. In similar situations we can recognize livestock's emphasis on deficiencies in sulfur when this element, measured in amounts by analytical procedures like ignition, duplicates phosphorus in soils and crops very closely.

VI. Potassium. Since potassium—a highly soluble, monovalent, alkaline element—is more intercellularly than intracellularly distributed in plant tissue and since within the cell it is found in the vacuole rather than within the living cytoplasm (protein), we would not be so prone to emphasize the animal's common discrimination in favor of forages growing on soils given extra potassium treatments to affect plant protein directly. But in experiments with legume forages on heavily-limed soils, with heavy infestations of root-rot on corn given phosphate but showing low exchangeable potassium by soil test, animals showed decided preference for forage grown with extra potassium.

In this case again, the legumes were improved through possible help in their synthesis of carbohydrates as forerunner compounds in the synthesis of proteins or other essentials preferred by animals rather than just by increased concentrations of inorganic alkaline potassium. Potassium gives indirect support to protein production.

VII. Nitrogen, Only a Symbol of Protein. The element nitrogen is normally a gas. It appears inorganically in combination with hydrogen as the positive ammonium ion and with oxygen as the negative ions of nitrites and nitrates. Organically, its extensive combinations with carbon and hydrogen make up 16% (as a mean) of the living protein tissues and fluids. It is readily transformed analytically and synthetically by microbial life, especially in the soil, from either organic to inorganic forms and vice-versa by many kinds of chemical and biochemical reactions, with its final simplification through oxidation or even by reduction to gas that returns to the atmosphere. In animal metabolism of proteins, its simplification results in its elimination in the form of organic urea.

In the struggle for nitrogen for proteins, only microbes (when amply supplied with energy foods) can combine elemental gaseous nitrogen with carbon and hydrogen to produce proteins. Hence, all higher forms of life, so far as we now know, ultimately depend on the lowly microbes for their combined nitrogen.

The fact that animal choices are determined so generally by proteins more than by other feed components is not evidence that nitrogen is the element characterizing the choice. Rather it is the organic nitrogenous compounds that are responsible. Animal choice is not guided by the ash element nitrogen.

In support of that contention, one needs only to see the spots of lush green grass in a pasture that mark the urinary droppings, which liberally fertilize the grass with nitrogen broken down by soil microbes into either ammonia or nitrate form. But livestock refuse to take this forage so heavily fertilized with nitrogen, though they crop closely around the distinguishing spots. In spite of animals' refusal of that much greener and more abundant growth, there is an increased concentration of nitrogen in it. Though measured by chemists as more "crude" protein and considered quality feed worth eating, it is simply "too crude" for the cow customers to buy.

That ignition analyses do not show us the determiners of animals' choices among the soil borne nutrient elements was shown by making bio-assays and chemical analyses of alfalfa hay grown with different soil treatments. The hay was grown on four plots given:

Animal Assays (Weanling Rabbits) and Chemical Analyses of Alfalfa According to Soil Treatments (Putnam Silt Loam)

Plot No. & Order of Choice	Consumption (ratios)	Chemical Analysis (Percent of dry matter)			
		Nitrogen	Calcium	Magnesium	Phosphorus
3 (1)	25.6	2.627(3)	1.100	.319	.237
1 (2)	22.2	2.523(4)	1.290	.325	.218
6 (3)	9.2	3.024(1)	.959	.256	.294
10 (4)	5.1	2.748(2)	1.260	.236	.299

Plot 3, no nitrogen; plot 1, 60 pounds of nitrogen in the spring and after cutting; plot 6, 100 pounds of ammonium sulfate and 200 pounds superphosphate annually, 60 pounds potassium chloride biannually; plot 10, 100 pounds ammonium sulfate, 200 pounds superphosphate, 60 pounds potassium chloride annually, plus 2 tons of limestone per six years.

1. No treatment, plot No. 3.

2. Nitrogen, 60 pounds per acre in spring and after each cutting, plot No. 1.

3. 100 pounds ammonium sulfate and 200 pounds superphosphate per acre annually plus 60 pounds potassium chloride in alternate years, plot No. 6.

4. 100 pounds ammonium nitrate, 200 pounds superphosphate and 60 pounds potassium chloride per acre annually, plus 2 tons of limestone every six years, plot No. 10.

The four lots of hay were offered for measured consumption by choice as supplements to a single lot of corn. Four weanling rabbits per pen in five pens were used in five trials—the equivalent of trials by 100 animals. The data is assembled in the table.

The test indicated that rabbits do not recommend chemical nitrogen as soil treatment for legume hay. Their consumption (quality evaluation) varies, because of those soil treatments, as widely as five to one. That variation was not correlated with order of nitrogen concentration, hence not with "crude" protein as measured chemically. Nor was it correlated with concentrations of calcium, magnesium or phosphorus. Nor were those correlated with each other. The rabbits simply preferred hay grown from nitrogen supplied by microbial decomposition of soil organic matter and by microbial nitrogen fixation on the roots of legume plants. Since the much larger portion of nitrogen in any legume crop comes from the former source rather than the latter, it is evident that the animals prefer protein nitrogen which comes from soil organic matter rather than from chemical salts. The animals distinguish between protein qualities to such a fine degree that they may separate them according to amounts of their components, *i.e.,* their values in terms of nutritional balance of the required amino acids.

In another similar assay by rabbits, using fescue hay grown with only increasing additions of ammonium nitrate to the soil, the first increment of fertilizer brought about first choice; no treatment was second choice; and, with higher increments applied to the soil, choice dropped more and more below no treatment. Daily total consumption by choice in two trials with this non-legume hay was but 6.1 and 10.5 grams of hay per rabbit. By contrast, in two trials with alfalfa the corresponding figures were 31.1 and 33.9 grams of hay per day per rabbit.

The one plant species was apparently chosen under duress of starvation when the ratios of amounts consumed were as wide as five to one in favor of alfalfa over fescue. The animals separated differences in quality by spreads far wider than any shown by chemical analysis. When animals are such capable connoisseurs of the rations we offer, why not cater to their choices? Why not use them in research to discover the criteria by which animals judge what they want?

3. the struggle for proteins and evolution of helpful body organs

Anatomical and functional designs of the alimentary canal and other organs helping it pass food through for preparation, digestion, absorption, chemical censoring, metabolism, excretion, *etc.,* impress one quickly with the complexities of body organs as they digest and conserve proteins in contrast to the relatively simple task of handling carbohydrates and fats.

I. Lower Life Forms. Anatomical arrangement of the digestive organs of sucking insects vary widely for separating low concentrations of proteins out of the plant saps and juices composing their diets. The protein-poor but carbohydrate-rich solution is not put directly into the stomach, arranged to attack proteins with strong acid (as is also the case with warmblooded bodies). Instead, the liquid diet is shunted first through auxiliary canals and pouch-like structures adept in filtering out the proteins and other nitrogenous materials, while the sugary liquids are moved on for excretion. But the proteins are returned to the major alimentary canal for digestion.

One needs only to cite the common aphid as an illustration, with its sugary excretions collecting like mist on auto windshields or serving the honey bee, in seasons of low nectar flow, with resulting honey of little use save for its fermentations and thereby self-clarified alcoholic solution.

II. Warmblooded Animals. All ruminants, as herbivorous feeders, are particularly unique examples of the modification of the anterior portion of the alimentary canal into three extra pouches for treatment of food before digestion by the true stomach. Increasing acidity in that sequence to the very acid condition of the stomach suggests more complete hydrolysis of proteins into amino acids and increasing rates of many other reactions. It lengthens the incubation time under warmer temperatures for microbial syntheses before treatment of the microbes themselves by strong acid. It favors many other chemical reactions under nearly anaerobic conditions for initial fermentations and other attacks on more stable carbohydrates. That succession of increasingly drastic treatments seems necessary to handle high-cellulose feeds as well as proteins.

The pig and the chicken, habitually close followers for the droppings of cattle, reveal wisdom and unique nutritional values in their choices. These two non-ruminants pay tribute to the microbial synthetic services performed in the ruminants' anatomy but not in theirs. They search out vitamin B_{12} (and possibly other essentials), because ruminants' symbiosis with intestinal microbes synthesizes it. The uniqueness lies in the fact that this vitamin is required in only micro-units. Hence, with bio-assays of that refinement, discriminations between quantities of amino acids should not be an animal endowment beyond our imagination when survival by evolution is considered.

Perhaps a more challenging evolutionary adaptation of body organs for more favorable management of the struggle for proteins is exhibited by the camel. Because it inhabits the deserts, our attention focuses first on shortage of water. But such arid ecological settings forcefully remind us that equally as (or more) hazardous is the shortage of proteins in any vegetation rooted in salt-saturated soils and with it tops in an atmosphere of maximum temperature and near-zero humidity.

That the camel can tolerate severe water depletion of its body by going without water for days is well known. It has been reported [by Knut Schmidt-Nelson in *The Physiology of the Camel, Scientific American,* 1959] that this domestic animal can deplete its body's water content to the extent of a body loss of one-fifth its weight. Then by drinking once it can restore that weight and body appearance to normal in a very short time.

But the camel's metabolism includes a practice of conservation rather than excretion and later intake as is true for water. Urea, as the end product of protein metabolism, is not sent from the liver to the kidneys for excretion. Instead, urea is retained in the system by recycling from the liver back into the rumen. Thereby this metabolite of previously ingested feed is merely the chemical nucleus passed up front again to be built into microbial protein, then later to be digested enroute through the alimentary canal and become urea again for the repeated process.

In this particular case, evolution has given us nature's practice of adding urea to the feed of the ruminant, possible at least under duress of near-starvation. This is done by a simple modification of the anatomy in the form of a vessel from the liver to the paunch. By that method, we believe urea is protected against the rapid changes to ammonium carbonate or to ammonia, carbon dioxide and water as moist urea salt does on atmospheric exposure. The amino nitrogen of the urea would remain linked to the carbon and save the synthetic costs of restoring that connection distinguishing the protein nitrogen and that is so costly in laboratory synthesis.

In the case of the camel, nature has long been feeding urea for maintenance of the body proteins. Much is yet to be learned about what the ruminant herds and flocks of primitive man may have been doing for survival of man in his closer connection, via the animals, with his own nutrition from the ground up.

4. summary

Much is yet to be learned by studying microbial metabolism in its primordial setting, namely, on rocks and within the soil. Nutrition of the microbes, as the first forms of life, started there on simple minerals mixed with microbes consuming their own dead bodies, all as life in single-cell stages. The success of that is well demonstrated by the growth of the lichen on rock faces as a cooperative struggle with algae, fungi and bacteria to begin soil development and its potential for creation of multi-celled bodies.

This discussion emphasized the simple fact that all life is dependent upon (or is in symbiosis with) microbes when the provision of nutritive substances, especially proteins for growth rather than carbohydrates for energy, is considered. The symbiosis so universal in the lower end of every alimentary canal has not occupied much of our thought (so shadowed by "fear of germs"), but it is a requisite for health to which life forms lower than man cling.

Biotic strata other than man are gifted in assaying their food intake according to different plant species and different degrees of rock development into soil on which plants grow. While this has been animals' means of survival by evolution during the ages, we humans are just beginning to recognize the ecological patterns of various strata in the biotic pyramid that reflect the soil as the major factor underlying food quality and, thus, health. According to the degree that such natural facts are accepted as essential truths by which man must also survive, so will the soil be more carefully and completely conserved; and thereby higher quality nutrition for more abundant health of all life will be the result.

35.

Most of the things brought to focus in Hidden Hungers Point to Soil Fertility *appear elsewhere in* The Albrecht Papers. *Still, this little wrap-up seems to belong under the heading of "natural" attention. "Even the microbes, the lowest forms of life within the soil, have their hidden hungers," wrote Dr. William A. Albrecht in this 1947* Victory Farm Forum *presentation. That microbes eat first, and crops eat at the second table is more than an aphorism. It is a fact of agriculture the farmer cannot ignore with impunity.* Hidden Hungers *tells why.*

Hidden Hungers Point to Soil Fertility

Hunger is not a new problem. Next to the sex instinct, it is the principal force driving man and beast into action. It projects one into areas where he had not previously ventured either in body or in mind.

Today we understand hunger as world wide in extent and importance. We are examining deeply enough into it to distinguish its "hidden" forms. We recognize these as due to shortages, not so much in the bulk of the food as in its nutritional qualities. We have not yet been able to tag all the different organic and inorganic compounds that provide these qualities, but have come to believe that many are *grown into* our foods. Consequently, we are thinking about deficiencies in the fertility of the soil as responsible for the failure of food to fully satisfy our body needs.

1. simplest forms of life hunger for fertility of the soil

Hidden hungers are not experienced by man only. Even the microbes, the lowest forms of life within the soil, have their hidden hungers. Organic matter of the soil, which is the source of their energy food, accumulates in some nutrient-mineral deficient, or acid, soils while the microbes literally starve. In the face of abundance, hidden hungers exist for nitrogen, calcium and other elements on the soil fertility list. Under such conditions there is a surplus of bulk and a shortage of protein producing, growth promoting compounds. Consequently, only limited supplies of energy foods serve, and then probably inefficiently.

Sweet clover, fed as a green manure to the soil bacteria, may cause hidden hunger for potassium. While this popular, soil-improving legume grows and feeds ravenously on calcium, it can make bulk despite a meagre supply of potassium. It grows well enough on a pile of crushed limestone suitable for fertilizer use. But it has manufactured little potassium into itself and to satisfy the microbes, decaying it in the soil, should be supplemented by potassium which they must take from the soil. Thus the corn crop, which is expected to benefit from this green manure as a supplier of nitrogen actually is robbed by it of potassium in the process. In such cases, the soil microbes, too, are struggling to cover their hidden hunger.

2. soil microbes pass on their hungers to our crops

Mature sweet clover residues of late summer, and straw left after the combine, plowed under before seeding a wheat crop represent hidden hungers of microbes for nitrogen. Whether these microforms of life so fed may not be suffering hunger for other elements of fertility has not yet been fully established. This soil condition also represents the hunger of the wheat crop for nitrogen, but this one no longer is hidden from us. However, we have not appreciated the fact that the wheat crop "eats at the second table," and that the microbes in their hunger for nitrogen are literally passing the hunger on to the wheat crop.

Since wheat manifests this nitrogen hunger in the autumn or at a season when soil moisture is ample, we have not been so prone to blame the drought for it as we are when corn shows these same symptoms in the summer. That crop's hidden hunger for nitrogen has been too readily interpreted as excessive thirst and consequently the weather—beyond our control—is the scapegoat while we do nothing about the deficiency in fertility. It is important to note that both the corn crop and the soil microbes are well supplied with energy—the one from the sun making photosynthetic compounds and the other from similar but decaying carbonaceous compounds. Both, however, are suffering hidden hungers for small amounts mainly of nitrogen, by which their surplus energy foods can be converted into proteins and their diets properly balanced. It is through difficultly synthesized substances like the proteins that cell multiplication is possible and by which the stream of life is kept flowing and shortages of them really provoke the hidden hungers.

3. fertile soils grow bigger fish

Even the lower forms of green plants, like the plankton in our fish ponds suffer hidden fertility hungers. In turn, the fish with their hunger for *grow foods* in more and better plankton do not multiply or grow so rapidly as when the fish ponds are properly fertilized. One dare not believe, however, if a little fertilizer in the fish pond is good, that more will be better. For then the plankton may have hunger for carbon dioxide, of which only limited amounts can go into solution in the pond water. Curing hidden hungers calls for an understanding of their physiological causes in even so simple a practice as feeding fish.

Wild animals well up in the biological scale have their hidden hungers, too, though the fact is not always associated with the fertility of the soil. Animals that are strictly herbivorous feeders are not commonly found on the highly leached soils of the tropics. Instead, buffaloes, elephants, antelopes and other grass-eating species are fund on the prairies and savannas. They subsist on vegetation produced in areas of lesser rainfall, on calcareous soils, where natural legumes are abundant, and on soils which under cultivation produce the hard, or high protein wheat. Soils given less to production of proteinaceous products and more to vegetation of carbonaceous contents give us forms of wildlife compelled to eat the seeds, the growing tips of branches, and other plant parts representing the maximum concentration of the plant's proteins. The ecological picture of wildlife is a pattern of placement where the animal is not destroyed by hidden hungers for the proteinaceous, mineral-rich foods that favor reproduction more than for those that serve mainly to lay on fat.

The roaming of wild animals, and their ravaging of farm crops, usually connotes an effort to satisfy hidden hungers. In leaving the forest to graze on fertilized land, the deer signals his recognition of better nutritive values in the feed growing there. When they break through the fence dividing the fertility depleted pasture from the virgin soil of the highway or railroad right of way, domestic animals likewise reveal their intuitive recognition of the dependence of the feed for nutritive quality upon the fertility of the soil. They are driven by particular hungers to risk their lives against the barbed wire just as the wild animals risk their lives in coming into the open for feeds grown on better soil.

In pointing out the animal's ability to detect differences in the grazing according to differences in soil fertility—almost beyond the capacity of chemical means of detection—we are apt to think of differences only in the ash constituents. We forget that the animal is not looking to the plants for service as haulers of minerals but rather as synthesizers of the many organic and organo-mineral complexes that build the animal body and supply energy to keep it in action. Some of these complexes have been catalogued as we consider them in making up a ration or a diet. Can we doubt that many yet remain to be listed? Their complete chemical nature and the many kinds of services they perform are still unknown facts. It is the still unlisted complexes that may be the main provokers of the hidden hungers.

4. prevention simpler than cure

Fortunately, we are better able to combat these hungers at the point of origin, namely in the soil, than at any later stage in the agricultural assembly line. At that point, the problem is no more complex, probably, than supplying one or more of a few simple inorganic elements. A little effort there cures the deficiencies that cause the hidden hungers of the soil microbes and the plants. Properly fed plants prevent deficiencies in their synthetic products that serve as animal feeds and human foods. Here are solved the problems of providing the hosts of essential chemical compounds, the required amino acids, the necessary vitamins and the specific fatty acids. These problems of provision in the diet are more nearly insurmountable than those of getting some dozen elements, both major and "minor," applicable as fertilizers on the soil. At any later stage the problem is more complex and the situation more prone to induce the micro-hungers.

Lespedeza hay grown after phosphate application and fed to sheep caused them to grow fleeces that were low in fat or yolk and that scoured out too poorly to be carded except as broken fibers. Yet the same plant species grown on soil given both lime and phosphate helped to grow fleeces of heavy yolk and wool that scoured well and carded out as fibers of good quality for spinning and weaving. Treating the soil to grow good quality wool was as simple as giving the soil some extra fertility in the form of calcium. Just what should have been chosen as the particular supplement to make this deficient lespedeza hay better sheep feed so as to make better wool is a problem not so simply and easily solved. It is clearly a case of hidden hunger, the *cure* of which is extremely perplexing, but the *prevention* of which is as simple as the practice of liming the soil.

In our thinking about diseases, both empirical and scientific knowledge are influencing us to think less about cure and more about prevention by ministrations to sick soil. Once the mind thinks soil fertility, observations come rapidly. Calves eating plaster, not the exposed first coat but the hidden last coat, in a fine barn prompted a farmer to ferret out a magnesium deficiency in his soils. Prompted by curiosity and intelligence to use some magnesium as a fertilizer he started a train of apparent miracles, including the curing of scours in calves, and some reduced mortality, less mastitis in the cows, better alfalfa, better corn, and other blessings in his farming program. When other major and minor elements given the cattle make them negative to the blood test for brucellosis, and when medical research is pointing to similar good suggestions of improvement of undulant fever patients, these are no longer hidden troubles. Attention to the soil fertility, the point of their origin as deficiencies rather than as diseases, is making them major hungers for major attention by more of us than those in the curative professions alone.

5. *by saving our soil we save ourselves*

It can truthfully be said that rapid progress is being made in recognizing hidden hungers. Many of them are now being prevented because they are being diagnosed as originating in our declining soil fertility. Foremost among the gross nutrient factors of serious decline are those connected with the synthesis of proteins by plants. Soil treatments are no longer appreciated only because they encourage production of greater bulk per acre. They are being made on increasing acreages because they add nutritional qualities to relieve the long chain of hidden hungers coming up from the soil through the entire biotic pyramid to torment man at the top.

For better reproduction of farm animals, and for the better health for them and for ourselves as well, we are becoming increasingly concerned to know more about the fertility of the soil as the means by which such good fortune can be guaranteed. The disturbing and perplexing micro-hungers are hidden mainly from our thought, our recognition, and our full appreciation of their origin. They are not hidden from our body physiology nor from our mental processes when as little iodine, for example, as a fraction of a grain coming from the soil up through the plants to us is all that "stands between us and imbecility." It is a good sign for the future that we are coming to realize that our hidden hungers are provoking deficiencies in mind as well as body. We are coming to think about keeping up the soil in order to keep us mentally able to realize that our hidden hungers are pointing to the soil fertility as ready means for their prevention.

36.

Soil and Nutrition *was read before the 11th Annual Meeting of the American Institute of Dental Medicine, Palm Springs, California, November 2, 1954. The political climate for that meeting must be appreciated in order to understand the depth of Dr. William A. Albrecht's thinking. First, the United States had emerged from a great war. The country was, perhaps, a bit uncertain as to the direction of the future. Very few farmers understood the public policies that had shaped their destinies, perhaps the destiny of the nation itself. In economics, the Keynesian Employment Act of 1946 had been signed into law. Farm parity was well into its slide toward a world price level. The temporary income tax withholding law of WWII had been made permanent, thus assuring government an unending stream of funds for bureau expansion. Finally, this bureau expansion had been made certain with the passage of the Administration Procedures Act, that broad-spectrum measure that would henceforth permit government agencies to write laws by the pound, not the page.*

That these several considerations added up to soil exploitation in the mind of Dr. Albrecht came clear as he spoke from the podiums of trade and professional assemblies. That nature would not comply with many of man's ill-considered designs was equally apparent. Albrecht warned that agriculture rested on biology, not on technology and economics. Fond hopes and fountain pen ink didn't make crops grow or nourish man so that he could account for a long series of correct decisions. Albrecht closed this paper by calling for "views of soils as the means for health rather than for only wealth."

Soil and Nutrition

"If we approach the subject of disease from the ecologic viewpoint, we shall be on much sounder ground than if we use the conventional approach and emphasize the immediate rather than the ultimate cause. From this viewpoint disease is the result of a breakdown on the part of the total individual." This is the view of Jonathan Forman, M.D., editor of the *Ohio Medical Journal.*

"While proper exercise, clothing and shelter are necessary for optimal health, the most important factor—far and away—is food. Food may be spoiled in seasoning, preserving, processing, storage or harvesting, but it must in any event have been good food in the first place if it is to produce optimal health."

"Food of high quality can only be raised on soil rich in all the essential nutrients. This is why the health of the nation rests primarily with the farmer and why I, as a physician," continues Dr. Forman, "interested in the health of the people, put so much emphasis on conquering disease by good food raised upon good soil; and so little upon the use of pills to patch up damaged bodies in which, as soon as we conquer one disease, another will straightaway appear."

In a similar vein, Icie G. Macey, Ph.D., emphasizes health based on nutrition in a broader pattern when he says, "The prerequisite to a good, healthy, buoyant and resilient life is the provision of a food supply furnishing all the nutrients necessary for structural and functional activities in an environment permitting full utilization of them."

As these quotations indicate, the curative profession, with the help of scientific research, is gradually giving more attention to nutrition as the foundation of health. It is coming to look at disease as due to present or past deficiencies in nutrition. That we should see nutrition not just as bulk of food, but in terms of both the required inorganic elements and the long and growing list of organic compounds—synthesized by microbes, plants and animals—that are connected with the soil fertility, is a highly gratifying mark of progress. It is testimony that we are putting professional and scientific sanction on the layman's concepts when he says, "Whatever we eat comes out of the soil. Therefore, sick people result from sick soil." We are more and more ready to believe that there is an ecologic pattern of human health and survival just as there is the ecology of plant survival, and animal survival according as soil is complete nutrition for them.

Our dwindling acres in relation to an increasing population portend problems already in providing the necessary quantity of food. However, for only a few minds is the dwindling soil fertility and consequent dwindling quality of food as nutrition for health a serious matter. Therefore it is proposed here to view soil and nutrition as cause and effect in terms of this latter category and under two headings, namely, the evidence from soil exploitation, and the services from soil treatments.

Such a proposal may seem bold when soil science has not yet outlined a complete concept of just how soils feed crop plants and microbes. But that degree of ignorance about soils finds consolation in a corresponding degree of it about nutrition. "Medical science," said Dr. John B. Youmans, Dean of the School of Medicine, Vanderbilt University, "is not certain as yet what constitutes optimum nutrition." Perhaps by taking a good look at nutrition via the soil from which all foods ultimately come, we shall learn more about each of these two categories of national concern in trying to feed our population.

1. evidence from soil exploitation

By considering man in his ecological setting of the world as a whole, over which his technologies of transport have now distributed him, it is high time that we see the exploitation of the soil fertility—the ultimate source of food—as the force that has been compelling him to move westward in the temperate zones; from there toward both the poles and equatorial tropics; or to be extinguished. Food as a basic factor in the ecology of man, as of all life forms, may be viewed in terms of the soil equivalent of it. Food, which is nutrition only via the soil, makes the latter, then the basic factor in the ecological pattern of man on his planet. Man then within larger areas may be expected to reflect in his nutrition—and all else determined by it—the soil that is the ultimate source thereof.

By deductive considerations of man on the larger scale of the soils of the world, we may possibly work out the integrated factors through which soil either feeds him or fails him. These are the rainfall, the temperature and other forces active in the development of soil from the magmatic rock as the productive power for growth of life forms. It is according to the degree of the weathering of the rock that the soil has been in control, even of man. The exploitation of fertile, virgin soils, not extensively developed to that particular degree, has long been a factor working against man's unbridled reproduction and against his survival in one place with the optimum of health there. Now that he is no longer a nomad led by the hunger instincts of his animal flocks assaying the soil, but is limited by fixed agricultural areas, he appreciates the need to manage the fertility of the soil under the fullest knowledge of the physiology of plants, animals and himself to make the soil his as well as their most complete nutrition.

Under no self-restraint, man multiplies his own numbers geometrically. The additional acres for food production in the future can be increased, at best only arithmetically. Shall he then not concern himself about making every acre more fertile and each volume of soil deliver more nutrition? His exploitation rather than wise management of his own sources of nourishment has only recently been recognized on a larger scale of world organization. Even then it has not been recognized as a basic factor in world problems now pitting east against west, too commonly viewed as matters of politics in place of problems of agronomics, or of production of ample nourishing food via the soil.

2. *history gives pointed suggestions*

In a hasty, sweeping scan of history—if one sees much of man's expansion starting from the Orient—can one visualize any other factor than hunger as the compelling force behind his constant westward march? Surely it was food and not air and water as biological requirements, with the first and not the latter two, connected with the soil. Movement under hunger pressure is regular and expectable by the nomad living by hunting and gathering. But movement westward from a fixed agriculture given to planting, cultivating, domestication and other practices founded by science and by accumulated experiences, should set us to thinking about soil exploitation as causative. Westward marches of later history connected with such names as Julius Caesar, the Apostle Paul, Genghis Khan, Christopher Columbus, and even very recently, Sir Winston Churchill, should bring soil exploitation and hunger pressure behind the westward marches and emigrations from the older East. Since these movements of peoples are within the larger land belt of the temperate zone in the northern hemisphere, there is suggestion that the climatic forces in soil development were outlining the ecological pattern of man's migrations to more virgin soils for his exploitation of them in his struggle to survive. The direction pattern suggests climatic forces of moderate, not excessive, degrees of soil development supplying protein-rich and mineral-rich food crops as giving the invitation. Man's early movements under less technological helps did not guarantee survival on soils offering only wood as fuel and shelter, or giving only crops for raiment, save along the sea as protein source. There is the strong suggestion that his migration and survival were guided by climatic settings making soils which were producing the proteins along with the carbohydrates, rather than producing carbohydrates mainly.

History deserves a restudy to examine soil exploitation as the factor behind the wars so commonly emphasized as history shaping events. Soils failing to supply proteins suggest areas of emigration. Soils providing them suggest areas of immigration. The struggle for meat and its prized nutritional values indicated by prices, controls and other economic indices, has been a factor in westward marches, wars and historical changes. Columbus in his search for spices came west for those preservatives of more dessicated meat. More spices expanded the European population by way of more and better sausage and bologna. It was a similar trouble with meat under spoilage in the Russian Navy at Odessa that started the Russian revolution in 1916 and sent the hungry hordes of Bolsheviks marching west, not only as far as Berlin, but into the western hemisphere as well.

Having probably had your own ancestors in the vanguard of the many past westward marches and having been blest ourselves with the products of seriously exploited soils giving the maximum height of living standards, we have not been looking back over the wake of our own ancestral migrants to see hunger in it as reason for their traveling in the direction they and we have gone. We have scarcely seen even the soils ahead. Our gaze has been too intent on concrete highways, or on landscapes from too much altitude to discern there the soils that create what grows on them to feed us.

The present tense moments in the history of our country and the mounting problems of the economics of agriculture suggest that our own westward march of escape is over. We cannot maintain high living standards by opening more acres, or by more high powered exploitation of those already catalogued. The present moment for ourselves in relation to world history should help us to see that exploited soils of the rest of the world under more years of use than our own have been driving the hungry hordes toward us. Those are apparently more murderous (though possibly more diplomatically so) than were those under Genghis Khan, when our western hemisphere is the last potential Marshall plan for them. Perhaps more data assembled by the Food and Agricultural Organization of the United Nations will convince the quantitatively minded folks and bring the realization of what should long have been qualitatively granted, namely, that the soil is nutrition.

That soil is nutrition for growing our seafoods holds true also. The sea along with the land has been exploited of its protein producing power. Like the American bison, the carrier pigeon, the wild turkey, the prairie chicken and present game animals, which were once abundant meat, or protein, supplies but have been taken at rates approaching the near extinction of the land animals providing them, so the exploitation of the proteins from the sea is bringing the various aquatic species of dwindling numbers. Even the large whale, once harvested for its oil value only in such numbers to demand protection, has been under limited take in the Antarctic by international agreement for its meat value to the victims of World War II. But when these protein crops of the sea are the result of fertility inwash from the land, we may well remind ourselves that even when we harvest the sea this, too, is a case where soil is nutrition.

3. extended soil use requires balance of crop load against level of fertility maintenance

Soils are scarcely ever exploited in the Old World to the point of no productivity or to that of abandonment, as may have been the case often in the United States. Old World soils are kept up in production by the maximum return of the residues of what they have grown, and by the limited supplement of purchased fertilizer which parsimonious farming can afford. Such old soils soon bring the loads they can carry into balance with their productivity under fertility maintenance. European farming has had the manure pile in the front yard as the index of the farm's prosperity. The mass of treasured nitrogen turnover, with other fertility accompanying, is the report of what the soil can do as nutrition or creation of the farm products. The amounts of those can be no larger and the delivery of them no faster than the amounts of fertility annually returned in the various manures and coming from the weathering of the reserve soil minerals will allow.

Nitrogen via organic manures was long the major single limiting factor in production. European agricultural output was determined by such nitrogen. Now that chemically fixed forms of it for fertilizer use are so plentiful, exploitation of the soil will not be hinged so simply to the turnover of this one element. Instead, it will be concerned with many inorganic ones from the soil to complicate the maintenance of soil productivity.

It was the soil fertility of the Old World, maintained at a fair level under the major limits set by nitrogen, that limited the population to where its numbers were almost holding steady. That was the case because emigration to colonies, or to new countries, removed the peoples that were surplus to the carrying capacity of the soil. Soils have thus held up long under the equilibrium state of peoples balanced against a soil productivity that was maintained at moderate levels. Such was true for France, Holland and several other countries. Now that colonies have revolted against absorbing the Crown's extra peoples and new countries are not to be taken over by immigrants, the equilibrium state between soil productivity and population numbers is being displaced against the soil for its further exploitation to give an agricultural peasantry at still lower living standards.

Soil exploitation has shown the connection between the soil fertility and plant nutrition in the declining yields per acre of any continuous crop, or even of a rotation of crops. Numerous experiment stations have now demonstrated these facts. The long-time soil fertility studies of Pennsylvania, Ohio, Illinois, Kentucky and Missouri testify that lands mined rather than managed for plant nutrition cannot continue in their high productivity. Instead there is a steady decline in crop yields with successive years of cropping unless nutrition for the plants is regularly supplied as soil treatments. It is this fact that gave rise to the fertilizer business. Both continuous and rotation cropping without restoration of fertility, according to Missouri data, reduced the soil productivity to non-economic levels in 30 years after the plowing out of the virgin sod. The six year rotation reduced the fertility supplies to levels lower then where continuous cropping was tested. Under moderate degrees of development of the soil by the climatic forces or with supplies of minerals added regularly by deposits from wind or inwash, or with other contributing factors of such nature the productivity period will be lengthened. Under extended time of cultivation, we soon learn that soils are not mines from which productivity may continually be taken. Instead they are only sites on which plant production may be managed through the feeding of the plants via the soil while nature supplies the meteorological contributions, namely, air, water and sunshine. We are slowly recognizing our past soil exploitation as responsible for declining crop yields when too often we blamed the weather.

4. mining the soil means nutrition of declining quality for scarcely one generation

Under continued soil exploitation there have come not only quantitative but also qualitative changes in the feed and food values of the vegetation for animals and man. Declining fertility as nutrition for the plants has contemporaneously given declining nutritional values of the crop. Finding crop substitutes has brought the introduction of new crops under acceptance for their production of vegetative bulk rather than for the high nutritional values delivered in it. This emphasis on the crop yield with no attention to the soil as nutrition for it brought a shift to physiological plant functions giving mainly carbohydrate delivery through photosynthesis rather than those bringing about biosynthesis of proteins. Consequently the crop yields have been kept up, or even increased, while the protein production has not gone up correspondingly and too often has gone down. This was a decline not only in total crude protein produced, but also in terms of the protein's complete array of amino acids required for good nutrition.

Crop breeding with recent emphasis on hybrid vigor in the case of corn brought more bushels per acre but lowered the concentration of protein already by 30% of that formerly common. Hybrid corn could not keep its own species surviving, if turned out on its own. The faith in plant breeding continues and the hope for some new creations from it seems undimmed. But to date no great advantage to the survival of the species on its own has resulted from the breeding work. Much has been done to explain how the hindrances to survival are transmitted, and how defects distribute themselves according to certain ratios in the future generations.

It is to be hoped that the studies in plant breeding will be combined with studies in plant feeding, whereby plant breeding will aim not only to offset the decline in plant values brought on by soil exploitation, but rather will demonstrate how this combination can raise soil as nutrition to much higher values both quantitatively and qualitatively.

Soil exploitation has been accompanied—and also covered against our observation—in a large measure by marked technological developments. More acres per man under power machinery have prohibited more universal recognition of soil depletion. High yields per acre on lowered fertility, especially lowered nitrogen, have demanded crops of high carbohydrate output. Those of protein-producing prominence and lower yields per acre have not been used so widely. Nor have they been improved so pronouncedly by either crop breeding or crop feeding by fertilizers. Calories in animal nutrition have been the major criterion of crop values. Feeds for fattening and rapid gains in weight have emphasized the castrated males as farm animals rather than the unaltered ones in good health and high fecundity. Speculative sales of livestock and the growing of feeds mainly for such rather than for high nutritional values for healthy animals and reproduction have become prominent. Declining soil fertility under technologies for high powered farming has emphasized these so thoroughly as to allow little concern about the biology responsible for the creative performances via the soil on which the performances of all agriculture ultimately depend.

5. *agriculture rests on biology not alone on technology and economics*

Our high standard of living with emphasis on technologies has left the biology of agriculture overshadowed or even hindered in its services of feeding various life forms. We have in some instances upset it rather than supplemented or supported it. Our teachings in agriculture are often given mainly to telling how *we* make plants grow or how *we* feed animals. Too little teaching time goes to observing the plants and animals for us to learn how they feed themselves when we do not interfere.

Research must often be given to deleting biological error in the agricultural practices of the past as well as to establishing wiser practices founded on biological principles. The research effort must fit man and his agriculture into the forces of nature rather than aim to modify her or make her accomodate herself to man's wants. The latter represents exploitation. The former is conservation.

Only slowly is soil research deleting the mistaken concepts of the last two decades about natural soil acidity. This soil irregularity is not due to the presence of the hydrogen on the clay complex so much as to the absence there of other positively charged ions rendering nutrient services to the plant which hydrogen there does not. The very activity of taking up nourishment from the soil by the plant root puts hydrogen, or acidity, into the soil via root respiration. Carbon dioxide, the by-product of that life process in the soil moisture, provides the hydrogen which is exchanging itself to the soil colloid for the nutrient cations like calcium, magnesium, potassium, copper, etc., held there to be taken up by the plants. The fight on soil acidity in agricultural practice by applying carbonates or alkalies was an error in our concepts about this soil condition until the soil was viewed as nutrition for the plants and the microbes.

Our erroneous beliefs that nutrition of a plant by the soil consisted of the same behaviors demonstrated by hydroponics in the botany laboratory even codified fertilizer values according as their fertility contents are *available* or soluble in water. Yet water soluble plant nutrients applied to the soil are not taken from there by plants because they are in aqueous solution. Were they kept there in that form they would soon be lost to percolating water. Should they stay there in that form in quantities sufficient to mature a crop, they would salt the seedlings to death. Soluble fertilizers are more serviceable because they react more readily with the soil colloids and the microbes to be held there against loss by leaching. *Nutrients in hydroponic service must be so dilute that they behave according to the gas laws rather than those of ionic, inorganic solutions.* Our concepts of plant nutrition built on such bases are anything but the facts about the soil processes serving for nutrition of our field crop plants.

In place of emphasizing the biology of agriculture, undue emphasis seems to have gone also to the economics of it. Now that one can purchase more fertilizers, especially nitrogen considered the key element of crude protein, and can increase crop yields decidedly thereby, bankers are given to the belief that now "we can substitute capital for land." Unfortunately, capital cannot substitute when the soil is so much more than a collection of "plant-ash equivalents" in a bag with the fertilizer label. Soil under agricultural production is an integration of many chemical, biochemical and biological reactions for which certainly dollars as credit cannot substitute. Economic manipulations have put technologies and speculation into agriculture so fully and biology out so far, that the banker is about to disregard the soil as the wealth creating force of nature that is agriculture. The handful of dust that must have the warm breath of creation blown into it is by no means "gold dust." All the capital in all the fertilizers cannot give agricultural production except through the soil.

The biological perversion by artificial insemination for dairy and beef cattle matings into what seems mainly technologies under the bad economics of excessively priced males, lifts high the faith in this breeding technique. But it seems to be pushing the biology of reproduction out on its very fringes for the survival of the species. Natural mating services by the bull in the herd usually bring the so-called "shy-breeder" cow into calf bearing even if his repeated servings are required over numbers of heat periods too large for the high costs of artificial insemination to duplicate. As a consequence the latter consigns such females to slaughter while the natural services by the bull keep them in the herd and in the procreation of the species.

Perhaps the natural mating plan is more than just an impregnation with the spermatozoa. Might it not also be a series of hormone treatments to encourage better ovulation through successive absorptions of semen in the larger doses? Would such not possibly contribute to the better survial of the species through more than sperm delivery from the male by the mating process? Might not natural matings thus be doing more than artificial insemination does to keep the species alive?

Exploitation of our soils by a youthful agriculture under a westward movement to deeper and more fertile soils of the midcontinent has given national crop yields representing some crop surpluses in the hands of the government and under economic manipulations by it. The cost of these surpluses has not been estimated in terms of the soil fertility exploited, dumped into foreign countries, or taken out of circulation to lower our own future food security. By that production of surplus as taken from the farms there have been built up also surpluses of urban population where food security depends on highly complicated economic entanglements. These are not readily circumvented in case of self-flattery because of our technologies and economics interpreted as high standards of living (more correctly high rates of spending our resources including those in the soil). We have moved so far from the soil and its biological support of us, that sudden regional irregularities in weather to disturb crop results of a single year are no longer viewed as acts of God as the pioneer saw them. Instead they are national emergencies considered for solution by a call on the national treasury. We have forgotten that our soils are nutrition and that they represent the basic force making possible much of what we call "high standards of living."

While at this moment in our agricultural history the biology of the soil's creative services is not shining much in contrast to the bright light from the technologies and economics, nevertheless the former is the living eternal hand that writes creative history of any area. The latter writes history of consumption and exploitation. The former under our exploitation has written "finis" for the agriculture of some areas. For other areas it is outlining their closing chapters while the neglected fertility sand in the hourglass is running low. Fortunately for us in our extensive geographical area under democracy, the lessons learned from exploitation and the knowledge so gained and put to soil fertility management in balance against population may save ourselves in the future on the remaining less exploited renewable areas. To this can be added the good fortune of deposits of mineral fertility reserves for fertilizer service, but demanding wiser and more thrifty use. All these coupled with the growing research that is gradually recognizing the soil as the basic source of our nutrition, suggest that we can do much to keep ourselves well fed in the future.

6. exploited soils emphasize principles of nature's conservation to guide the wiser management by man

The study of man's exploitation of soil under agricultural cropping reveals resulting chemical soil conditions similar to those in soils where nature's excessive development has resulted under higher rainfall. Our exploitation of virgin soils under higher rainfall but lower temperatures was less rapid than that of soils of similar rainfall but higher temperatures. The degree of natural development represented by virgin soils in our northeastern United States with clays of high exchange capacity and much total acidity suffered less speedy exploitation under cultivation when cleared of forests than that of virgin soils of equal clay content and less total acidity developed under higher temperatures and similar rainfall in southeastern United States. Soils farther west developed under the corresponding temperatures illustrated above but under lower rainfall like those near the midcontinent soon approach (under exploitation) the fertility array illustrated in soils farther eastward naturally more highly developed under higher rainfalls. When man uses the soil under rainfalls generous enough for higher crop yields, he depletes the fertility rapidly. He encourages leaching to enlarge its toll and he brings the soil to a much higher degree of development and at a higher rate than nature would in that climatic setting. His depletion of the soil fertility leads him to believe that the climate (average of weather) has become worse in terms of the quality of the present vegetation contrasted to what was virgin in the area. Consequently the vagaries of the weather are much more disastrous to his agriculture via poor soil as poor nutrition than they once were. Nutrition in the fullest sense must then be declining along with the fertility of the soil.

The successive stages of increasing soil development under higher rainfall to give decreasing soil fertility naturally, give successively more deficiencies in the plant compositions as related to animal (and human) nutrition. This fact establishes a principle illustrated also by the similarly changed plant compositions or increasing nutritional deficiencies. The increasing rainfall (temperatures constant) in the United States on going eastward to the Atlantic from the soils that once grew proteins in the bison body, give decreasing supplies of the essential nutrient elements in total coming from the soil as nutrition for microbes, plants, animals and man. However, the relative decrease in supply of calcium, which element the soils must give generously for the production of protein-rich forages (illustrated by lime for legumes and better grasses) is much greater than the decrease in potassium supply, serving in the plant's production of the carbohydrates. Thus with the increasing soil development as soil depletion the fertility supporting the plant's biosynthetic processes of converting the carbohydrates into protein is lost first as a factor in the ecological picture (calcium loss from the soil). The fertility supporting the photosynthesis of carbohydrates to give big crops of vegetative bulk (illustrated by sufficient potassium) remains. That shift in fertility, or ratio between amounts of calcium and potassium, shifts the ecological pattern of plants from the grasses and legumes of high nutritional value per acre for providing proteins along with carbohydrates under the original herds of bison, to the deciduous forests able to support only a few browsing animals and then to the coniferous forests on which such life cannot even survive. This is a principle of soil fertility in relation to the climatic pattern of natural soil development that controls the ecological patterns of different life forms. It is basic also for the interpretation of our exploitation of the soil. It tells us to expect the nutritional values of our soils to dwindle accordingly as we extend the use of soils without concern about maintaining their fertility for the plant composition we expect the soil to deliver.

7. services from soil treatments

Because we appreciate the technologies giving us nearly complete control of the processes required to accomplish certain productions by industry, one is apt to believe that equally as complete controls should be possible in agricultural production. Up to this moment man is more of an observer of the natural performances which we claim as agricultural production than he is director and controller of those biological manifestations. Factors and processes entering into production through technology are usually few, well understood, carefully catalogued, readily controlled and additive in their final summation. However, the factors and processes giving the biological performances of growing plants and animals are legion, not well comprehended, seldom controllable and integrative in their final summation. The latter can then be catalogued only as general natural principles. What control within those we exercise will be larger only as we gather more knowledge of the biology involved. Since the soil is the starting point of the biological processes of agricultural production, those can be directed to no small degree by what fertility is put at the disposal of nature at the point of takeoff for what constitutes the natural assembly line. The starting point represents the area of most decisive control of the direction taken by any creative process.

8. when surpluses are carbohydrate crops and proteins are the shortage, soils must become better nutrition

A chemical study of the ash contents of both crop and non-crop plants gives a clearcut illustration of the natural principles previously outlined and concerned with the variation in the chemical composition of the forage in relation to the climatic soil development. On those soils naturally but slightly developed and calcareous enough to grow nitrogen-fixing legumes to build deep, black, nitrogen-rich soils growing livestock generously, the ratio of the percent of calcium to the potassium (as oxides) in the dry matter is 6.8 to 1.9, or 3.5 to 1.0. On soils moderately developed where legume production requires soil fertility additions in limestone, phosphates and possibly others, the same ratio as a mean is 4.1 to 1.6 or 2.5 to 1.0. On the highly developed soils of our east and southeast, the calcium and potassium put into the plant tissues are in the ratio of 1.0 to 1.0. Thus with the increasing degree of soil development, the crop's composition declines rapidly in calcium and all that it represents in protein potential there. But the crop's composition does not decline so rapidly in potassium and all else associated with it in carbohydrate production. Thus most any soil may grow carbohydrates but not necessarily proteins along with them. This variable calcium in its ratio to the potassium in the ash of the crop and the corresponding variation in the nutritional organic compounds synthesized, namely, proteins to carbohydrates, moves our thinking about inorganic fertilizers to recognition of their equivalents in organic compounds synthesized. It moves our thinking of fertilizers to thinking of food. It lets the soil demonstrate itself as nutrition. Soil thus becomes creation. Agriculture becomes more of a biological performance than one of mainly technologies and economies.

Experimental studies using the colloidal clay as the medium for well controlled plant nutrition have demonstrated that differences in the ratios of the several inorganic fertility elements which are active on this soil fraction bring about differences in plant composition as proteins and carbohydrates. Thus by different ratios, or variable balance, of the calcium and potassium, we can grow either much plant bulk of high carbohydrate mainly and of low protein content, or less bulk of higher protein concentration and ratio to carbohydrates. By varying the fertility ration in the soil for the plant there are brought about not only different amounts of carbohydrates but also differences in the sugars, starches, hemicelluloses, etc., composing them. Likewise, the ratios of the different amino acids composing the proteins, and the amino nitrogen as part of the total nitrogen will be variable according as the ratios and amounts of inorganic fertility in action are varied for movement into the root as plant nourishment from the soil.

Up to this moment, the operation in agriculture of this basic principle of soil fertility in relation to crop production has exemplified itself in the introduction of crops for increased yields of carbohydrates, and for managed crop composition and improved yield under the program of testing of soils and the application of fertility accordingly. While the introduction of the corn hybrid covered itself with some semblance of glory, its yield increase comes by loss of its power to procreate itself by its own seed. It suffers also in terms of being of lessened value in animal nutrition because of its delivery of a crude protein highly deficient in certain required amino acids, notwithstanding its reputation as the crop giving high farm income. In the case of hybrid corn the nutrition of this plant for its exaggerated carbohydrate production over protein was a case where the introduction of the crop fitted well into the natural principle of the soil under declining soil fertility or increased soil exploitation, expressing itself through higher yields because of crop nutrition with mounting carbohydrate accordingly. This occurred even before the possible services from soil treatments in relation to this natural principle were recognized and ever considered for application in practice.

More particularly the principle of the ratios of the active inorganic fertility elements in the soil for control of the synthetic services by crops has offered decided promise when applied in relation to increased protein production along with the carbohydrates. Thus we can use the soil to grow more nearly balanced nutrition. It has given us the concept of specific ratios of active fertility elements as well as total amounts of them for production of plants of a certain nutritional value by their organic composition (proteins in balance with carbohydrates), and a generous yield per acre respectively. As a result of this view we see reasons for believing that 75% of the soil's exchange capacity might well be saturated by calcium; 7.5 to 10% by magnesium; and 2.5 to 5% by potassium for the growth of forages rich in protein and carrying also the quota of the many other inorganic elements, vitamins, enzymes, hormones and other essential compounds associated with the production of crops with nutritional purposes in mind becomes possible. This management moves soil into the food-creating category in place of keeping soil growing only filler feeds for fattening values.

9. diagnoses of the soils suggest relief for nutritional deficiencies of plants, animals and man

The testing of the fertility of the soil to obtain an inventory from which to plan its treatment, has now become a help in diagnosing the soil's production potential. Thus the nutrient store in it is measured so that by supplementing this with limestone, rock phosphate, and more soluble fertilizers the soil can become plant nutrition for certain physiological functions which we hope to use in growing the crop for our nutrition. Soil testing has become a much greater help to guide the services possible from soil treatments now that our fuller knowledge of the soil and the plant processes gives us newer concepts of the soil as nutrition.

Soil treatments are being used in terms of both the soil and the crop. Of some soil treatments we may say that they fertilize the soil in that their effects are more lasting there. This is true of limestone and other fertilizers supplying calcium and magnesium. Rock and superphosphate are similarly considered in providing the soil with phosphorus. Nitrogen and sulfur more truly fertilize the crop since they are not held so lastingly by the soil, but influence crop composition and growth pronouncedly. Guided by the soil test and by the fuller knowledge of the plant's synthetic services in response to the ratios of the nutrient elements exchangeable in the soil, we are now in better position to manage soils for better nutrition of all that must be fed by them and for an approach to conservative soil maintenance rather than soil exploitation.

Extremely wide variation in the plant's chemical composition, viewed as nutritional values, have come from research on soil fertility under that objective. Crop quality modified by soil treatments takes on new and larger meaning, not so much when the crude proteins are chemically measured but when the amino acid array composing them is revealed. Soil fertility as plant nutrition in terms of the crop's synthesis of the essential amino acids stands out all the more significantly when bioassays are made using laboratory animals and even insects, like the household cricket. According as more refined tools of research help us detect and measure the plant's organic creations of nutritional values for us resulting from soil treatments, shall we see more clearly the connections between soil and nutrition. That vision will probably suggest the ministrations for better nutrition and health which we can affect by what may be relatively simple treatments of the soil.

10. more problems ahead

It has taken a long time to realize that the ecological pattern of higher life forms (including man) is determined largely by the soil's provision of the essentially complete proteins in the crops grown. This realization established itself when nitrogen, the key element of proteins, fluctuated no more widely or suddenly in its soil supply than was provoked by natural phenomena. Whether our concepts of plant composition in relation to soil fertility so established will be helpful now that nitrogen can be added so generously to the soil, remains to be learned. Certainly we should be in better position to use this added fertilizer contribution toward more protein potential at least more wisely by the help of these concepts as guides than had we not established them. Thus the research challenge for fuller study of soil and nutrition is all the larger because of this technological possibility for application of much more nitrogen to the soil and to the biology of agriculture in making more and better proteins. Man's future will depend on his conservative use of the soil to create the living foods for him. That calls for research progress which views the soil as the means for health rather than for only wealth.

37.

Chemical studies tell us all too little about how the elements of the earth are shaped, hammered and bolted together into compounds. Often they tell us nothing when they purport to tell us all. This brace of realities caused Dr. William A. Albrecht and his associates to look to the biological assay for final determination of how well crop production nourished animals. This particular report is a classic. Its main content was given world-wide circulation in Andre Voisin's Soil, Grass and Cancer. *Basically, Albrecht and his co-workers observed that bins and bushels and tons did not properly describe the value of a farm crop. Only the biological assay—the actual running of that crop through the digestive systems of animals—could properly hint at what nature was all about. This paper appeared in the* Soil Science Society Proceedings *for 1943.*

Biological Assays of Some Soil Types Under Treatments

The soil, which is biologically altered geological materials, provides the support and nutrients for the plants so that life itelf may be sustained therefrom. Of the dry matter of the plant, 95% is combustible and 5% is ash. Four elements make up this 95% or combustible part. These are carbon, hydrogen, oxygen and nitrogen. They are supplied by water and air—uniformly distributed through their fluidity. They are available in unlimited quantities, and when one is utilized, there is more of the same to move in to replenish the supply. At least 12 elements make up the 5% that is ash in the plant. These elements are limited in supply for any given plant. Since they are many and present in the soil in small quantities, a limiting supply of one or another for maximum growth is more commonly the rule than the exception.

Plants utilize the nutrients delivered to them by the soil, as elements or molecules to be combined with the elements making up the combustible plant portion—carbon, hydrogen, oxygen and nitrogen—to form plant compounds.

Different compounds may be manufactured by the growing plant depending upon the rate and balance of delivery of these nutrients to the plant by the soil.

Our chemical studies of the plants have made use of methods that determine the relative concentrations of the elements in the plant. They tell little about how these elements are combined into compounds. It is true that tests are common for certain standard compounds such as fats and carbohydrates, but for protein the nitrogen figure multiplied by a factor is the common evaluation procedure. This tells nothing about the different amino acids making up the protein.

The appreciation of vitamins and the search for chemical tests for them has turned attention to the more complex organic compounds and away from the simple uncombined elements found by chemical procedures. Plant growth needs more emphasis as a process of fabricating compounds rather than one of siphoning soil- and air-borne elements into the plants. Yet the general tendency prevails to measure plant response to soil and soil treatment in terms of tonnages per acre in the field and to determine the concentration of elements in the dry matter rather than the organic compounds suitable or unsuitable for growth and energy for animals. Since quantitative chemical analyses for the elements are inadequate as measures of the synthesized plant compounds, and since the animal's nutrition depends on compounds rather than elements, the animal assay of plant differences caused by varied fertility delivery according to soil type and soil treatments seems most logical. It was in support of this hypothesis that the following study was carried out.

1. method of assay

The domestic Chinchilla rabbit was used in lots of 8 to 10 animals for the bioassays of different soil types and soil treatments reported herewith. The rabbit is herbivorous and thus lends itself readily to assays of forages. It has been compared with steers by Crampton, *et al.* [writing in ***Pasture Studies XIV. The Nutritive Value of Pasture Herbage,*** Scientific Agriculture, 1939; ***Pasture Studies XVII. The Relative Ability of Steers and Rabbits to Digest Pasture Herbage,*** Scientific Agriculture, 1940], and they conclude that the rabbit as a "pilot" animal digests protein in certain rations more thoroughly but crude fiber and cellulose less completely than do steers. They state that the rabbit digests pasture herbage less completely, but that there are close correlations between the two species in the cases of protein and lignin. They report that the rabbit as a pilot animal has much promise and that further work with it is justified.

The rabbits used in this bioassay were genetically quite similar. They were all of approximately the same age, and they were distributed at random with respect to weight, sex and litter. The feeding period started at weaning time and continued for six weeks. They were fed hays from different soil types without soil treatment, and from different soil types with soil treatment. The hay was fed ***ad libitum*** on wire-floored cages with facilities for collection of urine and feces. The voidings were measured and samples taken for analysis. The amount of hay fed and the amount wasted were measured so that actual consumption was determined. From these different values digestion and retention were determined.

One single species of plant, *viz.*, Korean lespedeza, was grown on five different soil types within the limits of of the state of Missouri. All hays were produced during a single season. The season varied only with locations of the soils. The hays varied widely in appearance chiefly because of soil differences, except in one case where the season had a distinct effect.

The soil types were Eldon sandy loam, Putnam silt loam, Clarksville gravelly loam, Grundy silt loam, and Lintonia fine sandy loam. They represent five distinctly different areas of the state, and were developed from distinctly different parent materials. They also vary considerably in age, topography and vegetation.

The Eldon sandy loam is an unglaciated border Ozark soil developed from sandstone and cherty limestone under the influence of prairie vegetation on level to slightly rolling topography. The Putnam silt loam is a level, prairie soil, thought to have been formed from glaciated material covered by a thin layer of loess. It is underlain by a very distinct claypan. The Clarksville gravelly loam is a very old, residual soil formed on cherty limestone under the influence of forest vegetation and very rolling topography. It makes up the bulk of the Ozark area. The Grundy silt loam was developed on glaciated material with a deep layer of loess under the influence of prairie vegetation and level to slightly rolling topography. The Lintonia fine sandy loam is a soil developed under forest cover upon sandy alluvial material laid down by the Mississippi river during the Coastal Plains period.

All soil treatments included lime and phosphorus. On the Eldon and Putnam soils potash deficiency appeared a few years after lime and phosphorus were first applied. As a result the treatments on these two soils included potassium as well as lime and phosphorus. Previous work has shown that where potassium was not deficient for crop growth, potassium as fertilizer did not increase the feeding value of forage.

Chemical assays showed slight differences in composition of the hays suggesting possible differences in the compounds synthesized by the plants. A constant type of animal with respect to age, weight, genetics and sex was used as bioassay of the plant differences.

2. results

The hays grown on these different soil types without treatment varied widely in their physical make-up. The hay on the Eldon was somewhat grassy and low in quality. That grown on the Putnam was somewhat grassy, but otherwise the quality of this hay was high. The hay from the Clarksville was very grassy and had small short stems. The quality of this hay was low as based on color and leafiness. The hay grown on the Lintonia soil was broken up and appeared very trashy because of the dry season during which it was produced. All other hays were grown where there was an abundance of rainfall during the growing season.

The hays from the different soil types with treatments were more similar than those grown without soil treatments. There was much less grass and more lespedeza on the Putnam, Clarksville and Eldon soils. However, there was a tendency for dodder to come into the Eldon plot.

When the hays from the untreated soils were fed to the rabbits for a period of six weeks, great variations occurred in the physical condition of these animals. Those on the hay from the Grundy were very large and thrifty. Their eyes were bright, and their coats sleek. The animals on the Eldon, Clarksville and Lintonia were all of fair size, but their coats were rough, their condition was poor, and they were unthrifty. The animals on the hay from the Putnam were smaller and more unthrifty in appearance than any of the others.

When the hays from the treated soils were fed to the rabbits during the same time, the differences in animal condition were not nearly so marked as those in animals on hays from untreated soils. All of these animals on treated hays except from the Putnam soil, were very thrifty. Their coats were sleek and they were in good condition. On the hay from Putnam soil, the animals were smaller and slightly less thrifty, but otherwise they were comparable to those on the hays from the other treated soils.

The yields of the untreated hays on the different soil types and the responses of the animals on these hays are summarized in the table. The differences between the lowest and highest figures expressed as per cent over the lower are possibly of greater interest than are the absolute values. These ranges of differences are as follows: For animal gains in 6 weeks, 101.8% for hay consumed per gram of gain 94.9%; for yields of hay per acre 380.8%; and for yields as pounds of gain per acre 507.7%.

With the yields of the treated hays on the different soil types, the maximum differences expressed as percentage were as follows: For animal gains in 6 weeks

Rates and efficiencies of animal gains and yields as hay and as animal gains per acre on different soil types without and with soil treatments.

Soil Type	*Rabbit gains grams*		*Hay consumed pr. gr. of gain**		*Yields of hay, lbs. per acre*		*Yields as lbs. of gain per acre**	
	Untr'td.	Treated	Untr'td.	Treated	Untr'td.	Treated	Untr'td.	Treated
Putnam	315.7	419.9	13.23	9.41	2,180	3,800	116	254
Clarksville	419.0	616.6	7.85	5.49	520	2,020	39	233
Eldon	504.7	674.8	6.79	5.88	2,500	4,500	241	471
Lintonia	561.4	741.9	6.95	5.06	2,250	2,800	180	316
Grundy	637.1	593.9	7.43	6.47	2,400	3,760	180	303
Average	487.6	609.4	8.45	6.46	1,970	3,376	151	315
Range of difference, %	101.8	79.1	94.9	86.0	380.8	122.7	507.7	101.2

**Assuming all gains from the hay*

79.1; for hay consumed per gram of gain 86.0; for yields of hay per acre 122.6; and for yields as pounds of gain per acre 101.2.

A comparison of these ranges in difference shows immediately that there is much less variation in the treated soils than on the untreated soils. On the untreated hay, the maximum difference figures range from 94.9% to 507.7%, while comparable figures on those treated ranged only 79.1% to 122.7%. The averages for each of the columns of data show that the animal gains are considerably greater, the hay consumed per gram of gain is less, and the yields as pounds of hay per acre and as pounds of gain per acre are greater on the treated than on the untreated hays. Thus, the treatments not only made differences less conspicuous, but improved the yields and feeding values of the hays produced.

The data of the routine laboratory analyses of the soil [fully reported in E. O. McLean's University of Missouri Master's Thesis, *Soil Treatments on Some Missouri Soil Types as Measured by Biological Assays,* 1943] makes obvious the wide differences in these soil types. Their exchange capacities vary from the lower value of 4.76 M.E. per 100 grams of soil in the Lintonia to the higher value of 23.28 M.E. in the case of the Grundy. The total carbon, total nitrogen, exchangeable calcium, and the contents of magnesium, strontium and hydrogen have the same general range and correlate closely with the exchange capacities of the soils. The potassium, phosphorus, and manganese vary at random with the exchange capacities of the soils. Since these chemical properties of the soil are so variable, it seems only reasonable that plants and animals should respond to the variable fertility delivery by these soils. However, that the small amounts of fertilizer as soil treatments should have such a pronounced effect upon reducing these differences in plant and animal response does not seem so logical.

The chemical analyses of the lespedeza hays grown on these soils and fed to the rabbits in this experiment offer some explanation of the animal responses. The concentrations of the nitrogen, phosphorus, and lignin were all increased by the soil treatments to give data with less range of difference than on the soil types without treatments. The concentrations of silica in the hays were reduced by soil treatment. There was a like tendency in potassium, manganese and strontium. Again, these data have less range of difference than on the untreated soil types. The calcium showed no definite trend according to soil treatments, except that the range of difference was greater for this element in the hay from the treated soil types. The calcium concentrations in the hay and the calcium saturation of the colloid of the soil correlate with animal response more closely than any other single character studied in the soil types with or without soil treatment, except for the Putnam silt loam. The Putnam produced somewhat irregular hays both with soil treatment and without treatment as evidenced by the low rates of gain, the inefficiency of feed utilized, and the low percent of dry matter digested as compared to the other four soil types.

It was postulated that there were greater differences in plant physiology as a result of different soil types and soil treatments than differences in yields as tonnages indicate. The hay was assayed with rabbits as an attempt to evaluate some of these physiological differences of the plant. If the physiology of the plant was not changed qualitatively as well as quantitatively by the soils and soil treatments, then all animal responses would be expected to be similar; but the rates of animal growth indicate that there were great differences in the plant physiology. The question then arises as to whether gains of animals are very accurate as indicators of these differences. Should they be much more accurate in showing the differences in plant physiology than are yields of hay for showing variable soil fertility delivery? Of the animal growth characters there are many that might be used as helps to evaluate these differences. Only one of these, the bone properties, was studied.

After the rabbits had been fed for six weeks, they were killed and the femur bone removed from the left leg. Physical tests were run on these bones and the data are summarized in appendix A. With the exception of the three items of specific gravity, calcium retained from the feed, and calcium-phosphorus ratio, the maximum differences were less for the animals fed hays from the treated soil types than for those fed hays from untreated soil types. The specific gravity of the bones from the animals on hay from the Putnam was less on the treated than on the untreated soil. This points again to some irregularity of the animal response on the hays from the Putnam soil type. Because of this reversed condition as compared to all the others, the maximum difference exceeded that of the hay from the untreated soil type. The calcium-phosphorus ratios show larger maximum differences because of the larger maximum differences of the calcium retained from the hay. That the calcium retained from the feed should show larger differences on the treated plot is of interest since this was also true of the calcium concentration within the plant. This immediately leads one to expect possible correlations between calcium concentration and grams of calcium retained from the feed. This correlation is very definite, but whether it has significance in this experiment is not known.

The bone data, in general, show very wide differences and are indicative of great differences in animal physiology on the hays from the different soil types without soil treatment. With certain exceptions the differences on the hays from the different soil types become less when soil treatments are used. As the average data indicate, the soil treatments resulted in hays which produced bones of greater weight, length, diameter, thickness, breaking strength, volume and specific gravity. There are individual cases where this was not true. All of these bone properties increased with soil treatments except for the decreased specific gravity in the case of the Putnam hay with soil treatment and for the bones of the animals fed hays from the Grundy soil. On the Grundy soil only the wall thickness and specific gravity of the bones were increased by soil treatments. It is possibly significant that where the breaking strength of the bone went beyond 30 pounds, the limit of the measuring mechanism, the assimilated phosphorus went above 58 grams per pen of 8 to 10 animals. When the strength of the bone went beyond 30 pounds, the retention of calcium was high, but this relationship was not as close as with the retention of phosphorus. Of no less significance is the fact that the smaller and weaker bones are all still more closely associated with lesser retentions of both calcium and phosphorus.

3. discussion

That small applications of lime and commercial fertilizer should reduce the differences in yields of crops from different soil types has been observed universally. Possibly, too, the effect of soil treatments upon eliminating many of the differences in composition of plants grown on different soil types has been a common observation. But that these small applications of fertilizer should tend to obscure the differences in the soil types in terms of nutritional qualities of these forages and make these feeds more efficient may not have been recognized.

It is logical to expect a greater increase in yield of hay from treatment put on infertile soils than on more fertile soils. Thus differences in yield between soil types are expected to be reduced when the soils are all treated. The studies reported herewith show that chemical compositions of hays from different soils become more nearly the same when the soils are treated. From this one would expect smaller differences in animal gains on these hays of more homogenous composition. But gains are only gross indicators of animal physiology. Gains, too, may be differentiated qualitatively and quantitatively. Tonnages of hay reflect physiological differences in the plant, but only the bioassay has succeeded in making a more complete measure of qualitative differences. An attempt was made to measure the varied physiology of the animal qualitatively by the bone study. The results of this also show that soil treatments tend to obliterate differences between the untreated soil types.

The major fertility variations in the majority of the soils are in the supply and availability of nitrogen, phosphorus, potassium and calcium. They become deficient because of low supply and losses through leaching, erosion or plant removal. In the final analysis production of crops is a matter of the provision of fertility. This provision depends upon the root surface exposed, the amount of clay surface available, the presence of all nutrients required by the plant in approximately the proportions utilized by it and the restoration of the clay saturation from the mineral reserve in the silt as the nutrients are removed. That the balance between nutrients available is important in plant physiology is borne out by previous work. Timothy hay grown on soil having an excess of nitrogen, alfalfa grown on soil with an excess of calcium or phosphorus, and soybeans grown on soil made deficient in potassium through excessive lime applications have all been lower in nutritive value than where no soil treatments were added.

The soil type is a classification to establish the morphological or "physical unit." Soil service calls for differentiations to give an edaphological or "functional unit" in terms of the production of plants, animals and humans. Classification comes first. It is followed by the detailed studies of physiology. The physical units of soil have done much in the understanding of soil science. The functional concept of soil, however, needs to be correlated with the physical units of soil classification. The results of this study show that the functional concept cannot always follow gross crop yields. Results in terms of soils as functional units should be used cautiously until enough data of this type are catalogued and evaluated. Then possibly definite physical units may be associated with the finer values of the functional units. Until the soil chemist and the soil surveyor become soil physiologists the chemical building blocks and the structural units separately will add little to the understanding of functions. It is only when these chemical building blocks and the structural units are correlated with their functions that progress in most complete understanding of soils will be possible.

4. summary

1. Soil treatments tend to obscure the effects of soil type upon plant yields and composition, and these in turn tend to obliterate differences as measured by animal response.
2. Production as animal products is a matter of balanced fertility delivery by the soil.
3. Gains in animals, like tonnages of hay, are not the entire picture of yields from the soils.
4. Varied physiology in plants is shown in part by chemical analysis, but more forcefully in terms of the bioassay.
5. Varied physiology of animals is indicated by varied rates of animal gains and efficiency of gains, but is more forcefully demonstrated by the bone differences.
6. There is a need for correlation of physical units of soil classification with functional units in terms of crop and animal production as a means of measuring the many factors operative in soils in their service in agriculture.

38.

With or without added fertility, soils determine human growth, achievements, history. How nature chose to weather the mineral content of rock into topsoil in fact cast the die for the game of life. Short grass in the semi-arid plains means more to a cow than belly-high grass in Mississippi. Certainly the point can be made that man has manipulated fertility for only a few seconds in the countdown of life on earth, and most of this manipulation has been done badly. This reality hands back to nature herself the delicate task of determining human growth. How Soil Determines Human Growth *tells about the process. It was first printed in the* Southwest Review *in 1945.*

How Soil Determines Human Growth

That the fate of men and civilizations is dependent on soil fertility is still a new concept to many of us. But it is unquestionably true. The soil and its mineral content of weathered rock control the quality of our food and thereby our health and productivity. The soil map of the United States indicates the future development of the regions of America. Soil fertility, far more than variations in climate, determines the fate of plants, animals and people.

The magnitude and geographical position of Texas make it possible for this state to serve as a replica of the United States as a whole. In fact, the range of soil possibilities in Texas is wide enough to illustrate the basic principles of soil development over a great part of the world. Texas is an empire in its range of meteorological conditions and the variety of its soils fashioned from the rocks by diverse climatic forces.

About the rainfall in Texas, you can say as much as Ring Lardner once said of the "civilization of Tennessee": "There are all possible degrees of it as one goes from east to west." Traversing Texas in the same direction, the annual rainfall ranges from almost fifty-five to five inches! Poised fanlike above Texas on the map, the rest of the Union has a rainfall of but little greater range. The longitudinal bands of rainfall in Texas, decreasing westward, extend themselves northeast, north and west over the entire United States. The rainfall pattern of Texas is thus duplicated in the United States as a whole, except that temperatures are higher in Texas.

The meaning of this pattern of climate and the resulting soil fertility is to be found in the nature of plant and animal growth. The carbohydrates—energy foods—can be produced on soils of limited fertility; but the proteins, vitamins and many other life-sustaining essentials must come from soils rich in minerals. The ten or more chemical elements that control the body-building processes are the dust of creation, and they determine whether the germinating seed shall die, as it does "on stoney ground," or whether it shall reproduce its kind a hundredfold.

1. soil fertility

Soil fertility, therefore, not only determines whether there is any vegetation at all, but whether the growing plant is a woody skeleton of fuel value only or a plant rich in the complexes that serve to build our bodies and enable them to reproduce.

Wood production makes no great demands on the soil for an extensive list of chemical elements. Forests grow on shallow and rocky ground that has not weathered enough to develop into what could be truly called a soil. Such lands would not produce much food for man or animals. Forests are common, also, in the tropical jungle where rainfall is great, temperature is high, but the soils are so leached of their fertility that they will not grow food crops well even when cleared.

Mixtures of sand, silt, clay and humus are commonly the productive soils, and the proper blending of these separates is basic in meeting the need for human food in grains and for good animal feed in grasses. A crop anywhere on the less fertile soils tends to be a producer mainly of wood and of foods having value only as fuel.

Western Texas and the Rocky Mountains of the United States in low rainfall areas are forest producers. Likewise, eastern Texas and the eastern half of the United States, as a whole, are covered with forests, or vegetation not consumed as food. In the east the rainfall, rather than the soil fertility, is the dominant factor.

The increasing degree of soil development under increasing amounts of rainfall, across Texas from west to east, serves well to demonstrate that it is the dominance of soil fertility in the west, rather than the weather, that controls the chemical composition of the vegetation.

In the midlands of the United States, in the regions of moderate rainfall, the grasses of the prairies abound. Here the great herds of buffalo roamed. Here the populations of grazing animals today are most concentrated. The great numbers of cattle and sheep bear a direct relation to the soil, according as it is a provider of the essential and properly compounded mineral elements required for body growth.

In mid-Texas the prairies have a rainfall high enough to develop the clay supply in the soil and to saturate the clay generously with exchangeable or available calcium, magnesium and other plant nutrient materials. But the rainfall is not so high that it leaches these elements out and leaves soils with an acid clay. In the prairies we have the maximum of soil construction and minimum of soil destruction. Prairie soils are the safety zones of life. They grow foods of health-giving qualities because the fertility is available for fabrication within the plants.

Prairie soils have their droughty seasons and consequent extreme losses of moisture, and it is this near balance of evaporation and rainfall that stocks the clays of the soil with fertility in readily usable forms. It is this belt that produces "hard" wheat because the soil provides the elements for producing the proteins that makes it "hard."

The extensive wheat fields of Texas, her enormous herds of cattle, and her dauntless and aggressive people are consequence, in part at least, of her great area of prairie soils. They are tokens of the soil's efficient provision of nourishment for plants, animals and men.

In the more arid areas of the western part of Texas, where the precipitation is less than the evaporation, the soils are being built. They are in the process of construction. The soils are only partially broken down, with some clay resulting. The soils are more sandy, with much nutrient mineral reserve in their silts and sands. Parts of the rock contents have been dissolved, but these have not been leached away. Instead, this mobile soil fertility is held by the clay as exchangeable plant nourishment. Sort grass occurs there, because the limited water supply does not permit a great bulk of tonnage per acre. But this short grass is crammed with the nutritional products induced by ample soil fertility from the soil minerals. It is, therefore, a body builder and a grower of animals rather than a fattener of them.

It is within the "short grass" belt extending northward out of Texas that the buffalo lived. It was there that this large quadruped was well supported on a strictly vegetarian diet. It is there that cattle are grown today "on the range." It is on that soil belt that they reproduce well and are relatively free from disease. Thence they are shipped to the east and north, or to the "tall grass" country with its moister, more highly developed soils, to be fattened. It is the soil in its different degrees of development, or its different stages of construction and destruction, that controls these great forces in agricultural production, seldom connected so directly with the soil.

In the more humid areas, like the eastern part, where precipitation is greater than evaporation, the soils are being broken down. They are in the process of destruction. Much clay has been formed in them. They are said to be "heavy," and require more plowing and tilling to make them produce. In making the clay, the weathering has broken down the softer, more soluble minerals to leave the harder, insoluble sands and silt made up mainly of quartz. These remaining, larger separates are not a nutrient mineral reserve to be readily broken down to pass their contents to the clay and from there to the plant root. The clay, too, has long been subjected to excessive rainwater and thereby has given up most of its fertility in exchange for hydrogen from carbonic acid to make itself sour or acid. Under the higher temperatures the redder clay resulting has less capacity for exchange or less capacity to hold either fertility or acidity. Thus the soil has less possibility of service in its clay as a "jobber" of fertility, and less in its silts and sands as mineral nutrient reserves.

Here, then, under higher rainfall, the "tall grass" and forests represent plants crammed with photosynthetic products of mainly energy value. There are ample supplies of water, air and sunshine to concentrate fuel and fattening values into sugars, starches, celluloses and woody products. There is potassium in the soil to serve as catalyst for this energy-concentrating process. But there is too little mobile soil fertility in the clay, in the mineral reserve and in the highly lignified humus, and the plants are therefore low in protein, and low in those complexes built from soil fertility synthetically.

2. *cotton and forests*

Eastern Texas and eastern United States are therefore the country of cotton and forests. The lesser delivery of soil fertility is in command. The crop products are of highly carbonaceous nature because they are made mainly of air and water.

We are just beginning to appreciate the fact that as we exhaust the fertility of our soils, the nutritional nature of the vegetation shifts downward. Both as different species and as change in the chemical composition within the same plant species, there is a shift from body-building values to mainly energy-providing values. As one crop declines in tonnage yields and we find a substitute making equal or more bulk per acre, this new crop is more of fuel and less of nourishing values. It may pack the paunch and stuff the stomach, but leaves hidden hungers. It is the equivalent of the tall grass that never lured the buffalo eastward away from his short grass, though both grasses were on a continuous and level plain and though he raomed widely north and south over his chosen areas under scant rainfalls and short grasses.

Excessive cropping of our soils and the consequent removal of soil fertility nas been basic to many westward movements and pioneer invasions. This represents invasions by crops as well as by peoples and politics. The forest flora of the east is seemingly going west. Our crops are becoming literally more woody as those of distinct food-fabricating services are passing out, and we take to those making their products mainly from the weather. Cotton, with its product of cellulose, has moved west. The prairie grasses are being replaced by the invading mesquite. Can the encroachment by this woody, non-nutritious growth be pictured as a case of this wood producer surviving under little soil fertility when as a legume it uses nitrogen from the air and as a deep rooter it can satisfy its scant need for minerals and thereby take over the soil where the shallower-rooted nutritious grasses cannot hold out?

"Soft" wheat or the more starchy grain common in Missouri and east thereof has been invading the "hard" wheat belt of Kansas. Unfortunately, the "hard" wheat has been decreasing in the relative share of this food grain output in that state. Four or five years of generous rainfalls and consequently four or five wheat crops of almost record-breaking production have lowered the soil fertility supply and with it the protein production in the grain. At the same time the increased photosynthetic action has pushed up the starch production. The "hard" wheat has become a "soft" wheat. Is it any wonder, then, that the bakers are decrying the quality of the grain while the farmers are boasting of its excellent yields measured as bulk of bushels per acre? "Soft" wheat, too, like our forests and other woody crops, is moving westward as many other movements have gone to follow after the soil resources that feed them.

The soil pattern of the empire of Texas invites detailed study as it influences not only agriculture but other wealth-creating industries as well. It challenges us to consider the soil as it sets the ecological pattern of various forms of life, not only in this one state, but in the United States and in similar climatic areas in Europe, Australia, South Africa, China and Argentina. Such study will push the soil fertility pattern into play in the great international struggle that we have to date seen mainly in terms of economics of food-bulk, rather than in the quality of food and in the fertility of the soils by which both are produced.

It may eventually be realized that the historic westward movements of peoples were accompanied by exploitations of soil fertility resources. Possibly the two hemispheres will eventually understand their troubles, when their respective inventories of soil fertility are weighed against each other. We shall probably understand these larger national and international problems when we learn to understand our soils and to manage them so as to feed ourselves properly. It is on our soils, their appreciation and care, that this hope depends.

39.

When Dr. Allen E. Banik of Kearney, Nebraska made a trip to Hunza Land, it all started out as an Art Linkletter publicity affair. Banik surprised everyone. When he returned he and Renee Taylor wrote Hunza Land, *and caused it to circulate widely. The subject intrigued Dr. William A. Albrecht. Dr. Banik's observations suggested that Hunzacutts represented a climax crop of human beings. They prac-*

ticed soil chemistry au natural *rather than* au science. *Albrecht produced several papers on this subject, but this entry in* The Journal of Applied Nutrition, *1962, must be rated as the best of them all.*

The Healthy Hunzas A Climax Human Crop

Because of a visit to "The Land of the Hunzas," or "The Fabulous Health and Youth Wonderland of the World," Dr. Allen E. Banik of Kearney, Nebraska reported [in *Hunza Land*] the significant facts which help us to understand why the people of Hunza Land live to such ripe old age and how they maintain themselves in good health. Certainly the latter, namely, their excellent health, is not the work of dentists and doctors. It results in almost complete absence of those professional groups.

Dr. Banik's report leads us to believe that the Hunzas are what might be called *A Climax Human Crop* because they conform to the laws of nature with reference to their management of the soils for the most nearly complete conservation of them. Thereby, their healthy soils seem to be good reasons for those healthy people.

When we find a climax crop of healthy plants, like the pure stand of trees in a forest, or of a grass on the prairies, or of animals like the bison was on the plains, we are reminded that any crop reaches its natural climax of development and survival because: the same species grows there persistently; and it is nourished well enough to protect itself from pests and diseases, and to reproduce regularly with healthy off-spring in goodly numbers. Those climax aspects result in the single area without additions to the soil, save for the dust-blown minerals dropped from the atmosphere or such brought possibly by inwash from nearby higher elevations. Much of the climax effect results from the complete annual return to the soil of all the organic matter it has grown. That serves to put energy and "life" into the soil. Apparently those are the same factors operating in the case of the healthy, long-lived Hunzas, and are those by which that climax crop of humans has resulted in the Himalaya Mountains for persistence during a score or more of centuries.

1. soil management, au naturel

Their soil management for its successful maintenance of productivity consists of two kinds of fertilization, each applied semi-annually: the application of the silt-size, or powdered, natural rock particles in the "glacial milk" which they use as the irrigation water to flood the terraces; and the regular applications of the organic manures made from the carefully collected and composted materials which those same soils have grown.

As for the powdered, natural rock fertilizer, "The Hunza fields (generally one-half to five acres in area) must be irrigated constantly, especially since two crops are grown each year," Dr. Banik reports. "The water comes from melting snows high in the mountains. The mountain water is rich in minerals, and carries to the fields a plant-nourishing silt which is invaluable in replenishing the growing properties of the soil. Early in the spring, and again before the second crops are planted, canal gates are opened and all the terraces are flooded. When the water has been absorbed by the soil, leaving the silt deposit, the terraces are plowed and fertilized."

In the Hunza's plan of soil and water management, their irrigation is inadvertently also a fertility treatment with natural minerals. By applying semi-annually the natural rock which was freshly pulverized and mixed by the glaciers in the high mountains, they are maintaining (and building up) the soils in all of Mother Nature's fertility elements. Those folks are not concerned about ratios of plant-nutrients, their degree of availability, their possible dangerous salt effects demanding placement in relation to the planted seed, or any of the other details of soil and fertilizer chemistry about which we have been so concerned in the late years. But they are concerned about their duplication of nature's processes regularly making more new soil by their help in weathering natural mixtures of rock dust in the surface soil. There the decay of organic matter speeds up those weathering processes and combines the inorganic and the organic essentials in ways that are coming more recently to be considered,though only partially comprehended.

Then when the Hunzas fertilize their plowed fields with organic manures, they again duplicate nature's build-up of nourishment of her climax crops of plants by the return to the soil of whatever it has grown. It is reported that "Every solitary thing that can serve as food for vegetables, field crops and fruit trees, is diligently collected, stored, and distributed in rationed equality over every square foot of the hundreds of terraces. Sunken compost pits are conveniently located, and into them go the ashes from cooking and heating fires, inedible parts of vegetables, pulverized animal bones, dead leaves, rotten wood and collected manure from animals."

This organic manuring is the simple return to the soil of what it has grown. It is the provision there of the many organic compounds—still unknown in their chemical composition—which serve in plant nourishment, but of which, unfortunately, the function is information we still lack and are thereby lead to believe, erroneously, that only the inorganic fertilizers are needed—and the more soluble in water the better. Thus, instead of encouraging the microbes to burn out the soil's natural supply of organic matter by treatments with highly soluble salt fertilizers, the Hunzas maintain, and even increase, the soil's organic matter supply and its services, not only as organic nutrition of the crops, but also for its values in hastening the weathering of the soil's inorganic reserves of silt minerals for their release of plant nutrient contents through the help of the decaying organic matter via its temporarily active organic acids. The Hunzas have been practicing soil chemistry *au natural* rather than *au science.*

2. a little science stifled our curiosity about nature

For academic purposes and educational conveniences, we have made two major divisions of our chemistry, namely, the inorganic and organic. But in natural phenomena, like plant growth, those are not separated kinds of chemistries. In fact, plant growth is such an intimate union of the two that they cannot be separated. Even though we, in our own bodies, are considered organic and are composed of highly combustible organic substances, we still contain about 5% of ash as inorganic and non-combustible matter. That is intimately combined with the organic, even in the bones and the teeth which we emphasize as "mineral" in composition.

Because the inorganic chemistry has been the easier to learn, the simpler to demonstrate, and the more ready to comprehend, we start learning about chemical facts concerned with inorganic matters first. Then, all too often, we do not arrive at learning much about the organic matters of nature. As a consequence, we have studied plant growth—in relation to the soil that brings that about—mainly in terms of matching the quantities of ash elements in the plants against the quantities of those many inorganic elements in the mineral supplies in the soil. By ignition first to get the ash sample which we analyze, we have destroyed the many organic-inorganic combinations in which the elements initially occurred.

Liebig, the foremost early agricultural chemist, started our thinking about plant nourishment from the soil when he began with that analytical, or inventory-taking kind of study of the ignited residues. Only very slowly have we come to appreciate the simple fact that our most rapidly growing food crop of commerce, the mushroom, is nourished almost wholly by its feeding on the organic compounds which it takes from its bed of barnyard manure brought beforehand to a certain degree of its natural decomposition. This crop grows so rapidly that the grower speakers about its mycelia "*running* through the bed." The mushroom takes inorganic nourishment in very small amounts from the thin layer of "casing" soil on top of the manure, into which tiny root-like growths are extended only just before the mushroom matures enough to develop its spores, or before the reproductive process is completed and the crop has fully matured. This most speedily growing crop is produced, in the main, from salvaged organic compounds that were parts of preceding crops. It is not a crop synthesized from the elements coming directly from air, water and soil. It is rather a synthesis of a new crop from organic compounds left over from the old and dead crops.

With so much concern about sales of inorganic salts as fertilizers, we have given little thought to the nutritional role played by the decaying organic matter of the soil *as it moves into the plant in the form of organic complexities,* making up a significant part of the plant's diet, and as it mobilizes the inorganic elements in their intimate combinations within them to serve more efficiently thereby. It is in this latter role that nature has her inorganic and organic chemistries working together in what would seem unique, at least in our delayed recognition of it.

3. "chelation," a new term for an old natural phenomenon

Before the organic fertility compounds of the soil can carry any inorganic element into the plant, the latter must be combined with the former in ways by which the inorganic is not ionized or is not acting on its own. This activity, or situation, in which the two, namely, the organic and inorganic behave as a unit or molecule, is now included under the term "chelation." It is a case in which the organic part takes over the command of the situation. In addition, a protein-like part of the organic complex seems to be involved often, or at least the nitrogen (the common indicator of protein) is a means of linking the inorganic element so firmly into the organic complex.

In our "ash," or inorganic, chemistry of agricultural crops we have been emphasizing the natural separation (in solution) of the elements of compounds and of molecules, or their ionization by which they become the chemically active parts of them. Nature works in the very reverse with those elements as we are now interpreting them in chelation. Many soil and plant interactions consist of the biochemistry of large molecular compounds holding the ionization to a minimum by the union of the inorganic with the organic, and thereby allowing literally no ionization by mainly gigantic molecules, so well illustrated by the growth of the mushroom.

Some of our pioneer agronomists, as able chemists, may have had a vision of this unique natural phenomenon of the mineral fertility and the chelation of it by the soil organic matter, when, as an example, about 45 years ago Professor A. W. Blair of New Jersey said: "It is well known, for example, that by judicious use of lime and vegetable matter on the soil, reserve or locked-up mineral plant food may be made available." Similarly, one of the scientists of Germany, O. Flieg, touched on this idea when he considered "Braun Kohl" (brown coal) an effective agent, in some unknown manner, for mobilizing the inorganic fertility from rock to crop. As far back as 1905, the Russian chemist Tschugaeff discovered that the inorganic salts of some of the trace elements combined with the organic compound known as dimethylglyoxime ($C_4H_8N_2O_2$) do what we would now call chelation. Since 1916, and the research at the University of Illinois that organic compound has been an important analytical reagent for several inorganic elements by a single organic molecular structure. In this case the nitrogen part is the portion active in chelation, much as during the past ages nature has used magnesium as the inorganic central—but not ionized—element of the chlorophyl molecule for service in photosynthesis by plants. Now that the commerical markets offer chelating compounds, like ethylenediamine tetraacetic acid (EDTA), we are looking into nature's processes using similar ones as parts of her organic compounds occurring naturally in the soils and within the plants.

That such natural processes of chelation are common has been confirmed in a report from the Macaulay Institute of Soils Research, Aberdeen, Scotland, in which P. C. deKock tells us that "we showed that the growth-promoting effect of lignite was the result not of its trace element content, but of humic substances that made iron in the nutrient solution readily available even in the presence of high phosphate. Iron chlorotic plants were found to have high concentrations of iron in the roots, probably immobilized as complex ferric phosphates. This indicated that the effect of the humic acids was not merely an ion exchange at the root surface, but that the translocation of iron to the leaves was being promoted. This could be demonstrated by the split-root technique.

"We have been able to confirm that EDTA, absorbed by the root in one compartment, caused translocation to the leaves of iron supplied in the presence of high phosphates from the roots of the other compartment. Within one month about two-thirds of the EDTA supplied had disappeared.

"If a water extract of peat was used instead of EDTA, the plants grew equally as healthily. Humus-like substances from 'A R' sucrose were equally effective in preventing iron chlorosis. The plants became chlorotic if the humic water was withdrawn."

Here, then, we learn that by feeding on these organic compounds which are serving as chelators (including protein-like substances) the plants increase their power of feeding themselves, both internally and externally, on the inorganic elements. *It suggests that if plants are nourished by the organic matter first, they will feed on the inorganic elements, second, rather than vice versa as we have commonly believed.*

The Hunzas, in adhering to their ancient practice of using only organic manures, have been practicing the natural principle of chelating their inorganic fertility—applied as pulverized granitic dust—for its more efficient service in nourishing their crop, long before we even caught the idea or principle involved. They have been practicing the art of natural chelation without waiting for the technology of it to urge them to adopt it for practice under sales pressure of artificials for that purpose.

4. human ecology, too, according to soil fertility

The Hunzas are a good illustration of human kind located with varied health and success in survival according as the varying soils feed them for such. They are an example of the unappreciated fact that the geography of mankind, like the biotic geography of other forms, reflects the fertility pattern of the soil in its production of protein foods according as the climatic forces develop the soil to feed, or to fail, in that respect.

The 25,000 or more Hunzas are apt to be considered as a very limited case in point, but it happens to be a very positive case. It is the exception—possibly the proof—to the prevailing rule of bad health, or its serious degeneration. The Hunza case has been cited numerous times in literature on health, but it is to the late report by Dr. Banik, telling of the soil, that we owe the knowledge about Hunza practices of their management of the soil which is the foundation supporting them as a climax crop of humans, just as it has been the soil which has determined any other natural growth like the plants and animals to a climax.

It is the positive nature of the Hunzas as a natural evolution that makes their case so worthy of public notice. The Hunzas illustrate survival of the fit humans under adherence to the natural laws of biology—supported on a soil maintained by pulverized rock and organic fertilizers—rather than a following of the dictates of human ingenuity in technologies or economics. That simple fact is especially pertinent now that we are beginning to substitute the word "degeneration" for the time-worn professional term "disease." More careful note of a case like the Hunzas can do much to shift bad health, or the prevailing negative, to good health, as the positive, by educational emphasis, and otherwise, as the consequence of the individual's more careful study of it. Man's survival too—in no wise different in biological principles than survival of other life forms—reflects the forces of nature growing nutrition as the soil's creation of foods for such.

We may well study the Hunzas more critically because they are a people who have been healthy by natural evolution and individual efforts, rather than one so maintained through professional design and business by any portion or group amongst them. As we learn more about the fertility of the soils which have been giving us the natural climax crops of plants and animals, we may well learn what treatments we need to give to our soils to bring human health, in general, more nearly to its climax when such can come via proper soil management for the Hunzas as a climax human crop, isolated as they have been so long in the Himalaya Mountains.

• • •

Young Soils Soon Get Old Sanborn Field Testifies

When the young soil in Sanborn Field was plowed out in 1888 and put under different treatment in the many plots by the Experiment Station, Professor Sanborn left several with continuous cropping and no soil fertility return. These had the entire crops, both grain and straw, removed.

We can get a picture of some of the indicators telling us that soils become old, or how productivity declines, by studying the wheat yields of Plot 9. Nothing but the wheat stubbles were put back as organic matter. The grain grown there was not used as seed. Consequently there was no fertility applied save some very small importation of it via that route. Early plowing in the fall also reduced the chance for weeds to serve as catch or cover crops.

The wheat yields during the last 30-some years, plotted as a graph and illustrated herewith, show that the mean grain yield has been going down slowly to indicate the declining soil fertility. But more significant is the evidence of the declining rate of the chemical dynamics within the soil for release of active nutrients from either the inorganic supplies in the mineral reserves or from the crop residues as organic ones. The rise of the yield curve in nearly alternate years fairly regularly, and its corresponding schedule of fall suggest that this soil's chemical and biochemical activities are not ample to get it ready by the October seeding when it gave a yield in July. This soil now requires a "barren" year, or low yield year, to rest and to get itself ready for a yield the next year. This behavior of the soil body is a clear illustration that it is getting old, more frail and is slowing down.

Had this soil been in common agricultural use, it would long have been abandoned since its yields would not have paid for the costs of farming it to wheat. It tells us clearly that many of our soils are not what we could call "productive pastures" to feed our crops rustling to feed themselves there, but rather they are nearly empty feed lots with no stored feed supply nearby. These are the facts emphasizing the need to learn how to feed plants properly. Only more research in soil fertility and plant nutrition will help us become proficient in that task.

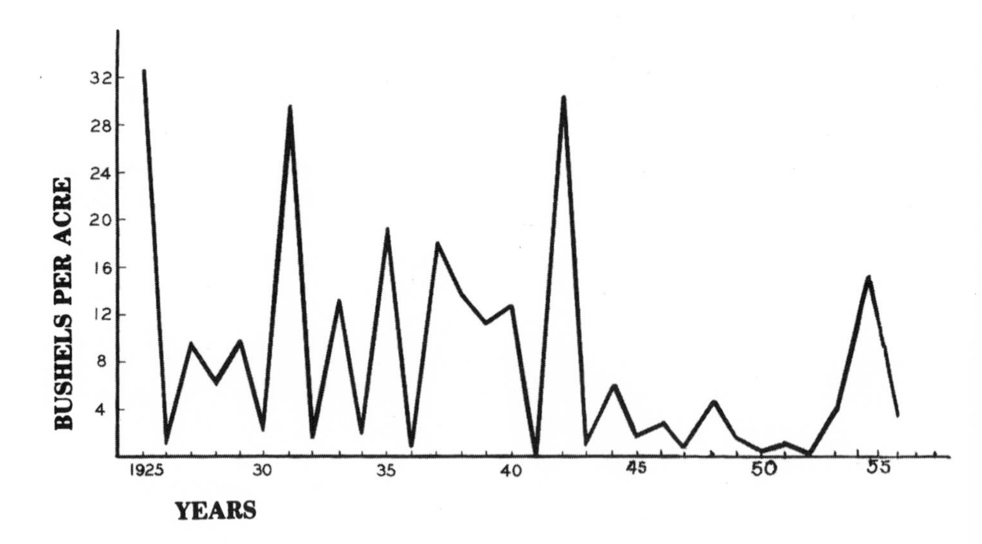

BACKGROUND

Thinking Naturally

"Are we not in danger of confusing education with training?" asked Dr. William A. Albrecht in the opening paragraphs of this paper. After all, "education is a life-long struggle under that self-discipline by the 'still small voice'." That small voice told Albrecht that nothing would ever be discovered in synthetics. Man was required to study nature so that the Creator's plan could reveal itself. "Those who teach must constantly hold up the challenge to study nature, not books," Albrecht often reminded those who would listen. This report explains both the philosophy of a man and the proper direction for inquiry into scientific matters. One of Albrecht's pet antipathies was the pseudo-researcher, the college drone who discovered no new knowledge, the one who takes the position that if he doesn't know about it, then it does not exist—"it" being findings recorded in detail in the literature. Science, Thinking Naturally *has not been published before. It was one of the many manuscripts by Dr. Albrecht that demand far wider circulation than they are likely to receive.*

Science, Thinking Naturally

One of our poets suggested, some years ago, that we should "Go forth unto the open sky and listen to nature's teachings, while from all around, the earth, the sea and the sky, comes the still small voice."

It is interesting to ponder just what the poet referred to when he said "the still small voice." Since he spoke of nature's teaching, it could be the voice of order about nature, in her regular, predictable and delicately controlled behaviors by which her activities are distinguished. It was in awe and respect of her teachings that man undertook his study of the natural order, the causes, the effects, and the many relations and interrelations of all that happens naturally. For many years before the term "science" was in common usage, the study of nature was called "natural philosophy." The accumulated knowledge from such study, arranged in order, is what is now included in the term *science*.

The word *science* comes from a Latin one, "scio—I know." When so much is said about science today, it might be well to clarify our thinking about that study.

1. education versus training

It was a bit of international jealousy—intensified by the first successful space missile, the Sputnik, in orbit—that has thrown our concern about our educational activities into a kind of temporary national hysteria. As the result, we are making extensive self-inspections of our teaching of the sciences. Some of us raise the question whether we believe we can now throw educational efforts into high gear when, in fact, the development of the mind is a growth, a biological matter, for which the fourth dimension, *viz. time,* is a requisite. Does man create his own body or his mind quickly? He does not so create crops or livestock. He waits for each of them to grow during its season. Are not mind and body also natural growths, too?

Will more classrooms, higher salaries for teachers, and the assembling of students into larger masses for instruction alter widely each separate mind's growth according to its own innate potentiality? Will more educational technologies and tricks of speedy memorizing improve the quality from natural mental development?

We need to ask ourselves, "Are we not in danger of confusing education with training? Training is a high-speed performance in the army, or navy, in time of war. Soldiers must be prepared quickly to take orders and to perform precisely as ordered, with speed on commands of the occasions when *"It is not to reason why, 'tis but to do or die."*

But that training makes slaves. Education has a far different objective if the fullest meaning of it is carried out. Education makes man free, responsible to his calling and mainly to himself; a fit citizen among fellow citizens and one possibly outstanding among all the others of the large human family. It is not the object of education to provide special privilege at the expense of others of the group. But it does aim to let each of us develop our special talents to the highest potential at the price of maximum self-discipline. It reminds each of us that every opportunity to learn carries with it that much more responsibility to do and to serve.

Education is a constant and life-long struggle under that self-discipline by the "still small voice." It is a constant emphasis on organizing the fragments of knowledge for better use in thinking in terms of the dominating and controlling principle, or principles, in each natural phenomenon we observe. Our learning, our reasoning and all mental behaviors fit into that category to give fuller understanding of the basic principles, or tenets, by which each of us reasons as the individual student, to behave according to high moral and religious teachings. Training comes via the strong voice of the classroom teacher; education via the still small voice from nature's teachings to the separate student.

2. *"study nature, not books"—agassiz*

But it is difficult to learn those moral and religious principles when one is separated from nature as environment, or is taken out of contact with all the biotic specimens lower than man himself. If one is to think naturally and live so, that thinking must follow mother nature first and human nature second. But one begins his responsible period of life with emphasis on gaining a living, keeping one's body healthy and active in the biotic struggle for shelter, raiment and food. Our moral and religious codes cannot be formulated far apart from the living as a body of flesh and blood. For that reason much for education in moral and religious matters can come out of education in the sciences. By contact with nature and the logical order of her behavior, the logical thinking in the moral and religious principles comes naturally, too. It avoids the phantom conflict so often mentioned between science and religion. The study of nature in basic sciences rather than as technology for gaining only a livelihood lays the foundation of self-discipline. The poet said more than we commonly appreciate when he suggested that we "Listen to nature's teachings," reminded us that man's survival demands that he fit into nature's larger pattern, rather than aim to conquer her and shape her performances to suit his economic and technological hopes and advantages.

The self-disciplines of one's body and mind, which make up education of ourselves, are apt to be considered as but a four-year session under training in a college community. In reality it is a life-long struggle. It begins with childhood and its disturbing and often perplexing questions by children. It should carry on during every day of one's life. It is a process of organizing our knowledge with emphasis on causes and effects (the eternal "why?" of the child) and with the correlation of every added item with similar ones that have come through and were organized in the mind previously.

When we concentrate our thinking on the mental organization of the facts and phenomena observed daily in the great out-of-doors, we call that knowledge "science." If our organized knowledge is about the truths of the thought processes, we call it logic and philosophy (Greek—*"phil"* love, *"soph"* truth, according to etymology of the word, or "love of truth"). When our organized knowledge concentrates on man's relations to his creator, as our God and Father of the human family, we call it religion. Thus, among the many categories of knowledge, "science" is only one of them. The same methods of logical organization and orderly thought sequence from causes to effects, which are included in science as natural thinking, are required in all other categories of organized knowledge.

It is the mental activity of organizing facts into a usable system of future reference while we are getting acquainted with them that constitutes what we mean by the word "study" in its fullest sense. By that lonely, quiet, mental, self-discipline our knowledge is a natural, quiet growth. The study of the sciences is basic for any later life practices of professions, and even theology is no exception. The study of sciences helps us to think naturally, i.e., to develop the mind more under nature's guidance. Sciences help us to hear the still small voice which we do not contradict without penalty or punishment.

3. the scientific method of research

The "high standard of living" is cited often to the credit of our knowledge of the sciences applied to harnessing the natural forces. That application has extended man's domain. It has given him extended scope in conquering distance, in locating and bringing resources or necessities to himself, and in extending his power over lower life forms as well as "to win friends and influence people." Much has been claimed for "the scientific method," and has placed scholars high in education and national thinking in what is termed "research."

Research is, first, a matter of a good fund of general knowledge. It requires more diligent study and more mental struggle toward fuller comprehension of nature's workings. It includes broader visions. The researcher's mind was, doubtless, of no different anatomical make-up at the outset than other minds. It was not born with a special label or a bright halo. The methods of effective learning are similar for all minds. But that of the researcher happens to be given more advantages and experiences in the gathering of facts and the organization of them.

Research uses logically organized thinking in that it builds the knowledge on nature's truths, by recognition of the relations between the facts operating naturally. Those relations were established and have been in action through the ages before man comprehended them, much less tabulated the facts of their orderly behaviors.

Credit for the existence of the facts is not due to research. That belongs to the creative pattern of the universe. Only late in its history, and much later in man's study of nature, is our comprehension of the numerous facts coming about. The order of causes and effects is initially established by nature. Man, understanding their relations, sees their combinations in serving to establish principles in certain thought areas to his advantage of their services for him. Only by accurate thinking can natural facts become the advantages we have in what we know as technology, and its application in labor-saving helps.

In accepting much of our technology, we fail to comprehend the principles operating in it, and forget about its coming to us mainly via commercialization for momentary profit. We do not see it as nature's laws serving us. It is apt to be emphasized as unusual because of man's ingenuity. For that reason, the patent laws were set up with protection of the patenter's claims for 17 years. Patents have concentrated on the technology, when it is accepted as a simple truth that laws of nature cannot be patented.

4. technology is merely natural laws in purposeful arrangements

As an example of a single technology, let us consider the facts and those arranged in their sequences and relations of causes and effects which we call the "internal combustion engine," used so widely in the automobile. That mechanical principle depends on a precise arrangement of many natural facts in sequence. But man's vision has put them together in a certain order of causes and effects to bring about specific results to his desires and advantages for him in his environment.

It has always been a natural fact that liquid matters evaporate into the atmosphere. In so doing, they become a mixture with the air. Gasoline is no exception to the laws of evaporation. But as a liquid consisting mainly of carbon and hydrogen and almost no oxygen, its two constituent elements burn explosively when the gaseous mixture of it with oxygen of the atmosphere meets with the high temperature of a lighted match or of an electric spark. It has always been a natural fact that the changes of carbon to carbon dioxide, and those of hydrogen to water, both as mixtures with oxygen, represent heat release on ignition with the production of powerfully expanding gases. If confined, those changes represent the exertion of tremendous power as pressure against the confining cylindrical chamber. If one wall is the end of a mobile piston in a cylinder, the planetary motion of that piston under power results; but in a limited single direction only. It is an additional natural fact that man's design of a crank on a shaft centered in a wheel to turn it, serves as a means of converting reciprocating, planetary motion into a rotary one of the wheel. Then by arrangements of valves to be opened and closed as intake and exhaust in relation to the reciprocal travels by the piston, gaseous mixture of gasoline and air can be brought into the cylinder, compressed, exploded and exhausted with an engine resulting if the timing of the explosions, by means of the electric spark, is also related to the positions of the piston in the cylinder. All these, and other natural facts, must, however, be correlated and coordinated to each other most refinedly to give power for high speed from the internal combustion engine by which technology we fly the air, sail the sea and travel the country.

Though we marvel at our technologies, shall we not remind ourselves that they have never brought forth any living bodies. They cannot reproduce themselves. They only consume, and destroy; and, thereby, dare we say they are a creation? We need that humiliating thought, the still small voice of our self-discipline, that constantly reminds us that *nature's creations, all about, transcend all that man ever envisions, contrives, invents or builds.*

We need to review the human experience, recorded in a drama of four centuries B.C., when the victim of many misfortunes told us to "Speak to the earth and it shall teach thee." Also the great Roman orator, Cicero, some 50 years B.C., raised his resounding voice to tell us that "To rebel against nature is not that to fight like mere giants against the mighty gods?" Isn't our science but the slow process of our learning to think on a higher plane and to comprehend but a small part of the wisdom of nature in her designs and methods of thinking and creating? Shall we boast of our thinking? Shall we not think via nature, or naturally, if we are to fit into nature's laws rather than think and work contrary to them?

5. science: a mental venture sailing the seven c's— curiosity

The development of any phase of knowledge in the logical order of arriving at the truth must follow the natural process of mental discipline by study. It is given to the child who soon struggles to make the first step in learning naturally when it exhibits the first factor of learning by study and inquiry, namely, its curiosity.

Curiosity begins with the child's innumerable questions and struggles for answers with understanding of what it observes. That is the start of the scientific method. That is the "take-off into research."

Equipped at birth with eyes to see and ears to hear, as two of the five senses given us early to comprehend and measure, the insatiable curiosity is exhibited at the very outset of life. It is a nature-given asset for acquiring knowledge, though at times almost deadly in nature's price for some self acquired knowledge.

But mere experiences by way of the visual and auditory senses are not education unless they result in mental records marking in memory their significance in relation to past and future experiences. Seeing may be a thrill for the eyes, but for education's sake it must be an observation. In that, the seeing brings to mind the comprehension of all that gives the properties and behaviors of what we behold. What we see must arouse the mental curiosity and set up the mental rearrangements to include that particular experience for future reference.

It was the seeing without the observing that the ancient teacher was distinguishing when he said: "They have eyes but they see not; ears and they hear not." In modern translation, we would say, "They see and they hear, but they do not study, comprehend and organize the report in their minds." The radio and television talk too rapidly to allow that.

The pre-school mind with its curiosity and continued asking "Why?" is commonly recognized as making the child's first steps in its education (not training). It is the beginning of the mind's self-guided struggle to get knowledge about what it sees, what it hears, and what it touches, tastes or smells. By the five senses, the mind is observing, measuring and analyzing objects, motions, properties and phenomena to be catalogued promptly for future reference. Childen's minds assemble some knowledge on their own and soon multiply requests to parents for help in accumulating more.

But in that parental opportunity—and responsibility—to help and encourage, we are prone to miss the obligation in the child's self-education. We are apt to seize the opportunity to exhibit parental authority to train the child. Unfortunately, when school systems seem to operate as an industry, as imparters of certain sets of information in specific curricular arrangements as requisites of an educational hierarchy, the youthful curiosity has little opportunity to develop, or remain the motive for study. Mere memorizing for test passing may replace diligent study which brings mental organization and sound reasoning as the mind's equipment for all the years thereafter. With no more mental development than the simple memory—so capable in the young mind—there later results frustration.

The fact is well illustrated by the junior university student about to discontinue his studies because, as he said, "I can't learn any more. I can't remember for the quiz what the teacher told us in the lecture." More appropriately, he should have said, "I have not been organizing by relations, causes, effects, *etc.*, of what the teacher discusses in connection with the reading assignments, or what I previously learned; consequently, I cannot think and write on the assignments of the quiz. My attempts at simple memorizing are failing. My memory alone is not equal to what a maturity in education demands." There are needs for training in some aspects of learning. But the fuller development of the mind requires the self-discipline of study, if true education is the objective. In that, it is curiosity that is excellent stimulation.

6. the second "c"— collection

The second step in education is the making of orderized mental collections. The mother's examination of a boy's pants pockets at night soon reveals that instinctive behavior of daily collections. But in that performance, the collection of materials does not proceed without the mind's cataloguing the properties and behaviors associated with the items collected. Closer observations result from what seems but a collecting instinct. That collecting is a precursor of the mental recordings of the differences between the properties of the items collected. Rocks and rainworms in the boy's same pocket are the beginnings of differentiations, shocking as they may be to the mother discovering them. That instinct is the beginning of collecting letters as the differentiations between words; between words in collections as sentences; and collections of sentences to transmit visions, and thought relations, or mental behaviors in all their varied potentials of originality in thinking.

For one interested in soils, the collecting of rocks is basic. Rocks are the starting of the Creator's work of making soil for the handful of dust. By blowing the warm, moist breath of the climatic forces of rainfall and temperature on the rocks, they are weathered into soil with a part remaining as land, and a small part going into solution and into the sea. On that land, as soil, the lowly rainworm is a noble inhabitant under our feet to surpass, in its delicately balanced relations to the soil that feeds it, any managerial abilities man has ever demonstrated toward the domesticated plants, animals and even himself for healthy survival with a fitness comparable to that in nature or the Creator's patterns. We need to catch a vision of how we might cooperate more successfully with nature's creative agricultural art before we boast that our scientific agricultural management is oversuccessful economically, and while we are forgetting that we are not overproductive in pest free, disease free and highly fecund, as well as decidedly healthy, self protecting plants, animals and man as nature had them in her climax crops when man first arrived.

7. the third "c"— classification

Classification is the third step in our mental venture into the scientific method, or in our thinking naturally. The angle worms in the boy's pockets are in all probabilities still alive. The rocks are not. The boy soon classified his collected objects according to the criterion of presence and absence of life. That is his mental step of meeting the chaos of the collection in his pocket and making a mental order for himself out of it, possibly notwithstanding contradiction by his mother.

It was the Swedish student and scholar of plants, Carolus Linneaus, who—as an ambitious collector satisfying his curiosity—realized that he had collected so many different plants to compel his grouping of them into smaller lots. He did that according to anatomical differences in their body parts, or such as means of reproduction, as the criteria. Classification, or taxonomy, of that kind was the height of botanical science as late as the 18th century. Form of the body alone was the first step in classification. Visions of body parts, as functions, waited to make their advent into the scientific method of thinking naturally until the 19th century.

Later science has taken to many other criteria in classification, according as we learned more about nature. We classify soils of the surface crust of the earth first according to what a vertical cut through the soil, as a profile, permits us to observe. From observations there, we apply, as criterion, the degree of soil development, or degree of change by original rocks to leave clay with its creatively active inorganic elements and as they have collected organic matter from previous vegetations, all making the soil in place, while other parts of the rocks have been transformed by weathering to be carried away.

Every science uses classification as an early step in its organization of knowledge. It is one so requisite and regular in the scientific method of thinking that classification is not emphasized. Nevertheless, we still have herbaria with their hundreds of plant specimens as a kind of fossil illustrations of the early emphasis on classification as a basic sector in the evolution of our botanical science.

8. the fourth "c"— correlations

When one classified objects, like the chemical elements composing the dead rocks and minerals of the universe, or the living matters of the earth, like the microbes, plants, animals and man, the properties by which they are classified suggest themselves as being correlated. This prompts our taking the fourth step, namely in correlations in properties and behaviors of objects, or in thinking naturally. In these correlations we note that different properties seem connected and behaviors linked together. By mathematical helps, we calculate the degree or regularity of certain associations, and we envision that such can scarcely be only an accidental one.

Relations between them was the reason for arranging the known chemical elements into the several periodic tables, one of which has been so commonly used that students of chemistry, before the atomic energy age, labored under the belief that there was but the one periodic table of Dimitri Mendeleyef, the Russian held in high honor by the Soviets. Similar tables, by a Frenchman and a Britisher contributed to the evolution of that scientific step. But now in the atomic energy era of science when the atom is no longer our concept of the smallest bit of matter or energy (vice versa, if you choose), the periodic table has become a much more complex one with the correlations still more refined, even though the arrangements by Mendeleyef and his predecessors is still the core of it.

Correlations, a kind of intensified study of properties, serve as challenges to our closer observations and more critical examinations. They were an early help in the belief that the form or anatomy is an end in itself, but is the means by which function, or physiology, is made possible. The structures of the cell of the microbe, the plant, and the insect or animal make possible their behaviors, including those which make life possible. Correlations are, then, in no small way refined classifications showing the objects more similar. They move our minds into extended visions and challenge our continued study just for more knowledge of nature.

9. the fifth "c"— causes and effects

While we may find two properties, or two behaviors, so closely correlated that the mathematical measure of that says there is an agreement of 100%, even then we may be in error. There is place then for righteous doubt. Such situations in the scientific method bring us to the fifth step in natural thinking, namely, the necessity of establishing the relations of cause and effect by experimentation to establish that relation by demonstration. Even when two properties or two behaviors always occur simultaneously, and also when the value mathematically established of their correlations is 100%, the one may not necessarily be the cause of the other. They may have a common cause, and hence are perfectly correlated. This, the cause and effect stage in scientific thinking, represents the test of truthful thinking, or doing so in accord with nature in the case of the sciences. It is the stage where true research enters. All the preceding steps, or stages, in the scientific thought are preliminary to what is truly research, namely, establishing proof of the cause and effect relation.

For the establishment of a case of the cause and effect more is demanded than careful observations, accurate classifications and closest correlations. Situations must be arranged where one factor is varied while all the others remain constant. Only by that set-up can the single variable be demonstrated as the cause of a certain effect. Research begins with a new vision or a new theory. It then proves the theory. Such carries out the ancient writer's injunction when he said, "Prove all things. Hold fast to that which is true."

Much that may have erroneously been called research does not fit the above limited sense of that term, namely, starting with a new theory and establishing new truths to add new knowledge to our present supply of it. *Much that is claimed to be research is merely more search in that it establishes no new knowledge. Scientific research increases the sum total of it. The pseudo-researcher may demonstrate facts new to him, but previously reported and recorded in detail in the literature, which has escaped his attention or notice.* Or the pseudo-researcher may test a theory new to him, but one previously established by a predecessor, hence the work is not research in the strict meaning of the scientific term, *research.* In the correct sense of the term, it is research that builds our thinking to its highest level of doing so naturally.

10. the sixth "c"— conclusions

The real service, in the broader respects, from research comes in the sixth step of scientific thinking when the proper conclusions are drawn from research, namely, the test of a theory resulting from progressive thinking within one area. With sound logic in the conclusions drawn, new facts set themselves up to be added to our satisfaction in knowing more and comprehending a larger portion of nature's behaviors.

11. the seventh "c"— contributions

When once the research performed reports its conclusions, then the contributions to other thinkers follow closely as the last step and the reward in the scientific method of proper thinking and doing so correctly. Any new fact of research soon finds itself organized with other facts into principles underlying new practices and technologies. It serves in new predictions or new theories. Those then call for more testing in more new research and more contributions which are all parts in mental growth in total. But only the correct thinking in accord with nature's laws will give mental growth offering contributions to service of human kind.

12. nature emphasizes the survial of the individual, not the crowd

In the study of nature, one sees the emphasis on the survival of the individual, multiplied to compare the survival of the crowd, or the species. It is a common saying that "self-preservation is the first law of nature." Yet nature operates with group survival, by division of protective labor, in the case of the family under an intense union as parental protection until the period of infancy has passed. There are matings for life in some species. Large numbers in a group find many illustrations as a gregariousness for advantage to the survival of the individuals. There is even division of the duties within the group as each contributes to survial in terms of a single factor.

In the swarm of bees as a distinct colony, there is single parenthood in the queen bee. In a city of prairie dogs, it may be several families. The same is true of a herd of bison, wild goats, wild sheep, zebras, giraffes, elephants, monkeys and other species. In those cases, it is significant to note that relative to food they are mainly all herbivorous. When one considers the pack of wolves as a grouping, there is no consideration of one individual for others since all are brought together by dire hunger and are carnivorous. If the prey is lacking, then cannibalism within the pack breaks out. It is significant to note that the prey of the carnivora consists mainly of the herbivors. Between the predator and its prey, the natural laws are constantly adjusting the number of the one against the other without the complete extinction of the other.

But among the omnivora, among which man is grouped as those nearest the apex of the biotic pyramid, the natural laws of relations of prey to predator no longer hold, and both groups serve to extinguish other lower species, or strata completely. Man has demonstrated that disregard of the fact that thereby he—at the apex of the biotic pyramid—is destroying other segments, or population strata, by means of each of which in its proper place only he can be maintained at the top.

Man fails to consider the facts of nature reminding him that his behavior must fit him, not only into his single species of homo sapiens for its and his survival, but, in turn, his species too must fit its behavior into those of all the lower ones of the pyramid for which the soil is the creative foundation of all the units or strata maintaining the entire living structure. "Can man save himself?", a dramatist of two dozen centuries ago inquired, and would paraphrase it today by asking, *can man be at the apex unless all other species beneath him act as their separate supports for the entire living pyramid?*

It was by intense natural thinking, or by well-organized knowledge of nature's laws that the pioneer found his better health, and greater strength for survival in cooperation with nature. But as we moved to apply science to modify and attempt to manage our environment for the many purposes included in attaining our "high standards of living," we have also collected the individual survival by cooperation with nature while demanding competition with human nature. Can we visualize the latter as a move away from the gregariousness common to the herbivora looking to the soil directly for survival, and move toward the collection into hungry packs of carnivora, and grant the propriety of the philosophy of "dog eats dog?"

Dare we venture the thought that as we think less naturally, less according to the laws of nature, as illustrated by sciences for knowledge's sake, we move counter to even the concept of the family with its biological fatherhood and go strongly counter to any concept of the human family under spiritual fatherhood? If the term "science" includes only phenomena of nature, dare we classify studies of human behaviors under organized knowledge as reliable as that of natural sciences? Studies of nature's laws in the sciences give humility and guidance under natural forces as a property of human behavior. One lives by cooperating with nature and by laws which transcend legal classification in courts. Study of human behavior for wisdom in behavior's advantage among others put us at a loss for the immutable standard accepted without argument of contradiction when thinking naturally. Behavior studies tempt one to emphasize living by our ingenuity to collect it from other humans. On this motive for living, the vision of a spiritual union or family is not so readily caught as when we are thinking naturally.

In terms of our problems of living our own lives, it appears that we need to practice more than conservation of things natural, more than matter of food, raiment and shelter. We need the creation of the spirit of natural conservation, the more elements of real worship of the power of creation, apt to be a power taken over, in part, by man with his scientific knowledge applied to nature's control under technology. Perhaps we are making a turn, now that degeneration of our bodies is recognized and it is more widely agreed that we are the agents of our own degeneration, no longer ascribed to "disease."

Perhaps the shades of the poet are about to return, and we shall hear him say, but with a new meaning for us, "Go forth unto the open sky and listen to Nature's teachings while from all around, from the earth, the sea and the sky comes the still small voice."

LESSON NO. 3

Life

40.

George Washington Carver once said that we ought to appreciate people who could do simple things quite well. Growing food, he pointed out, rated as one of man's finest pursuits. Dr. William A. Albrecht would have agreed. All phases of farming were important to the great Missouri scientist. No subject was too mundane for his attention or pen. Thus a long series of articles on everything from mulches to mychorriza. The Use of Mulches *was printed as a* Bulletin *of the Garden Club of America, 1952.*

The Use of Mulches

In considering the use of mulches in gardening and field practices, it is necessary—at the outset—to define the word "mulch." For the purpose of this discussion, let us agree that *a mulch is a particular arrangement of, or addition to, the surface of the soil to simulate, or to provide, a cover.*

According to this definition, then, a mulch may be made, either by nature or by man, from the upper or surface portion of the soil itself. It may also be the result of applying on the soil some cover consisting usually of organic materials, particularly plant residues. These may be imported and applied as a direct effort for that service. They may be grown in place as a by-product of some other crop and be managed in connection with that crop to accomplish the particular purpose. In addition, it is possible to get mulching services, not only from the residues of predecessor crops, but also from companion crops under conditions where one crop is growing for its direct values while, as a by-product value, it is serving as the mulch for another growing crop. The mulching services can then be a by-product of the management rather than the result of a direct effort or a specific investment for the sake of only the mulch.

Already you recognize the use of the word, mulch, as both a noun and a verb. By specific definition of it as the former, we consider it a kind of cover put on the soil or a simulated cover consisting of a modification of the surface soil itself. In discussing the use of mulches, then, one might begin to classify them into different kinds and to describe their appearances, sources, costs and other aspects of the different materials composing them. It is not the aim here to discuss the kinds of mulches and mulching materials. But when the word, mulch, becomes a verb, naturally we ask: "What are we doing to the soil when we mulch it?"

In dealing with the functions of the mulch, we must consider not only its direct effects on the soil by being a cover to reduce evaporation but also those effects via the soil on the crops growing on it. It is this complicated relation of the mulch to the soil body, to the soil fertility or chemical contents, to the microbes, and the crops growing on the soil that calls for discussion here. It calls for full understanding of these relations if the use of mulches is to be a wise practice in place of disappointment or even disaster.

1. the soil mulch—au naturel

When a soil has been watered generously by rain and when the drying of its surface starts, it is by that rapid surface drying that nature herself makes a soil mulch. In some cases, this is a soil crust of one or two inches of depth. In others it is a granular, non-crusting cover. Both of these dried soil layers are mulches to help to reduce evaporative loss of water up through them. That the cultivation of the moist soil to hasten its surface drying into a cover or mulch could hold down water loss from the lower depths of the soil was emphasized by the followers of Professor F. H. King's experiments of about 50 years ago. At that date the water in the soil was emphasized in the use of mulch while the fertility of the soil was disregarded. Unfortunately, King's followers forgot that his demonstrations were on soils with their water table within three feet of the mulch. This was an experimental situation and one not common in soils under cultivation.

While the crust mulch functions to cut down water loss by evaporation, it fails in the second function of the mulch-soil-water relations; namely, it does not facilitate ready entrance of the rain water into the soil. Aiding infiltration of the rain is the really significant function of the mulch or the soil cover in relation to water. For this function alone, the breaking of the soil crust and the mechanical maintenance of the granular mulch are well considered practices.

It is the higher organic matter and fertility contents of the soil that bring about the naturally granular in place of the crust mulch. The soils of the midwest and the semi-humid plains granulate themselves so readily that the winds pick up the soil. Strong winds make dust storms of it. This ready granulation depends on the fertility of the soil.

These are the lime-laden soils. They are well stocked with other nutrient elements. Their virgin vegetation of many legumes built up their content of nitrogenous organic matter. Other complex organic compounds suggesting the long-chain chemical structure of the commercial product, Krilium, are also abundant. It is the chemical aspects afforded these soils by the climatic conditions controlling their particular fertility development that makes them take water so readily. Here granular soil mulches have always been natural because of the fertility of the soil. These situations raise the question whether the crust mulch might not be a suggestion that the soil needs more fertility and more incorporated organic matter for a naturally induced mulch rather than more tillage treatments for a mechanically created one. Shall we not consider the serious crusting a serious fertility shortage?

2. the organic mulches—by application

The use of straw, leaves, sawdust, and other organic materials as mulch for growing plants is a well-known and good gardening practice. But the relations of these materials to the soil fertility and via that to the crops of microbes within, and of plants above the soil is less well known. Were one to incorporate these organic substances into the soil, their woody nature offered as fuel foods to the microbes would be disturbing to the crop. It would compel these single-cell performers, using these carbon-rich compounds as energy foods, to balance their deficiency in nutrient elements by drawing on those in the soil. Thus the microbes, more profusely distributed through the soil than are any plant roots, would be serious competitors with the plants for the nitrogen, phosphorus, calcium and other soil-borne fertility elements. The microbes eat at the first table.

The application of such woody materials in a mulch sets up this competition at the zone of mulch-soil contact. Decay moves the partially changed organic matter from the mulch down into the soil to do slowly and without serious irregularity what the incorporation of these materials into the soil would do all too rapidly. Consequently, the use of these mulch materials assumes that the soil under them has been brought up in its fertility level to the point of discounting any possible disastrous effects by microbial competition with the crops for the nutrients in the soil's supply. It is around this simple principle that the wisdom of the use of the mulch turns for failure or success.

Naturally as an alternative, and for the avoidance of this possible competitive hazard, the use for the mulch of legume hays and other materials, themselves rich in fertility elements, is most logical. The higher feed values for animals of these mulching materials suggests that they are also well balanced microbial diets. Consequently they decay rapidly as mulches without injury to any crop they mulch, but they do not last long in place. Some thought about this simple relation of the composition of the mulch materials to the fertility of the soil may repay well in the more effective results from the wise management or use of the mulches.

3. the mineral mulches—by application

More recently there have become available mineral materials of very little weight per unit volume which serve as mulches, particularly in potted plantings. The most prominent of these, known by various commercial names, are the expanded micaceous materials. Almost completely inert chemically, as they are—save for possibly the contribution of some potassium—they are long-lasting and bring about an open structure when incorporated into the soil. This effect is particularly advantageous on soil of more clayey nature. Such soils are benefited highly by the mulches in preventing crusting and cracking and then still more by the successive incorporations of these flakey mineral materials into the body of the soil itself. This repeated incorporation brings on the improved granular condition for self-mulching, so much desired.

4. the organic mulches—by growth in place

That the mulch might be a by-product of the crop itself or that extra fertility might be added to the soil for growing one crop that is to be the mulch for another, is a newer concept in the use of mulches. Unfortunately for the weed crops, we have given little thought to their value as mulches. Little consideration has gone to fertilizing the soil well enough so that with the minimum of tillage for weed destruction during the early life of the desired crop there would be enough fertility in the soil to grow an abundance of it and yet enough of the weeds or unwanted crop later to let them serve as soil cover or as the mulch. It may be difficult to consider weeds as friend when so long by tradition they have been considered as foe.

Corn has been condemned and labeled as an "erosive crop" largely because our soils have had insufficient fertility to grow the corn crop and along with it enough weeds to provide protection against serious runoff of water and damaging erosion. During the early use of the cornbelt soils the weeds coming in after the corn roots were well down into the soil were a cover crop serving to retain the soluble nutrients elaborated in the surface layer from which the corn roots were already excluded by tillage, to reduce erosion, and to return to the soil a mineral rich organic matter for manurial values when plowed under for the succeeding crop. But since our cornbelt soils have been so depleted of their fertility under intensive cultivation for corn production, they are not only failing to produce a paying corn crop, but are also failing to produce enough weeds to be a sequel crop to the corn and to be winter cover against soil erosion.

Now that we have remaining but few upland soils, in Missouri for example, which will produce enough corn to pay the costs of production unless fertilizers are used to increase the acre yields; and now that we are treating the soils to grow a hundred bushels of this cereal per acre, the weed problem is almost eliminated by the vigorous and dense growth of the higher populations of corn stalks required for such yields. These stalks as crop residues are so much more abundant that they are an excellent soil mulch. Here we are growing not only a good corn crop but, as a by-product of this crop, we are simultaneously providing a dense mulch lasting over winter. In addition, this organic matter worked into the soil with extra fertilizers at the time of being plowed under decays readily in such an underground composting procedure for artificial manure production from it. It gives the surface soil an increased self-granulation or brings about the natural production of a soil mulch when the young corn crop is benefitted most by this condition of extra aeration, rapid delivery of available nutrients and deeper infiltration of the rainwater.

Some recent trials, using chemical fertilizers as soil treatments by which the corn stalk residues might be used as a means of this crop's building up the soil organic matter under itself, are telling us that, even with continuous cropping to corn, this way of building up soil organic matter is no myth or idle dream. With increasing amounts of nitrogen in the fertilizer mixture applied, the annual crop residues going back into the soil were increased so much that now from 2½ to 4 tons of dry residues per acre are left over winter. This procedure going forward for four successive crops is suggesting that it was a starving corn crop that went with an erosive soil, and that if the crop is well nourished at the outset with chemical fertilizers it can grow its own mulching materials to reduce erosion and simultaneously build up the organic contents to provide ample organic fertilizers within the soil.

As a crop combination, corn is now producing well in more widely spaced rows with even a grass or a legume growing as an interrow or companion crop. This pattern of cropping was published and copyrighted some eight or more years ago and put into practice by M. J. Spivack, of Bayonne, New Jersey. It is now getting more attention in the cornbelt states with some marked reception and claims for it. There is the growing recognition of the necessity for soils to be fertile; at least there must be a significant amount of fertility coming out of the soil—possible only when much is going back in chemicals applied or in the organic matter grown in place.

5. *nature offers suggestions*

While the soils in the midwestern United States form a granular mulch naturally under sparse grass vegetation, nature has been applying an organic matter mulch in the form of either heavy prairie grasses or forest leaves and litter on the more humid soils in the eastern United States. Here is nature's suggestion that the degree of development of the soil, according to the differing degrees of weathering under the climatic forces with the resulting different levels of soil fertility, points out the mulching procedure that would be wise. If the soils of the eastern United State are to be self-mulching by their own granulation in place of requiring applied mulches, then they must be brought up in their fertility to the level duplicating that of the soils in the midwestern states.

Only by building up the organic matter and the fertility contents in the more highly weathered humid soils will they become granular and mulch themselves effectively. With the naturally more granular midwestern soils, the high fertility and the organic matter contents make each unit of rainfall more effective. Mulching artificially seems to have come into vogue because the soils were less fertile. But the emphasis went to the water rather than to the creative power of the soil. If such is the case, then, by reasoning conversely and building up of the soil fertility to enable the soil to grow into itself more organic matter, the need for extra mulching should be less or the extra mulching should be so much more effective.

Nature's climatic patterns of the soil are giving their suggestions by which we can make better use of mulches. Our soil will find itself undergoing conservation much more extensively and will be used more efficiently when we see nature's pattern of natural mulching with its benefits according to the levels of soil fertility concerned. Mulching alone, as a mechanical ministration, cannot offset completely the shortage of fertility in the soil. Conversely, however, building up the fertility can be all the more reason for mulching also, a combination with doubled benefits because of the more efficient use of both the soil and the mulch that covers it.

41.

Few plots of ground on spaceship earth have revealed as many of nature's secrets as Sanborn Field, University of Missouri. When Dr. J. W. Sanborn, in 1888, outlined plans for use of the test plots that bear his name, he discerned that short term experiments meant all too little. It would take 25, 50, 75, even 100 years to determine whether farming practices fit into the curve of history. Dr. William A. Albrecht was on hand to read the results as they arrived. This report first appeared in Soil Science Society Proceedings, 1938, *and again in* Bio-Dynamics, *1950-1951.*

Variable Levels of Biological Activity in Sanborn Field After 50 Years Of Treatment

Sanborn Field, where the predominant soil treatment has been manure additions and some fertilizers in continuous cropping and different rotations, provides an opportunity to study the properties and the performances of the soil organic matter as an index of the biological activity in relation to soil management practices. After 50 years of treatment the organic matter fraction of the soil might well be expected to have some earmarks of stability in nature and behavior which could be recognized by analytical study.

The hypothesis for undertaking such a study is the belief that organic matter returned to the soil is a greater source of nutrients than commonly considered. Organic matter is readily recognized as the source of energy for the majority of soil chemical changes. As the source of carbon dioxide, we grant it a prominent place, and have been led to believe that in setting free this acidic agent through microbiological struggles, it serves to break the bases, or cations, out of the crystal minerals to be absorbed or held less firmly by the clay complex, and thus changed from the unavailable to usable forms for the plants. Skepticism toward such a belief in the mineral breakdown as the main source of nutrients for plants should arise from the acquaintance with the high ash content of the so-called "humus" or organic fraction taken from the soil. Also, the stability of the colloidal mineral, or clay fraction, under rather drastic chemical treatment in the laboratory, indicates its insignificant change during a single growing season. When organic matter decay reduces soil acidity, isn't such possibly the result more from oxidized alkali or alkaline earth residues left as ash than from basic residues liberated by rock and mineral breakdown under carbonic acid attack? By determining the more exact nature of the carbonaceous and nitrogenous fraction of the soil, and then by learning its rate of breakdown—particularly during the growing season—it will be possible to learn whether or not this fraction of the soil is the main, or almost sole, contributor of nutrients to the plant during the growing season.

Sanborn Field is of the Putnam silt loam type in the Putnam-Vigo-Clermont association, whose profile consists of a silty surface underlain by an impervious horizon of compact clay. Its climatic equilibrium locates its surface acre nitrogen content at approximately 2,500 to 2,600 pounds. Its exchange capacity amounts to 13-15 M.E. per 100 grams soil, and its degree of base saturation is about 50%.

The cropping systems on Sanborn Field include continuous crops and rotations while the list of soil treatments contains manure, ammonium sulfate, sodium nitrate, superphosphate, complete fertilizers and limestone. Some of these treatments have extended through only the last 25 years, while some have been regularly followed since the establishment of the field 50 years ago. Attempts have been made to learn something not only about the changes in the supply of the organic matter, but also about the nature of the organic matter, particularly the differences in its composition as shown by the carbon and nitrogen contents, by the differences in the extracted humus, by the degree of lignification (woodiness), and by the differences in rates of its nitrification. The studies on the nature of the organic matter of this old field were undertaken to learn more about the biochemical activity within a soil as related to its treatment in the hope that such knowledge might contribute to more wise soil management.

1. changes in supply of organic matter

Careful analyses of the plots for nitrogen and carbon were made at the end of 25 years and again after 50 years. Not all the plots will be cited here. Only a few of them have been selected for citation as they lend themselves for common use through the other phases of this study. Table 1 gives plots in continuous wheat and continuous timothy to illustrate the extremes in organic matter changes in the field.

Outside of the plots in continuous sods, the depletion of the organic matter is the characteristic not only of the plots cited in the table but for most of the plots. Since no analyses were made of the original soil, the change in the organic matter content, as based on carbon, is determinable for only the last 25 years. During this brief period the changes amount to as much as a loss of more than 10 to 20% in the original supply with continuous cultivated cropping and a gain of 20% under continuous sod cropping.

2. nature of the organic matter

The lowest supply of organic matter after 50 years occurs in Plot No. 2 under continuous wheat given heavy additions of commercial fertilizer. This seems to suggest that the addition of the nitrogen and minerals has supplied a microbiological deficiency, and that fertilizers served thus to deplete the organic matter further here than on any other plot. With over 750 pounds of mixed fertilizers, of

Table 1
The Lignin Content of the Organic Matter After 50 Years with the Changes in the Organic Matter Stock in the Soils of Sanborn Field During the Last 25 Years

Plot No.	Crop and Treatment	Organic Matter Content % After 25 Years	After 50 Years	Change - or +	Lignin (%)
2	Wheat—Fertilizer	1.94	1.75	-.19	39.5
5	Wheat—Manure 6T, 25 yrs. Manure 3T, 25 yrs.	2.62	2.18	-.44	48.9
29	Wheat—Manure 6T, 25 yrs. Ammonium sulfate, 25 years	2.37	1.84	-.53	46.1
30	Wheat—Manure 6T, 25 yrs. Sodium nitrate, 25 yrs.	2.77	2.24	-.33	50.5
23	Timothy—No treatment	2.29	2.49	+.20	40.5
22	Timothy—Manure 6T	2.91	3.51	+.60	48.8

which no small part was mineral nitrogen, the carbon supply of the soil has been extensively burned out. That the phosphate and potassium are essential contributions toward increased microbiological activity is testified to by the higher content of organic matter in the continuous wheat plots given ammonium sulfate and sodium nitrate. These are still higher at this time in organic matter than the plot given complete fertilizer. The ammonium sulfate treated wheat plot which is now very acid, is rapidly moving toward the lower level of organic matter found in the plot given complete fertilizer. If the minerals are deficient here the nitrification process may be liberating some from the mineral fraction of the soil.

These facts suggest that organic matter decay in the soils of many of the Sanborn Field plots is a much slower process because of the deficiency in phosphorus and possibly in potassium. The addition of mobile or soluble nitrogen is also instrumental in decomposing the organic matter or carbon as based on the determinations of this latter element. Thus, we may reason that the organic matter in the soils like those on Sanborn Field must be approaching a very stable condition. This stability would seem to rest on its highly carbonaceous nature, or its supply of immobile nitrogen, which does not serve microbiological life for its digestion or decay unless balanced or supplemented by soluble nitrogen and minerals. If this is the nature of the organic matter then its accumulation under the timothy would suggest such an occurrence because of a deficient supply of nitrogen, deficient phosphorus, and deficient minerals, or it may be simply remaining there because these deficiencies prohibit the microorganisms from decomposing or removing it.

3. changes in the nature of organic matter with time

A study of the carbon-nitrogen ratios of these sample plots also points further to a wide difference in the nature of the organic matter, and consequently, a wide variation in microbiological activities. In table 2 are presented the carbon-nitrogen ratios along with percentage contents of these elements for a few of the interesting plots.

The shift to a wider carbon-nitrogen ratio as a result of the nitrogen additions indicates a more active form of organic matter in the soil as a result of nitrogen fertilization. Even though the total supply in this soil was decidedly lowered as measured in terms of either carbon or nitrogen, yet this lower supply with its wider carbon-nitrogen ratio suggests less stable or more active compounds. These changes which occurred in these nitrogen treated plots within 25 years suggest that the stage of activity was very low already 25 years ago, when the carbon-nitrogen ratio was very narrow, and that more activity is now prevalent in the material of wider ratio. On the wheat plot, given complete fertilizer, the nature of the organic matter changed little during 25 years as indicated by the stability of the carbon-nitrogen ratio, or its similarity in the two intervals. Manure additions, likewise, maintained a stable carbon-nitrogen ratio, in the face of a declining total supply of organic matter. Under timothy without manure, the total carbon increased while the total nitrogen decreased, thus giving a wider ratio. Where manure was applied on the timothy soil, the carbon-nitrogen ratio widened at the same time while the total organic matter increased. This suggests a tendency to hoard carbon. In four of the plots, whether manured or not, the carbon-nitrogen ratios were narrow but similar. They were much wider after 25 years where nitrogen was applied, even though the total organic matter supply was decidedly lowered during that same interval.

Table 2

Total Carbon and Nitrogen Contents Together with Their Ratios at 25 Year Intervals on Sanborn Field

Plot No.	Crop and Treatment	After 25 Years Carbon %	After 25 Years Nitrogen %	After 25 Years C/N	After 50 Years Carbon %	After 50 Years Nitro gen %	After 50 Years C/N	Changes in Carbon	Changes in Nitrogen
2	Wheat—Fertilizer	1.13	.107	10.5	1.02	.100	10.3	-.11	-.07
5	Wheat-Manure 6T 25 yrs. Manure 3T 25 yrs.	1.52	.140	10.8	1.27	.119	10.6	-.25	-.21
29	Wheat-Manure 6T 25 yrs. Ammonium sulfate, 25 yrs. . . .	1.38	.145 9	9.5	1.07	.081	13.2	-.31	-.64
30	Wheat-Manure 6T 25 yrs. Sodium nitrate, 25 yrs.	1.61	.171	9.4	1.30	.094	13.8	-.31	-.77
23	Timothy-No treatment	1.33	.141	9.4	1.45	.135	10.7	+.12	-.06
22	Timothy-Manure 6T	1.69	.177	9.5	2.04	.195	10.4	+.35	+.18

4. nature and quantity of the extracted humus

As a partial attempt to differentiate the organic matter on the different long continued plots, humus extracts were made of six of them, and analyzed for their contents in calcium, phosphorus and silica in addition to their carbon and nitrogen. These data are presented in table 3.

Table 3
Analyses of Extracted Humus from Some Sanborn Field Plots

Crop	Treatment	Soil Content %	Humus Composition: Nitrogen %	Carbon %	C/N	Calcium %	Phosphorus %	Silica %
Corn	Manure	3.280	7.40	14.15	1.91	2.24	.710	6.08
	No Manure	3.218	3.45	8.76	2.54	2.12	.448	3.34
Timothy	Manure	4.712	7.09	14.11	1.99	1.71	.648	3.44
	No Manure	3.314	5.52	16.51	2.99	1.71	.842	2.82
Rotation	Manure	3.958	5.34	18.09	3.38	2.74	.724	5.70
	No Manure	3.322	4.99	17.07	3.42	1.56	.808	2.16

The total quantity of humus was lowest in the cultivated soil, followed closely by the soil in the rotation and that in timothy, both where no manure was used. They take this same order, but at a higher figure, where manure was added.

The nitrogen content of the humus varied from 3.45 to 7.40%, with the high figures under corn with manure, and sod given the same treatment. The lowest nitrogen figure occurred under corn that had no treatment. Under rotation, the manure addition gave the least variation in nitrogen content of this extract. These rotated plots and the sod plot without manure fluctuated closely around the nitrogen content of 5% customarily considered as the average for humus.

The carbon content of the humus fluctuates from the high figure under the rotated soils to the low one under the corn without manure. In each case, the extract from unmanured soil is lower in carbon than from the manured soil. The carbon-nitrogen ratio is also wider for unmanured than for the manured soil, and points to relatively more nitrogen in this humus form as a consequence of manure additions. Manure nitrogen then must be retained, in part at least, in the more stable humus fraction of the soil.

The mineral contents of the humus are also widely different with the different soil treatments. Except for the timothy soil where the two analyses are the same, the calcium contents are higher for the soils with manure than for those with no manure. That the lowest calcium content should be in the untreated soil with rotation may not be causally related to the fact that this is the only soil in the group attempting to grow a legume. The silica content was higher in every case of manure treatment agreeing in general in this respect with the calcium content. The phosphorus showed the reverse in the case of the timothy and the rotation, but paralleled the calcium and silica with reference to manure for the continuous corn. The lowest phosphorus content in the humus extract occurred under corn given no manure. Manuring seems to have increased the calcium and silica contents in the extracted humus and raises the question whether the extra silica means merely more silica as ash from manure decay, or a contribution from the mineral fraction to the organic fraction by more interaction between the colloidal organic matter and the colloidal mineral matter of the soil.

The extracted humus reflects the soil treatments and follows trends in amounts and in composition as one might surmise their influence. The widest spreads occur between manure and no manure additions under continuous corn, for all the tested properties of the humus. The differences between these two soil treatments were less significant under timothy sod and even less so under rotation. Manure gave the humus with the highest content of nitrogen, with the narrower carbon-nitrogen ratio, and the larger total amount of humus in the soil. It also gave the higher calcium content, and silica content to this extracted fraction. It suggested, however, the reverse relation for phosphorus.

5. *humus as a possible supply of minerals for the crop*

If we were to assume that this humus is the sole supply of soil nitrogen to the growing plants and that, during the liberation of this nitrogen, the humus is broken down to liberate also its calcium and phosphorus, it is interesting to calculate the crop possibility in terms of calcium and phosphate so released. Taking the continuous corn plot without manure, as an illustration, of which the soil contained 3.2% humus, there would be 64,000 pounds of humus per acre, carrying 3.45% or 2,208 pounds of nitrogen. This plot has been producing 20 bushels of corn per acre as an average requiring 30 pounds of nitrogen as an estimated figure. The release of this amount of nitrogen would break down 1.36% of the total humus in the soil. If at the same time, the calcium and phosphorus in this corresponding amount of humus were liberated, these amounts would be 18.4 pounds and 3.91 pounds, respectively, or the calcium needed for a corn yield of 50 bushels, and the phosphorus for 25 bushels, both of which are yields above that being produced per plot. If this is a correct assumption, then the organic matter or humus could supply not only the nitrogen, but the calcium and phosphorus needs of the crop as well.

6. *degree of lignification of the organic matter*

As an attempt at further differentiation of the nature of the organic matter in relation to soil treatment, tests were made of the degree of lignification according to the method suggested by S. A. Waksman. By the treatment of the soil with 72% sulfuric acid it is assumed that only those carbonaceous compounds approaching lignin remain unhydrolyzed and insoluble. The percentage of the organic matter found by this method to be in the lignin form is given in table 1, which gives also the changes in total organic matter as based on the carbon.

There are no excessively wide differences in these lignin figures. The lowest percentage of this inactive fraction in the organic matter occurred where complete fertilizer had been applied heavily on continuous wheat. The highest percentage occurred where the sodium nitrate was used and has apparently been responsible for an organic matter that is highly lignified. Manure addition seems to have had an effect similar to that of nitrate additions.

These few determinations suggest that nitrogen additions alone, or manure with its high nitrogen content tend to give an organic matter of high lignin percentage and of a stable or less inactive nature, while under phosphate-potash fertilizers the organic matter remaining is much lower in this decay resistant fraction.

7. variable nitrification activities as related to soil treatments

Laboratory studies were undertaken to evaluate the nitrate accumulating capacities by these soils under different long continued cropping and soil treatments as a measure of the influence of these on the microbial behavior, within the soils. The samples were brought into the laboratory in late February, handled at optimum moisture into jelly tumblers, with treatments including lime and organic matter, and tested for their accumulated nitrate content at fortnightly intervals during 10 to 12 weeks. Twenty-eight plots were included and the data assembled to include the effects of different cropping systems and rotations, and then for the different soil treatments including manure, superphosphate, fertilizer, ammonium sulfate, sodium nitrate and limestone.

The nitrifying activity under all the soils with continuous crops was relatively low. The low level for corn, oats, wheat, and timothy or for all of these crops rather than a differentiation amongst them is the significant fact about the separate crop influence on this microbial soil process.

The rotations were not, however, without their influence. The nitrate accumulations were regularly higher for the three year rotation, followed by the four year and then in order by the six year rotation. Except for the six year rotation, they were higher in nitrate activity than the highest of the continuous crops, namely, timothy. Since clover occurred only once in each of these rotations, it suggests that the greater number of recurrences of this crop in the shortening of the rotation may be the reason for their taking their particular order in levels of nitrate accumulating activity.

The long continued soil treatments manifest their influences markedly on nitrate accumulating activities in the soil. The manured soils were especially responsive with nitrate production when limed in the laboratory, pointing out that these soils are not nitrifying their stock of organic matter rapidly because of lime deficiency. That this effect by lime is not so much a matter of acidity as that of supplying calcium, is demonstrated by the influence of the superphosphate treatment. Soils, receiving superphosphate for 25 years, given nitrifiable organic materials in the test were almost as active in nitrate production as when given both lime and organic matter. Where the latter combined treatments produced 9.7 mgms. of nitrate, the former single addition gave 8.08 as the increase during six weeks of

Table 4
Nitrate Nitrogen Levels in Soils Under Different Laboratory Treatments as Influenced by the Past History of Cropping and Soil Treatments (Soils from Sanborn Field)

Cropping History and Field Treatment		(Pounds Nitrogen per Acre) — Laboratory Treatments: No Treatment	Limestone	Organic Matter	Lime & Organic Matter	Mean of all treatment	Increase over Lowest Item
Continuous	Corn	25	83	55	122	71	—
Crops	Oats	30	26	121	190	92	21
	Wheat	59	102	52	172	96	25
	Timothy	48	104	110	183	111	40
Rotation	Six-year	34	52	74	176	84	—
	Four-year	56	75	109	208	112	28
	Three-year	44	97	182	219	135	51
Soil	None	40	61	92	179	93	—
Treatments	Manure	49	93	120	196	114	21
	Phosphate	50	106	184	282	155	62
	Fertilizers	43	94	149	222	127	34
	Ammonium Sulfate	60	113	150	193	129	36
	Sodium Nitrate	62	87	174	216	134	41
	Lime	36	74	223	267	150	57

incubation. The higher general nitrate level by this phosphated soil when given something to nitrify was the distinguishing feature.

The effects by fertilizers were similar to those by phosphate as might be expected, since the latter is their main constituent.

In the plots treated with ammonium sulfate and sodium nitrate, the former has lowered the activity in the soil as a nitrifier. Lime additions were decidedly effective on nitrate production where ammonium sulfate had been used, though it was not without significant effect on soils treated with sodium nitrate, suggesting again the need of the soil for calcium. Where the field treatment of limestone had been used, the soil was at a lower nitrate level than where such had not been used. This was particularly noticeable by the failure of the soil to respond to liming in the laboratory in contrast to the marked response by the unlimed soils when given lime. However, the limed soil given something to nitrify was almost on a par in nitrate accumulation with the unlimed soil given both lime and green manure as laboratory treatments.

The general levels of biological activities as demonstrated by nitrate accumulation in these soils can be measured from the data in table 4, giving the average nitrate contents during 10 weeks of laboratory study and particularly the mean of all treatments.

8. summary

From these various observations of the organic matter in the soils under different treatments on Sanborn Field it is evident that there are wide differences in the biological activities now after 50 years. Where the organic matter remains at a relatively high level it has failed to be consumed by the microorganisms very probably because of its deficiencies as a bacterial ration. Some of the deficiencies suggested are nitrogen, calcium, phosphates and possibly potassium in the mobile forms to serve in the nutrition of the microorganisms. Where these have been added as soil treatments, the total supply of organic matter has been reduced. The increased crop returns, running parallel with organic matter reduction in the soil, suggest strongly that where the organic matter fraction of the soil is not breaking down, it is being retained at modest levels because these mineral elements are too deficient in the soil for the microbiological processes. They are then consequently deficient in the soil for the plants' activity on a high level. Such conditions suggest that when these deficient nutrient items are delivered by the soil to the plant, they may come in the main from the more slow decay of the more stable organic matter.

42.

During the last decade of his life, Dr. William A. Albrecht wrote out some of his most readable papers. These appeared in "small" magazines and journals, and they circled the globe. No one could possibly recapture the distribution pattern, nor would there be any great reason for doing so. Dr. Albrecht's objective, always, was to stimulate activity between the ears. The "Half-Lives" of Our Soils *drew nourishment from dozens of papers in the scientific literature. It was published in 1966 in* Natural Food and Farming.

The "Half-Lives" of Our Soils

It is significant to note that when our virgin soils, with their climax crops, were put under cultivation, the decline in productivity was at a rapid rate first, and then at a decreasing one. It is helpful to learn that the life-curve of a productive soil duplicates the death-rate curve, or loss of energy, of a radioactive chemical element from the moment forward of its creation by atomic explosion. Some illustrations of measured "part-lives" of soils will help us to appreciate the soil problems ahead for agriculture, to say nothing of those for increasing population, so recently come into the political limelight.

The unit of measure of the radiating atom's loss of half of its initial energy in a given time period is called its "half-life." The loss then in the second period of that same number of years, days or hours is but one-half of the energy remaining in the element. In the succeeding equal-time periods, the energy loss during each is again and again one-half of what is left. Thus, the radiating energy declines toward a zero supply but never arrives at complete exhaustion.

The decline in the productivity of a virgin soil put under agricultural use and its rate pictured by measured yields, or remaining fertility supply, against time follows a similar graphic line, or curve. Accordingly, we can now take the data from long-continued experiments by some of the agricultural research stations and translate them into "half-life" records for soils. Thereby, we can predict their death so far as economic production is concerned.

1. sanborn field speaks appropriately

Sanborn Field of Missouri, established in 1888, with its carefully kept crop records and periodically measured changes in soil fertility for now over 75 years offers helpful data. They report the "part-lives" of soil under treatments starting with the virgin soils. Those data include continuous cropping singly to each of corn, wheat, grass and also crop successions or rotations.

Also on the same soil type at some distance from Columbia, Missouri, additional studies of normal farming of Putnam silt loam, and of that same soil type maintained in virgin prairie conditions nearby, permit the fertility contrasts from data of the "part-life" or the "half-life" of this extensive soil type, not only in Missouri, but also in nearby prairie states.

The Putnam soil illustrates the development of soil in the mid-continental belt where the annual rainfall is near 40 inches. It is in the temperate zone. It has its prevailing winds from the southwest, and where about a half-ton of scarcely weathered, wind-blown mineral dust is deposited per acre annually for maintenance of the reserve fertility supply in the surface horizon.

The Missouri Agricultural Experiment Station has been fortunate for its soil science studies because of this particular geo-climatic setting in the glaciated soil area where such soil records are nature's contribution. This setting speaks for the wise choice in the location of Sanborn Field now listed as an historical landmark along with the plots which gave the first original data from soil erosion studies under Professor F. L. Duley and the late Dean M. F. Miller [In June, 1965, Sanborn Field and the First Erosion Plots, Columbia, Missouri, were recognized for their far-reaching contributions to the science and practice of agriculture, and were, therefore, listed among the landmarks in the history of the United States.]

2. soil "half-life" in years

Just how long might we expect a virgin prairie soil to remain economically productive under continuous cropping to corn with no concern about conservation or exploitive soil management? Growing corn continuously under such practice during the first 40 years on Sanborn Field spoke up in behalf of the soil to tell us that by 1928 the yield of corn had dropped 40%. The average number of bushels per acre for the first ten years went down by the above percentage when compared with the average acre yield during the last 10 of the 40 years (1888 to 1928, inclusive).

That was a "two-fifths" life in 40 years, namely, a fertility loss of two-fifths, 40% of corn yield potential in the first 40 years under cultivation. In the next 40 years the soil would be expected to lose 40% of the remaining 60%, or a net loss of 24% of original fertility supplies in the second "two-fifths" life. By that time, or after 80 years, the fertility loss would be a total of 64%. By 120 years, which is about the maximum age of farming on most land of Missouri, the third "part-life" period would lose 40% of the remaining 36 of original fertility, *i.e.,* 14.4% to make a total loss of 78.4% of virgin soil fertility which was turned over to our stewardship when the prairie was broken out. Would it have been economically possible to continue cultivation of that soil for even that period of three generations or three such "part-lives" had not both nature and man been putting back the inorganic and the organic fertility, respectively, to support continued cultivation of that soil area?

As confirmation of those data from the first years of the soil on Sanborn Field, there are the results from the studies by Professor Hans Jenny, reported in 1933 [*Soil Fertility Losses Under Missouri Conditions,* Missouri Agriculture Experimental Station Bulletin 324, 1933]. He reported his measurements of the fertility remaining after 60 years of exploitive farming of a virgin prairie soil (Putnam silt loam) for 60 years. Corn, oats, and wheat were grown on nearly level land with no erosion. For comparison of this soil's remaining fertility potential with a standard, he measured the stock of fertility simultaneously of a virgin prairie (Tucker Prairie) across the road. The latter had never been plowed; was put to occasion pasturing only; and had some hay removed during the same 60 years.

The measurements by Professor Jenny showed that for the land under cultivation for 60 years, there was a decrease in the soil's virgin supply of nitrogen by one-third, 33%. This was a loss of 1,360 pounds/acre from an original supply of 3,940 pounds/acre or a drop to 2,580 pounds/acre in 2 million pounds of soil per plowed acre-layer of near seven inches deep.

According to these data, only 60 years are the "one-third life" of a virgin prairie soil. The next 60 years would show a loss of one-third of what nitrogen remains, and so on. But what are 60 years in the life of a family or a nation, or what are even two "one-third life" periods, if during those very few years more than half (56%) of the life-supporting assets in food resources have been squandered? What can the future output be in quality food products for healthy life forms by successions of crop substitutions required by declining soil productivity and under no stricter criterion than merely growing some vegetative bulk?

3. *"half-life" shortened by commercial nitrogen fertilization*

The above two reports gave data gathered before 1928 and 1933. They resulted from soils under no chemical fertilizer treatments. What happens to the rate of fertility decline when the chemical, or salt, fertilizers are put into the soil? Apparently Dr. J. W. Sanborn had anticipated that question already when in 1888 he decided on the plans of managing the soils of the experiment field bearing his name. For the answer to that question, commercial nitrogen fertilizers were put to a detailed study, or test, during the second quarter of a century of the field's history.

Two plots under continuous cropping to wheat, after the previous 25 years had them under additions of barnyard manure (6 tons/acre), were continued under wheat, but were given nitrogen fertilizers only. One plot was given ammonium sulfate. The other was given sodium nitrate. Each treatment added but 25 pounds of the element nitrogen per acre per annum.

Soil analyses at the outset, and again at the close of the second quarter of a century, showed losses of nitrogen of 44.1% and 45.0%, respectively, from the supply of the soil's nitrogen at the start of the test. When the 625 pounds of nitrogen, added as chemical salts, are included in the initial totals for the calculations, then the losses were 54.0% and 53.5%, respectively. In simplest words, the data represent roughly one "half-life" of about 25 years for this soil under such treatments.

4. soils must first support other life supporting man

In facing the double problem of more people and soil of less acres with less than half of its original productivity nutrition-wise, the future half-lives of these better soils do not hold forth much promise. They are in the more favorable climatic settings of the United States. Those locations, reported as results of white man's settlements, include only moderately developed soils. They are in neither excessively high nor low annual rainfalls, and those moderate ones have their lower limits near 15-20 inches per annum. They are in the temperate zone giving us largely the favorable temperature factor at the figure 53 F. as the desired annual mean, or a range from 35-60 F. Those values are also the naturally favorable ones for grain crops.

In these facts there is the serious reminder that the geo-climatic setting must first be favorable for all the other life forms in the biotic pyramid, since it is they, which in simpler bio-chemical elaborations from the soil, determine whether any geo-climatic setting is favorable for man. Of course, man can tolerate some ranges in temperature which we note in the simple natural fact that in the northern hemisphere the optimum climatic setting for the wheat crop is also near-optimum for white settlements. But man makes the accommodation to move to the southern hemisphere with wheat where present optimum for it is on the arid and warmer side of that for white folks [Griffith Taylor, *The Distribution of Future White Settlement, The Geographical Review,* 1922, University of Sydney]. Our domestic animals, supplying much of man's food and raiment, require geoclimatic settings factor-wise within ranges of those including their master's. The minimum mean temperatures are 30 F. for sheep, 35 F. for cattle and 40 F. for pigs and goats, while the maximum mean of all livestock is 70 F.

It is fortunate for man that three of the four kinds of livestock are ruminants, or grazing animals depending on grass for their nutrition. It is the grasses that are "natural" to moderately developed soils which have unweathered mineral reserves of the nutrient elements in their pronounced organic matter accumulations preventing erosion and hastening rock weathering in the surface soil on top of mineral-rich lower soil horizons. Also, grasses may die back in drought and regrow quickly with return of moisture, because their meristematic (viable) tissue is preserved deep down in the crown of the plant to survive the dry summers and cold winters.

The higher fertility of less-weathered soils makes grazing deliver mineral-rich and protein-rich feeds of nutritional values far above those in browsed terminal branches of forest trees. The latter grow on soils under higher rainfalls required as permanent, not intermittent, soil moisture, for trees of which their meristematic tissues are in the terminal ends of the branches. We need to remember that man's early migrations were from one grassy area to another with those emphasizing a relation to nutritional needs but also to surface water supplies which often registered serious risks on the lower rainfall prairies. The risk for water would not have been such in the higher rainfall forests, but the greater risks nutrition-wise were avoided since man's herds and flocks of grazers were not apt to lead him there.

5. geoclimatic settings include rock, climate and soil

The term "geoclimatic setting" includes three factors. The first is the rock that represents the supply of mineral elements, including much silicon, the former to be weathered out and moved to the sea, and the latter, to remain longer as soil. The second factor is rainfall as source of water enough to break the rock down slowly and remove excess sodium (disturbing to growth of food plants) and also to make enough silicate clays to hold the several cationic (positive) inorganic, nutrient elements against loss by leaching yet available to plant roots. The third is temperature to increase speed of chemical reactions seasonally through its ranges favorable for the soil's biodynamic processes creating, not only man himself, but before him also the wild and domestic life forms attuned to those same many factors by which all those life strata result. That occurs because of their interrelations of healthy growth and survival controlled, in the main, by the microbial soil life forms. That, too, is premised on the soil by which plants are the food source of all. Complete summation is pictured when we say the ecological pattern of the earth's surface is premised on the many geoclimatic settings outlining the varied patterns of what we are because of where we are on the rocks weathered into soil.

Can we understand well enough, that natural pattern of the past to fit our management of the soils for extension of their "half-lives" beyond their short ones which our recent experiences suggest they will have? Dare we continue their future management along the same methods followed during the history of our recent but short stewardship? Dare we continue very long under the limited political foresight in an economy management for agricultural production of surpluses at low prices which exports our future food-potential, even now so limited, in order to maintain favorable position in international politics, including a cold war? These are some of the questions the "half-lives" of our soil have raised.

43.

Soil microbes do get their food first. They have the first place at the table. This reality caused Dr. William A. Albrecht to explore the merit of putting fertilizers down deep where plants, rather than microbes, could eat first. This little report was published by the Victory Farm Forum *in 1948.*

Soil Microbes Get Their Food First

It was less than three generations ago that Pasteur's work in France suggested the bacterial causation of disease. Even though we are coming to see that the bacterial entry into the body may be encouraged by weakness induced by deficiencies of many kinds, yet the fear of microbes, germs and bacteria is almost universal. Everybody is afraid of getting germs. Pasteur told us that heat is the best weapon for fighting these microscopic life forms and we have been heating, boiling, steaming and sterilizing in the fight against microbes.

Now that the science of microbiology has brought us penicillin, streptomycin and other similar microbial products as protection for our bodies against the microbes, it is a question whether we want to continue the universal attack on microbes, particularly since we are learning to live with them more for our benefit than for our harm. We are coming to see that microbes are a foundational part of the pyramid of life forms, of which we are the topmost. If we are to live complacently with them, we must remember that they are next to the soil in that pyramidal structure. They are between the soil and the plants. They either cooperate with, or compete with, the plants for the creative power in the form of nutrients in the soil. Hence, they are a part of the biotic foundation on which animal and human life depend. Microbes are now recognized as important because they eat more simply than all other life. They also eat first of the fertility of the soil.

1. struggle for calories

Microbes are less complex in their anatomy and in many respects are less highly developed than plants. Unlike plants, the microbes cannot make their own energy-food compounds by the help of sunlight. On the contrary, sunlight kills microbes. By the process of photosynthesis, plants build their own carbohydrates for body energy from carbon dioxide in the air and from hydrogen and oxygen in water from the soil. Plants make many carbonaceous complexes from these three simple elements which they build into intricate energy-giving compounds of high fuel value and as deposits above the soil or as additions within its surface layer. Plants work in the light. Microbes work in the dark. Unable to derive energy directly from the sun, they must get it from these chemical compounds passed on to them by the death of the plants.

As a means of getting energy for heat and work, the microbes burn or oxidize organic compounds, just as we do in our bodies. Microbial life depends on just such compounds as make up dead plant and animal bodies. It simplifies them. It tears them apart. It is the wrecking crew taking over dead plant and animal tissues to return the separate elemental parts back to the air, water, soil or other points of origin. It is working in the dark and sending back to simplicity all that the plants built up to complexity.

This microbial struggle is what we call decay. The process of rotting organic matter is the result of microbial processes of digestion and metabolism of the organic matter, by which the energy initially put into chemical combination through plant photosynthesis is released again for microbial life service.

As humans, we too use organic compounds such as sugars, starches, proteins, fat and other food components to provide our energy. This occurs as part of the process by which we break down these compounds into carbon dioxide, water, urea and other simple substances eventually thrown off as body excretions. Humans, like the microbes, are struggling for calories. In humans we call it digestion and metabolism. For the microbes, it means decay, or the simplification process which the different substances are undergoing when we commonly say "They are rotting."

2. competition with crops

Plowing under some organic matter in the garden or field is a good way of disposing of crop residues because the microbes "burn" or oxidize them. They do it slowly, however. Yet the process of microbial combustion of such materials may have disastrous effects on a crop planted soon after plowing, when we say we "burned out" the crop.

Microbes need more than energy "go" foods. They need the "grow" foods, too, just as we do. They do not demand that their nitrogen be given them in the complete proteins or the more complex compounds of this element as we do. Nevertheless, they are just as exacting in their needs for nitrogen, at least in its simpler forms. This is a "grow" food necessary to balance their energy foods in the proper ratio just as we demand the balance in speaking of our own nutritive ratio, or the balance of carbohydrates against proteins in our own diets, or in the ration of feeds for our domestic animals.

So when we plow under any woody residue of stalks, leaves or other parts of plants that have given up their protein contents for seed making, these residues are an unbalanced microbial diet. They do not permit the microbes to grow rapidly on them. They are too much carbohydrate. As a diet they are deficient in "grow" foods. They are short in proteins, or nitrogen, and in minerals, hence decay very slowly.

Woody crop residues, like straws, have long been used for roof covers in the Old World. They last well but need to be replaced more often at the ridge top than over the entire roof. It is at the ridge tops that birds sit more often to leave their droppings, which are rich in urea nitrogen. When this soluble nitrogen—along with the mineral salts of the bird droppings—is added to the straw, the first rain hastens its decay. This decay, however, is limited to the ridge of the roof, or to the area in which these supplements of nitrogen balance the microbial diet originally consisting of straw. Until this balance was brought about the straw was too carbonaceous to decay, and was good thatch. Microbes require little of the "grow" foods but without it they do not carry out their decay processes.

When strawy crop residues or sawdust, for example, are plowed into the soil, the soil microbes are offered a diet that is high in carbon, or energy, and low in body-building foods. Since the microbes are well distributed throughout this plowed soil, they are in such intimate contact with the clay that they make colloidal exchanges with it for its available nutrients. They can take ammonia nitrogen, potassium, phosphorus, calcium and other nutrients for their own growth from the clay to balance the sawdust as a more adequate diet.

It is unfortunate for the plants when woody residues are plowed under. When the microbes are more intimately in contact with the soil than are the plant roots, the microbes eat first of the available fertility elements. While the microbes are balancing their sawdust diet by taking the fertility of the soil into their own body compounds, we do not appreciate the production of the microbial crop, nor the proportion of the available fertility which they appropriate for their own needs. Instead we see how poorly the corn crop or other plants grow when planted soon after straw, heavy weeds or sawdust are plowed under. We say "The crop is burned out," when it is extra fertility and not water that is needed. Yes, the microbes eat first. This disaster follows inevitably when the soil is too low in fertility to feed both the microbial crop within and the farm crop above the soil.

But fortunately the disaster is only temporary. While the energy compounds are being consumed, the excessive carbon is escaping to the atmosphere as carbon dioxide. The nitrogen and inorganic nutrient elements are kept within the soil. Thus while the carbon supply in the soil is being lessened by volatilization, the ratios of the carbon to the nitrogen and to the inorganic elements are made more narrow. These ratios approach that of the microbial body composition—more nearly that of protein.

Thus by decay the straw with a carbon-nitrogen ratio of 80 to 1 leaves microbially manipulated residues going toward what we call "humus" and toward a carbon-nitrogen ratio of nearly 12 to 1. This resulting substance is then more nearly like the chemical composition of the microbes themselves. So when no large, new supplies of carbonaceous organic matter are added to the soil, new microbes can grow only by consuming their predecessors or the humus residues of their creation.

Humus residues, used as food by the microbes, comprise a diet low in energy values, but high in body-building values. Humus is also unbalanced, but unlike straw, it is unbalanced in the opposite respect. It is not badly unbalanced, because "grow" foods, like proteins, can be "burned" for energy. Man can live by meat (protein) alone, as Steffanson and other Arctic explorers have demonstrated. It is a bit costly, however, so we use carbohydrates to balance the protein. In that case the proteins are going for tissue building rather than to provide energy. The microbes also can use protein-like compounds for energy and very effectively. We encourage them to do this when we plow under legumes. Here again they balance their own diets but with benefit to the crop above the soil, rather than with disaster which follows the plowing under of straw.

When we plow under proteinaceous organic matter, such as legumes, with not only a high content of nitrogen but also a high content of calcium, phosphorus, magnesium, potassium and all the other inorganic nutrient elements, the microbes are placed on a diet of narrow carbon-nitrogen ratio. The ratio of carbon to the inorganic nutrients is also narrow. It is like an exclusively meat diet would be for us, or like a tankage diet would be for a pig. The energy foods in such a ration are low in supply. Conversely, the nitrogen and minerals are a surplus. This surplus is not built into microbial bodies. Instead, it is liberated in simpler forms which are left in the soil as fertilizers for farm crops.

What we plow under determines what we have as left-overs for the crops. The microbes always eat first. The crops we grow "eat at the second table." In wise management of the soil we must consider whether the composition of the organic matter we plow under is a good or poor diet for the microbes. If the soil is so low in fertility that it grows only a woody crop to be plowed under, then there can be little soil improvement for the following crop. It gives the microbes only energy foods. They must exhaust still further the last fertility supply in the soil to balance their diet and consequently the crops starve.

But if the soil is high in fertility so that it grows legumes, and if we then plow these protein-rich, mineral-rich forages under, the microbes receive more than energy foods. Given the nitrogenous, fertility-laden green manures plowed under, they pass this fertility back to the soil. Here their struggle is for energy, a struggle by which they are not in competition with the crop, the energy for which comes not from the soil but from the sunshine instead.

Microbes eat first. On poor soil with little humus and inorganic fertility, this spells disaster to the farm crop if we plow under only the poor vegetation which such soils produce. Growing merely any kind of organic matter to let it go back to the soil is not lifting the soil to higher fertility, any more than one lifts himself by pulling on his bootstraps. On soils that are more fertile in mineral nutrients, the idea in growing cover crops to turn under is to help the farm crop. It helps them if we plow under the more proteinaceous and leguminous cover vegetation which fertile soils produce.

While we have been mining our soils to push them to a lower level of fertility, the microbes that originally were working *for* us are now working *against* us. They are eating first, not only so far as the plants are concerned, but indirectly so far as even we and our animals are concerned.

It is in this competition with the microbes that inorganic fertilizers and mineral additions to the soil can play their role by balancing the microbial diet. Such minerals are taken by both the plants and the microbes. But if the fertilizers are put deeper into the soil, they may be below the layer where they affect the microbes, either favorably or unfavorably. They will serve the plants, which send their roots down there, under the power coming from the sunshine. They will not affect the microbes unless they are mixed into the humus-bearing surface soil. *Putting the fertilizers down deeper puts their nutrient contents where the plants, rather than the microbes, eat first.* This is fertilizing, by means of inorganics, the fertilizing crop that combines them with organics to serve the microbes when this fertilizing crop is turned under for true soil improvement. This is a way of composting the inorganics within the body of the soil itself.

44.

Agassiz said, "Study nature, not books." Dr. William A. Albrecht chose that statement as his theme for Organic Matter for Plant Nutrition, *a* Journal of Applied Nutrition *entry in 1964. Earlier, in correspondence with Andre Voisin of France, Albrecht had objected to the "ash" mentality so evident in the republics of learning. Here he explained his reasoning, and he punctuated his reasoning with reference to the famous Dr. F. M. Pottenger experiments in California. Gentle in his criticism, lethal in his logic, Albrecht never lets his readers forget that "nature's part is still the major unknown factor in technological agriculture."*

Organic Matter for Plant Nutrition

It was the famous zoologist and geologist, Agassiz, who suggested that we "study nature, not books." Then, with similar suggestions that we put "nature before man," there was a poem in the ancient Latin, by the title (in translation) *For Nature Will be Conquered Only by Obeying.* That thought set up in rhyme was widely quoted several decades ago as *The Riddle of the Sphinx.* Science started as, and in truth still is, the organization of knowledge of natural phenomena. Unfortunately, we study technology, now, more than "natural" science.

The so-called "natural" gardeners and farmers are given to practices founded on empiricism or on the knowledge that those practices readily serve. Why their methods succeed is not yet interpreted by either science or sales literature. Natural gardeners use organic matter for the return of plant nutrients and the mobilization by decomposition and "chelation" of both the active and the reserve mineral elements as nutrients, to say nothing of newly synthesized organic compounds for higher qualities in the vegetable crops. Because by those gardening and farming practices, they try to duplicate the natural environment through which climax crops are grown in man's absence rather than under a synthetic environment by his technological managements, they have been ridiculed by techniques which in public debate are called *Argumentum ad hominum,* a personal attack on the opponent and not on the subject.

Adherence to the belief that organic soil compounds are essential in plant nutrition is claimed by pseudo-scientists and their sophisticated science to be a case of "clinging to a myth." Those claims are often prompted by some bias for sales promotion. They forget that much that is "professional" in practice and of extensive serve is no more than empiricism. It is by starting with empirical knowledge, however, that we are prompted to search for causes, and become scientific. Hence, we can still wisely use empiricism. We still do so extensively. The physiology of relief by taking aspirin, a drug consumed annually by the tons, is still a scientific *unknown.* Its use in medical practice may be also called a "myth" when it is founded on no more than the grinding of swamp-willow bark and using the extract of this as medicine.

Consequently, when the inorganic or mineral parts of plant nutrition are so widely emphasized and promoted, but the organic parts of the nutritional support are neglected, it might be well to tabulate some of the reasons for the undue emphasis on the inorganic and to present some of the unfamiliar "organic" facts of natural plant growth. It would be helpful to prevent some of the unfortunate tactics of useless debate about plant nutrition as commercially practiced. It might be more helpful to study crop growth in nature without man's management, but on fertile soils and according to particular climatic settings responsible for these.

1. the dead ash, not the living organic substance is emphasized

When some of the pioneer botanists studied plant growth in the very dilute aqueous solutions of highly-ionized and chemically active salts, they were aiming at an accurate determination of the different elements required for plant growth. Research today is still using that method to determine the essentiality of various micro-nutrients, or "trace" elements not yet so catalogued. But even that method is, seemingly, not refined enough to measure the small amounts of chemical substances to which the life processes of plants respond.

But nutrient solutions will supply plants over but a short period of time. Their use cannot enlighten us on the ratios of the amounts of elements taken in, or what amounts in combination of them would be considered a "balanced" plant diet. Such solutions require frequent changes or supplementation by a few elements. Unlike the soil growing plants naturally, they cannot supply, at the outset, the growing season's total needs within the volume of root reach.

As for the ratios of the dozen or more inorganic elements in any nutrient solution, those relations are determined by solubilities and the necessary prevention of their precipitation out of solution as insolubles and unavailables. Hence, solutions do not enable one to determine the effects on the plant's chemical composition by variable ratios between two, three or more nutrient elements. Those effects by natural soil variations in such ratios are widely demonstrated by differences in the amounts or ratios of elements adsorbed on, and exchanged to, the root by the soil's clay or humus fractions. Those effects can now be readily demonstrated and measured by the colloidal technique which uses organic compounds both natural in the soil and synthesized in commerce, as well as by the finer fractions of colloidal clay.

The nutrient solution technique of the laboratory, or its commercial application as "hydroponics," does not permit variation in ratios offered to meet the needs of different plants. Instead, it demands a dilution to the degree of expecting the inorganic ions to behave more nearly according to the laws of mixtures of gases (behaving independently of each other) rather than according to interchanges and reactions with precipitation out of solution complete enough to duplicate conditions accepted for quantitative chemical analyses. When the use of only the soluble inorganic salts for plant nutrition as a technology falls so far short of duplicating the growth of crops rooted in the soil as were nature's methods during the ages of plant evolution for their healthy survival—and when the major crops are grown where rainfall exceeds evaporation to wash soluble salts out of the soil—shall we chide the unsophisticated gardeners and farmers for trying to fit their soil and crop practices more nearly to what they call the "natural," and what they have found successful during the centuries rather than what is only a recent technology? For them, the nutrients for all life coming from the soil are still "insoluble but available" to the crops by natural methods. They still speak of "*a living soil*" and according to their successes, "natural" farming is no myth.

2. *knowledge about nature comes slowly*

The development of our scientific information about plant nutrition and crop production naturally depended on the advances of chemical science. That consisted of only inorganic chemistry for many years. That early science was given to "ash" analyses. By combustion, those procedures eliminated, at the very outset, what is organic and what makes up about 20 times as much of the crop bulk as the inorganic parts do.

Then, also, we learn chemistry by beginning with the inorganic aspects including now about 100 elements. Even that beginning phase of chemical science proves highly lethal to any further interest of a high percentage of students of that science. All too many of them fail to arrive at the organic phase of it. They do not learn of its many synthetic processes representing at this date about a half million different known compounds resulting mainly in connection with the life processes dependent initially on the organic synthetic processes of plants.

The study of soil as plant nutrition began by matching plant ash with its list of inorganic elements against the similar list of the ignited soil's composition. When organic chemistry came along so late to become a synthetic science in place of mainly an analytical one in only recent decades, it should not be surprising that agricultural practices have not become concerned about organic compounds taken from soils by plants for their nutrition.

But it should be surprising (to say nothing of poor faith to his students) to see the reported claim of a chemist-agronomist of an experiment station that "Before the plant foods contained in compost, manure and other organic matter can be used by plants, they must be broken down by bacteria into simple *mineral* compounds which the plants assimilate, and there is no difference between these minerals and those processed in fertilizer factories" [reported in *Reader's Digest,* July 1962].

For one content with such a naive concept of plant nutrition which limits "plant foods" to the inorganic substances of mineral (rock) origin even in the use of composts, the sight of a mushroom crop and its rapid growth would not be impressive, even when it grows by feeding on the organic compounds released by decomposing manure at only a certain stage in that natural process. Perhaps the teachings of the above type and the neglect to teach so many other basic and natural truths are reasons why the crop's growth on soil by means of many organic compounds, as well as of the inorganic elements—both assimilated by the roots—remains a myth in the classroom where it might better be interpreted as nature's contribution through which man attempts to manage plant nutrition.

Seemingly, our teachings are limited to thinking only of inorganic salts and their plant service, or only as the very beginnings of chemical science envisioned them. Apparently the botanical science of plant physiology also is not studied extensively in agricultural production to learn the natural plant processes of organic compounds involved, when the tests of the use of only chemically derived inorganic elements in plant nutrition on almost any soil are undergirded by commercial grants to subsidize them.

3. fertilizer inspection uses duel criteria

Inspection of commercial fertilizers by states also abets our ignorance of the role of soil organic matter in plant nutrition, while it helps the manufacturers see the chemical compositions of their competitors' goods determined by the state inspection. That is a special service to the fertilizer manufacturers, even though it is not claimed to be such to the farmers who pay the costs of it. That inspection uses water solubles as the criteria for the amounts of nitrogen and the potassium in the erroneous belief that such a quality *outside* of the soil is an index of their availability to, or absorption by, the plant roots for fertilizers *within* the soil.

But for inspecting the phosphorus guaranteed in the fertilizer, which element is taken by plants from mineral compounds too insoluble for solution in water to approach their availability to plants from soil, the criterion is solubility in an organic solution *viz.* ammonium citrate. This suggests that it is a chelating solution of the phorphorus by means of an organic substance and, unwittingly, makes a close approach to the methods by which we now know plants mobilize the inorganic nutrients both from the soil and within themselves. By accident, not by science, one-third of the service, namely, the test for one of the three inspected elements, is unknowingly using what is *natural* and has only recently been recognized within that aspect by our thinking.

4. chelation, nature's mobilization of the inorganic elements by the organic molecules

Within the last decade or two the significance of the soluble salt aspect and the ionization of its inorganic elements in plant nutrition has been reduced decidedly. Instead, their role in non-ionized union within larger organic molecules has become magnified and extensive in nature. Now the chemistry of plant growth (and of other life forms) is spoken of as "molecular biochemistry." Reactions are between large molecules, as illustrated by magnesium in chlorophyll during the past ages. That inorganic element within the immense chlorophyll molecule is not rendering service in photosynthesis by ionic behaviors of the magnesium as we have been comprehending most inorganic behaviors. We do not know just how the magnesium serves.

Now that some organic compounds, produced in the laboratory, take from solutions and hold the inorganic elements by what seems a duplication of their adsorption (and exchange) by the inorganic clay minerals, we speak of that organic behavior toward the inorganic as "chelation" of the latter by the former. The commerical organic compound ethylene diamine tetra-acetic acid (EDTA) is an illustration used extensively in recent experimental work. It adsorbs, or chelates, many different inorganic parts with improved nutritional services by the latter as a consequence of the uptake by the root and activities within the plant of that combination of the inorganic as an integral and non-ionizing part of the larger organic molecule. This is a service by the soil organic matter, or humus, in ways even more complex than those corresponding helps rendered by the clay fraction of the soil when it adsorbs and exchanges to the roots, the calcium, magnesium and other ions which are insoluble yet available because of that natural phenomenon.

One of the significant demonstrations of this natural phenomena was made by one of the naturalists among scientists, Professor Midgeley of Vermont, years ago. He studied the effects on the crop of phosphates and barnyard manure applied separately to the soil, in contrast to mixing them before their application. The latter gave much better plant growth and made much more phosphorus available—as shown by plant composition—than the former method.

Drs. Hopkins and Whiting of Illinois, gave similar demonstration of the higher availability of phosphorus from rock phosphate when that was plowed under along with red clover as a leguminous green manure in a kind of sheet composting performance within the soil.

But more recently Vernon E. Renner of the Missouri Experiment Station, introduced radioactive phosphorus into the soil for use by young barley. That crop was harvested, carefully analyzed and used as an organic manure for a soybean crop on a soil high in inorganic phosphorus, according to soil test. *The application of the organic manure at the rate of a ton per acre mobilized its phosphorus into the soybean crop with an efficiency just a hundred times that of the movement of of the inorganic soil phosphorus along the same biochemical course for crop production.* (This seems to add to the myth.)

5. methods of teaching may need modification

The separation of inorganic chemistry from organic chemistry, in our methods of learning, has magnified unwittingly the former phase, as if it were more important than the organic one in crop production. This has been unfortunate, when in nature there is no such separation. The research into the biochemistry of photosynthesis by Melvin Calvin [*The Path of Carbon in Photosynthesis, Science,* 1962], which won the Nobel prize recently, emphasizes phosphorus active in the first carbon compound in the production of sugar by photosynthesis. That first compound is not a six-carbon sugar. Instead, it is a three-carbon compound containing phosphorus. Then, two of those unite by splitting out the phosphorus to result in the six-carbon sugar—formerly considered the first stage in photosynthetic action—and releasing the phosphorus to repeat its service in synthesizing the three-carbon compounds into six-carbon ones.

Thus, there are natural exhibitions in which the so-called anionic phosphorus coming from the soil is chelated and combined into a vastly improved molecular biochemistry of growth. We are gradually appreciating the similar services by cationic elements such as magnesium in chlorophyll, iron in hemoglobin of blood and copper in the similar life fluid of the crustaceans, like the lobster. We are viewing similarly many other enzyme-like performances in which the extensive list of inorganic macro- and micro-elements of the soil as either cations or anions are doing wonders within large molecules with a ratio of their ash to organic part that duplicates the ratios of those in living tissues. The arts of the pioneer practitioners of empiricism, *i.e.,* the trial and error methods aiming to duplicate nature's behaviors, have long preceded what is now technology and the science supporting it. The latter is, unfortunately, coming along all too slowly when potential commercialization rather than curiosity about the natural is the major stimulus for research.

6. agronomic science becoming more organic-minded

Textbooks of botany and bulletins of some years ago reported the uptake of organic compounds for their nutritional service. Among those reported as improvers of plant growth were such complex ones as coumarin, vanillin, pyridine, quinoline, asparigin, nucleic acids and, in fact, some in each group of carbohydrates, organic acids and nitrogenous compounds. In entomological work, commercial research by the Boyce-Thompson Institute of near two decades ago reported that some 300 and more organic compounds—aimed to be "systemic" insect poisons within the plants—were taken into the plant roots from the soil. *Apparently such reports have not registered as cases of soil organic matter serving as large organic molecules moving into the plants from the soil.* Research at the Missouri Experiment Station by Dr. George Wagner, working with seedlings under sterile conditions, has been demonstrating sugars and proteinaceous substances taken as nutritional values for them.

But it remained for P. C. de Kock of Scotland to combine science and empiricism for attention by both the agronomists and the so-called "natural" practitioners, when in 1955 he reported his exhibition of the curative effects on chlorotic plants offered and taking up the commercial organic compound, ethylene diamine tetra-acetic acid (EDTA, molecular weight 380.20) and thereby mobilizing iron from the soil; and when he duplicated the same beneficial effects by substituting for the EDTA the organic matter extracted from the soil. Thus natural organic matter demonstrated its activities in moving the essential mineral nutrient for plants, namely, iron in chelation, just as was demonstrated for the commercial organic EDTA [*Influence of Humic Acids on Plant Growth, Science,* 1955].

Only very recently was it shown that an inorganic element fed to plants through the roots in chelated union with EDTA is no longer under control of that organic chelating agent by the time it is moving up through the stem. When samples of the exudate from the cut stem were examined, the inorganic element initially controlled by EDTA was chelated with other compounds, mainly the simple malic (apple) acid (molecular weight 158.11) and malonic acid (molecular weight 104.06) [see L. O. Tiffin and J. C. Brown, *Iron Chelates in Soybean Exudates, Science* 1962]. This indicates the possibility that many organic compounds within the plant are operating to reduce the ionic activities of inorganic elements and are moving them about within the plants as parts of larger organic molecules. It is the latter, then, that dominates the former. The "ash" contributions from the soil come under control of *molecular biochemistry*.

Now that we find the natural behavior duplicating what we have done by laboratory techniques, we as agronomists are more ready to accept possible functions in plant nutrition by soil organic matter with more credence. The way is now open for many forthcoming research projects on soil organic matter for plant nutrition. This highly neglected half or more of past plant nutrition bids fair to be elucidated in the near future. But when the essentiality of inorganic elements is still an unfinished task that involves a list of no more than a hundred in total, then by what names might he not be called who will suggest a research project even under federal funds to determine the essentiality of the many organic compounds, of which but a half million are characterized or catalogued to date?

7. discriminating animals may aid research in organic compounds

The use of animals as bioassayers of feeds demonstrates that they recognize the dangerous, or beneficial, effects within what they will consume because of organic compounds taken directly by the plants from the soil. Animals respond to effects also from organic compounds in the soil as manures or organic fertilizers of the crops fed. Livestock has not taken readily to pasturing green sweet clover. If confined to a field of such, the various animals will first clean up the fence rows, water courses and areas of vegetation other than the sweet clover. Their reluctant taking of any sweet clover suggests such as only an act of desperation. They apparently recognize in the dicumerol, synthesized by that crop, its anti-coagulating effects on the blood and the resultant bleeding to death on injury or surgical treatment of the animals.

Hogs, given choice of corn grain grown with sweet clover as the preceding crop, turned under for leguminous organic manure on plots given increasing combination treatments of calcium in lime, phosphorus and potassium, discriminated sharply among four simultaneous offerings of the grains. Those choices differed according to whether the sweet clover was turned under green in spring, or merely as the residue of a crop grown for seed the preceding year. In case of the organic manure of sweet clover residues from its use as seed crop on the four plots, the hogs chose to eat more of the grain according as more inorganic fertilizers were applied, or as more yield of corn per acre resulted. However, when sweet clover was turned under green ahead of the corn planting, the hogs decided to choose exactly in the reverse order. For them the less green, sweet clover used to fertilize the corn, the better. They preferred no green sweet clover as organic fertilizer for the production of their corn (maize) as feed.

No chemical data were taken to tell us whether dicumerol, or organic compounds suggesting it, were in th corn grain. The hogs merely reported that organic matter from green sweet clover, used to fertilize corn, carried organic effects in the grain which they refused. On the contrary, sweet clover used as dried, matured residue carried organic effects into the grain of their highest choice. While many folks may be deriding organic farming, the hogs vote for it. But they are not speaking of it in general. Even they report that judgment must be exercised as to the kinds of organic farming one is talking about, and that associated with attention to the inorganic essentials.

It is not common knowledge that the organic compounds of well known, specific, chemical structure giving the fecal odor, namely, indole and skatole, will be taken up by a plant like the white dwarf bean of Michigan fame. But it was demonstrated by Dr. F. M. Pottenger, Jr., of California, that those odorous organics as taken my be either stored in the seeds with the scent emanating from them, or may be converted into the well known growth-hormone, indole-acetic acid. That was suggested when fertilization of the soil by dung from cats fed raw milk, converted the dwarf beans to pole beans in two cat pens, but corresponding dung or manure from cats fed cooked milk in six pens did not violate the growth behavior claimed by the beans' pedigree as dwarf plants.

8. composting pulverized minerals is natural

Nature's management of soil for extended maintenance of its high productivity of crops as nutrition for all other life forms, consists of two major practices. The first is the regular application by wind (loess, dusts, etc.) and by water (alluvium, inwash, etc.) of deposits of unweathered, finely pulverized rock-mixtures on the surface of the soil. The second is the regular covering of the soil surface by organic matter as crop residues.

By the latter, as energy and sustenance for the microbial flora, its life processes decomposing the finely pulverized minerals represent an active development of a new stratum of surface soil where the earth and the atmosphere with its meteorological forces meet. The surface phenomena there are the dynamics by which nature's composting processes are combining the organic matter with the rock-mineral fertilizers to make those insoluble inorganic soil elements become available for plant nutrition (not salts), through their union with the conserved organic nutrient compounds, or through nature's blending in support of microbes and plants. Unfortunately, our inorganic salt concepts, magnified by their commercial potential, have kept us blinded from nature's more efficient management of production to support all that lives.

The reported "Healthy Hunzas," isolated in the high Himalayas, have not been viewed as a case of a climax human crop dependent on their complete adoption of the practices by which the soil productivity is naturally maintained for climax crops (and livestock) with the Hunzas as a corresponding climax human crop included.

9. nature's part is still the major unknown factor in technological agriculture

Of course, we talk about and teach only what we know. Hence, if we don't know that crops are nourished by organic compounds taken from the soil—a fact demonstrated as early as the first climax crops in the course of their evolution before domestication—we are just naturally content to believe that crops are nourished only by what has been "broken down by bacteria into simple mineral compounds." Then, too, "we are just naturally down on what we are not up on." With limited information of nature's performances, our discussions and debates about our management of those in what we include in agriculture may miss the truth widely and become illustrations of "*argumentum ad hominum.*" We need to study nature, not only books.

45.

Much of the drive for sound eco-agriculture became lost in the battle over semantics, the term "organic" becoming both a cudgel for defense and a weapon for attack. Dr. William A. Albrecht always tried to sidestep this fruitless battle. He tried to structure nature's own syllogism, usually by retelling the premises he had told before. Soil Management by Nature or Man? *is such a paper. It appeared in* Natural Food and Farming *in 1965.*

Soil Management by Nature or Man?

In our studies of how Mother Nature was growing crops which were able to protect themselves against pests and disease to survive the ages, and to be available for domestication by man when he took over the soil and crop management, we find that two basic requirements had always been met or fulfilled.

In the first place, rock minerals were weathering in the soil to remind us of the poetic claim that "The Mills of God must grind." In the second place, the organic matter grown on the soil was naturally put back in place on top or within the soil for its decay there. That served to put microbial life into the soil. It generated the carbonic acid there (and other acids of decay) to break some of the nutrient elements out of the rock more rapidly for them to be caught up and held, or adsorbed, by some of the more stable, weathered, non-nutrient elements like the silicon of the clay. That adsorption holds them for plant services when the plant uses the same kind of carbonic acid to take those nutrients off by trading the hydrogen, or acid, for them.

1. nature's self protection

By means of grinding fresh rock regularly as natural mineral fertilizers in the soil, and by conserving the organic matter to go back to maintain the soil's humus at higher levels, nature had protected her crops so they grew annually from their own seeds. By a unique self protection they were doing well when man came along to take over what we call "scientific" crop management and "scientific" soil management. Certainly we are not now duplicating those practices in which nature was more successful than we appreciate.

When we began the study of how soils feed different plants naturally, we started by chemical analyses of the elements we could separate out of the plant's ash. We listed the kinds and amounts our analytical ingenuity let us specify. We used the ashing procedure to do the same to the soil. Then we tried to match the elements given for the soil against those specified for the plants.

Among those in the plant, especially wood ashes, the potassium was the highest one. Phosphorus was next in order and calcium, or lime, was the third at the top of the list. There was often some sodium. Then there were many other elements of lesser amounts.

In the ash of the human body and other warm-blooded animals, there appear again calcium, phosphorus and potassium—the same three elements as in plants. But calcium, not potassium, comes first in the largest amounts. Phosphorus is second and potassium is third, or quite the reverse of the positions in plants. As vegetarians building bodies, we ask our physiology to juggle those elements, eaten in one order of amounts in our food and into another decidedly different order for building our bodies.

2. protein supplements

That chemical composition of warm-blooded bodies, demanding high calcium and phosphorus, gives reason for growing and eating legume plants (peas, beans and other pulses) which are protein-rich as well as mineral-rich, to be protein supplements to the non-legume, low-protein or fattening foods we eat. We use legumes also in animal rations as protein supplements to the pasture grasses for prevention of deficiencies in essential mineral elements and in nitrogen—the element by which we distinguish, chemically, the proteins or living substances.

Legumes can take nitrogen from decaying organic matter in the soil. But when supplied with generous mineral fertility, especially calcium and magnesium as carbonates in limestone and other substances, legumes can use some of the 70 million pounds of the atmospheric nitrogen (over every acre) to build their own protein.

Putting legumes into pasture grass mixtures was nature's way of growing good grazing for the American bison on the prairies. That provided a more nearly balanced animal ration by the bison's choice among the 60 to 75 species of plants offered by the prairies and plains. Legumes were nature's synthesizers of protein by using atmospheric—not chemical-salt—nitrogen. They supplemented the non-legume plants in their nitrogen demands from the soil. They also prepared the combined nitrogen for giving more complete protein supplies in the nutrition of other life forms calling for that food supplement.

We now have commercial chemical salts of nitrogen so widely available to be applied as fertilizer to pasture grasses to make them rich in nitrogen, and "crude" protein, as the chemist measures it. Since nitrogen is the element distinguishing the protein, it is apt to be claimed that "we can now grow more grasses per acre and make them richer in nitrogen and, therefore, in protein, to be the equal of legumes. Consequently, we do not need to grow the legumes, which are so hard to grow anyway."

3. specific compound

But proteins are compounds very specific in more than just nitrogen. Nitrogen in some plant compounds may be poison, as it is in regrowth of sorghums with cyanide nitrogen. ***That pasture grasses fertilized heavily with nitrogen are not the equal in feed value of grasses growing simultaneously on soils with legumes as companion crops for animals grazing them*** is told us when the latter in that crop combination serves to make them as feed grow more young animal weight than the former does alone.

Fertilization of pastures by nitrogen only is by no means the equal of grasses limed and phosphated to give plenty of calcium, magnesium, phosphorus, potassium and all else and with legumes grown along with them. The latter crop combination, coupled with the fertile soil, exhibits the higher creative power for the living substance—protein—which demands the higher amounts of ash elements and nitrogen from the air to grow a quality of feed supplement which grasses cannot do even if they can take more nitrogen than usual from the soil.

Consequently, we need legumes to grow our bodies of the separate cells. Those have the *grow* power. *Non-legumes* make the carbohydrates, *the "go" power, and the power for hanging on the fat.* But a fattened animal with power to "gain" in weight cannot go very far nor very fast. We need soils high in calcium and other fertility elements to supply "grow" foods. We are still able to grow many "go" foods easily. But only "grow" foods build bodies, protect them against diseases and pests and enable life forms to reproduce the species to populate the earth.

For growing legumes and better non-legume crops which are mineral-rich and protein-rich, the soils must be well-stocked with calcium, magnesium and potassium, usually accompanied by sodium, some hydrogen or acid, all as positively charged elements in their electrical behaviors. There must also be present some of the trace elements in that same classification of the positive. In order to be available to the plant roots, they must all have been broken out of the rock reserves and adsorbed, or held, on the clay surface against loss in percolating soil water. Yet they must be exchangeable to the plant root, trading hydrogen or acidity for them.

4. soil's holding capacity

According to our knowledge to date, the soil's total capacity to hold electrically positive nutrients in available form should have about 60-75% for calcium, 6-12% for magnesium, 3-5% for potassium, and not more than that much of sodium and also all the needed trace elements and non-nutrient hydrogen, or acidity.

Those figures represent the soil's content of positively charged elements in what, to date, we may consider a *balanced plant ration* of that portion of the list of required elements for growing legumes. Simultaneously, more nutritious grasses are grown, to make for better grazing—a point well proven when animals have a choice. It may not be so complete as nutritious feed in baled hay or in mechanically pelleted feeds.

In our preceding remarks, we have not spoken about the soil's organic supplies of nitrogen, sulfur and phosphorus in the required plant's ration. We have not mentioned some of the trace elements also connected more actively with the supply of organic matter than with the reserve minerals.

We need to look to the organic matter of the soil to make these last three more essential major nutrient elements available to the crops. We need to remind ourselves that it is the organic matter that makes the surface layer the "living soil" and the "handful of dust" with its power for creating life.

We must not forget that *microbes are what make a living soil "alive."* And far more important, we must remember that soil microbes, like all other microbes, eat at the first sitting, or first table. Plants eat at the second. Microbes go first for energy food, since they cannot use the sunshine's energy directly. Plants go first for "grow" food, since they can use sunshine energy that way.

A sprouting seed "roots" for a living, or for "grow" food first. It puts up its advertising of growth by showing its leaves above the soil in the sunshine second.

Microbes are the decomposers of the organic matter and the conservers of the inorganic fertility, of the nitrogen, of the sulphur and of the phosphorus. Those three elements do not escape so much from a soil which has plenty of organic matter and growing crops to conserve those elements. We need to consider organic matter to conserve, to mobilize and to increase the nitrogen, the sulfur and the phosphorus of the soils, if those are to be fully productive.

Soil microbes oxidize carbon, nitrogen, sulfur and phosphorus to get energy thereby. It is in their oxidized forms that those elements are taken into the plant. Carbon is taken into the leaves. The others are taken into the plant root and, thus, all are in cycles of re-use.

5. symbiotic activity

It was by that more complete recycling for conservation that nature built up the soils in organic matter which we are compelling our microbes to burn out so rapidly when we return primarily chemical salts and little carbon of organic matter by which in this combination for microbial service, these fertility elements must be held in the soil. Plants and microbes must be in symbiotic activity and not in competition for fertility if our productive soils are to be maintained.

Carbon, nitrogen, sulphur and phosphorus are the negatively charged elements with which the positively charged hydrogen, calcium, magnesium, potassium and sodium combine to make *the readily soluble inorganic salts.* But in those combined forms they are not held by the soil as such. They are *ionically injurious to plant roots.* They are leached out by percolating rainwater. It is the clay-humus part of the soil which filters the positively charged ions, or elements, out of those salts, much like the household water softener takes the calcium, or lime, hardness out of the water supply. The clay-humus holds them as insoluble, yet available, to plant roots which are trading acid, or hydrogen, for them.

The negatively charged, soluble nitrates, sulfates, phosphates, so oxidized by the microbes, serve as nutrition for them and for the plants to be reduced into the organo-molecular states of living tissue where they are insoluble but functional in large organic molecules and not as salts. On death, they are oxidized again for microbial energy and repeat the cycle.

It is in this natural plan of soil management where we must recognize the real service by the fertility elements of soil, air and water playing their roles in creation before we can take over for wiser management of nature's part in crop production. Her two phases of management stand out. Nature returned the organic matter as completely as possible, in that she held many of the fertility elements and kept them available. *She grew crops where she also added unweathered mineral salts and dusts through winds with their storms of such and by overflowing waters with their inwash of deposited minerals.*

By that simple, two-phase procedure of fertility management, nature had many different crops of healthy plants here for man when he arrived. But each crop was on its own particularly suitable soil in its specific climatic, geo-chemical and balanced fertility setting with man and warm-blooded animals on the high-calcium soils. *We have not yet included calcium as the foremost fertility element* when we list the contents of commercial fertilizers, for the inspector, even though we lime the soil to combat its acidity and, thereby, work against the very mechanism by which the plant roots feed our crops.

With a threatening population explosion, and the threat of pollution of our entire environment in our war on pests and diseases, to say nothing of the wide use of hydrocarbon fuels, will we eventually repent as prodigals in spending our creative substances of the soil, and return to nature which we must finally admit is still the greatest creator the earth has known in her quiet patience and transcending wisdom? Let us hope that we return from our prodigality of the soil before it is too late.

46.

Chemurgy is a term that was coined during the early part of this century. It has since all but passed from the scene. Basically the chemurgic people sought to have a flow of by-products from agriculture move into the industrial society, thus assuring farm prosperity. This flow would start with farmers' alcohol and ultimately involve plastics, medicines, fabrics, all sorts of quasi-synthetics, and—of course—food products. Dr. William J. Hale of Dow Chemical led the fight for farmers' alcohol during the 1930s. The general drive was taken over by the National Farm Chemurgic Council. When the Council sought the advice of Dr. William A. Albrecht, he wrote out this paper, Reconstructing the Soils of the World to Meet Human Needs. *This report was first published as No. 5 of the 1951 series,* Chemurgic Papers.

Reconstructing the Soils of the World To Meet Human Needs

Man's lofty position at the top of the biotic pyramid is apt to give the impression that, since he is put over all other life, he would surely have an answer in the positive and would outline a *modus operandi* in detail for the question assigned to us for this occasion, namely, "Can we reconstruct the soils of the world in their productivity to meet human needs?" Unfortunately, even though the growth of conservation thinking has been phenomenal up to this moment, one of the features about it that emphasizes itself most is the fact that man has not yet demonstrated much success in the conservation of other life forms and living products below him in that pyramid. The loftiness of his position is not one of regality in which he may gloat. On the contrary, it is one of hazard because these life forms beneath him may readily refuse their support. The declining soil fertility below all life may topple him from it.

All too slowly is this great fact being recognized. While man may rightfully boast of his technologies of construction and reconstruction in engineering accomplishments, we can scarcely subscribe to the belief that man is capable of participating in the processes of creation to the extent of reconstruction of the soils of the world to meet human needs. Can man save himself? The subject under discussion here must be approached, then, mainly for the analysis of the problem, but as yet, for no plan guaranteeing the full and positive answer in favor of soil reconstruction for man unlimited. On the converse, nature's great forces will, in all probability, reconstruct man to fit the soil.

1. needs, not wants

As a partial clarification of the subject, let us note that the statement of it specifies human *needs* and not *wants* as the objective of soil reconstruction. "Need is a state of circumstances requiring something," [wrote John D. Detweiler of the Canadian Conservation Association]. If we consider only the needs, that is, the bare requirements for human survival, those are far below human wants. But when so much is said about the "standard of living" let us remind ourselves that when our soils must meet that requirement they must satisfy our wants, not our needs. Those wants include, so often, extensive phases of jealousy, greed, selfishness, and similar attributes emphasizing man with disregard of all else.

Our wants are too often mistaken for our needs. They are readily interpreted as demands and come into the picture of economics matched against supply. Demands of such nature with their flow of dollars and other monetary equivalents submit to measurements by cash register recordings. It is by such specifications, then, that our standards of living are listed to include wants and desires but by no means according to the criterion of need. One needs, for example, only 70 grams of protein per day, according to the calculations of the scholars of nutrition. But any one of us may want a T-bone steak or a filet mignon or more than three times that many grams for consumption at one sitting. Wants and demands are not constants. Hence the common views of economics may well be laid aside. We shall do well even to think about rebuilding the world's soils for human needs. It was man's wants and not his needs that brought on the disastrous soil exploitation.

It is man's needs and not his wants that must guide soil conservation and reconstruction. In this discrimination between the needs of the human and the wants claimed by him, there is much to clarify our thinking about conservation of our resources. According to our wants there can certainly be no hope in the face of mounting population. According to our needs, there remains at least a challenge to our thinking about the problem, the wish for its solution, and the effort at least by some of us, to cling to the faint hopes for the positive answer to it in limited localities.

2. our major needs

The human needs may be listed as numbering mainly three. When put in the order of increasing effort to obtain them in the struggle for survival as a human—reasonably sociable and amenable to the laws and behaviors of good society—they are shelter and fuel, raiment or clothing and food. In considering the reconstruction of the soils of the world to meet these, the following three simple questions are posed. How much of a task will it be to have the soils of the world extensive enough and fertile enough to provide our needs for shelter? What must be done to assure that the soils will supply fiber crops sufficient for clothing and fabrics? Then, finally, can we reconstruct our soils to grow the food in sufficient amount and of required nutritional quality for the mounting numbers of our world population?

3. *the shelter problem*

The problem of shelter has commonly been disposed of either by facing and solving it or by escaping from it. The migration of many folks from the colder to the more moderate and warm climates is, and always has been, the escape from the needs for fuel and extensive shelter. In the past, the soil has been the productive source of most of our sheltering materials. The forests were plentiful. The pioneer's erection of the cabin was almost incidental to the removal of the trees in clearing the land for cultivation and food production. The call by Gifford Pinchot for conservation of the forests has not been heeded extensively because shelter can be had from many substitutes for wood in home construction. Even today, the efforts in forest conservation and the program of reforestation are not so much a cry for means of shelter as they are a cry for pulp for paper and industrial uses other than for lumber in building houses.

The reconstruction of our soils for growing wood for shelter does not represent much of a problem for several reasons. In the first place the growing of wood, which as a chemical product is lignified cellulose, makes no call for a particularly fertile soil. Instead it is a call on the air and water for the major chemical elements of its construction. These are meteorological contributions. They are not soil borne. They are fabricated into combustible products by sunshine energy. Consequently, they give that energy or heat back on burning as fuel wood or as fossil wood in coal.

In the second place, very little of soil fertility, or relatively small amounts of chemical essentials from the soil, enter into the wood. Those used to make the seasonal growth of a tree are returned to the soil annually to a large share in the regular drop of leaves. Even for the non-deciduous trees the growth is so scant that the drain on the soil's essentials is very small.

In the third place, the growth of a tree so far as the soil is concerned is the result of continued root extension. This is one going not only horizontally over larger areas but also vertically through greater soil depths. As a consequence, wood as a shelter product is possible on soils of fertility level far below that required to meet other human needs.

The need for shelter is not a call so much for reconstruction of our soils but rather for a reallotment of soils now in no crops to trees for future crops of wood. It is a call for more abandoned areas to be planted to trees. Our foresters are demonstrating clearly that planting trees in the once forested but cleared and burned areas under abandon is still a great opportunity for large crops of wood. This will be good reconstruction of vegetation if we are only farsighted enough for each of us to do more planting of this slowly ripening crop where it will be no competition for a shorter-lived one. More acres planted rather than more soil reconstruction is the solution. Growing our shelter does not invoke serious pessimism about the future. Trees require so little fertility that they are almost the first crop before the rocks are scarcely developed into a soil and are also near the last vegetation holding forth on soils developed so completely as to have been moved nearly into solution and on to the sea.

Even if we could not grow shelter, the soil itself and the rocks that might make it will serve as shelter. This was demonstrated by the sod houses of the western pioneers and the shelters of the cave dwelling primitives. Modern home construction has gone forward while the role of wood in it has almost passed out. We are making buildings completely fireproof. Soil scientists have not given much thought to reconstruction of the soil to meet the needs for growing our shelter. They have escaped that responsibility in the substitutes which do not call on the soil for their creation by growth. It would be no insurmountable difficulty to reconstruct our soils to meet the human needs if shelter were the only one in that category calling on the soil.

4. fiber crops pose serious soil problems

Division of our fiber needs into those of vegetable, animal and technological origins makes the problem of soil responsibility for their provision less complex. By no means, however, can one escape the necessity for soils and their reconstruction to meet these needs. Technology has exhibited what may be some of the most outstanding applied research in giving us the synthetic fibers. Even then, of those still in the minds of the research men and in the prospect of creation, there is a good number. However, in seeing the metallic spineret replace the corresponding anatomical equivalent of the silk worm, we must remind ourselves that both are fed by digestion of vegetable matter grown on the soil, either recently or in the distant past. But here again, as in the case of shelter, the chemical composition of synthetic fibers calls for mainly carbon, hydrogen and oxygen which are delivered gratis as air and water over extensive land areas. Then, too, with cellulose serving as the raw materials for the synthetic chemist, such fiber production does not demand the most fertile soils on the list of those serving human needs. Technological creations of fibers for clothing and plastics offer consolation in the problem of growing fibers, skins, *etc.* as body cover and comfort. While such helps in fiber production lessen the soil's responsibilities and push the day of soil exhaustion under this need into the distant future, nevertheless, we must not forget that the carbon and the nitrogen in the coal come from what is now fossil crops but grown once upon a time by means of soil fertility before it escaped to the sea.

More significant in relation to soil is the observation that our choice synthetic fibers are not those consisting wholly of the *carbohydrate* equivalents, namely carbon, hydrogen and oxygen. The cellulose-acetate fabrics were originally a welcome creation and fill a significant human need. But the more purposeful fibers coming later at more cost and greater synthetic complications approach the *pro-*

teins in chemical composition. They must have nitrogen in their molecular structure as well as the more common constituent elements of vegetable matter. Nylon, vicara, orlon and other fibers, perhaps not yet commercially born, suggest that, like the proteins in agricultural production, they are the deficiency in supply and are hard to produce or grow except under the complex chemical combinations composing the most fertile soils. This is quite in contrast to the rayon fibers, suggesting carbohydrates, which are a crop that is easy to grow, and is plentiful. They give us big yields on acres which are still equal to that production load from the remnants of their original fertility supply.

5. growing fiber crops demands that we grow protein seed crops too

Vegetable fibers like cotton, flax, hemp and others, commonly considered for fabric use, bring the soil problem more sharply into focus. These crops cannot be grown in disregard of the level of fertility in the soil. Abandoned acres are numerous which these crops have exhausted of fertility. Growing the cotton *fiber* is a matter of growing also the *seed* to which the fiber is attached. It is a matter of a soil fertile enough to produce first the proteins in the seed and then the fibrous, cellulosic cover enshrouding it. Growing a cotton crop calls for the high level of fertility needed for production of any seed, or any protein-rich crop. In the case of fibers taken from the plant stems, the maturation of the seed is a part of the plants' total performances of making the fiber. The making of the cellulosic fibers, even if they themselves contain little that was brought up from the soil, cannot result unless the plant also carries forward the physiological load of producing the proteins. It is that latter performance which makes heavy demands on the soil for liberal production of them.

The growing of cellulose for fibers calls for more soil construction and reconstruction than the growing of cellulose as wood for shelter. In the coniferous trees, the physiological load of seed production is reduced to the very low level of a fungus spore. In the fiber crops the corresponding physiological demand is much higher, calling on the soil for more help. Fiber crop production is production of cellulose but also of protein along with it. This is possible only on soils fertile according to the protein output rather than the delivery of cellulose.

Soils for fiber crops need not be as highly fertile as soils for production of protein in foods and feeds when we recall that cotton seeds are protein but one not complete enough to serve all our domestic animals. They serve the cow. Her safety factor in this connection lies apparently in the cooperating help of the microbial flora in her paunch through which she can use cottonseed proteins, insufficient as they are for other domestic animals. Soils growing cotton fibers and the proteins associated with them call for reconstruction in their fertility if those soils are asked to grow the more complete proteins for human consumption.

If we are to grow more fiber crops for more folks, such crops are no escape from the necessity of either finding more soils inherently fertile in certain specific respects, or reconstructing soils accordingly by means of added fertility materials. Fortunately, fibers are not a perishable crop. Also, their services are long lasting and would be much more so were we not such addicts to fashions, changing with the season's demands rather than remaining undisturbed according to human needs.

6. wool is a protein crop

When we consider soils in relation to wool, our favorite fiber of animal origin, the problem of reconstructing soils to grow the sheep to make the wool incidentally, is far more complex than at this moment we appreciate. The wool fiber itself is a protein. It is bathed in a particular fat during its production by the sheep. Wool production is a physiological performance of high order. It calls for the provision of proteins in the food which the sheep eats. Proteins grown into the feeds are a call for soils fertile to degrees much higher than required for a plant's production of cellulose and other carbohydrates.

Animals cannot synthesize proteins from the elements. They only assemble them from the amino acids as parts of the protein synthesized by the microbes and the plants, and in completeness of all those required only as the fertility of the soil supports the conversion by the plant of its carbohydrates into amino acids. Wool production is a question of soils fertile enough for protein production to build sheep bodies. Is it too much stretch of the imagination to see human bodies of highly similar physiology in the same picture? Protein production is the major call for the reconstruction of many soils too low for that. If we are to produce wool, this demands regular maintenance of the fertility of any soil, too long taken for granted when after exploitation in one generation we escaped the responsibilities of soil reconstruction by going west. When the nutrition and the physiology of the sheep approach those of the human so closely, any consideration of soil reconstruction for feeding sheep for wool and meat may well carry its implications for the nutrition of man, too. In thinking of feed for the sheep we are then thinking simultaneously about reconstruction of the soil for food for ourselves.

We trust you will not deem it unkind to the sheep loving flock-masters when we believe that wool production (and mutton production) in the past may have resulted more from the instincts of the sheep than from the knowledge of animal nutrition and physiology—much less in relation to the soil fertility—on the part of the shepherds. For the pioneer, the sheep were the chemists that went ahead of him and assayed the vegetation for its quality of protein to make wool and to support reproduction of the flock. This bioassay for good sheep nutrition was cataloging simultaneously those more fertile soils into which the plow could be put for good crops as nutrition for man. Sheep have spread over the land and multiplied because they do so with a fuller knowledge of their soil security under them than can be said of their owners. Along the same line of thought, we must acknowledge the fact that sheep have become of major importance, not under closer domestication but out on the range where they search out their feed from the native and virgin vegetation. We have not yet set up the complete fertility inventory required of a soil to grow forages that will produce sheep of good health, prolific lamb crops and incidentally big weights of wool. So far, the soil has not come into the picture for its fuller significance in wool production. This crop of protein (and fat) is still much a matter controlled by those supposedly weak animals or dumb beasts themselves. To date the growth of our protein crops, whether in wool fiber, in the carcass of the sheep and in the bodies of our other animals as meat and choice food, has been mainly at the expense of ravaged soils, and not reconstructed ones.

7. *protein-producing soil areas are limited*

Consideration of the two preceding human needs, shelter and raiment, in regard to reconstruction of soils has already separated out the two parts into which the third need, namely, food for any living body, divides itself. Foods serve to provide the body with energy and to build the body, *i.e.,* to be parts of the construction or to be tools in this process. Carbohydrates, of photosynthetic and meteorological origin, illustrate the former. Proteins and all that is associated with their fabrication make up the latter. For our crops' delivery of carbohydrates in sugars, starches and cellulosic bulk of large yields per acre little soil fertility and thereby little soil reconstruction is required. Merely going forth to sow with unbounded faith in the pedigree of the seed but with no attention to the good ground on which it must fall is about all one requires. Natural cover on most any soil illustrates this contention. But for the production of cell-multiplying, body-building, species-reproducing proteins along with those carbohydrates, the areas of fertile soils have always been limited. Soil surveys have measured acres, not fertility, much less protein production. Statistics of crop yields represent bulk as the criterion for agricultural output. Delivery of nutritional values has not yet been included in that category.

Our sheep and other animals ahead of us, while we were trailing them for benefits unappreciated, have been going west to more proteins but with exploited soils in our wake. We are now being turned back in this journey by the soils under us, and by our animals, too. We prefer to dodge the responsibility of reconstructing our soils to produce sufficient protein. We are content with a superficial thinking that has not yet come to believe that shortages of proteins find their causes in failing soil fertility. It is instead a contentment with legislation that will roll back the prices on meat to please the majority and disregard the food-producing minority. Great facts are not necessarily established by majority votes.

With reference to protein as feed and food coming from our soils the sheep population deserves the same critical consideration we are about to give to human population. Sheep population has given its curves of increasing numbers from the earliest records until the maximum was reached in 1942. Since then, the numbers have declined so sharply that today they are below the figure of the first count taken. No reason can be found in the economic demands for such a decrease, when wool prices and meat prices, offered by those able to pay them, are also the highest in history by several times. We may well ask whether it is not the decreasing soil fertility that is rolling the sheep population back in spite of increasing demands for wool and mutton as the economist uses them? Isn't the human population apt to be rolled back, too, eventually, in relation to the soil fertility that must feed it?

An interesting correlation, suggesting failing nutrition of the sheep because of a deficiency in the soil fertility's trace element copper, deserves mention in this connection. It was in 1942 when phenothiazine, the organic compound for killing worms in sheep (one of the many deadly ring compounds), was announced as replacement for the inorganic copper sulfate that had formerly been used. Might it not have been possible that by drenching the sheep with copper salts regularly under the guise of killing internal parasites of an undernourished animal we were feeding copper to cover a deficiency in the soil and were making well nourished, healthy animals within which even worms do not survive? Such hypothetical consideration of what may seem to be only a correlation ought to push research farther. When it does we may discover that it only magnifies the task of reconstructing our soils with copper to the sheep's needs for feed, to say nothing of the magnitude for the human needs of foods with respect to all the other trace elements known and still unknown.

8. soil fertility pattern suggests reconstruction pattern

In the soil pattern of the United States, the major production of proteins along with the carbohydrates to make these more nearly a balanced diet for healthy bodies of animals and man has been on the soils in the climatic region of moderate to low rainfall in the narrow longitudinal belt bordering the 97th meridian. That is where our high protein wheat is grown. When our meat and wool animals range and rustle for themselves out there they are healthiest, longest-lived, and most fecund in reproduction. Protein production is favored by the less weathered soils where wood and water were not so plentiful for the pioneer. Soils too dry for massive annual crops are still rich in their stores of inorganic essential nutrients. We speak of them as soils of the prairies and the plains with grass as good feed for growing (but not for fattening) animals. We then conclude that grass must always be good feed for growing our meat- and milk-producing animals because of its pedigree. We fail to realize that grass is good feed (when grown out there) because of its high protein content, its high concentration of inorganic bioelements and its location where the periodic droughts prohibit forests but permit grass that can grow by stops and starts during the season according as the rainfall moisture in the soil permits. Soils developed under such a climatic setting are fertile. They grow protein-rich grass. They grow their own nitrogen-fixing, protein-rich legumes naturally. They make every mouthful of forage going into the animal a case of real feeding and not one of merely filling and nutritional fooling. The choice of the sheep and cattle for their best health from this ground up in the mid-continent was previously confirmed by the numbers of bison delineating the same soils for his high production of bone and brawn to say nothing of prolific reproduction.

9. production of food protein poses still bigger soil problems

The provision of complete proteins is the major food problem for both man and beast. It is more serious in other parts of the world than in parts of the United

States, of Argentina, of Australia, of Canada and of South Africa, for example. This is a problem for which nothing but more acres of more fertile soils can offer a solution. Technologies cannot be called upon to deliver synthetic proteins. Unlike fossil energy compounds, no fossil protein compounds, except for a little nitrogen in coal, have been unearthed. Animals of prehistoric times may have needed protein supplements much as our domestic and wild animals demonstrate their needs for them when they break through the fences and become marauders in their struggle for them. Now that we have had man's nomadism, undergirded by technologies overcoming distance to cover the earth with his population, we are face to face with the problem of peacefully feeding that mounting crowd on decreasing acreages under tillage and dwindling fertility in those shrinking volumes of soil.

Perhaps it will be sufficient to consider the protein problem (or the meat problem) by matching the human needs for this food portion against the possibilities of reconstructing the soils to meet them. This is no new mental pass-time. Thomas Robert Malthus of England indulged in it as early as 1798. He discussed "The principle of population as it affects the future improvement of society." He pointed to the fact that the increase of population is a geometric function in which the number doubles every 25 years. The rate of increase in the earth's production is an arithmetic function which could never keep pace with the geometric rise of numbers of people to be fed. But while folks laughed at Malthus' idea because the day of doom he predicted was delayed by man's migration to the western world for increased food by soil exploitation of new acres there, Malthus' spirit is now coming back to enjoy a chuckle as he says, "I told you so long ago."

We have now moved over just about the entire potential world's surface for the purpose of mining the soil's resources, for seining the possible proteins out of the seven seas from pole to pole, and for collecting nature's savings from the deeps everywhere. Technology has lengthened the food life lines beyond their elastic limits. Many of them are breaking. Most of them are being shortened. Man's reaching hand is being cut off by the failure of other life forms which he robs by his reach. We have not yet come to think in terms of our individual land allotments, their limitations and our responsibilities in their conservation.

According to recent figures, our world population is about 2.2 billion people [1951]. The usable land for food production is 2.4 billion acres. As a mathematical mean, this is slightly more than one acre for each of us. In the United States, the recent population figure is 151 million people. We had 345 million acres under cultivation in that census year. For purposes here in the U.S. you can imagine, then, that you are managing a bit more than two acres of agriculture to guarantee your keep. On a world basis, you are limited to farming only one acre.

In order to simplify the problem, let us remind you that your protein food requirements are 70 grams per day, so with your allowances of food fats near the ratio approaching that of protein and fat in beefsteak, your annual needs in only these two as beef would call for 320 pounds of this meat. At a dressing percentage of 56%, this requirement put into beef alone is the equivalent of 570 pounds of live beef weight that you must grow on your land allotment.

One acre of good soil in grass for a season will produce 300 pounds of beef. Hence one and one-half acres are needed to grow the protein (meat) and fat. A half acre remains in your allotment in the U.S. for the production of 200 pounds of cereals, 250 pounds of potatoes, 50 pounds of sugar, to say nothing of fruits for other carbohydrates and accessory foods you might desire to produce on that limited area. It is immediately evident why the world as a whole is not on beefsteak, and why our growing population in relation to soil acreage and productivity in protein potentials is rapidly taking many of us off that excellent diet and desired high standard of living.

That we dare not assume continuation of the past increases in production into the future so far as protein is concerned, is suggested by the records before us. The crop acreage of the U.S. in 1930 was 359 million. In 1950 it was only 345 million in the face of higher prices. Increased acreage without costly reconstruction of the soil is out of the question. Decreased acreage is the inevitable prospect when erosion is cutting it so rapidly that within 100 years, they tell us, we shall have a total of only 100 million acres left for crops. That area for even our present population would cut down your allotment from two acres to two-thirds of an acre. Your allowance of all protein as beef equivalent would be cut from 70 grams per day to but 25 grams. It would leave no acreage for growing other foods. That situation shifts our standard of living downward seriously when it suggests that we like Nebuchadnezzar shall eat the grass like the ox (in place of eating the ox) knowing that our soil fertility kingdom is departing from us.

Not only the shrinking acreage but the declining fertility in any acre comes into consideration for soil reconstruction as experiment fields at Missouri and Illinois emphasize it. They tell us that even in the cornbelt and its glacial soils, a time longer than 50 years of cultivation will exhaust the fertility below the point of paying the costs of working them, much less paying the taxes on them. But you say, "We can replace the fertility taken out in the produce by means of fertilizers." Already these inorganic mineral resources were producing 25% of our crops in the year 1950 with a maximum mixed fertilizer consumption of 18 and ⅓ million tons. This was an increase by 12% over 1949 when the increases in yields per acre in none of our crops were equally large. In the last ten years the fertilizers' share in crop production, in contrast to that of virgin fertility, increased from 20 to 25%. This says nothing of the thousands of tons of fertilizer materials beside the mixed fertilizers used on our soils. Even then, the total food production has not increased since 1944.

Here is the evidence that our food curve is no longer going up. The curve of population is. With such increased reconstruction of our soil fertility but with no corresponding increase in yields per acre and with no more than "holding our own" in total yields for the cultivated acreage of the country as a whole from such fertilizer increase, certainly these artificials on the soil will not offset the food needs of the population increase by 13 million people during that same period when production was already a constant. Soil construction cannot meet the mounting needs of such population increase. The falling curve of soil resources for protein food production under all efforts is coming to cross the rising curve of food needs by more people. "It is estimated we shall this year consume 148 pounds of meat per person, but that the effective demand at parity prices would be 160 pounds. It is estimated that (in 1951) we will grow enough feed grains to produce 138 pounds of meat per person. We will draw on our feed grain reserve for 10 million tons of grain necessary to bring our 1951 meat supply up to 148 pounds per person."

To produce these 10 million tons of feed grains which we are drawing from reserve this year we would need another 7 million acres of land. If we produced enough feed to equal the *demand* for meat, it would require 20 million more acres of land. To meet our increase in population and to maintain only our meat supply. . .we would need to add three million acres annually. In this next decade, we would need to find 30 million acres. . .another state like Iowa."

As partial relief from the problem of finding more acres, we can turn to the alternative of fertilizers to make the present number of acres produce more proteins. "One ton of nitrogen in fertilizers equals 14 acres of good farm land. . .We use an average of seven pounds per cultivated acre. In Holland the average application is 50 pounds. We will need to balance the nitrogen with phosphate and potash." In terms of nitrogen, then, ". . .to get our emergency need of 7 million more acres of feed grains we must produce 500,000 more tons of nitrogen as fertilizers. . .We are producing over twice that amount now to get our present production. Even if we get this 500,000 tons (more nitrogen) to solve our (present) emergency in 1951, we must add 100,000 additional tons every year to keep up with our population." What, then, can we hope for from more soil construction when the curve of productivity is leveling off (if not falling) under even so much present soil building? What hope is there when during the last two decades tractors using fossil crops as fuel have replaced the feed crops in the protein of 20 million horses displaced, and when the protein produced in soybeans was increased from 28 to 280 million bushels of these, both of which cannot be repeated?

Attention to soil building on its broader scale can be relief to needs in limited localities. Any discovery of new fertilizer resources brings the world to pounce on it immediately. This occurred for Gafsa in North Africa, and for Nauru, an island in the Pacific, both phosphate fertilizer suppliers, but objectives of whole armies in World War II, or as occurred for the fixed nitrogen on the market today. Technologies have put us on a world war basis. They put us into a United Nations with scarcely no nation any longer independent but submerged into the group and held there by the veto. On such great dimension and under such world political pattern, some folks would boast of our opportunity for world leadership in it. But they are failing to see that we have taken over the responsibility—and the attending dangers —of world feedership.

The virgin soils of the western hemisphere were a relief to human needs for the interval extending from Malthus' day to our recent time. Technologies made relief from population pressure on the soil resources most extensive, as man overran to consume what he could and to collect from everywhere. By such means of feeding itself, the population shifted its curve from what then was near a straight line, either level or slowly rising, to one rising geometrically with populations doubling every 25 years.

But as we look ahead there are handwritings on the wall again like those of Malthus, to say that the curve of population will take a fall since the curve of support, coming from the soil or the sea (into which soil has washed), are not only failing to rise commensurately, but are turning to a relative decline. Even food as bulk per person cannot be increased at rates of present population increase. The human needs as outlined by FAO (Food and Agricultural Organization) and matched against world crops show the wide disparity of these two sets of figures. More significant is the fact that in such data the needs as food protein show still greater disparity between these and the supplies of them.

In the older parts of the world, the population pressures on the fertility-exhausted soils have been so heavy and already for so long a time that our imagination cannot picture this as reason behind seven Nazi generals, recently hanged at Landberg, Germany, for the crime of exterminating near hundreds of thousands of folks by starvation, thousands of anti-communists recently eliminated in ever-hungry China through what we call a "Communist purge," and thousands of communist soldiers thrown into the murderous cannon fire of the United Nation's armies. Technologies once pushing the world population upward are now turned to slaughter, apparently to pull the population downward to fit the food resources of the soil. Is it beyond belief, that underneath all these disturbing manifestations there is the soil and with it the controlling factor of insufficient fertility as food?

10. we must still hope

Perhaps just at this moment our despondency might overwhelm us and shake severely our faith in the soil sciences. But the human species is quickly reduced from social stature to animal nature by hunger. The hidden hungers as an earlier stage in the reduction procedure are the most dangerous. *It would appear as if the major pattern of the world's population is suffering mainly from this hidden form of hunger which represents our living in a mental health too poor to exercise noble human judgment, but in a body condition still well enough to fight brutally to survive.* It suggests that we are not yet starved down to the degree of resignation to the forces of fate.

Soil reconstruction may be a part of the struggle under goad of hidden hungers for protein. In our humble opinion, soil construction cannot hold up the world's multiplying population. Those numbers will eventually be pulled down until they are balanced against the fertility flowing as food from the land and from the sea.

Such is the world picture as we see it. Fortunately, none of us is a single individual needs to assume that world responsibility. There is the place yet for each of us to make his own local soil support him the best he can. We need not be swept with the indifferent horde, made up of those unwilling to struggle independently. Soil conservation is still an individual responsibility. Soil conservation is still an individual opportunity. By multiplying the individual soil conservationists, we can meet the hunger needs in this segment of population, at least, yet for a while.

Only in a struggle on such a democratic basis can a democracy survive. We shall survive only according as our needs are reduced to come into balance with the possible reconstruction of the soil. This view of the future for each of us, puts real meaning into the words, "Soil Conservation," and calls for more folks to become Friends of the Land in the fullest significance of that title.

SUMMARY

Wastebasket of the Earth

"Man is upsetting much of his natural environment by his contamination of air, water and particularly soil, reducing nature's ability to safely dispose of wastes. He seems to be creating his own destruction." So read the cover blurb for a little report by Dr. William A. Albrecht, Wastebasket of the Earth, *which was circulated in the Bulletin of the Atomic Scientists, 1961. It has been reprinted a number of times since then. Here, as in many of the Albrecht papers, the biotic pyramid comes first. As a prelude to the last paper in this "Albrecht collection," it asks questions and stays on to answer them in terms of nature.*

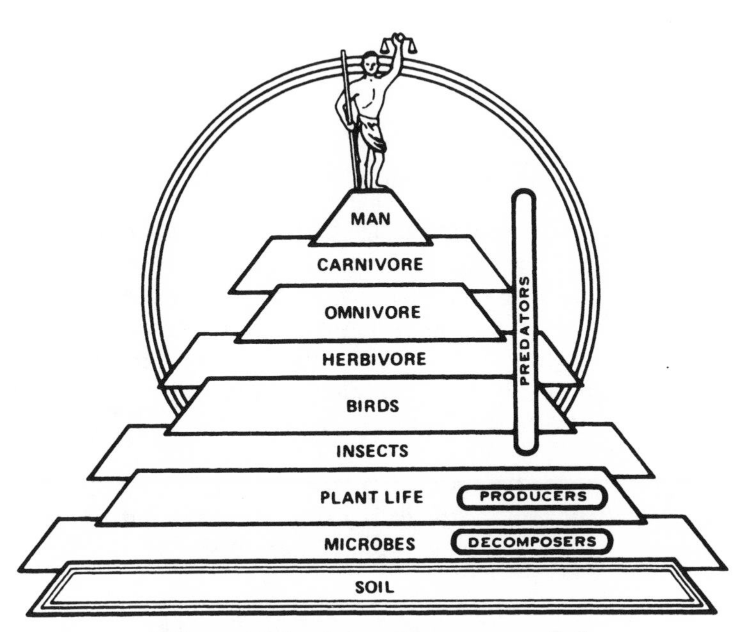

The Biotic Pyramid

Wastebasket of the Earth

Contamination in the food we eat, the water we drink, and the air we breathe suggests close connection with the soil. Usage such as "your hands are dirty," serves as a reminder that the thin surface layer of the soil is the wastebasket of the earth, the collector for the disposal of all matter that has once lived and moved.

1. role of the soil

The thin layer of the earth's surface is an intensive transformer of all the waste it collects. In that shallow stratum elements are separated out of combinations and reunited into other compounds, effecting vast changes in every kind of matter. These activities include transformations of energy by such processes as oxidation, reduction, hydration, hydrolysis, and molecular rearrangement. Oxidation, like combustion, dissipates energy in the form of heat, which escapes from the earth. Reduction concentrates it in compounds of high heat and fuel values. This is illustrated most significantly by the plant's reduction of carbon dioxide by water, storing the sun's energy in the resulting carbohydrates.

These transformations of matter and energy create, nourish, protect, and maintain all living creatures. These creatures vary in physiological complexity from the simplest microbial cell to man, whose high state of evolution has equipped him with a mind to comprehend—and to modify—his environment, even as far as unwittingly contaminating it.

2. term with many meanings

The term *contamination* has many meanings. The wastes of one population may be either the poison or the life support of another. We consider populations of different forms of life to see whether their coexistence is cooperative or competitive, and whether their respective wastes may be of benefit or harm to each other.

A high degree of cooperation through evolution and adaptation characterizes nature's management of environment. But when man manages environment, competition stands out—with frequent examples of contamination. As managers, we are biased toward aims and benefits for man alone. We have concerned ourselves little with how, in modifying the environment for ourselves, we may be disrupting it for all the other populations of the biotic pyramid—microbes, plants, animals and man—all supported by the one creative foundation—the soil.

3. man upsetting environment

Instead of speaking of the soil as "dirt" to emphasize its contamination of ourselves, the converse might be more appropriate. Man is upsetting much of the natural environment by his contamination of the air, the water, and particularly the soil, reducing its ability to dispose of wastes safely, and to nourish healthy populations of men and all other living things. It is a dangerous boldness to believe that we can manage environments completely by technologies designed for our economic advantage.

Given to a belief in the homocentric purpose of the earth, we have come to take our soil for granted. This view is quite the opposite of that of the pioneer—living mainly by agriculture—who respected and studied his environment and struggled to be naturally fit for his evolutionary survival there. He considered the seasons, the annual amount and distribution of the rainfall, the degrees of heat and cold, the winds and storms. He did not consider land as a commodity. The pioneer

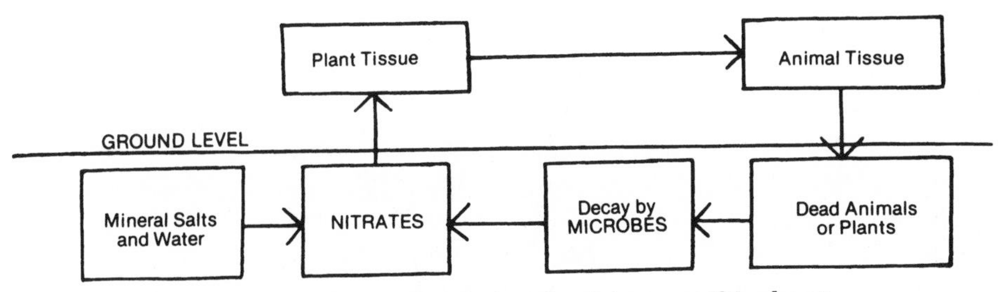

Nature's pathways for continued existence on this planet.

appreciated the fact that the soil had been built by nature during the ages predating him, brought about by the climatic forces breaking down the rocks, growing the microbes, the plants, the many other kinds of life. For him, those were the natural forces of soil construction, and he knew he must maintain soil productivity if he were to survive. The pioneers were truly agrarian people. For them the soil was holy ground. Too, it was living soil.

4. man forgets nature

Because of the scientific organization of our recently increased knowledge about the soil, we forget that the decomposition of rocks, the growth of vegetation, and the complete return of that organic matter in place—all under nature's management—are what brought about productive soils. At first, these soils were not contaminated against the healthy coexistence of a specific, but limited, set of species. The early balance represented an evolutionary set, each form unique in relation to the others, to the soil's particular geologic-climatic setting, and its degree of development in a given area. Examples of this limited balance would include the particular plant and wildlife forms of virgin forests, or the American plains or prairies. All of these conditions were major determinants of the coexistent species; the survival of each was dependent on the survival of the others.

Man's management of agricultural crops and livestock has not been directed by knowledge of the limitations of soil fertility, nor by knowledge of the required climatic-geologic setting for crops and livestock with each in its natural ecological climax. Instead, we have been given to transplanting any species from anywhere to everywhere for economic gain, ignoring biological benefits or dangers to the species involved. We have brought in higher (or lower) species while depleting the very soil fertility support required to grow them in health.

5. evolution in reverse

We have thrown natural evolution into reverse. We struggle to nurture species we have made unfit for the environment because creative forces there cannot offer the required quality of food and energy support. We now need to view the pampered species as contamination against all other lower species which would otherwise arrive at their natural climax. We must accept the fact that the soil, with its dynamics of producing and accumulating organic substances through plants with the support of climatic forces, is still the only energy supply on which all kinds of life depend. Transformation is the major role played by the soil.

Granting that the entire biosphere is dependent on the inflow of that bit of the sun's energy fixed in organic substances by photosynthesis, it is helpful to note that as Nelson G. Harston, Frederick E. Smith, and Lawrence B. Slobodkin said in their article, "Community Structure, Population Control and Composition," in *The American Naturalist,* last year, populations divide themselves appropriately into three trophic levels: the decomposers, the producers, and the predators.

6. the decomposers

The decomposers are represented by the varied kinds of microbial life which live by degrading organic debris through processes giving the microbes their energy and growth substances. The slow accumulation (at great depths below the earth's surface) of these remnants as fossil fuels of organic origin and a high degree of oxygen removal indicates that this heterotrophic group did not have much in the way of energy-giving "leftovers." Nor can these remains be considered contamination when they are far beneath the soil surface. But when they are brought into the atmosphere, and into the highly aerobic surface soils, after laboratory work has turned them into products such as antiseptics, pesticides, and herbicides, they are the most extensive and powerful biochemical contaminations we have yet known. Their range of disturbances covers the entire biotic pyramid.

Hydrocarbons, which can disrupt the transformations in the soil's surface zone, were buried by nature at great soil depths. These high concentrations of energy are now our industrial fuels. But even such deep burial of the wastes resulting from the use of atomic fuels will not serve as safe removal. Atomic fuels, with their lingering rays for a lingering death of all living cells, are not respecters of the beneficial portions of the soil's microbial flora.

7. source of nutrition

Nutritional requirements of microbial decomposers are met by the contents of the debris and the soil. The essential elements remain very much in the cycle of use and reuse, since sulfur, phosphorus, nitrogen, and carbon occur as major elements in the leftovers. Oxygen is almost absent there, but carbon, linked to hydrogen, occurs in high concentration.

Ever since the work of Pasteur, our fear of microbes has singled them out as our environment's major contamination. We have made them the victims of vengeance and we boil them under steam pressure at every opportunity. Their disrepute has been shared recently by dusts, fungus spores, pollen grains of trees, grasses, weeds, and other particles. Today the professional allergist devotes his attention to atmospheric contaminants and to other substances disturbing the mucous membranes and similar tissues of the human body. But when microbes are viewed as disposers and transformers they become major benefactors for other populations. Microbes in surface soil serve as wrecking crews and salvage agencies. Simplifying the residue of past populations for energy release and reuse, these decomposers make the upper stratum of soil the real living foundation of the entire biotic pyramid. They keep open the sewage disposal systems of all the population levels.

8. uniquely equipped

Microbes are uniquely equipped to maintain their own populations. They reproduce at the rate of one generation per hour or even faster. They synthesize their own extra- and intra-cellular compounds—some of which we term antibiotics—for protection against competitors for their environment. The antibiotic quality may be merely an evolutionary accident that served to bring death to competitor cells. Nature's conservation practices use wastes from some life forms to make the environment serve its own survival more completely.

Benzene rings characterize the chemical structure of the antibiotic terramycin, for example. Modified ring structures with substitutions of nitrogen and sulfur for some of the carbon are found in penicillin, aureomycin and other microbial products developed for their bacteriocidal effects on the human body. That ring structure represents highly reduced organic compounds such as those found in crude oil and coal. The chemical structures and biochemical energy potential represent the opposite of that of natural organic wastes dumped on the surface soil. While natural organic wastes offer much as energy through microbial oxidations, crude oil and coal are too stable for biochemical transformation and energy release, even though they rank high as industrial fuels. They are seldom broken down by digestion. They overload the liver, the chemical censor of the human body. They are leftovers from anaerobic microbial populations, and are well removed as serious contaminants by their natural placement far below the surface soil and by disposal well beyond the entire biotic pyramid.

9. acts as deadly poison

The benzene ring, in a simpler compound distilled from coal tar—carbolic acid—was an early antiseptic. But now, long-chain compounds and ring structures of carbon, sulfur or nitrogen substitutions, or in chlorinated, nitrated, and sulfonated forms as synthetics from the industrial chemistry laboratory, are being distributed extensively, acting as deadly poisons against the populations in the biotic pyramid. They come into the atmosphere in "smog" and carbon monoxide, and in herbicides, pesticides, and the like. Those microbial wastes welling upward from the depths of the earth, in their natural forms and in our more poisonous alterations of them, must be considered contamination by man of his own environment and of the environments of all populations that support him. His efforts to so completely destroy microbes are contributing to his gradual destruction by his own hand. Decomposers are not respecters of man when they release their own wastes as contaminants.

The second group among the populations, or trophic levels, of the earth are the plants, which produce organic compounds carrying the chemical and other energy transferred from the sun. They are the only means of storing and distributing that supply. Energy is collected by photosynthesis, the unique process whereby the chlorophyll of the leaves binds it into compounds of carbon, hydrogen, and oxygen in the molecular arrangement of carbohydrates. Plants are the source of energy for all biochemical processes, and of their own starter compounds into which they synthesize nitrogen, sulfur, and phosphorus to yield the different amino acids of proteins and living tissues which grow, protect, and reproduce. Plants are the only producers since they are the sources of food energy and growth potential synthesized directly from the chemical elements and flowing through all the other trophic levels.

10. significant stratum

Plants root themselves first into the soil and then extend their tops into the atmosphere. They may be vulnerable to contaminants from both directions. Their unique position and special processes bring both inorganic and organic decompositions from the soil into biochemical union with water, and carbon dioxide, and nitrogen. Plants, representing that limited zone where earth and atmosphere meet, act as a kind of interface for concentrations of different kinds of matter, becoming the significant stratum for the creation of all that lives. All life is possible because the synthetic power of sunlight operates through the plant enzyme, chlorophyll. It is a mobilizer, or chelator, of the chemical elements, unique because the plant itself creates it. Its chemical structure consists of the inorganic element magnesium as the core, combined with nitrogen, linked with carbon, and all three connected with hydrogen. Even this chelator's composition represents a chemical union between the soil, which furnishes the magnesium, and the atmosphere, which yields the nitrogen, carbon, and hydrogen.

Because plants support themselves by their own capacity for combining elements and using solar energy to create compounds, they surpass all other populations in the struggle for survival. Microbes can use the elements in synthesis but must decompose the organic compounds synthesized by plants to provide the necessary energy. Plants, in turn, profit in their extended survival because the microbes simplify the accumulated organic matter to keep the soil's inorganic elements and the atmosphere's organic elements—carbon and nitrogen—in cycles of reuse. Otherwise the accumulated products of plants would contaminate their own environment.

11. symbiosis for survival

Plants and microbes may be considered in a symbiosis for survival independent of more complex populations. But even that symbiosis may be disrupted by one or the other symbiont acting as a competitor, a parasite, or a predator. Either may even produce contamination by its waste products: plant compounds may be poisons for microbes and microbes may be poisons for plants. By competing for essential inorganic elements—calcium, magnesium, phosphorus, potassium, nitrogen, sulfur—and the several "trace" elments, one may limit the other via the soil.

All trophic levels above the plants must live by the compounds of the latter's synthesis—carbohydrates—for energy; proteins for growth of tissue, protection, and reproduction; and inorgano-organic combinations associated with the proteins. Proteins in plants result, not from photosynthesis, but from the plant's biochemical processes, which require expenditures of stored energy and assembly of inorganic and organic requirements by the roots. The roots penetrate only a limited volume of soil, living there largely through activities of the decomposers. Plants, like microbes, use carbon dioxide waste—from the roots—to produce active hydrogen in the resulting carbonic acid to mobilize the soil fertility elements for plant survival. These two populations, the producers and the decomposers, as contaminants or as transformers, determine the environmental support of all other populations.

12. the predators

All the heterotrophic populations crowd each other to get to the food delivered to them by that mundane team of autotrophs—the symbiotic team of microbes and plants. That crowding, under the deficiencies provoking it, makes the wastes of each a contamination for the other whenever the natural processes of the producers and decomposers are disrupted. All populations above plants and microbes are included in the third category: the predators. They prey upon the plant population, and upon each other to an increasing degree, since the plants are not nourished by the living soil completely enough to be ample prey for man and all the other predators and parasites.

Predators—other than man—do not destroy their prey completely before the numbers of predators drop down so far that the numbers of prey mount to domination again. Increase in prey favors an increase in predators, and then, in turn, a decrease in prey, to yield naturally alternating dominations but not the extinction of either. But man is not merely the predator of one trophic level; he also aims his technology at complete extinction of many populations. Thus he breaks the law of predator-prey relations and of survival. When he causes extinction, he reduces the number of basic segments by which the biotic pyramid supports man.

Interpopulation predations increase as the soil becomes less able to grow the vegetation of nutrition in required quality and quantity. When the soil's inorganic elements have been depleted, then the exploited soils must eventually register their damage on all trophic levels. The most serious effects fall first on man, raising the question of whether the baffling degeneration of our bodily health, now increasing our concern about what was once commonly called "disease," may not suggest patterns of hidden causes connected in some way with the climactic-fertility pattern of the soil. Thus, human ecology may develop into the most important science.

13. population time chart

When the numbers in any population are charted as a graph against a base of time, they give a sigmoid curve. Its shape suggests the top of the letter "S" pulled to the right while the bottom remains attached. The introduction of a single living microbe into a given volume of medium results in a slow increase in numbers with the curve moving along the near-horizontal. But soon the move turns toward the vertical, suggesting a population explosion. Then the increase lessens, ceases, and finally, there is a population decrease followed by eventual extinction. Contamination by its accumulated non-transformed wastes, coupled with exhaustion of nutritional and other environmental supports, eliminates the population from its limited setting.

Populations of older countries, similarly charted, suggest the final third of the biotic curve. France, Scandinavia, Italy, and Spain, starting as far back as 1900, are illustrations. The United States illustrates the lower two-thirds, or the beginning of the chart. Predictions from 1920 onward pointed to the falling percentages of increase, if immigrations were limited.

14. biological liability

Since the multiplication of man still conforms to the natural laws of biological phenomena, and since human multiplication has by no means managed to give biological advantages for improved survival, we must characterize man, at this stage, as the main biological liability, not only to himself, but to the other populations supporting him. ***He is the contamination in the environment.***

Man's disregard of his ecological limitation to the temperate zone—by the required supply of proteins alone—should be cited as only one powerful factor. As one moves from the temperate to the torrid and humid tropical zone, the desperate struggle for proteins at any trophic level is evident in the carnivorousness and cannibalism that are common. Here, life forms are mainly predators. The same is true as one moves from the temperate to the frigid zone.

15. climatic extremes

The migrations into the frigid zone, into climatic-soil-fertility settings of little or no soil construction, are badly handicapped by the paucity of producers and decomposers. The predators win support by their carnivorousness or through lifelines reaching back into the temperate zone. In the tropics, high rainfall and temperatures have caused excessive decomposition of the rocks to the point of destruction of soils which then fail to fully nourish plant life. Even proteins are there replaced by poisonous compounds of many producers. Carnivorousness and cannibalism characterize the survival of decidedly limited populations. The dry tropics are in the same category. Populations at climatic extremes are limited because the producers, operating in combination with the decomposers, are not providing the necessary proteins in complete array, nor with their natural accompaniments.

The epoch of man is but a minute segment in the paleontological column of the earth's populations as they have come, gone, or remained. Man has shifted away from the rugged individualism of open country and its diversity of agriculture according to the laws of nature. He has collected himself into congested cities which have been said to require monocultures and chemicalized agricultures; he has controlled the environment of those cities according to the dictates of technology and economics, disregarding nature's laws and even his own biochemistry.

During the latest part of the brief epoch of man's existence, his technologies and their political complications have served to harvest the natural living resources, to exploit soil fertility and to compel the human march from east to west. The march has rolled on until the resources of most recent possession—those of the western hemisphere—are dwindling rapidly under the political demand for coexistence of the western world with those bringing up the rear in the march—a total world population approaching three billion predators.

16. man ignores nature

Because they must behave according to the natural laws controlling biological bodies, each population below man has—through evolution—exhibited itself as a climax crop for a limited time in its limited ecological setting. But man, with mental capacity transcending that of the others, has used most of the other levels to his advantage and their disadvantage so that, in general, he does not conform to the pattern of evolution. His technological powers disrupt the pattern and destroy the very conformers that support him. Unwittingly, his development and management of environmental control over materials and energy have increased the contamination of the soil and the atmosphere. He has destroyed the decomposer populations to the extent that the natural basic support of the producers is weakening; the entire biotic pyramid is tumbling because of man's dominance at the top.

In managing her contaminations, nature either transforms them through biotic disposers in the surface soil, or buries them safely at greater depths. Man, managing his contaminations as a helpless novice, seems to be on his way to his own destruction by and amongst them. In all probability, nature, not he, will determine the final outcome.

Index

E

F

G

H

I

R

S

T

U

V

Appendix A

Properties of the bones of rabbits fed on hays from different soil types without and with soil treatments.

Soil Type	Weight grams	Length cm	Diameter cm	Thickness mm	Breaking strength lbs.	Volume cc	Specific gravity	Retained from feed in grams per pen of animals animals		
								P	Ca	Ca/P
Without Soil Treatments										
Eldon	2.26	6.88	0.520	0.70	22.8	3.11	0.71	25	163	6.4
Putnam	2.02	6.77	0.525	0.59	18.9	2.65	0.76	36	171	4.6
Clarksville	2.09	6.77	0.520	0.58	22.1	3.03	0.69	20	165	6.3
Grundy	3.07	7.24	0.570	0.84	30.0*	3.94	0.77	58	225	3.8
Lintonia	2.59	6.95	0.530	0.73	27.6	3.31	0.78	43	22.7	5.2
Average	2.40	6.92	0.533	0.69	24.3	3.21	0.74	36	190	5.3
Maximum difference, %	52.0	6.9	9.6	44.8	58.7	48.7	13.6	189	39	66.6
With Soil Treatments										
Eldon	2.96	7.20	0.560	0.72	25.2	3.48	0.85	44	261	5.8
Putnam	2.35	6.88	0.540	0.77	26.4	3.21	0.72	34	179	5.2
Clarksville	2.63	6.93	0.548	0.71	24.2	3.29	0.79	37	165	4.3
Grundy	2.85	7.20	0.530	0.95	30.0*	3.63	0.78	60	201	3.3
Lintonia	3.26	7.33	0.565	0.82	30.0*	3.79	0.86	79	379	4.7
Average	2.81	7.11	0.549	0.79	27.2	3.48	0.80	51	237	4.7
Maximum difference, %	38.7	6.5	6.6	33.8	19.1	18.1	18.4	131	129	74.0

*In these cases the bones withstood the limit of pressure possible on the testing machine.

Also from Acres U.S.A.

Eco-Farm: An Acres U.S.A. Primer

BY CHARLES WALTERS

In this book, eco-agriculture is explained — from the tiniest molecular building blocks to managing the soil — in terminology that not only makes the subject easy to learn, but vibrantly alive. Sections on NP&K, cation exchange capacity, composting, Brix, soil life, and more! *Eco-Farm* truly delivers a complete education in soils, crops, and weed and insect control. This should be the first book read by everyone beginning in eco-agriculture . . . and the most shop-worn book on the shelf of the most experienced. *Softcover, 476 pages. ISBN 978-0-911311-74-7*

Weeds: Control Without Poisons

BY CHARLES WALTERS

For a thorough understanding of the conditions that produce certain weeds, you simply can't find a better source than this one — certainly not one as entertaining, as full of anecdotes and homespun common sense. It contains a lifetime of collected wisdom that teaches us how to understand and thereby control the growth of countless weed species, as well as why there is an absolute necessity for a more holistic, eco-centered perspective in agriculture today. Contains specifics on a hundred weeds, why they grow, what soil conditions spur them on or stop them, what they say about your soil, and how to control them without the obscene presence of poisons, all cross-referenced by scientific and various common names, and a new pictorial glossary. *Softcover, 352 pages. ISBN 978-0-911311-58-7*

Science in Agriculture

BY ARDEN B. ANDERSEN, PH.D., D.O.

By ignoring the truth, ag-chemical enthusiasts are able to claim that pesticides and herbicides are necessary to feed the world. But science points out that low-to-mediocre crop production, weed, disease, and insect pressures are all symptoms of nutritional imbalances and inadequacies in the soil. The progressive farmer who knows this can grow bountiful, disease- and pest-free commodities without the use of toxic chemicals. A concise recap of the main schools of thought that make up eco-agriculture — all clearly explained. Both farmer and professional consultant will benefit from this important work. *Softcover, 376 pages. ISBN 978-0-911311-35-8*

To order call 1-800-355-5313

or order online at www.acresusa.com

Hands-On Agronomy

BY NEAL KINSEY & CHARLES WALTERS

The soil is more than just a substrate that anchors crops in place. An ecologically balanced soil system is essential for maintaining healthy crops. This is a comprehensive manual on soil management. The "whats and whys" of micronutrients, earthworms, soil drainage, tilth, soil structure and organic matter are explained in detail. Kinsey shows us how working with the soil produces healthier crops with a higher yield. True hands-on advice that consultants charge thousands for every day. Revised, third edition. *Softcover, 352 pages. ISBN 978-0-911311-59-4*

Hands-On Agronomy Video Workshop

Video Workshop

BY NEAL KINSEY

Neal Kinsey teaches a sophisticated, easy-to-live-with system of fertility management that focuses on balance, not merely quantity of fertility elements. It works in a variety of soils and crops, both conventional and organic. In sharp contrast to the current methods only using N-P-K and pH and viewing soil only as a physical support media for plants, the basis of all his teachings are to feed the soil, and let the soil feed the plant. The Albrecht system of soils is covered, along with how to properly test your soil and interpret the results. *80 minutes.*

The Biological Farmer

A Complete Guide to the Sustainable & Profitable Biological System of Farming

BY GARY F. ZIMMER

Biological farmers work with nature, feeding soil life, balancing soil minerals, and tilling soils with a purpose. The methods they apply involve a unique system of beliefs, observations and guidelines that result in increased production and profit. This practical how-to guide elucidates their methods and will help you make farming fun and profitable. *The Biological Farmer* is the farming consultant's bible. It schools the interested grower in methods of maintaining a balanced, healthy soil that promises greater productivity at lower costs, and it covers some of the pitfalls of conventional farming practices. Zimmer knows how to make responsible farming work. His extensive knowledge of biological farming and consulting experience come through in this complete, practical guide to making farming fun and profitable. *Softcover, 352 pages. ISBN 978-0-911311-62-4*

To order call 1-800-355-5313

or order online at www.acresusa.com